D0118894

A SOURCE BOOK IN GEOLOGY

SOURCE BOOKS IN THE HISTORY OF THE SCIENCES

EDWARD H. MADDEN, GENERAL EDITOR

Source Book in Astronomy, 1900–1950
Harlow Shapley, Harvard University

A Source Book in Greek Science
Morris R. Cohen, College of the City of New York and University of Chicago, and I. E. Drabkin, College of the City of New York

A Source Book in Animal Biology
Thomas S. Hall, Washington University

A Source Book in Chemistry, 1400–1900
Henry M. Leicester, San Francisco College of Medicine and Surgery, and Herbert M. Klickstein, Edgar Fahs Smith Library in the History of Chemistry, University of Pennsylvania

Source Book in Chemistry, 1900–1950
Henry M. Leicester, San Francisco College of Medicine and Surgery

A Source Book in Physics
William Francis Magie, Princeton University

A Source Book in the History of Psychology
Richard J. Herrnstein and Edwin G. Boring, Harvard University

Source Book in Geology, 1900–1950
Kirtley F. Mather, Harvard University

From Frege to Gödel: A Source Book in Mathematical Logic, 1879–1931
Jean van Heijenoort, Brandeis University

A Source Book in Mathematics, 1200–1800
D. J. Struik, Massachusetts Institute of Technology

A SOURCE BOOK

IN

GEOLOGY

1400–1900

by

Kirtley F. Mather, Ph.D., Sc.D.

Professor of Geology, Emeritus, Harvard University

and

Shirley L. Mason, Ph.D.

Geologist, Houston, Texas

Harvard University Press
Cambridge, Massachusetts

Library of Congress Catalog Card Number: 67-12100

SBN 674-82277-3

PRINTED IN THE UNITED STATES OF AMERICA

"*If I saw farther, 'twas because
I stood on giant shoulders.*"

ISAAC NEWTON

GENERAL EDITOR'S PREFACE

The Source Books in this series are collections of classical papers that have shaped the structures of the various sciences. Some of these classics are not readily available and many of them have never been translated into English, thus being lost to the general reader and frequently to the scientist himself. The point of this series is to make these texts readily accessible and to provide good translations of the ones that either have not been translated at all or have been translated only poorly.

The series was planned originally to include volumes in all the major sciences from the Renaissance through the nineteenth century. It has been extended to include ancient and medieval Western science and the development of the sciences in the first half of the present century. Many of these books have been published already and several more are in various stages of preparation.

The Carnegie Corporation originally financed the series by a grant to the American Philosophical Association. The History of Science Society and the American Association for the Advancement of Science have approved the project and are represented on the Editorial Advisory Board. This Board at present consists of the following members.

The series was begun and sustained by the devoted labors of Gregory D. Walcott and Everett W. Hall, the first two General Editors. I am indebted to them, to the members of the Advisory Board, and to Joseph D. Elder, Science Editor of Harvard University Press, for their indispensable aid in guiding the course of the Source Books.

Edward H. Madden, General Editor

Department of Philosophy
State University of New York at Buffalo
Buffalo, New York

A SOURCE BOOK IN GEOLOGY

Preface

The literature of geological science is extraordinarily voluminous, partly because of the wide ramifications of the science itself
and partly because of the tendency displayed by geologists,
especially many of the eighteenth and nineteenth century writers
in this field, to set forth their ideas *in extenso* and to give lengthy,
detailed descriptions of the phenomena with which they dealt.
The modern student of geology is much more likely to get his
information from textbooks, which in turn depend upon earlier
compilations, than from the original writings of the classical
contributors to geological lore. To provide in a single volume a
representative series of excerpts from such writings is, therefore, a
task well worthy of the arduous labor that it necessarily involves.

In making our selections we have tried to keep in mind the
needs of the reader interested in the historical development of the
science of geology as well as the needs of the student who wants to
secure first-hand information concerning the origin of the principles upon which he depends in his own research. We have also
given thought to the probable availability of source material in the
various libraries throughout the country and have given preference, where other considerations were nearly equal, to the publications which are likely to be least accessible for the majority
of geologists. In this connection, the resources of the Library of
Congress, the New York Public Library, and the libraries of
Lehigh University and of the Museum of Comparative Zoology, as
well as the Widener Library at Harvard University, proved to be
unusually rich.

Excerpts have for the most part been arranged in chronological
order according to the dates of birth of the contributors. For
ready reference, a Guide to Subject Matter has been provided, in
which the articles dealing with a score and a half of the major

concerns of the geologist have been listed in the order of the dates of publication. The brief biographical notes do little more than identify the time, country, and major activities of the authors. No systematic attempt has been made to evaluate the theories, now discarded or radically emended, which in the past were of great significance in the development of the science. The student of the history of geology will continue to depend upon such treatises as Zittel's *History of Geology and Paleontology*, Geikie's *The Founders of Geology*, Merrill's *Contributions to the History of American Geology*, and Adams' *The Birth and Development of the Geological Sciences*.

Contributions originating since 1900 have not been considered, nor has the work of living geologists been included. The aim has been to provide the material which must serve as the basis for a picture of seventeenth, eighteenth, and nineteenth century geology. Although the current development of the science has its roots in the past, its contemporary fruits in many instances depend upon ideas and techniques that were scarcely glimpsed by the earlier workers. On the other hand, many of the modern methods of geological research, such, for instance, as the use of the portable seismograph, were clearly envisioned by them.

In each instance the style of the original article has been retained, even though frequent inconsistencies in punctuation and spelling have resulted. Footnotes, unless otherwise indicated, are from the original publications. Our own comments, whether in the text or in footnotes, are enclosed in square brackets. A few of the titles given for the articles are our own, not those of the original authors.

Translations not credited to others were prepared by Mr. Mason, with the exception of the excerpts from the works of Perrault, Logan and Hunt, Surell, and de la Noë and de Margerie, which were translated by Helen S. Pattison, who also rendered able assistance in much of the editorial work involved in the final preparation of the manuscripts.

To ensure as wise a selection as possible from the large number of contributions considered for inclusion in this volume, more than a hundred geologists were consulted during the progress of our work. Especially valuable suggestions were made by William C. Alden, William Bowie, Walter H. Bucher, Kirk Bryan, Rollin T. Chamberlin, Reginald A. Daly, Douglas Johnson, Alfred C. Lane,

Chester R. Longwell, Rudolf Ruedemann, Robert W. Sayles, Clinton R. Stauffer, Allyn C. Swinnerton, Lewis G. Westgate, and Alfred O. Woodford. Professor Woodford also read critically a near-final draft of the entire manuscript and supplied one of the translations as indicated in the text. It should be noted, however, that not all the suggestions from these colleagues have been followed and that the final responsibility for determining the content of the volume is ours.

In addition to acknowledging our debt of gratitude to those mentioned in the foregoing, we wish to express our thanks to the many librarians upon whom we have relentlessly imposed ourselves. In supplying us with biographical data, David C. Mearns, of the Library of Congress, and Fletcher Watson, of the Harvard College Observatory, have rendered great assistance. Leonice Cook Seyfert, Doris Mann Kennedy, and Constance Grady have worked long and patiently to prepare the manuscript for publication and attend to innumerable details incident to its completion. The illustrations were prepared by Edward A. Schmitz.

We also acknowledge with thanks the courtesy of the Empire State Book Company in granting permission to use passages from *Leonardo da Vinci's Note-Books*, and of Professor John Garrett Winter and the University of Michigan Press in permitting the use of the excerpts from *The Prodromus of Nicolaus Steno's Dissertation*. Professor Winter very kindly checked the text that we used and made a few minor alterations from that which he published in 1916.

KIRTLEY F. MATHER,
SHIRLEY L. MASON.

CAMBRIDGE, MASSACHUSETTS,
May, 1939.

Contents

PAGE

Leonardo da Vinci (1452–1519)
THE EARTH AND THE SEA 1

Agricola [Georg Bauer] (1494–1555)
THE STRUCTURE OF THE EARTH AND THE FORCES WHICH CHANGE
THE EARTH. 7

George Owen (1552–1613)
THE COURSE OF ROCK SEAMS IN PENBROKSHIRE 12

René Descartes (1596–1650)
RATIONAL BASIS FOR HYPOTHESIS OF CREATION OF SOLAR SYSTEM 14
THE COMPOSITION OF THE EARTH. 14

Athanasius Kircher (1602–1680)
THE SUBTERRANEAN WORLD 17

Pierre Perrault (1611–1680)
THE SOURCE OF WATER IN SPRINGS AND RIVERS 20

Bernhard Varenius (1622–1650)
THE FIRST PHYSICAL GEOGRAPHY. 24

Robert Boyle (1627–1691)
CONDITIONS AT THE BOTTOM OF THE SEA. 27

Robert Hooke (1635–1703)
EARTHQUAKES 28

Nicolaus Steno (1638–1687)
OF SOLIDS NATURALLY CONTAINED WITHIN SOLIDS 33

Gottfried Wilhelm Leibnitz (1646–1716)
THE EARTH ORIGINALLY MOLTEN. 45

Benoit de Maillet (1656–1738)
STRATA FORMED BENEATH THE SEA 47

PAGE

John Woodward (1665–1722)
    EARTH HISTORY AND THE DELUGE   49

John Strachey (1671–1743)
    OBSERVATIONS ON THE STRATA IN THE SOMERSETSHIRE COAL FIELDS   53

Lewis Evans (c. 1700–1756)
    THE PHYSIOGRAPHY OF THE MIDDLE BRITISH COLONIES IN AMERICA   55

Georges Louis Leclerc, Comte de Buffon (1707–1788)
    THE COLLISION HYPOTHESIS OF EARTH ORIGIN.   58
    EPOCHS OF THE HISTORY OF THE EARTH.   65

Giovanni Targioni-Tozzetti (1712–1784)
    VALLEYS FORMED BY STREAM EROSION.   74

Giovanni Arduino (1713–1795)
    ORIGIN OF MARBLE AND DOLOMITE   76

Jeanne Etienne Guettard (1715–1786)
    MAP SHOWING THE ROCKS THAT TRAVERSE FRANCE AND ENGLAND   77
    THE FLUVIATILE ORIGIN OF THE PUDDINGSTONES [CONGLOMERATES]
    OF THE PARIS BASIN.   78

John Michell (1724–1793)
    THE NATURE AND ORIGIN OF EARTHQUAKES   80
    THE EARTH COMPOSED OF REGULAR AND UNIFORM STRATA.   84

Immanuel Kant (1724–1804)
    A NATURALISTIC HYPOTHESIS OF EARTH ORIGIN   88

Nicolas Desmarest (1725–1805)
    THE VOLCANIC ORIGIN OF BASALT.   90
    AGE DETERMINATION IN A VOLCANIC REGION.   91

James Hutton (1726–1797)
    THEORY OF THE EARTH   92

Lazarro Spallanzani (1729–1799)
    ON BASALTIFORM LAVAS AND THE ROLE OF GASES IN VOLCANIC
    ERUPTIONS. 101

Henry Cavendish (1731–1810)
    EXPERIMENTS TO DETERMINE THE DENSITY OF THE EARTH. 103

Jean Baptiste Louis Rome de l'Isle (1736–1790)
    LAWS OF CRYSTALLIZATION. 108

Rudolph Eric Raspe (1737–1794)
    COLUMNAR JOINTING IN BASALTIC LAVA 111

PAGE

Horace Benedicte de Saussure (1740–1799)
ALPINE GEOLOGY. 114

Peter Simon Pallas (1741–1811)
MOUNTAINS OF VARIOUS AGES 123

Antoine Laurent Lavoisier (1743–1794)
LITTORAL AND PELAGIC BEDS. 126

Abbé René Just Hauy (1743–1822)
A BASIS FOR A SYSTEM OF MINERALOGY. 129

John Playfair (1748–1819)
PROOFS OF THE HUTTONIAN THEORIES. 131

Abraham Gottlob Werner (1749–1817)
THE AQUEOUS ORIGIN OF BASALT. 138
THE NEW THEORY OF VEINS, AND OF THE MODE OF THEIR FORMATION 140

Pierre Simon, Marquis de Laplace (1749–1847)
THE NEBULAR HYPOTHESIS 143

Guy S. Tancrède de Dolomieu (1750–1801)
RHYOLITIC LAVAS. 151
RELATIONS BETWEEN THE AUVERGNE VOLCANICS AND THE GRANITE
ON WHICH THEY REST. 152
ON THE DISTINCTIONS BETWEEN DOLOMITE AND LIMESTONE 154

Jean Louis Giraud, Abbé Soulavie (1752–1813)
METHODS OF DETERMINING THE CHRONOLOGICAL SEQUENCE OF
ROCKS. 155

Sir James Hall (1761–1832)
MARBLE FROM LIMESTONE IN THE LABORATORY. 158
RESULTS OF THE SLOW COOLING OF MELTED ROCK 161
THE ROLE OF HEAT IN THE CONSOLIDATION OF STRATA 165

William Maclure (1763–1840)
OBSERVATIONS ON THE GEOLOGY OF THE UNITED STATES. 168

Ernst Friederich, Baron von Schlottheim (1764–1832)
ON THE USE OF FOSSILS IN GEOLOGICAL INVESTIGATIONS. 174

William Nicol (c. 1768–1851)
THE NICOL PRISM. 176
THE PREPARATION OF THIN SECTIONS. 177

Friedrich Heinrich Alexander, Baron von Humboldt (1769–1859)
VOLCANIC ACTIVITY IN MEXICO. 179
EARTHQUAKES AND VOLCANOES IN THE AMERICAS. 182

PAGE

Léopold Chrétien Frédéric Dagobert Cuvier (1769–1832)
REVOLUTIONS AND CATASTROPHES IN THE HISTORY OF THE EARTH. . 188

Léopold Chrétien Frédéric Dagobert Cuvier (1769–1832) and *Alexandre Brongniart* (1770–1847)
STRATIGRAPHY OF THE PARIS BASIN. 194

William Smith (1769–1839)
THE STRATA OF ENGLAND 201

John Macculloch (1773–1835)
METAMORPHIC AND IGNEOUS ROCKS OF THE WESTERN ISLANDS OF SCOTLAND 205

Leopold von Buch (1774–1852)
THE IGNEOUS ORIGIN OF BASALT 208
THE RAISED COASTS OF SWEDEN 209
THE UPHEAVAL OF VOLCANOES. 210
THE IMPORTANCE OF FOSSILS. 212

Pierre Louis Antoine Cordier (1777–1862)
THE CRYSTALLINE NATURE OF VOLCANIC ROCK. 214
THE TEMPERATURE GRADIENT OF THE EARTH'S INTERIOR. 216

Jean Baptiste Julien d'Omalius d'Halloy (1783–1875)
THE SYSTEMATIC CLASSIFICATION OF GEOLOGIC FORMATIONS 220

Adam Sedgwick (1785–1873)
THE METAMORPHISM OF SEDIMENTARY ROCKS 222
ON THE CLASSIFICATION AND NOMENCLATURE OF THE LOWER PALEO-ZOIC ROCKS OF ENGLAND AND WALES 227

Louis Constant Prévost (1787–1867)
A NEW VOLCANIC ISLAND NOT UPHEAVED 230

Francis Walker Gilmer (1790–1826)
THE NATURAL BRIDGE OF VIRGINIA A RESULT OF EROSION. 233

Jean René Constant Quoy (1790–1869) and *Joseph Paul Gaimard* (c. 1790–1858)
THE FORMATION OF CORAL ISLANDS. 235

Charles Babbage (1790–1871)
THE TEMPLE OF SERAPIS. 237

Sir Roderick Impey Murchison (1792–1871)
THE SILURIAN SYSTEM. 244
THE PERMIAN SYSTEM. 247

CONTENTS

PAGE

Edward Hitchcock (1793–1864)

    FOOT MARKS ON NEW RED SANDSTONE IN MASSACHUSETTS 250

    METAMORPHISM OF THE NEWPORT CONGLOMERATE. 252

Bernard Studer (1794–1887)

    METAMORPHIC MESOZOIC ROCKS DISTINGUISHED FROM PALEOZOIC. 257

Henry Thomas De la Beche (1796–1855)

    THE DRAFTING OF GEOLOGICAL SECTIONS 260

Sir Charles Lyell (1797–1875)

    UNFORMITARIANISM. 263

    SUBDIVISIONS OF THE TERTIARY EPOCH 268

George Poulett Scrope (1797–1876)

    VOLCANOES 274

    THE ORIGIN OF VALLEYS. 278

Abraham Gesner (1797–1864)

    THE GEOLOGY OF GRAND MANAN. 280

Ebenezer Emmons (1798–1863)

    THE TACONIC SYSTEM. 284

Jean Baptiste Armand Louis Leonce Élie de Beaumont (1798–1874)

    THE NATURE AND CAUSE OF MOUNTAIN BUILDING 288

Sir William Edmond Logan (1798–1875) and *Thomas Sterry Hunt* (1826–1892)

    GEOLOGIC OUTLINE OF CANADA. 291

Sir William Edmond Logan (1798–1875)

    HURONIAN AND LAURENTIAN DIVISIONS OF THE AZOIC ROCKS . . . 295

    A PRE-CAMBRIAN FOSSIL. 296

Heinrich Georg Bronn (1800–1862)

    THE PLAN OF CREATION IN THE SERIES OF GEOLOGICAL AGES . . . 299

George Perkins Marsh (1801–1882)

    INFLUENCE OF THE FOREST ON FLOODS 305

Hugh Miller (1802–1856)

    A QUARRY LABORER AS A GEOLOGICAL OBSERVER. 313

Sir Richard Owen (1804–1892)

    THE ARCHEOPTERYX FROM SOLENHOFEN. 317

Daniel Sharpe (1806–1856)

    THE RELATION BETWEEN DISTORTION AND SLATY CLEAVAGE 321

PAGE

Reinhard Bernhardi
    AN HYPOTHESIS OF EXTENSIVE GLACIATION IN PREHISTORIC TIME . 327

Louis Jean Agassiz (1807–1873)
    EVIDENCE OF A GLACIAL EPOCH. 329

William Barton Rogers (1805–1881)
    FOSSIL EVIDENCE OF THE EARLY PALEOZOIC AGE OF CERTAIN ROCKS
    IN NEW ENGLAND. 336

Henry Darwin Rogers (1809–1866) and *William Barton Rogers* (1805–1881)
    MOUNTAIN BUILDING FORCES EXEMPLIFIED BY THE APPALACHIAN
    SYSTEM 338

Henry Darwin Rogers (1809–1866)
    AN INQUIRY INTO THE ORIGIN OF THE APPALACHIAN COAL STRATA,
    BITUMINOUS AND ANTHRACITIC. 346

Charles Robert Darwin (1809–1882)
    The Origin of Coral Reefs and Islands 354
    THE SETTLING OF CRYSTALS IN MOLTEN LAVA 357
    THE NATURE OF GRANITE CONTACTS 359
    RECENT UPLIFT OF THE SOUTH AMERICAN COAST 361
    THE IMPERFECTION OF THE GEOLOGIC RECORD OF LIFE DEVELOPMENT 363

Robert Everest (c. 1805–c. 1875)
    A QUANTITATIVE STUDY OF STREAM TRANSPORTATION 365

Andrew Atkinson Humphreys (1810–1883) and *Henry Larcom Abbot*
    (1831–1927)
    THE AMOUNT OF SEDIMENT CARRIED BY THE MISSISSIPPI 367

William H. Sidell (1810–1873)
    ON CERTAIN GEOLOGICAL PHENOMENA OF THE MISSISSIPPI DELTA. . 369

Thomas Oldham (1817–1878)
    STREAM EROSION AND TRANSPORTATION 371

Alexandre Surell (1813–1887)
    TORRENTIAL STREAMS AND THEIR CONTROL 372

Joseph Beete Jukes (1811–1869)
    DISTINCTION BETWEEN CONTEMPORANEOUS AND INTRUSIVE TRAP. . 378

Robert Wilhelm Eberhard von Bunsen (1811–1899)
    GENETIC RELATIONS OF IGNEOUS ROCKS. 381

Robert Mallet (1810–1881)
    THE DYNAMICS OF EARTHQUAKES 384

PAGE

John Henry Pratt (c. 1811–1871)

THE ATTRACTION OF THE HIMALAYA MOUNTAINS UPON THE PLUMB-
LINE IN INDIA 393

THE DEFLECTION OF THE PLUMB-LINE IN INDIA AND THE COMPEN-
SATORY EFFECT OF A DEFICIENCY OF MATTER BELOW THE HIMA-
LAYA MOUNTAINS 395
SPECULATION ON THE CONSTITUTION OF THE EARTH'S CRUST 398

George Bedell Airy (1801–1892)

AN HYPOTHESIS OF CRUSTAL BALANCE. 401

James Hall (1811–1898)

CORRELATION OF PALEOZOIC ROCKS OF NEW YORK WITH THOSE OF
EUROPE. 406
THE RELATION OF MOUNTAINS TO REGIONS OF THICK SEDIMENTARY
ACCUMULATION. 406
PHYSICAL CONDITIONS IN PALEOZOIC SEAS 413

James Dwight Dana (1813–1895)

ORIGIN OF THE MINERAL CONSTITUTION OF IGNEOUS ROCKS 416
ON THE ORIGIN OF CONTINENTS AND MOUNTAIN RANGES. 419
ON THE NATURE OF VOLCANIC ERUPTIONS 423

Sir Andrew Crosbie Ramsay (1814–1891)

MARINE DENUDATION IN SOUTHERN WALES 430
ON THE PROBABLE EXISTENCE OF GLACIERS AND ICEBERGS IN THE
PERMIAN EPOCH 435

Gabriel August Daubrée (1814–1896)

EXPERIMENTS ON THE ACTION OF SUPERHEATED WATER IN THE
FORMATION OF SILICATES 440
FRACTURES RESULTING FROM TORSIONAL STRAINS. 443
EXPERIMENTS ON THE FOLDING AND FAULTING OF STRATA 445

Alphonse Favre (1815–1890)

EXPERIMENTAL REPRODUCTION OF MOUNTAIN STRUCTURES BY
LATERAL COMPRESSION 448

J. Peter Lesley (1819–1903)

APPALACHIAN STRUCTURE AND TOPOGRAPHY 450
PHYSIOGRAPHIC PROVINCES OF EASTERN UNITED STATES. 457

John William Dawson (1820–1899)

MARINE ALLUVIAL SOILS. 461
CARBONIFEROUS FOSSILS OF THE JOGGINS SECTION 462

John Strong Newberry (1822–1892)

CIRCLES OF DEPOSITION IN AMERICAN SEDIMENTARY ROCKS 465

PAGE

Joseph Le Conte (1823–1901)
    ON THE ORIGIN OF NORMAL FAULTS AND OF THE STRUCTURE OF THE
      BASIN REGION 467

Sir William Thomson, Lord Kelvin (1824–1907) .
    ON THE SECULAR COOLING OF THE EARTH 472

Hermann Sternberg (1825–1885)
    STREAM PROFILES. 477

Thomas Henry Huxley (1825–1895)
    THE GEOLOGICAL HISTORY OF THE HORSE 479

Henry Clifton Sorby (1826–1908)
    ON THE ORIGIN OF SLATY CLEAVAGE 484
    ON THE MICROSCOPICAL STRUCTURE OF CRYSTALS, INDICATING THE
      ORIGIN OF MINERALS AND ROCKS. 488
    THIN SECTIONS AND THEIR VALUE TO THE GEOLOGIST. 492

Ferdinand André Fouqué (1828–1904) and *Auguste Michel-Lévy* (1844–1908)
    THE ARTIFICIAL PRODUCTION OF FELDSPARS 494
    THE ARTIFICIAL PRODUCTION OF A CRYSTALLINE IGNEOUS ROCK . . 495

William Henry Brewer (1828–1910)
    THE DEPOSITION OF CLAY IN SALT WATER. 497

Carl Ochsenius (1830–1906)
    THE DEPOSITION OF ROCK SALT 499

Eduard Suess (1831–1914)
    ON MOUNTAIN STRUCTURES AND THE FORCES WHICH FORM THEM. 503
    STRUCTURAL FEATURES OF THE EARTH. 507

Thomas Mellard Reade (1832–1909)
    THE IMPORTANCE OF SOLUTION AS A FACTOR IN EROSION. 509

Baron Ferdinand von Richthofen (1833–1905)
    A CLASSIFICATION OF VOLCANIC ROCKS 511
    MARINE ABRASION AND TRANSGRESSION. 512

John Wesley Powell (1834–1902)
    A CLASSIFICATION OF STREAMS AND VALLEYS. 518

Sir Archibald Geikie (1835–1924)
    ON DENUDATION NOW IN PROGRESS 523

Gaston de la Noë (1836–1902) and *Emmanuel de Margerie* (1862–
    CONCERNING THE FASHIONING OF SLOPES 529

Ferencz Posěpný (1836–1895)
    GROUND WATER AND THE DEPOSITION OF ORE BODIES. 537

PAGE

Raphael Pumpelly (1837–1923)

   THE RELATION OF SECULAR ROCK-DISINTEGRATION TO LOESS, GLACIAL DRIFT, AND ROCK BASINS 542

Richard Anthony Proctor (1837–1888)

   METEORIC SYSTEMS AND THE ORIGIN OF THE EARTH. 547

James Geikie (1839–1915)

   ON CHANGES OF CLIMATE DURING THE GLACIAL EPOCH 549

Nathaniel Southgate Shaler (1841–1906)

   THE CHANGE OF SEA LEVEL 551
   ORIGIN OF FIORDS AND ESTUARIES 553

Clarence Edward Dutton (1841–1912)

   THE GREAT DENUDATION 555
   ISOSEISMAL LINES AND EARTHQUAKE INTENSITY. 560
   ISOSTASY. 566

Karl August Lossen (1841–1893)

   THE DISTINCTION BETWEEN CONTACT AND REGIONAL METAMORPHISM. 569

Alexander William Bickerton (1842–1914)

   COLLISION AND ACCRETION HYPOTHESIS OF EARTH ORIGIN. 574

Charles Lapworth (1842–1920)

   THE ORDOVICIAN SYSTEM 578
   THE SECRET OF THE HIGHLANDS 579

Grove Karl Gilbert (1843–1918)

   AN ANALYSIS OF SUBAERIAL EROSION 581
   LAND SCULPTURE. 586
   LACCOLITHS AND THEIR ORIGIN. 592
   HISTORY OF LAKE BONNEVILLE. 596
   THE TOPOGRAPHIC FEATURES OF LAKE SHORES. 600

Thomas Crowder Chamberlin (1843–1928)

   THE METHOD OF MULTIPLE WORKING HYPOTHESIS 604
   ON THE INTERIOR STRUCTURE, SURFACE TEMPERATURE, AND AGE OF THE EARTH 612
   THE PLANETESIMAL HYPOTHESIS 618

Marcel Alexandre Bertrand (1847–1907)

   OVERTHRUSTING IN THE ALPS. 631

Archibald Robertson Marvine (1848–1876)

   PROCESSES OF EROSION IN THE COLORADO FRONT RANGE 637

PAGE

Israel Charles White (1848–1927)
    THE ANTICLINAL THEORY OF GAS ACCUMULATION. 639

Albert Heim (1849–1937)
    ON ROCK DEFORMATION DURING MOUNTAIN BUILDING. 642

William Morris Davis (1850–1934)
    THE EROSION CYCLE IN THE LIFE OF A RIVER 649
    PLAINS OF MARINE AND SUBAERIAL DENUDATION. 656

W J McGee (1853–1912)
    THE COASTAL PLAIN AND THE FALL LINE 661

Joseph Paxson Iddings (1857–1920)
    PHENOCRYSTS 663
    RELATION BETWEEN MINERAL COMPOSITION AND MODE OF OCCUR-
    RENCE OF IGNEOUS ROCKS 663

Charles Richard Van Hise (1857–1918)
    PRE-CAMBRIAN GEOLOGY. 666
    THE MEANING OF ROCK FLOWAGE 673

GUIDE TO SUBJECT MATTER 683

INDEX. 693

A SOURCE BOOK IN GEOLOGY

A SOURCE BOOK IN GEOLOGY

DA VINCI

Leonardo da Vinci (1452–1519), Florentine painter, sculptor, architect, and naturalist, made a profound impress upon art, engineering, and science but contributed little to the literature of any science. His *Literary Works* are not much more than fragmentary notes. In geology, his fame rests securely upon his ideas concerning the nature of fossils, ideas which were revolutionary in his day.

THE EARTH AND THE SEA

From *Leonardo da Vinci's Note-books*, pp. 95–112, arranged and rendered into English by Edward McCurdy, New York, 1923.

Of the Sea Which Girdles the Earth

I perceive that the surface of the earth was from of old entirely filled up and covered over in its level plains by the salt waters, and that the mountains, the bones of the earth, with their wide bases, penetrated and towered up amid the air, covered over and clad with much high-lying soil. Subsequently, the incessant rains have caused the rivers to increase, and by repeated washing, have stripped bare part of the lofty summits of these mountains, leaving the site of the earth, so that the rock finds itself exposed to the air, and the earth has departed from these places. And the earth from off the slopes and the lofty summits of the mountains has already descended to their bases, and has raised the floors of the seas which encircle these bases, and caused the plain to be uncovered, and in some parts has driven away the seas from there over a great distance.

Why Water Is Salt

. . . The saltness of the sea is due to the numerous springs of water, which, in penetrating the earth, find the salt mines, and

1

dissolving parts of these carry them away with them to the ocean and to the other seas, from whence they are never lifted by the clouds which produce the rivers.   So the sea would be more salt in our times than it has ever been at any time previously. . . .

Changes of Earth and Sea

The destruction of marshes will be brought about when turbid rivers flow into them.   This is proved by the fact that where the river flows swiftly it washes away the soil, and where it delays there it leaves its deposit, and both for this reason, and because water never travels so slowly in rivers as it does in the marshes of the valleys, the movement of the waters there is imperceptible. But in these marshes the river has to enter through a low, narrow, winding channel, and it has to flow out over a large area of but little depth; and this is necessary because the water flowing in the river is thicker and more laden with earth in the lower than in the upper part; and the sluggish water of the marshes also is the same, but the variation between the lightness and heaviness of the upper and lower waters of the marshes far exceeds that in the currents of rivers, in which the lightness of the upper part differs but little from the heaviness of the part below.

So the conclusion is that the marsh will be destroyed because it is receiving turbid water below, while above, on the opposite side of the same marsh, only clear water is flowing out; and, consequently, the bed of the marsh will of necessity be raised by means of the soil which is being continually discharged into it.

The shells of oysters and other similar creatures which are born in the mud of the sea, testify to us of the change in the earth round the centre of our elements.   This is proved as follows:—the mighty rivers always flow turbid because of the earth stirred up in them through the friction of their waters upon their bed and against the banks; and this process of destruction uncovers the tops of the ridges formed by the layers of these shells, which are embedded in the mud of the sea where they were born when the salt waters covered them.   And these same ridges were from time to time covered over by varying thicknesses of mud which had been brought down to the sea by the rivers in floods of varying magnitude; and in this way these shells remained walled up and dead beneath this mud, which became raised to such a height that the bed of the sea emerged into the air. And now these beds are of so great a height that they have become hills or lofty mountains, and the rivers, which wear away the sides of these mountains, lay bare the strata of the shells, and so the light surface of the earth is continually raised, and the antipodes draw nearer to the centre of the earth, and the ancient beds of the sea become chains of mountains.

Origin and Meaning of Fossils

Of the Flood and of Marine Shells. If you should say that the shells which are visible at the present time within the borders of Italy, far away from the sea and at great heights, are due to the Flood having deposited them there, I reply that, granting this Flood to have risen seven cubits above the highest mountain, as he has written who measured it, these shells which always inhabit near the shores of the sea ought to be found lying on the mountain sides, and not at so short a distance above their bases, and all at the same level, layer upon layer.

Should you say that the nature of these shells is to keep near the edge of the sea, and that as the sea rose in height the shells left their former place and followed the rising waters up to their highest level:—to this I reply that the cockle is a creature incapable of more rapid movement than the snail out of water, or is even somewhat slower, since it does not swim, but makes a furrow in the sand, and supporting itself by means of the sides of this furrow it will travel between three and four braccia in a day; and therefore with such a motion as this it could not have travelled from the Adriatic sea as far as Monferrato in Lombardy, a distance of two hundred and fifty miles in forty days,—as he has said who kept a record of that time.

And if you say that the waves carried them there—they could not move by reason of their weight except upon their base. And if you do not grant me this, at any rate allow that they must have remained on the tops of the highest mountains, and in the lakes which are shut in among the mountains, such as the lake of Lario or Como, and Lake Maggiore, and that of Fiesole and of Perugia and others.

If you should say that the shells were empty and dead when carried by the waves, I reply that where the dead ones went the living were not far distant, and in these mountains are found all living ones, for they are known by the shells being in pairs and by their being in a row without any dead, and a little higher up is the place where all the dead with their shells separated have been cast up by the waves, near where the rivers plunged in mighty chasm into the sea.   So it was with the Arno, which fell from the Gonfolina near to Monte Lupo and there left gravel deposits, which deposits are still to be seen welded together and forming one concrete mass of various kinds of stones from different localities and of varying colour and hardness.   And a little further on, where the river turns towards Castel Fiorentino, the hardening of the sand has formed tufa stone; and below this it has deposited the mud in which the shells lived; and the mud has risen by degrees as the floods of the Arno poured their turbid waters into this sea. So from time to time the floor of the sea was raised, and this caused these shells to be in layers.

This is seen in the cutting of Colle Gonzoli, which has been made precipitous by the action of the Arno wearing away its base, in which cutting the aforesaid layers of shells are plainly to be seen in the bluish clay, and there are also to be found other things from the sea.

As for those who say that the shells are found over a wide area and produced at a distance from the sea by the nature of the locality and the disposition of the heavens which moves and influences the place to such a creation of animal life,—to them it may be answered that, granted such an influence over these animals, they could not happen all in one line, save in the case of those of the same species and age; and not one old and another young, one with an outer covering and another without, one broken and another whole, nor one filled with sea sand, and the fragments great and small of others inside the whole shells which stand gaping open; nor the claws of crabs without the rest of their bodies; nor with the shells of other species fastened on to them, like animals crawling over them and leaving the mark of their track on the outside where it has eaten its way like a worm in wood; nor would there be found among them bones and teeth of fish which some call arrows, others serpents' tongues; nor would so many parts of different animals be found joined together, unless they had been thrown up there upon the borders of the sea.

And the Flood could not have carried them there, because things which are heavier than water do not float high in the water, and the aforesaid things could not be at such heights unless they had been carried there floating on the waves, and that is impossible on account of their weight.

Where the valleys have never been covered by the salt waters of the sea, there the shells are never found.

Since things are far more ancient than letters, it is not to be wondered at if in our days there exists no record of how the aforesaid seas extended over so many countries; and if moreover such record ever existed, the wars, the conflagrations, the deluges of the waters, the changes in speech and habits have destroyed every vestige of the past. But sufficient for us is the testimony of things produced in the salt waters and now found again in the high mountains far from the seas.

Shells and the Reason of Their Shape. The creature that resides within the shell constructs its dwelling with joints and seams and roofing and the other various parts, just as man does in the house in which he dwells; and this creature expands the house and roof gradually in proportion as its body increases and as it is attached to the sides of these shells. Consequently the brightness and smoothness which these shells possess on the inner side is somewhat dulled at the point where they are attached to the creature that dwells there, and the hollow of it is roughened, ready to receive the knitting together of the muscles by means of which the creature draws itself in when it wishes to shut itself up within its house.

When nature is on the point of creating stones, it produces a kind of sticky paste, which, as it dries, forms itself into a solid mass together with whatever it has enclosed there, which, however, it does not change into stone but preserves within itself in the form in which it has found them. This is why leaves are found whole within the rocks which are formed at the bases of the mountains, together with a mixture of different kinds of things, just as they have been left there by the floods from the rivers which have occurred in the autumn seasons; and there the mud caused by the successive inundations has covered them over, and then this mud grows into one mass together with the aforesaid paste, and becomes changed into successive layers of stone which correspond with the layers of the mud.

Of Creatures Which Have Their Bones on the Outside, Like Cockles, Snails, Oysters, Scollops, "Bouoli" and the Like, which are of Innumerable Kinds. When the floods of the rivers which were turbid with fine mud deposited this upon the creatures which dwelt beneath the waters near the ocean borders, these creatures became embedded in this mud, and finding themselves entirely covered under a great weight of mud they were forced to perish for lack of a supply of the creatures on which they were accustomed to feed.

In course of time the level of the sea became lower, and as the salt water flowed away this mud became changed into stone; and such of these shells as had lost their inhabitants became filled up in their stead with mud; and consequently, during the process of change of all the surrounding mud into stone, this mud also which was within the frames of the half-opened shells, since by the opening of the shell it was joined to the rest of the mud, became also itself changed into stone; and therefore all the frames of these shells were left between two petrified substances, namely that which surrounded them and that which they enclosed.

These are still to be found in many places, and almost all the petrified shell fish in the rocks of the mountains still have their natural frame round them, and especially those which were of a sufficient age to be preserved by reason of their hardness, while the younger ones which were already in great part changed into chalk were penetrated by the viscous and petrifying moisture.

Of Shells in Mountains. And if you wish to say that the shells are produced by nature in these mountains by means of the influence of the stars, in what way will you show that this influence produces in the very same place shells of various sizes and varying in age, and of different kinds?

AGRICOLA

Agricola [Georg Bauer] (1494–1555), Saxon physician and professor of chemistry at Chemnitz. His *De re metallica* was the basis for all later metallurgy.

THE STRUCTURE OF THE EARTH AND THE FORCES WHICH CHANGE THE EARTH

From *De ortu et causis subterraneorum*, 1546, in the translation of *De re metallica*, by Herbert C. and Lou H. Hoover, 1912.

Vein Systems of the Earth

I now come to the *canales* in the earth. These are veins, veinlets, and what are called "seams in the rocks." These serve as vessels or receptacles for the material from which minerals (*res fossiles*) are formed. The term *vena* is most frequently given to what is contained in the *canales*, but likewise the same name is applied to the *canales* themselves. The term vein is borrowed from that used for animals, for just as their veins are distributed through all parts of the body, and just as by means of the veins blood is diffused from the liver throughout the whole body, so also the veins traverse the whole globe, and more particularly the mountainous districts; and water runs and flows through them. With regard to veinlets or stringers and "seams in the rocks," which are the thinnest stringers, the following is the mode of their arrangement. Veins in the earth, just like the veins of an animal, have certain veinlets of their own, but in a contrary way. For the larger veins of animals pour blood into the veinlets, while in the earth the humours are usually poured from the veinlets into the larger veins, and rarely flow from the larger into the smaller ones. As for the seams in the rocks (*commissurae saxorum*) we consider that they are produced by two methods: by the first, which is peculiar to themselves, they are formed at the same time as the rocks, for the heat bakes the refractory material into stone and the non-refractory material similarly heated exhales its humours and is made into "earth" generally friable. The other method is common also to veins and veinlets, when water is collected into

7

one place it softens the rock by its liquid nature, and by its weight and pressure breaks and divides it. Now, if the rock is hard, it makes seams in the rocks and veinlets, and if it is not too hard it makes veins. However, if the rocks are not hard, seams and veinlets are created as well as veins. If these do not carry a very large quantity of water, or if they are pressed by a great volume of it, they soon discharge themselves into the nearest veins.

Underground Waters

Besides rain there is another kind of water by which the interior of the earth is soaked, so that being heated it can continually give off *halitus* (steam), from which arises a great and abundant force of waters. *Halitus* rises to the upper parts of the *canales*, where the congealing cold turns it into water, which by its gravity and weight again runs down to the lowest parts and increases the flow of water if there is any. If any find its way through a *canales dilatata* the same thing happens, but it is carried a long way from its place of origin. The first phase of distillation teaches us how this water is produced, for when that which is put into the ampulla is warmed it evaporates (*expirare*), and this *halitus* rising into the operculum is converted by cold into water, which drips through the spout. In this way water is being continually created underground . . . And so we know from all this that of the waters which are under the earth, some are collected from rain, some arise from *halitus* (steam), some from river-water, some from sea-water; and we know that the *balitum* is produced within the earth partly from rain-water, partly from river-water, and partly from sea-water.

Source of Minerals

I will now speak of solidified juices (*succi concreti*). I give this name to those minerals which are without difficulty resolved into liquids (*humore*). Some stones and metals, even though they are themselves composed of juices, have been compressed so solidly by the cold that they can only be dissolved with difficulty or not at all. . . . For juices, as I said above, are either made when dry substances immersed in moisture are cooked by heat, or else they are made when water flows over "earth," or when the surrounding moisture corrodes metallic material; or else they are forced out of the ground by the power of heat alone. Therefore, solidified juices originate from liquid juices, which either heat or cold have condensed. But that which heat has dried, fire reduces to dust, and moisture dissolves. Not only does warm or cold water dissolve certain solidified juices, but also humid air; and a juice which the cold has condensed is liquefied by fire and warm water. A salty juice is condensed into salt; a bitter one into soda; an astringent and sharp one into alum or into vitriol. Skilled workmen in a similar way to nature, evaporate water which contains juices of this kind until it is condensed; from salty ones they make salt, from aluminous ones alum, from one which contains vitriol they make vitriol. These workmen imitate nature in condensing liquid juices with heat, but they cannot imitate nature in condensing them by cold. . . .

Having now refuted the opinions of others, I must explain what it really is from which metals are produced. The best proof that there is water in their materials is the fact that they flow when melted, whereas they are again solidified by the cold of air or water. This, however, must be understood in the sense that there is more water in them and less "earth";* for it is not simply water that is their substance but water mixed with "earth." And such a proportion of "earth" is in the mixture as may obscure the transparency of the water, but not remove the brilliance which is frequently in unpolished things. Again, the purer the mixture, the more precious the metal which is made from it, and the greater its resistance to fire. But what proportion of "earth" is in each liquid from which a metal is made no mortal can ever ascertain, or still less explain, but the one God has known it, Who has given certain sure and fixed laws to nature for mixing and blending things together. It is a juice (*succus*) then, from which metals are formed; and this juice is created by various operations. Of these operations the first is a flow of water which softens the "earth" or carries the "earth" along with it, thus there is a mixture of "earth" and water, then the power of heat works upon the mixtures so as to produce that kind of a juice.

We have spoken of the substance of metals; we must now speak of their efficient cause. . . . We do not deny the statement of Albertus Magnus that the mixture of "earth" and water is baked

---

* One of the four "elements" of the ancients.—EDITORS.

by subterranean heat to a certain denseness, but it is our opinion that the juice so obtained is afterward solidified by a cold so as to become a metal. . . . We grant, indeed, that heat is the efficient cause of a good mixture of elements, and also cooks this same mixture into a juice, but until this juice is solidified by cold it is not a metal. . . . This view of Aristotle is the true one.   For metals melt through the heat and somehow become softened; but those which have become softened through heat are again solidified by the influence of cold, and, on the contrary, those which become softened by moisture are solidified by heat.

Construction and Destruction of Mountains

Hills and mountains are produced by two forces, one of which is the power of water, and the other the strength of the wind. There are three forces which loosen and demolish the mountains, for in this case, to the power of the water and the strength of the wind we must add the fire in the interior of the earth.   Now we can plainly see that a great abundance of water produces mountains, for the torrents first of all wash out the soft earth, next carry away the harder earth, and then roll down the rocks, and thus in a few years they excavate the plains or slopes to a considerable depth; this may be noticed in mountainous regions even by unskilled observers.   By such excavation to a great depth through many ages, there rises an immense eminence on each side.   When an eminence has thus arisen, the earth rolls down, loosened by constant rain and split away by frost, and the rocks, unless they are exceedingly firm, since their seams are similarly softened by the damp, roll down into the excavations below.   This continues until the steep eminence is changed into a slope.   Each side of the excavation is said to be a mountain, just as the bottom is called a valley.   Moreover, streams, and to a far greater extent rivers, effect the same results by their rushing and washing; for this reason they are frequently seen flowing either between very high mountains which they have created, or close by the shore which borders them. . . .

Nor did the hollow places which now contain the seas all formerly exist, nor yet the mountains which check and break their advance, but in many parts there was a level plain, until the force of winds let loose upon it a tumultuous sea and a scathing tide.   By a similar process the impact of water entirely overthrows and flattens out hills and mountains. But these changes of local conditions, numerous and important as they are, are not noticed by the common people to be taking place at the very moment when they are happening, because, through their antiquity, the time, place, and manner in which they began is far prior to human memory.

The wind produces hills and mountains in two ways: either when set loose and free from bonds, it violently moves and agitates the sand; or else when, after having been driven into the hidden recesses of the earth by cold, as into a prison, it struggles with a great effort to burst out. For hills and mountains are created in hot countries, whether they are situated by the sea coasts or in districts remote from the sea, by the force of winds; these no longer held in check by the valleys, but set free, heap up the sand and dust, which they gather from all sides, to one spot, and a mass arises and grows together. If time and space allow, it grows together and hardens, but if it be not allowed (and in truth this is more often the case), the same force again scatters the sand far and wide. . . .

Then, on the other hand, an earthquake either rends and tears away part of a mountain, or engulfs and devours the whole mountain in some fearful chasm. In this way it is recorded the Cybotus was destroyed, and it is believed that within the memory of man an island under the rule of Denmark disappeared. Historians tell us that Taygetus suffered a loss in this way, and that Therasia was swallowed up with the island of Thera. Thus it is clear that water and the powerful winds produce mountains, and also scatter and destroy them. Fire only consumes them, and does not produce at all, for part of the mountains—usually the inner part—takes fire.

OWEN

George Owen of Henllys (1552–1613). His book was completed in 1603 and circulated in manuscript copies until printed in 1795.

THE COURSE OF ROCK SEAMS IN PENBROKSHIRE

From *Description of Penbrokshire*, 1603; reprinted in Cymmrodorion Record, No. 1, 1892.

Now that I have brefflie overrune the maner of tillinge of the lande, I will speake somewhat of the naturalle helpes which are found in the countrie to better the lande, and to make it more fruitful and apt to beare corne and grasse: the cheffest thereof I recon the lyme for that it is most commonlie and most used and founde to be lesse chardge then the marle which I take to be the best kinde of those naturalle helpes yealded by the soile it self: and first you shall understand that the lymestone is a vayne of stones runninge his course for the most parte right east & west: althoughe sometymes the same is found to wreath to the Northe & southe. yet is the mayne course thereof as I take that all other vaines of this Realme are also found to be, from west to east; of this lymestone there is found of ancient two vaynes, the one smale & of noe greate breadthe: the other verie broade both these roninge estward as I shall declare unto you. the least of these which I will first speake of is first found in the sea Cliffes at *galtopp* in the parishe of *Talbenye* and beinge there verie deepe is noe more founde above ground untill it come to *Johnston* ground which lieth East of *galtop*. . . . from Johnston it runneth farther eastward and sheweth it self in the Cliffe at haroston somewhat south of the ould church there and so crosseth over there. the first braunche of *Milford* to *Boulston* grounde where it is found and burned, and goeth on to *Picton* land and *Sleabech*, and thence crosseth the other braunche of Mylford, and holdinge eastwarde appeareth at *Muncton* by the woodd, and so estwarde to *Ludchurche* & so eastward to the sea and passeth out of *Penbroksheere*. this veyne is not of breadth above a but lengthe or stons Cast, & therefore whosoeuer seaketh sowthwarde or Northwarde over that breadth misseth it, but eastward and westward is founde to continue althoughe not in every place appearinge by reason of his deepe lyeinge in the grounde in some places. . . .

The seconde vayne of lymestone & cheeffest of the two beginneth at the Mouthe of Millford havon, west of the *Nangle*, at a place called *west pill:* where the one side of the pill you shall perceave the lymestone, and the other a red stone, which kinde of Red stone for the most parte accompanieth this veyne of lymestone allmost throwe out, as it were a cognizaunce of the lymestone being hott & fyrie & therefore this redstone is in collor & substance like a stone burned with fire. this veyne of lymestone is verye broade, for southward as yt goeth yt reacheth to the sea both in *Penbrokshire: Glamorgan shire & Monmouthshire* and therefore will I followe the Norther lymitt thereof & so followe on estward as his naturall course Runeth. this veyne is about seaven Myles distant from the former more southerlie then it, & so or neere they contynue together as shall be declared. . . .

---

ffor the vayne of Coales which is founde between these two vaines of lymestone as a benefitt of nature, without the which the profitt of lymestone were neere lost, thoughe in some place they burne it with wood, I will deferr to speake of till hereafter, where I meane to speake of the severalle sortes of fuell in *Penbrokshire;* onlie this I thinke fitt to say in this place that betweene the said two vaynes, from the beginning to the endinge there is a vayne if not severalle vaynes of Coales that followeth those of the lymestone and serveth for a principall fuell, in most countries, where it is found and carried into foreine partes, alsoe if the comodiousness of the sea doe so permitt. This veine of Coale in some parts ioyneth close to the first lymstone vayne as in *Penbrokshire* and *Carmerthenshire,* and in some partes it is found close by the other vayne of lymestone as in *Glamorgan, Monmouth,* and *Somercetshires,* therefore whether I shall saie there are two vaines of Coales to be founde betweene these two vaynes of lymestone, or to ymagine that the coale should wreth or turne it self in some place to the one and some places to the other, or to thinke that all the lande betweene these two vaynes should be stored with Coales, I leave it to the judgmente of the skillfull Myners, or those which with deepe judgement have entred into these hidden secreates.

DESCARTES

René Descartes (1596–1650), great French philosopher, was probably more influential than any other individual in raising European thought from the medieval level.

RATIONAL BASIS FOR HYPOTHESIS OF CREATION OF SOLAR SYSTEM

Translated from *Discours de la méthode*, Leyden, 1637; reproduced, Paris, 1902.

I did resolve to leave all this world here to their disputes, and to speak only of that which would take place in a new, if God should now create in some region, in the imaginary spaces, enough material to compose it, and that he agitate diversely and without order the divers parts of this matter, so that he form of them a chaos as confused as the poets could imagine, and that, thereafter, he do nothing other than lend his usual aid to nature, and allow it to act according to the laws which he has established.

I shall show how the greatest part of the matter of this chaos, as result of these laws, ought to be disposed and arranged in a certain fashion which would render it similar to our heavens; how, meanwhile, some of its parts ought to compose an earth, and some, planets and comets, and some others a sun and fixed stars.

THE COMPOSITION OF THE EARTH

Translated from *Les Principes de la philosophie*, 4th ed., Paris, 1681.

That the Earth and the Heaven are made of but one same matter

Finally, it is not difficult to infer from all this, that the earth and the heavens are all made of one same matter, and that although there were an infinity of worlds, they would be made only of this matter. . . .

How this fourth body is broken into many pieces

But there being many crevices in the body *E*, which enlarge more and more, they are finally become so great that it cannot be longer sustained by the binding of its parts, and that the vault

14

which it forms bursting all at once, its heaviness has made it fall
in great pieces on the surface of the body *C*. But because this
surface was not wide enough to receive all the pieces of this body
in the same position as they were before, some fall on their sides and
recline, the one upon the other. . . .

How the mountains, the plains, the seas, etc. have been produced

As a result, we may think of the bodies *B* and *F* as nothing other
than air, that *D* is the water and *C*, a very solid and very heavy
crust upon the earth's interior, from which come all the metals, and

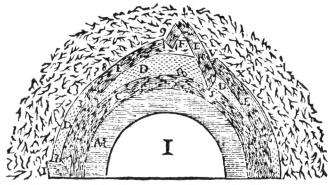

Fig. 1.—Diagram used by Descartes to illustrate his ideas concerning the
structure of the earth. (*From Les Principes de la philosophie*, 1681.)

finally that *E* is another, less massive, crust of the earth, composed
of stones, clay, sand, and mud. We see clearly the fashion in which
the seas are made above the pieces 2, 3, 6, 7, and the like, and that
those of the other pieces which are not covered with water nor
much higher than the rest, have made plains; but that which has
been more raised and strongly sloped, like 1, 2, and 9, 4, V, has
made mountains.

How the metals come to be in the mines; and how vermilion is made there

Thus the vapors of the quicksilver which ascend through the
small cracks and the wider pores of the inner earth, take with them
also accessories of gold, silver, lead, or some other metal, which
remain there thereafter, even though the quicksilver often does
not stop there, because being very fluid it passes beyond, or even
redescends. But it happens also that it does stop there sometimes,
to wit, when it meets several exhalations of which the very loosely

joined parts envelope its own and by this means change it into vermilion.   For the rest, it is only the quicksilver which can carry
with it the metals from the interior of the earth to the exterior, the
spirits and exhalations are alike as regards some others, as those of
copper, iron, and antimony.

Why metals are not found except in certain places of the earth

And it should be noted that these metals are rarely able to
ascend save at those points of the inner earth, touched by the
pieces of the exterior which have fallen upon it.   As, for example,
in the figure they ascend from 5 towards V. . . .

But it must not be hoped that, by dint of delving, one can ever
attain this inner earth which I have said to be entirely metallic.
For, more than that the exterior, which is above, is so thick that
the force of men could hardly suffice to dig through it, one would
not fail to meet various springs there, from which the water would
issue with as much more impetuosity as they were opened lower;
so that the miners could not avoid being drowned.

KIRCHER

Athanasius Kircher (1602–1680), Thuringian Jesuit, professor of languages mathematics, and natural history. His *Mundus Subterraneus*, quoted here, was the standard geological treatise of the seventeenth century.

THE SUBTERRANEAN WORLD

Translated from *Mundus subterraneus*, Amsterdam, 1678.

The Igneous Network

This drawing portrays the compartments of heat or of fire, or what is the same thing, the fire cells, throughout all the bowels

FIG. 2.—"Ideal System of Subterranean Fire Cells from which Volcanic mountains arise, as it were, like vents." (*From Kircher's Mundus Subterraneus*, 1678.)

of the Geocosm, the wonderful handiwork of GOD! These are variously distributed so that nothing be lacking which is in any way necessary for the preservation of the Geocosm. Also it is not believed that the fire is located exactly in the way the drawing shows, nor the channels placed exactly in this order. For who has

Fig. 3.—"Ideal system by which is portrayed the propulsion of waters from the sea into mountain water-chambers, by canals and subterranean channels." (*From Kircher's Mundus Subterraneus*, 1678.)

examined this? Who among men ever penetrated down there? By this drawing we only wished to show that the bowels of the earth are full of channels and fire chambers, whether placed in this way or in another. In the end we trace fire from the center through all the paths of the subterranean world clear to those volcanic mountains out on its surface. The letter *A* indicates the central fire. The rest are glory-holes of Nature, marked *B*. Fire-conducting channels *C* are not conduits; they are fissures of the earth through which the gusts of fire make their way.

The Underground Water System

The central fire, *A*, pours out surging and burning exhalations to each and every part by fire-carrying channels. Striking the water-chambers, it forms some into hot springs. Some, it reduces to vapors which, rising to the vaults of hollow caves, are there condensed by cold into waters which, released at last, give rise to fountains and rivers. Among others, some, drawing the juice of various minerals from the source matrices, coalesce into metallic bodies or into new formations of combustible material destined to nourish the fire. Here you will also see the manner in which the sea, by pressure of air and wind or movement of the tide, pushes the waters through subterranean passages to the highest water-chambers of the mountains. But the Figure will show you all this better than I could describe it with a profusion of words. You see also that the subterranean globe conforms to sea and to lands on the uppermost extent of the earth, and these to the atmosphere, as the drawing shows. The rest will be more clear from the very description of the effects and from reasoning.

PERRAULT

Pierre Perrault (1611–1680), French naturalist, dedicated his treatise on the origin of springs to "Monsieur, Mr. Huguens de Zulichem." It not only contains a brilliant deduction concerning the water running on the surface of the earth but also describes one of the earliest attempts to apply quantitative tests to a geological theory.

THE SOURCE OF WATER IN SPRINGS AND RIVERS

Translated from *De l'origine des fontaines*, pp. 196–207, Paris, 1674.

If the springs and rivers are engendered, as Aristotle says, from air condensed and resolved into water in the caverns of the earth, that is to say, as Lydiat explains it, from vapor that exhales its humidity when it is heated: and if this humidity comes from rain which it absorbs, as Aristotle says in another place, there must then be sufficient rain to give the earth enough moisture to make the vapor which can supply water to the springs and rivers throughout the entire year. In this case, according to his opinion, rain water must not only equal the greatness of the earth but greatly surpass it, since it is true that that water is subjected to the great waste which we have observed.

Concerning the idea expressed in all ancient and modern philosophy, I believe there is greater probability in attributing the source of springs and rivers to rain and snow water than there is in attributing it solely to internal distillation in the earth; and that common sense will never prefer a medium as obscure as this distillation, whose consequences seem rather feeble, to a medium as apparent as the rains, whose consequences are so great and so well-known. But as these reasons only make for the destruction of the contrary opinion, we must try to give other reasons which can establish what I maintain and show that rain water is sufficient to make springs and rivers run throughout the entire year.

Although it is no more necessary for me to prove this affirmation than for those who make objection to prove their negative, one being as difficult as the other, I shall nevertheless attempt to do so, by making rough estimates of the quantity of rain relative to the basins of rivers. . . .

. . . It is necessary for the success of our plan, to measure or estimate the water of some river as it flows from its source to the place where some stream enters it and to see if the rain water which falls around its course, being put in a reservoir, as Aristotle said, would be sufficient to make it run a whole year. I have chosen the Seine River and have examined it with sufficient precision in its course from its source to Aynay le Duc, where a stream enters and enlarges it: that is why I shall take it as the subject for the examination which I wish to make.

The course of this river from its source to Aynay le Duc is about three leagues, and the lands along its course extend to right and left about two leagues on each side, where there are other streams which go elsewhere; granting that those streams must have rain water for their subsistance as well as the Seine, I want to consider only half of this region of slopes and to say that the basin that the Seine occupies from its source to Aynay le Duc is three leagues long and two leagues wide, and consequently I make the following deductions.

If a reservoir of this length and breadth were made, it would be six square leagues in area, which reduced to *toises*,* following the measurements previously accepted, would be 31,245,144 square *toises* of surface.

In this reservoir one must imagine that during a year there has fallen from the rain a height of 19⅓ *poulces*, which is the height of an ordinary year, as we have observed. This height of 19⅓ *poulces* gives 224,899,942 hogsheads of water or thereabouts, following the measure on which we have agreed.

All this water thus collected must serve to keep this river running during a year, from its source to the place we have indicated, and must serve also to provide for all that it must lose for the nourishment of trees, plants, and herbs, and for evaporation, and for useless drainage into the river, which only enlarges it for a time and while it rains, and for wandering streams which may take another course than to this river because of irregular and contrary slopes, and other such wastes, losses, and impairments.

It would be difficult to ascertain exactly the measurement or estimation of the water of this small river, and to tell what quantity it contains. Nevertheless, as far as I can judge, it cannot have more than 1000 or 1200 *poulces* of water constantly running,

---

* A *toise* is about six feet.—EDITORS.

balancing the least that it has at its source with the most that it has at Aynay le Duc. . . .

All these things thus supposed, I say that following the measures on which we have agreed, 1200 *poulces* of water, on the basis of 83 hogsheads of water for a *poulce*, amounts in twenty-four hours to 899,600 hogsheads of water; during a year, which is 366 times more, this will total 36,453,600 hogsheads. The quantity of water flowing between the banks of this river from its source to Aynay le Duc during a year amounts only to that quantity of 36,453,600 hogsheads. But if I draw this quantity of water from 224,899,942 hogsheads, which are in this reservoir we have just imagined, there will remain 188,446,342 hogsheads, which amounts almost to five times more, and which serves to provide for the losses, diminution, and wastes which we have noticed. Therefore, only about the sixth part of the rain and snow water is necessary to make this river run continually throughout the year.

I well know that this deduction has no accuracy; but who could give one which would be precise? I believe, however, that it ought to be more satisfactory than a simple negative like Aristotle's or the premise of those who hold, without knowing why, that it does not rain enough to furnish the flow of rivers. At any rate, until someone makes more exact computations, by which he proves the contrary of what I have advanced, I shall remain convinced and shall content myself with this feeble light which gives me the observation that I have made, not having anything greater.

If then this water suffices for one river, it will suffice for all the other rivers of the world in proportion, considering especially the large margin for waste and the small area for the basin and course of the river which I allow. This is only a league on each side, for rivers are not usually closer to each other than two leagues. There is therefore some reason for saying that rain and snow water are adequate for all the rivers in the world.

I might say that there are regions where it rains very rarely, and others where it does not rain at all, which nevertheless contain fairly large streams. But the streams of these countries, where it rains only occasionally, are not continual; they are large only in winter and they dry up almost entirely in the summer. Because they are near high mountains, from which they come, abundant snow which falls on these mountains and which afterwards melts there can fill their beds as long as it lasts. When the supply is exhausted, they are left to the dryness of summer.

There are hardly any countries in the world where it never rains. The torrid zone, where this is more nearly true than anywhere else, is watered abundantly twice a year, possibly more than France is in the summer, and at least with greater abundance at certain times. But when we speak of regions where it never rains, we do not deny the possibility that there are large rivers there which may have their sources in other regions where it does rain, as, for example, the Nile, which flows through Egypt where it does not rain. There are countries in the world which do not produce wine, where not much of it could be produced, and business and commerce bring it from afar; similarly the great rivers make a kind of commerce of their waters to irrigate the provinces not watered ordinarily from heaven.

VARENIUS

Bernhard Varenius (1622–1650), German physician and geographer, laid the foundations for geography as a science other than the simple description of towns, nations, and rivers.

THE FIRST PHYSICAL GEOGRAPHY

From *Geographia generalis*, translated by Dugdale as *A Compleat System of General Geography*, London, 1734.

GEOGRAPHY is that part of *mixed Mathematics*, which explains the State of the Earth, and of it's Parts, depending on Quantity, *viz.* it's Figure, Place, Magnitude, and Motion, with the Celestial Appearances, &c. . . .

We call that *Universal* [general] *Geography* which considers the whole Earth in general, and explains it's Properties without regard to particular Countries. . . .

Of the Changes on the Terraqueous Globe, viz. of Water into Land or Land into Water

Proposition I.   To enquire how much of the Surface of the terra-queous Globe, the Earth and Water severally take up.   IT is impossible to know this accurately, because we are ignorant of the Situation of the Earth and Ocean, about the North and South Pole, and because their Superficies are terminated by irregular and crooked Lines, not easily computed or measured.   But so far as we can guess, from a bare Inspection of the Globe, it seems that the Super-ficies of the Earth and Water are nearly equal; each taking up half of the Globe's Surface.

Proposition II.   The Surfaces of the Earth and Waters, are not always equally extended, but sometimes more, and sometimes less; and what the one loses the other gains.   THE Sea frequently breaks in upon the Land in several Places and overflows it, or wastes it by degrees, and washes it away; by which means it's Superficies is enlarged to the bigness of the Plane of Earth it overflows.   But the greatest that we know of have made no sensible Alteration in the

Surface of Globe, tho' it is possible that, some Time or other, there will happen such as may. . . .

Proposition V.   Rivers leave their Shores (or part of their Chanels) dry, and form new Parcels of Ground in many Places.   1. IF their Water bring down a great deal of Earth, Sand, and Gravel out of the high Places, and leave it upon the low, in process of Time these will become as high as the other, from whence the Water flows: Or when they leave this Filth in a certain Place on one side of the Chanel, it hems in and raises Part of the Chanel which becomes dry Land.

2. IF a River take another Course, made by Art, or Nature, or some violent Cause, as the Wind, or an Inundation, it leaves it's former Chanel dry.

Proposition IX.   The Ocean in some Places forsakes the Shores, so that it becomes dry Land where it was formerly Sea.   THIS is caused by these Means: 1. If the Force of the Waves dashing against the Shore, be broken by Cliffs, Shoals, or Rocks, scattered here and there, under Water, the earthy Matter contained in the Water, as Slime, Mud, &c. is made to subside, and increase the Height of the Sand-Banks, whereby the Violence of the Ocean is more and more resisted, which makes it yield more Sediment; so that at length the Sand-Banks, being raised to a great Height and Bulk, entirely exclude the Ocean and becomes dry Land.   2. It contributes much to heightning the Shores if they be sandy and rocky, for when the Sea dashing against them, and withdrawing, carries little or nothing away from them, but every Time it approaches them it brings Dregs and Sediment, whereby they are increased in the Manner aforesaid.   3. If some neighbouring Shore consist of light, mouldring, porous Earth, which is easily washed away by the Flux of the Sea, it is mixed with the Water, and left upon some other adjacent Shore that is harder; besides, when the Sea encroaches upon one Shore, it relinquishes another not far off.   4. Large Rivers bring down vast Quantities of Sand and Gravel to their Mouths, (where they exonerate themselves into the Sea) and leave it there, partly because the Chanel is wider and shallower, and partly because the Sea resists their motion; but this is chiefly observed in Countries, whose Rivers annually overflow their Banks.   5. If frequent winds blow from the Sea to the Shorewards, and the Shore itself be rocky or of rough Earth without Sand, it gathers Slime and Mud, and becomes higher. 6. If the Tide flow quick, and without great Force, but ebb slowly, it brings a great deal of Matter to the Shore, but carries none away. 7. If the Shore descend obliquely into the Sea for a great Way, the Force of the Waves are broke and lessened by Degrees, and the Sea leaves it's Filth and Slime upon it.

BOYLE

Robert Boyle (1627–1691), the English natural philosopher, was the discoverer of "Boyle's Law," one of the most fundamental principles of physics.

Conditions at the Bottom of the Sea

From *The Works of the Honourable Robert Boyle*, Vol. III, pp. 352–354, London, 1772.

Another thing observed in the bottom of the sea is the tranquillity of the water there, if it be considerably distant from the surface. For though the winds have power to produce vast waves in that upper part that is exposed to their violence; yet the vehement agitation diminishes by degrees as the parts of the sea, being deeper and deeper, lie more and more remote from the superficies of the water.

The above-mentioned calmness of the sea at the bottom, will (I doubt not) appear strange to many, who admiring the force of the stormy winds and the vastness of the waves they raise, do not, at the same time, consider the almost incomparably greater quantity and weight of water that must be moved to make any great commotion at the bottom of the sea, upon which so great a mass of salt-water, which is heavier than fresh, is constantly incumbent. Wherefore, for the proof of the proposed paradox, I will set down a memorable relation, which my inquiries got me from the diver, elsewhere mentioned, who by the help of an engine could stay some hours under water.

This person then being asked, whether he observed any operation of the winds at the bottom of the sea, where it was of any considerable depth? answered me to this purpose, the wind being stiff, so that the waves were manifestly six or seven feet high above the surface of the water, he found no sign of it at 15 fathom deep; but if the blasts continued long, then it moved the mud at the bottom, and made the water thick and dark. And I remember he told me, which was the circumstance I chiefly designed, that staying once at the bottom of the sea very long, where it was considerably deep, he was amazed at his return to the upper parts of the water to find a storm there, which he dreamt not of, and which was raised in his absence, having taken no notice of it below, and having left the sea calm enough when he descended into it.

HOOKE

Robert Hooke (1635–1703), the English physicist and mathematician.

EARTHQUAKES

From *The Posthumous Works of Dr. Robert Hooke*, London, 1705.

That Earthquakes Change the Level of Strata

To proceed then to the Effects of Earthquakes, we find in Histories Four Sorts or *Genius's* to have been performed by them.

The first is the raising of the superficial Parts of the Earth above their former Level: and under this Head there are Four Species. The 1st is the raising of a considerable Part of a Country, which before lay level with the Sea, and making it lye many Feet, nay, sometimes many Fathoms above its former height. A 2d is the raising of a considerable part of the bottom of the Sea, and making it lye above the Surface of the Water, by which means divers Islands have been generated and produced. A 3d Species is the raising of very considerable Mountains out of a plain and level Country. And a 4th Species is the raising of the Parts of the Earth by the throwing on of a great Access of new Earth, and for burying the former Surface under a covering of new Earth many Fathoms thick.

A second sort of Effects perform'd by Earthquakes, is the depression or sinking of the Parts of the Earth's Surface below the former Level. Under this Head are also comprized Four distinct Species, which are directly contrary to the four last named.

The *First*, is a sinking of some Part of the Surface of the Earth, lying a good way within the Land, and converting it into a Lake of an almost unmeasurable depth.

The *Second*, is the sinking of a considerable Part of the plain Land, near the Sea, below its former Level, and so suffering the Sea to come in and overflow it, being laid lower than the Surface of the next adjoining Sea.

A *Third*, is the sinking of the Parts of the bottom of the Sea much lower, and creating therein vast *Vorages* and *Abysses*.

A *Fourth*, is the making bare, or uncovering of divers Parts of the Earth, which were before a good way below the Surface; and this either by suddenly throwing away these upper Parts by some subterraneous Motion, or else by washing them away by some kind of Eruption of Waters from unusual Places, vomited out by some Earthquake.

A Third sort of Effects produced by Earthquakes, are the Subversions, Conversions, and Transpositions of the Parts of the Earth.

A Fourth sort of *Effects*, are *Liquefaction. Baking, Calcining, Petrifaction, Transformation, Sublimation, Distillation*, etc.

That Water Counteracts These Effects

Another Cause there is which has been also a very great Instrument in the promoting the alterations on the Surface of the Earth, and that is the motion of the Water; whether caus'd 1*st*. By its Descent from some higher place, such as Rivers and Streams, caus'd by the immediate falls of Rain, or Snow, or by the melting of Snow from the sides of Hills. Or, 2*dly*. By the natural Motions of the Sea, such as are the Tides and Currents. Or, 3*dly*. By the accidental motions of it caus'd by Winds and Storms. Of each of these we have very many Instances in Natural Historians, and were they silent, the constant Effects, would daily speak as much. The former Principle seems to be that which generates Hills, and Holes, Cliffs, and Caverns, and all manner of Asperity and irregularity in the Surface of the Earth; and this is that which indeavours to reduce them back again to their pristine Regularity, by washing down the tops of Hills, and filling up the bottoms of Pits, which is indeed consonant to all the other methods of Nature, in working with contrary Principles of Heat and Cold, Driness, and Moisture, Light and Darkness, etc. by which there is, as it were, a continual circulation. Water is rais'd in Vapours into the Air by one Quality and precipitated down in drops by an other, the Rivers run into the Sea, and the Sea again supplies them. In the circular Motion of all the Planets, there is a direct Motion which makes them indeavour to recede from the Sun or Center, and a magnetick or attractive Power that keeps them from receding. Generation creates and Death destroys; Winter reduces what Summer produces: The Night refreshes what the Day has scorcht, and the Day cherishes what the Night benumb'd. The Air impregnates the Ground in one place, and is impregnated by it in another. All things almost circulate and have their Vicissitudes. We have multitudes of instances of the wasting of the tops of Hills, and of the filling or increasing of the Plains or lower Grounds, of Rivers continually carrying along with them great quantities of Sand, Mud, or other Substances from higher to lower places. Of the Seas washing Cliffs away and wasting the Shores: Of Land Floods carrying away with them all things that stand in their way, and covering those Lands with Mud which they overflow, levelling Ridges and filling Ditches. Tides and Currents in the Sea act in all probability what Floods and Rivers do at Land; and Storms effect that on the Sea Coasts, that great Land Floods do on the Banks of Rivers. *Egypt* as lying very low and yearly overflow'd, is inlarg'd by the sediment of the *Nile;* especially towards that part where the *Nile* falls into the *Mediterranean.* The Gulph of *Venice* is almost choak'd with the Sand of the *Po.* The Mouth of the *Thames* is grown very shallow by the continual supply of Sand brought down with the Stream. Most part of the Cliffs that Wall in this Island do Yearly founder and tumble into the Sea. By these means many parts are covered and rais'd by Mud and Sand that lye almost level with the Water, and others are discover'd and laid open that for many Ages have been hid. . . .

The Movement of Continents

But to proceed to the last Argument to confirm the 6th Proposition I at first undertook to prove, namely, that very many parts of the Surface of the Earth (not now to take notice of others) have been transform'd transpos'd and many ways alter'd since the first Creation of it. And that which to me seems the strongest and most cogent Argument of all is this, That at the tops of some of the highest Hills, and in the bottom of some of the deepest Mines, in the midst of Mountains and Quarries of Stone, etc. divers Bodies have been and daily are found, that if we thoroughly examine we shall find to be real shells of Fishes, which for these following Reasons we conclude to have been at first generated by the Plastick faculty of the Soul or Life-principle of some animal, and not from the imaginary influence of the Stars, or from any Plastick faculty inherent in the Earth itself so form'd; the stress of which Argument lies in these Particulars.

First, That the Bodies there found have exactly the Form and Matter, that is, are of the same kind of Substance for all its sensible

Properties, and that the same External and Internal Figure or Shape with the Shells of Animals.

Next, That it is contrary to all the other acts of Nature, that does nothing in vain, but always aims at an end, to make two Bodies exactly of the same Substance and Figure, and one of them to be wholly useless, or at least without any design that we can with any plausibility imagine. . . .

Next therefore, Wherever Nature does work by peculiar Forms and Substances, we find that she always joins the Body so fram'd with some other peculiar Substance. Thus the Shells of Animals, whilst they are forming are join'd with the Flesh of the Animal to which they belong. . . .

Fourthly, Wherever else Nature works by peculiar Forms, we find her always to compleat that form, and not break off abruptly. . . .

Further, if these be the apish Tricks of Nature, Why does it not imitate several other of its own Works? Why do we not dig out of Mines everlasting Vegetables, as Grass for instance, or Roses of the same Substance, Figure, Colour, Smell? etc.

The Seventh Proposition that I undertook to make probable, was, That 'tis very probable that divers of these Transpositions and Metamorphoses have been wrought even here in *England:* Many of its Hills have probably been heretofore under the Sea, and divers other parts that were heretofore high Land and Hills, have since been covered with the Sea. Of the latter of these I have given many Instances already, and that which makes the first probable, is the great quantities of Shells that are found in the most Inland Parts of this Island; in the Hills, in the Plains, in the bottoms of Mines and in the middle of Mountains and Quarries of Stones. . . .

Now 'tis not probable that other Mens Hands, or the general Deluge which lasted but a little while, should bring them there; nor can I imagine any more likely and sufficient way than an Earthquake, which might theretofore raise all these Islands of Great *Britain* and *Ireland* out of the Sea, as it did heretofore, of which I have already mention'd the Histories; or as it lately did that Island in the *Canarys* and *Azores*, in the sight of divers who are yet alive to testifie the Truth and Manner of it: And possibly

England and *Ireland* might be rais'd by the same Earthquake, by which the *Atlantis*, if we will believe *Plato*, was sunk. . . .

But as to those vast tracts of Ground that lye very far from the Sea, it may perhaps to some seem not impossible, that the Center of Gravity or Method of the attraction of the Globe of the Earth may change and shift places, and if so, then certainly all the fluid parts of the Earth will conform thereto, and then 'twill follow that one part will be cover'd and overflow'd by the Sea that was before dry, and another part be discover'd and laid dry that was before overwhelm'd. . . .

From all which Propositions, if at least they are true, will follow many others meer Corollaries which may be deduced from them.

First, That there may have been in preceding Ages, whole Countries either swallowed up into the Earth, or sunk so low as to be drown'd by the coming in of the Sea, or divers other ways quite destroyed; as *Plato's Atlantis*, etc.

Secondly, That there, many* have been as many Countries new made and produced by being raised from under the Water, or from the inward or hidden parts of the Body of the Earth, as *England.*

Thirdly, That there may have been divers Species of things wholly destroyed and annihilated, and divers others changed and varied, for since we find that there are some kinds of Animals and Vegetables peculiar to certain places, and not to be found else-where. . . .

Fourthly, That there may have been divers new varieties generated of the same Species, and that by the change of the Soil on which it was produced; for since we find that the alteration of the Climate, Soil and Nourishment doth often produce a very great alteration in those Bodies that suffer it; 'tis not to be doubted but that alterations also of this Nature may cause a very great change in the shape, and other accidents of an animated Body. . . .

---

* Probably a misprint for "That there may have been."—EDITORS.

STENO

Nicolaus Steno (1638–1687), Danish physician, churchman, and professor of anatomy, lived for many years in Florence. His *De solidarum intra solidum* marked a great advance in the knowledge of geology, although it was nearly a hundred years before the science reached the stage where this treatise should have placed it.

OF SOLIDS NATURALLY CONTAINED WITHIN SOLIDS

From *The Prodromus of Nicolaus Steno's Dissertation concerning a Solid Body Enclosed by Process of Nature within a Solid*, translated into English by John Garrett Winter, The Macmillan Company, New York, 1916.

The Problem of the Fossils

Do not be surprised, therefore, Most Serene Prince, if, for a whole year's time, and, what is more, almost daily, I have said that the investigation for which the teeth of the shark had furnished an opportunity, was very near an end. For having once or twice seen regions where shells and other similar deposits of the sea are dug up, when I observed that those lands were sediments of the turbid sea and that an estimate could be formed of how often the sea had been turbid in each place, I not only over-hastily fancied, but also dauntlessly informed others, that a complete investigation on the spot was the work of a very short time. But thereafter, while I was examining more carefully the details of both places and bodies, these day by day presented points of doubt to me as they followed one another in indissoluble connection, so that I saw myself again and again brought back to the starting-place, as it were, when I thought I was nearest the goal. I might compare those doubts to the heads of the Lernean Hydra, since when one of them had been got rid of, numberless others were born; at any rate, I saw that I was wandering about in a sort of labyrinth, where the nearer one approaches the exit, the wider circuits does one tread.

———————

The Dissertation itself I had divided into four parts, of which the first, taking the place of an introduction, shows that the inquiry concerning sea objects found at a distance from the sea, is old, delightful, and useful; but that its true solution, less doubtful in the earliest times, in the ages immediately following was rendered exceedingly uncertain.   Then after setting forth the reasons why later thinkers abandoned the belief of the ancients, and why, although one may read a great many excellent works by many authors, the question at issue has hitherto been settled by no one anew, I show, returning at length to you, that besides very many other things which under your auspices have in part been recently discovered, and in part freed from old doubts, to you is due our trust that the finishing touch shall soon be put upon this investigation also.

He, therefore, who attributes to Nature the production of any thing, names the universal agent which appears in the production of all things; he who calls the sun to share, limits that agent a little more; he who names the soul or the particular form, mentions a more limited cause than the rest: but one who nevertheless duly weighs the answers of all, finds nothing known, seeing that Nature, the sun's rays, the soul, and the particular form, are things known only by name.   But since, besides the agent, matter and place ought to be taken account of in the production of substances, it is clear that the answer ("produced by Nature") is not only more unknown than the very thing under investigation, but altogether incomplete; as, for example, mollusks found in the earth are said to have been produced by Nature, since those that grow in the sea are also Nature's work.   Nature indeed produces all things, seeing that the penetrating fluid has a place in the production of all things; but one may also say with truth that Nature produces nothing, since that fluid by itself accomplishes nothing; its determination depends upon the place and the matter to be moved. We find an illustration in man: he can produce anything if all the necessary things are at hand, but if they are wanting, can produce nothing.

If a solid body is enclosed on all sides by another solid body, of the two bodies that one first became hard which, in the mutual contact, expresses on its own surface the properties of the other surface.*   Hence it follows:

---

* Steno probably meant to write "that one first became hard which, in the mutual contact, displays the surface forms characteristic of itself."—Editors.

1. That in the case of those solids, whether of earth, or rock, which enclose on all sides and contain crystals, selenites, marcasites, plants and their parts, bones and the shells of animals, and other bodies of this kind which are possessed of a smooth surface, these same bodies had already become hard at the time when the matter of the earth and rock containing them was still fluid. And not only did the earth and rock not produce the bodies contained in them, but they did not even exist as such when those bodies were produced in them.

2. That if a crystal is enclosed in part by a crystal, a selenite by a selenite, a marcasite by a marcasite, those contained bodies had already become hard when a part of the containing bodies was still fluid.

If a solid substance is in every way like another solid substance, not only as regards the conditions of surface, but also as regards the inner arrangement of parts and particles, it will also be like it as regards the manner and place of production, if you except those conditions of place which are found time and again in some place to furnish neither any advantage nor disadvantage to the production of the body. Whence it follows:

1. That the strata of the earth, as regards the place and manner of production, agree with those strata which turbid water deposits.

2. That the crystals of mountains, as regards the manner and place of production, agree with the crystals of niter, although it is not therefore essential that the fluid in which they were produced should have been aqueous.

3. That those bodies which are dug from the earth and which are in every way like the parts of plants and animals, were produced in precisely the same manner and place as the parts of the plants and the animals were themselves produced.

What has been said concerning shells must also be said concerning other parts of animals, and animals themselves buried in the earth. Here belong the teeth of sharks, the teeth of the eagle-fish, the vertebrae of fishes, whole fish of every kind, the crania, horns, teeth, femurs, and other bones of land animals; since all these are either wholly like true parts of animals, or differ from them only in weight and color, or have nothing in common with them except the outer shape alone.

A great difficulty is caused by the countless number of teeth which every year are carried away from the island of Malta; for hardly a single ship touches there without bringing back with it some proofs of that marvel.   But I find no other answer to this difficulty than:

1. That there are six hundred and more teeth to each shark, and all the while the sharks live new teeth seem to be growing.

2. That the sea, driven by winds, is wont to thrust the bodies in its path toward some one place and to heap them up there.

3. That sharks come in shoals and so the teeth of many sharks can be left in the same place.

4. That in lumps of earth brought here from Malta, besides different teeth of different sharks, various mollusks are also found, so that even if the number of teeth favors attributing their production to the earth, yet the structure of these same teeth, the abundance in each animal, the earth resembling the bottom of the sea, and the other sea objects found in the same place, all alike support the opposite view.

Others find great difficulty in the size of the femurs, crania, and teeth, and other bones, which are dug from the earth.   But the objection, that an extraordinary size makes it necessary to conclude the size to be beyond the powers of Nature, is not of so great moment, seeing that:

1. In our own time bodies of men of exceedingly tall stature have been seen.

2. It is certain that men of unnatural size existed at one time.

3. The bones of other animals are often thought to be human bones.

4. To ascribe to Nature the production of truly fibrous bones is the same as saying that Nature can produce a man's hand without the rest of the man.

There are those to whom the great length of time seems to destroy the force of the remaining arguments, since the recollection of no age affirms that floods rose to the place where many marine objects are found to-day, if you exclude the universal deluge, four thousand years, more or less, before our time.   Nor does it seem in accord with reason that a part of an animal's body could withstand the ravages of so many years, since we see that the same bodies are often destroyed completely in the space of a few years.   But this doubt is easily answered, since the result depends wholly upon the diversity of soil; for I have seen strata of a certain kind of clay which by the thinness of their fluid decomposed all the bodies enclosed within them. I have noticed many other sandy strata which preserved whole all that was entrusted to them. And by this test it might be possible to come to a knowledge of that fluid which disintegrates solid bodies.

The Strata of the Earth

The strata of the earth are due to the deposits of a fluid:

1. Because the comminuted matter of the strata could not have been reduced to that form unless, having been mixed with some fluid and then falling from its own weight, it had been spread out by the movement of the same superincumbent fluid.

2. Because the larger bodies contained in these same strata obey, for the most part, the laws of gravity, not only with respect to the position of any substance by itself, but also with respect to the relative position of different bodies to each other.

3. Because the comminuted matter of the strata has so adjusted itself to the bodies contained in it that it has not only filled all the smallest cavities of the contained body, but has also expressed the smoothness and lustre of the body in that part of its own surface where it is in contact with the body, although the roughness of the comminuted matter by no means admits of similar smoothness and lustre.

Sediments, moreover, are formed so long as the contents in a fluid fall to the bottom of their own weight, whether the said contents have been carried thither from some other where, or have been secreted gradually from the particles of the fluid, that too, either in the upper surface, or equally from all the particles of the fluid.

Concerning the matter of the strata the following can be affirmed:

1. If all the particles in a stony stratum are seen to be of the same character, and fine, it can in no wise be denied that this stratum was produced at the time of the creation from a fluid which at that time covered all things; and Descartes also accounts for the origin of the earth's strata in this way.

2. If in a certain stratum the fragments of another stratum, or the parts of animals and plants are found, it is certain that the said

stratum must not be reckoned among the strata which settled down from the first fluid at the time of the creation.

3. If in a certain stratum we discover traces of salt of the sea, the remains of marine animals, the timbers of ships, and a substance similar to the bottom of the sea, it is certain that the sea was at one time in that place, whatever be the way it came there, whether by an overflow of its own or by the upheaval of mountains.

4. If in a certain stratum we find a great abundance of rush, grass, pine cones, trunks and branches of trees, and similar objects, we rightly surmise that this matter was swept thither by the flooding of a river, or the inflowing of a torrent.

5. If in a certain stratum pieces of charcoal, ashes, pumicestone, bitumen, and calcined' matter appear, it is certain that a fire occurred in the neighborhood of the fluid; the more so if the entire stratum is composed throughout of ash and charcoal, such as I have seen outside the city of Rome, where the material for burnt bricks is dug.

6. If the matter of all the strata in the same place be the same, it is certain that that fluid did not take in fluids of a different character flowing in from different places at different times.

7. If in the same place the matter of the strata be different, either fluids of a different kind streamed in thither from different places at different times (whether a change of winds or an unusually violent downpour of rains in certain localities be the cause) or the matter in the same sediment was of varying gravity, so that first the heavier particles, then the lighter, sought the bottom.   And a succession of storms might have given rise to this diversity, especially in places where a like diversity of soil is seen.

Concerning the position of strata, the following can be considered as certain:

1. At the time when a given stratum was being formed, there was beneath it another substance which prevented the further descent of the comminuted matter; and so at the time when the lowest stratum was being formed either another solid substance was beneath it, or if some fluid existed there, then it was not only of a different character from the upper fluid, but also heavier than the solid sediment of the upper fluid.

2. At the time when one of the upper strata was being formed, the lower stratum had already gained the consistency of a solid.

3. At the time when any given stratum was being formed it was either encompassed on its sides by another solid substance, or it covered the entire spherical surface of the earth. Hence it follows that in whatever place the bared sides of the strata are seen, either a continuation of the same strata must be sought, or another solid substance must be found which kept the matter of the strata from dispersion.

4. At the time when any given stratum was being formed, all the matter resting upon it was fluid, and, therefore, at the time when the lowest stratum was being formed, none of the upper strata existed.

As regards form, it is certain that at the time when any given stratum was being produced its lower surface, as also its lateral surfaces, corresponded to the surfaces of the lower substance and lateral substances, but that the upper surface was parallel to the horizon, so far as possible; and that all strata, therefore, except the lowest, were bounded by two planes parallel to the horizon. Hence it follows that strata either perpendicular to the horizon or inclined toward it, were at one time parallel to the horizon.

Hence it could be easily shown:

1. That all present mountains did not exist from the beginning of things.

2. That there is no growing [*vegetatio*] of mountains.

3. That the rocks of mountains have nothing in common with the bones of animals except a certain resemblance in hardness, since they agree in neither matter nor manner of production, nor in composition, nor in function, if one may be permitted to affirm aught about a subject otherwise so little known as are the functions of things.

4. That the extension of crests of mountains, or chains, as some prefer to call them, along the lines of certain definite zones of the earth, accords with neither reason nor experience.

5. That mountains can be overthrown, and fields carried over from one side of a high road across to the other; that peaks of mountains can be raised and lowered, that the earth can be opened and closed again, and that other things of this kind occur which those who in their reading of history wish to escape the name of credulous, consider myths.

The Shape and Growth of Crystals

A crystal consists of two hexagonal pyramids and an intermediate prism likewise hexagonal. I call those angles the *terminal solid angles* which form the apexes of the pyramids, but those angles the *intermediate solid angles* which are formed by the union of the pyramids with the prism. In the same way I call the planes of the pyramids *terminal planes*, and the planes of the prism the *intermediate planes*. The *plane of the base* is the section perpendicular to all the intermediate planes; the *plane of the axis* is a section in which lies the axis of the crystal, which consists of the axes of the pyramids and the axis of the prism.

The place where the first hardening of a crystal begins, whether it be between a fluid and a fluid, or between a fluid and a solid, or even in a fluid itself, may remain in doubt; but the place in which the crystal grows after it has already begun to form, is a solid in that part where the crystal is supported on it, whether the place be a stone or another crystal already formed.

———————

The following propositions will show what can be determined concerning the place of the crystal to which new crystalline matter is being added:

I

A crystal grows while new crystalline matter is being added to the external planes of the crystal already formed. No room at all is here left for the belief of those who affirm that crystals grow, plantlike, by nourishment, and that they draw their nourishment on the side where they are attached to the matrix, and that the particles thus received from the fluid of the rock, and transmitted into the fluid of the crystal, are inwardly added to the particles of the crystal.

II

This new crystalline matter is not added to all the planes but, for the most part, to the planes of the apex only, or to the terminal planes, with the result:

1. That the intermediate planes, or the quadrilateral planes, are formed by the bases of the terminal planes, and hence the

intermediate planes are larger in some crystals, smaller in others, and wholly wanting in still others.

2. That the intermediate planes are almost always striated, while the terminal planes retain traces of the matter added to them.

III

The crystalline matter is not added to all the terminal planes at the same time, nor in the same amount.   Hence it comes to pass:

1. That the axis of the pyramids does not always continue the same straight line with the axis of the prism.

2. That the terminal faces are rarely of a size, whence follows an inequality of the intermediate planes.

3. That the terminal faces are not always triangular, just as all the intermediate planes are not always quadrilateral.

4. That the terminal solid angle is broken up into several solid angles, this being the case frequently also with the solid intermediate angles.

IV

An entire plane is not always covered by crystalline matter, but exposed places are left sometimes toward the angles, sometimes toward the sides, and sometimes in the centre of the plane.   Hence it happens:

1. That the same plane, commonly so-called, does not have all its parts located in the same plane, but in different planes extending above it in different ways.

2. That a plane, commonly so-called, in many places is seen to be not a plane but a protuberance.

3. That in the intermediate planes inequalities rise like the steps of stairs.

V

The crystalline matter added to planes upon the same planes is spread out by the enveloping fluid, and gradually hardens, with the result:

1. That the surface of the crystal comes forth the smoother the more slowly the added matter has hardened, and is left wholly rough if the said matter has hardened before it has spread sufficiently.

2. That the manner in which the crystalline matter is added to the crystal can be distinguished, since where it has hardened suddenly, it reveals a surface full of small tubercles like variolar postules, as it were, just as small drops of oily fluid are wont to float upon an aqueous fluid; sometimes it shows also trilateral and depressed pyramids, if it has hardened somewhat more slowly. The tortuous fringes of the descending matter show now the place to which the fluid matter was being added, now the place toward which it was being advanced, now the arrangement of the matter added, that is, which came first, and which last. And in this way certain roughnesses always appear in the crystals of mountains, nor have I ever seen a crystal whose still unbroken surfaces possess the lustre which the rent sides of the same crystal show after it has been broken, however prolix writers on subjects relating to nature become in extolling the lustre of the crystal which is extracted from the mountains.

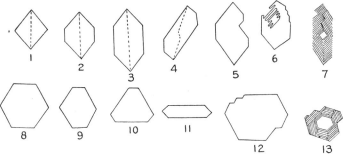

FIG. 4.—Figures 1 to 13 on the plate of illustrations accompanying Steno's *Prodromus*, 1671.

Explication of the Figures

Quoted verbatim from the "H. O. version" of the Prodromus, translated by Henry Oldenburg and published in London between 1671 and 1673.

The *thirteen* first figures, designed to explain the *Angular* Bodies of Chrystal, are reducible to two Classes;

The *first* contains *seven* differences of a Plane, in which is the *Axis* of the Chrystal. In the 1^t, 2^d and 3^d, the *axes* of the parts, out of which the body of the Chrystal is composed, do constitute one straight line, but by an intermediate column, which in the 1st figure is wanting, but is seen shorter in the 2^d, and longer in the 3^d. In the 4th figure the *axes* of the parts constituting the body of the Chrystal, do not make one straight line. The 5ᵗʰ and 6ᵗʰ figure are of the kind of those, of which I could have produced innumerable, to evince, that in the Plane of the *axe*, both the number and length of the sides are variously changed without change in the angles, and that in the very midst of the Chrystal there are left various cavities, & formed various plates. The 7ᵗʰ figure doth shew in the plane of the *axe*, how from the new Chrystallin matter laid upon the planes of the pyramids, both the number and length of the sides are variously sometimes increased, sometimes lessen'd.

The *Second* Classis contains *six* differences of the *Basis* of a Plane. In the 8ᵗʰ, 9ᵗʰ, 10ᵗʰ, and 11ᵗʰ figures, there are only six sides; yet with this difference, that in the 8ᵗʰ figure all the sides are

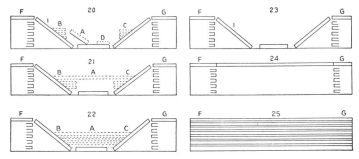

Fɪɢ. 5.—Figures 20 to 25 on the plate of illustrations accompanying Steno's *Prodromus*, 1671.

equal; in the 9ᵗʰ and 11ᵗʰ, not all but only the opposite sides are equal; but in the 10ᵗʰ, all opposite sides are unequal. In the 12ᵗʰ figure, the Plane of the *base*, which should be hexagonal, contains twelve sides. The 13ᵗʰ figure shews, how, by laying a new Chrystallin matter upon the planes of the Pyramids, sometimes the length of the sides, and the number also are variously changed in the plane of the base, without changing the angles.

The *six* last figures do *both* shew, how from the present face of *Etruria* we may collect the six distinct faces of the same Country, above discoursed of, *and* serve also for the more easy understanding of the particulars, we have deliver'd concerning the *Beds* of the Earth. The *pricked* lines represent the *Sandy* beds of the Earth, so nominated from their *main* matter, there being mix't with them divers both Clayie and stony beds.   The other lines represent the *Stony* beds, likewise so called *a potiori*, seeing there are Beds found in them that are of a softer substance. . . .

The 25th figure exhibits the perpendicular Plane of *Etruria*, at the time when the Stony Beds were yet entire, & parallel to the horizon.

The 24th shews the vast cavities, eaten out by the force of Fire and Water, without any breach in the upper Beds.

The 23th represents, how Mountains & Vally's came to be made by the ruine of the superior Beds.

The 22th, that by the Sea new Beds were made in the said Valleys.

The 21th, that of the new Beds the lower ones were consumed, the uppermost remaining untouch't.

The 20th, that by the breach of the superior sandy Beds *there* were produced Hillocks and Vallys.

LEIBNITZ

Gottfried Wilhelm, Baron von Leibnitz (1646–1716), the German mathematician and philosopher.

THE EARTH ORIGINALLY MOLTEN

Translated from Bertrand de Saint-Germain's translation, Paris, 1859, of
Protogaea, 1680(?).

In agreement with these views, some savants* developed an
hypothesis that serves to explain more clearly the order of the
world.   They suggest that vast globes, self-luminous like the fixed
stars or our sun, or projected from a luminous body, after reaching
the last stages of ebullition, were covered with scoriae like bubbling
foam.   It is as if the spots by which the ancients suspected the sun
could be altered and some day obscured, and which our optical
instruments have allowed us to see, should increase in amount until
they veiled the face of that orb.   For the greater part they believe,
in effect (as the sacred writers, in their way, tend to imply), that
*there are raging fires at the center of the earth from which flames can
break forth in eruption.*

Existing traces of the primitive aspect of nature support these
conjectures, for *all scoria* resulting from fusion is *a sort of glass.*
And the crust which covered the molten matter of the globe and
which hardened from the fused condition would be like scoria, as
happens with metals in the furnace.   If the great framework of the
earth, the exposed rocks, the imperishable silicates, are almost
entirely vitrified, does that not prove that they arose from fusion
of the bodies, brought about by the powerful action of nature's
fire on still soft material?   And as the action of that fire infinitely
surpasses that of our furnace, both in intensity and duration, is it
a matter of astonishment that it led to a result that men cannot
attain now? . . . It is everywhere true that the most simple and
primitive material in the composition of the earth, that which
represents most accurately the true nature of rock, is that which

---

* Descartes and his followers.

most resists fire, which melts under an excessive heat, and finishes by vitrifying. . . .

. . . At the same time it is readily believed that at the origin of things, before the separation of the opaque material from the luminous, *when our globe was incandescent*, the fire drove the humidity into the air, acting like a distillation. That is to say, as a result of the lowering of the temperature, it was converted into aqueous vapors. These vapors, finding themselves in contact with the chilled surface of the earth, condensed to water. The water, working over the debris of the recent conflagration, took up the fixed salts, giving rise to *a sort of lixivium*, which soon formed the sea. . . .

Finally it is credible that the consolidation of the crust of the globe, on cooling—as takes place with metals and other bodies which become more porous with fusion—has left *enormous bubbles* accordant in grandeur with the planet; that is to say, *cavities* enclosing water or air were formed under its immense vaults. It is also probable that other parts stretched out in the form of beds and that, by the diversity of material and the [irregular] distribution of heat, *the masses were not equally compressed and have burst, here and there, so that certain portions subsided to form the trough of valleys, whereas others, more solid, have remained upright like columns and, for that reason, constituted the mountains.*

To these causes would be added the action of the waters, which by their weight tended to furrow stream beds in the still soft surface. Then the vaults of the earth breaking, either from the weight of the material or from the explosion of gases, the water would be forced from the depths of the abyss across the wreckage, and, joining that which was flowing naturally from the high places, would give rise to vast inundations which would leave abundant sediment at divers points. These sediments would harden, and with repetition of the same condition sedimentary beds would be superimposed. The face of the earth, still only slightly firm, has thus been often renewed until, the causes of disturbance having been exhausted or balanced, *a more stable state was finally produced.* These facts should make us understand *the double origin of solid bodies*, first by their chilling after igneous fusion, and then by new aggregations after their solution in the waters.

MAILLET

Benoit de Maillet (1656–1738), French consul in Egypt, paradoxical philosopher. Because of its unorthodox ideas, he arranged to have *Telliamed* (anagram for de Maillet) published only after his death. The book consists of "Discourses between an Indian Philosopher and a French Missionary on the diminution of the sea, the formation of the earth, and the origin of men and animals." The excerpt contains what is probably the first recognition of the fact that older rocks contain fewer fossils of existing species than do the younger rocks.

Strata Formed beneath the Sea

From *Telliamed*, pp. 32, 34, 1748; translated anonymously, London, 1750.

My Grandfather, who has studied the various Works performed in the Bottom of the Sea, especially on the Coasts, easily discovered this Truth: He found in these two Kinds of Stone, the same Composition which the Sea daily produces, almost every Moment, in fixing to stony Bottoms or small Rocks, which it still washes with the Extremity of its Waves, the Substances which its Waters contained, or which were carried to them from the adjacent Mountains. The Position of these Quarries of Pumice and Rockstone, had the same Aspect as those Places where the Sea had formed similar ones upon the Coast. Thus the superficial Quarries in the large Mountains; which he had found very near their highest Summits, were to him new Proofs, both of the long Continuance of the Sea in such elevated Places, and of the prodigious Diminution of her Waters.

These two Kinds of Quarries are however much less frequent and thick near the Tops of high Mountains, than about the Middle; and still less so about their Middle, than at their Feet, and in Places at present near the Sea. The Reason of this is obvious; the Rock and Pumice Stones are composed of the Wrecks of certain Mountains, of small Stones which the Sea detaches from them: of small Flints which she contains, of Shells and Impurities which she brings along with her. Now nothing of all this existed at the Time of the Discovery of the first Soils. The Sea could not break them, nor convey their Wrecks to their Feet till after they had appeared:

47

Its Waters at first contained but very few Shells, since these are only found near the Shores, which were but of a small Extent at first.

In general, my Grandfather, in this Petrifaction in the Surface of our Soils, found numberless Shells, some known, others absolutely unknown, or such as are very rare on the adjacent Coasts. He found in particular a great many Corneamons,* which are very frequent in the Stones of *France*, tho' none of them are found on the Coasts of *France*.   He also observed that the unknown Shells were more deeply sunk in these Compositions, whereas those which are frequent on our Coasts were situated nearer their Surfaces.

---

* Ammonites(?).—EDITORS.

WOODWARD

John Woodward (1665–1722), professor of physick, Gresham College, London, founded, by bequest, the Woodwardian professorship at Cambridge.

Earth History and the Deluge

From *An Essay towards a Natural History of the Earth*, 1695; 3d ed., London, 1723.

My principal Intention indeed was to get as compleat and satisfactory *Information* of the whole *Mineral Kingdom* as I possibly could. To which End, I made strict *Enquiry* wherever I came, and laid out for Intelligence of *all Places* where the *Entrails* of the *Earth* were *laid open*, either by Nature, (if I may so say) or by Art, and humane industry. And wheresoever I had Notice of any considerable *natural Spelunca* or *Grotto:* any sinking of *Wells:* or digging for *Earths, Clays, Marle, Sand, Gravel, Chalk, Cole, Stone, Marble, Ores* of *Metalls,* or the like, I forthwith had recourse thereunto; where taking a just Account of every observable *Circumstance* of the Earth, Stone, Metall, or other Matter, from the *Surface* quite down to the *Bottom* of the *Pit,* I enter'd it carefully into a *Journal,* which I carry'd along with me for that Purpose. And so passing on from Place to Place, I *noted* whatever I found *memorable* in each particular *Pit, Quarry,* or *Mine:* and 'tis out of these *Notes* that my *Observations* are compil'd.

I likewise drew up a *List* of *Queries* upon this Subject; which I dispatch'd into all Parts of the *World,* far and near, wherever either I my self, or any of my Acquaintance, had any Friend resident to transmitt those *Queries* unto.

The Result was, that in time I was abundantly assured, that the *Circumstances* of *these Things* in *remoter Countries* were much the *same* with those of ours *here:* that the *Stone,* and other *terrestrial Matter,* in *France, Flanders, Holland, Spain, Italy, Germany, Denmark, Norway,* and *Sweden,* was distinguished into *Strata,* or *Layers,* as it is in *England:* that those *Strata* were divided by *parallel Fissures:* that there were enclosed in the *Stone,* and all the other denser kinds of *terrestrial Matter,* great Numbers of *Shells,* and

49

other Productions of the Sea; in the same Manner as in *that* of *this Island.* To be short, by the same Means I got sufficient Intelligence that *these Things* were found in like Manner in *Barbary,* in *Egypt,* in *Guiney,* and other Parts of *Africa:* in *Arabia, Syria, Persia, Malabar, China,* and other *Asiatick* Provinces: in *Jamaica, Barbadoes, Virginia, New-England, Brasil, Peru,* and other Parts of *America.*

I shall distribute them into *two* general *Classes* or *Sections,* whereof the *former* will comprehend my Observations upon all the *Terrestrial Matter* that is naturally disposed into *Layers,* or *Strata;* such as our *common Sand-Stone, Marble, Cole, Chalk,* all Sorts of *Earth, Marle, Clay, Sand* with some others.

Of this various Matter, thus formed into *Strata,* the far *greatest Part* of the *Terrestrial Globe* consists, from its *Surface* downwards to the *greatest Depth* we ever dig or mine. And it is upon my Observations on *this* that I have grounded all my *general Conclusions* concerning the *Earth:* all that relate to its Form: all that relate to the *Universal* and other *Deluges:* in a Word, all that relate to the several *Vicissitudes* and *Alterations* that it hath yet undergone. . . .

For upon the *particular* Observations of the said *Metallick* and *Mineral Bodies,* (which are the Subjects of the *second Section,*) I have not founded any thing but what purely and immediately concerns the *Natural History of those Bodies.*

A DISSERTATION CONCERN*ing* Shells, *and other* marine Bodies, *found at* Land: *Proving that they originaly generated and formed at* Sea: *that they are the real Spoils of once* living Animals: *and not* Stones, *or natural* Fossils,* *as some late Learned Men have thought.*

There being, I say, besides *these,* such vast *Multitudes* of *Shells* contained in *Stone,* &c. which are *intire, fair,* and absolutely free from such *Mineral Contagion:* which are to be *match'd* by others at this Day found upon our *Shores,* and which do not *differ* in any Respect from them; being of the *same Size* that those are of, and the

---

* Apparently, Woodward is here using the term "fossil" in accord with the custom of his time to designate minerals or rocks rather than in the modern sense of the term.—EDITORS.

same Shape precisely: of the *same Substance* and *Texture;* as consisting of the *same peculiar Matter,* and this constituted and disposed in the *same Manner,* as is that of their respective Fellow-kinds at *Sea:* the *Tendency* of the *Fibres* and *Striae* the same: the *Composition* of the *Lamallae,* constituted by these *Fibres,* alike in both: the same *Vestigia* of *Tendons* (by Means whereof the Animal is fastened and joyned to the *Shell*) in each: the same *Papilae:* the same Sutures, and every thing else, whether *within* or *without* the *Shell,* in its *Cavity,* or upon its *Convexity,* in the *Substance,* or upon the *Surface* of it. Besides, these *Fossil Shells* are attended with the ordinary *Accidents* of the *marine ones, ex-gr.* they sometimes grow to one another, the *lesser Shells* being fixed to the *larger:* they have Balani, Tubuli *vermiculares, Pearls,* and the like, *still* actualy *growing* upon them. And, which is very considerable, they are most exactly of the same *Specifick Gravity* with their Fellow-kinds now upon the *Shores.* Nay farther, they answer all *Chymical Tryals* in like Manner as the *Sea-Shells* do: their *Parts* when dissolved have the same *Appearance* to View, the same *Smell* and *Taste:* they have the same *Vires* and *Effects* in *Medicine,* when inwardly administred to Animal Bodies: *Aqua fortis,* Oyl of *Vitriol,* and other like *Menstrua,* have the very same *Effects* upon both.

When therefore I shall have proved more at large, that *those* which we find at *Land,* that are not matchable with any upon our *Shores,* are many of them of *those very Kinds* which the forecited *Relations* particularly assure us are found no where but in the *deeper Parts* of the *Sea:* and that as well those which we can match, as those we *cannot,* are all *Remains* of the *Universal Deluge,* when the *Water* of the *Ocean,* being boisterously turned *out* upon the *Earth,* bore along with it *Fishes* of all Sorts, *Shells,* and the like moveable Bodies, which it *left behind* at its *Return* back again to its *Chanel;* it will not, I presume, be thought *strange,* that amongst the *rest,* it left some of the *Pelagiae,* or those Kinds of *Shells* which naturaly have their Abode at *Main-Sea,* and which therefore are now never flung up upon the *Shores.* And it may very reasonably be concluded, that all these *strange* Shells, which we cannot so *match,* are of these *Pelagiae:* that the several Kinds of them are at *this Day* living in the *huge Bosom* of the *Ocean:* and that there is not any one intire *Species* of *Shell-fish,* formerly in Being, now *perish'd,* or *lost.*

So that I shall only proceed to make *Inferences* from them; which *Inferences*, in *this Part*, are all *affirmative*. Of these, the first is,

That these *Marine Bodyes* were born forth of the *Sea* by the *Universal Deluge:* and that, upon the *Return* of the *Water* back again from off the *Earth*, they were left behind at *Land*.

That during the *Time* of the *Deluge*, whilst the *Water* was *out* upon, and covered the *Terrestrial Globe*, all the *Stone* and *Marble* of the *Antediluvian Earth:* all the *Metalls* of it: all *Mineral Concretions:* and, in a Word, all *Fossils* whatever, that had before obtain'd any *Solidity*, were *totaly dissolved*, and their constituent *Corpuscles* all *disjoyned*, their *Cohaesion* perfectly ceasing.

That at length all the *Mass*, that was thus born up in the *Water*, was again *precipitated*, and *subsided* towards the Bottom. That this *Subsidence* happened generaly, and as near as possibly could be expected in so great a *Confusion*, according to the *Laws* of *Gravity*. . . .

. . . That the *Matter*, subsiding thus, *formed* the *Strata* of *Stone*, of *Marble*, of *Cole*, of *Earth*, and the rest; of which *Strata*, lying one upon another, the *Terrestrial Globe*, or at least as much of it as is ever displayed to view, doth mainly consist.

STRACHEY

John Strachey (1671–1743), English naturalist, was inducted into the Royal Society in 1717.   His geological cross section, reproduced here, was one of the first attempts at drawing a structure section.

Observations on the Strata in the Somersetshire Coal Fields

From *Philosophical Transactions of the Royal Society*, No. 360, pp. 972–73, 1719.

This is all I can say in relation to the different veins of coal and earth in the coal-works in these parts; wherein all agree in the oblique situation of the veins; and every vein hath its cliff or clives lying over it, in the same oblique manner.   All of them pitch or rise about twenty-two inches in a fathom, and almost all have the same strata of earth, malm, and rock over them, but differ in respect to their course or drift, as also in thickness, goodness, and use.

Now as coal is here generally dug in valleys, so the hills, which interfere between the several works before mentioned, seem also to observe a regular course in the strata of stone and earth found in their bowels: for in these hills . . . we find on the summits a stony arable mixt with a spungy yellowish earth and clay; under which are quarries of lyas, in several beds, to about eight or ten feet deep, and six feet under that thro' yellowish loom, you have a blue clay, enclinable to marle, which is about a yard thick: under this is another yard of whitish loom, and then a deep blue marle soft, fat, and soapy, six feet thick; only at about two feet thick, it is parted by a marchasite about six inches thick. . . .

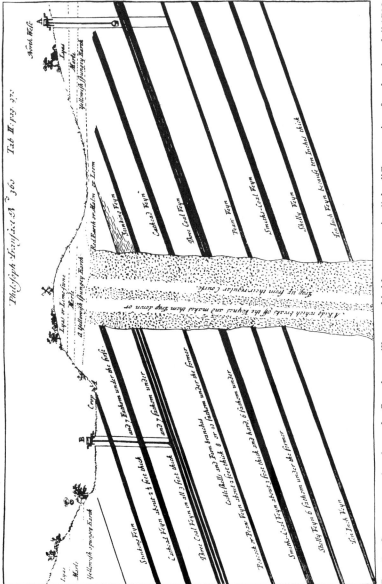

Fig. 6.—Structure section drawn by Strachey to illustrate his ideas concerning "the different veins of coal and earth," 1719.

EVANS

Lewis Evans (*c.* 1700–1756), Pennsylvania surveyor. His maps of the Eastern Colonies were the standard of the time.

The Physiography of the Middle British Colonies in America

From *Analysis of a General Map of the Middle British Colonies in America,* Printed by B. Franklin and D. Hall, Philadelphia, 1755.

The Western Series. The land, South Westward of Hudson's River, is more regularly divided, and into a greater Number of Stages than the other. The first Object worthy Regard, in this Part, is a Rief or Vein of Rocks, of the Talky or Isinglassy Kind, some two or three, or Half a Dozen Miles broad; rising generally some small Matter higher than the adjoining Land; and extending from New-York City South Westerly by the Lower Falls of Delaware, Schuylkill, Susquehanna, Gun-Powder, Potapsco, Potomack, Rapahannock, James River and Ronoak. This was the antient maritime Boundary of America, and forms a very regular Curve.

First Stage or Lower Plains. The Land between this Rief and the Sea, and from the Navesink Hills South Westward as far as this Map extends, and probably to the Extremity of Georgia, may be denominated the *Lower Plains,* and consists of Soil washt down from above, and Sand accumulated from the Ocean. Where these Plains are not penetrated by Rivers, they are a white Sea-Sand, about twenty Feet deep, and perfectly barren, as no Mixture of Soil helps to enrich them. But the Borders of the Rivers, which descend upon the Uplands, are rendered fertile by the Soil washt down with the Floods, and mixt with the Sand gathered from the Sea. The Substratum of Sea Mud, Shells and other foreign Subjects, is a perfect Confirmation of this Supposition. And hence it is, that for 40 or 50 Miles inland, and all the Way from the Navesinks to Cape-Florida, all is a perfect Barren, where the Wash from the Upland has not enriched the Borders of the Rivers; or some Ponds and Defiles have not furnished proper

55

Support for the Growth of White Cedars. There is commonly a Vein of Clay seaward of the Isinglassy Rief, some three or four Miles wide; which is a coarse Fullers Earth, and excellently fitted, with a proper Portion of Loam, to make Bricks of.

Second Stage of the Upland. From this Rief of Rocks, over which all the Rivers fall, to that Chain of broken Hills, called the South Mountain, there is the Distance of 50, 60 or 70 Miles of very uneven Ground, rising sensibly as you advance further inland; and may be denominated the *Upland.* This consists of Veins of different Kinds of Soil and Substrata, some Scores of Miles in Length, and in some Places overlaid with little Ridges and Chains of Hills. The Declivity of the Whole gives great Rapidity to the Streams; and our violent Gusts of Rain have washt it all into Gullies, and carried down the Soil to enrich the Borders of the Rivers in the *Lower Plains.* These Inequalities render half the Country not easily capable of Culture; and impoverishes it, where torne up with the Plough, by daily washing away the richer Mould that covers the Surface.

Third Stage, or Piemont. The *South* Mountain is not in Ridges like the *Endless* Mountains, but in small, broken, steep, stony Hills; nor does it run with so much Regularity. In some Places it gradually degenerates to Nothing, not to appear again for some Miles, and in others spreads several Miles in Breadth. Between the South Mountain and the hither Chain of the Endless Mountains, (often for Distinction called the North Mountain, and in some Places the Kittatinni, and Pequílin,) there is a Valley of pretty even, good Land, some 8, 10 or 20 Miles wide, and is the most considerable Quantity of valuable Land that the English are possest of; and runs through New-Jersey, Pennsylvania, Mariland and Virginia. It has yet obtained no general Name, but may properly enough be called *Piemont*, from its Situation. Besides Conveniencies always attending good Land, this Valley is every where enriched with Limestone.

Fourth Stage or the Endless Mountains. The *Endless Mountains,* so called from a Translation of the Indian Name, bearing that Significance, come next in Order. They are not confusedly scattered, and in lofty Peaks overtopping one but stretch in long uniform Ridges, scarce Half a Mile perpendicular in any Place above the intermediate Vallies. Their name is expressive of their Extent, though no Doubt, not in a literal Sense. In some Places, as towards the Kaats Kill, and the Head of Ronoak, one would be induced to imagine he had found their End, but let him look a little on either Side, and he will find them again spread in new Branches, of no less Extent than what first presented themselves.

Allegeny Mountains. The *further* Chain, or Allegeny Ridge of Mountains, keeps mostly on a Parallel with the *Isinglassy* Rief, and terminates in a rough stony Piece of Ground at the Head of Ronoak and New River. The more Easterly Chains as they run further Southward, trend also more and more Westerly; which is the Reason that the *Upland* and *Piemont* Valley are so much wider in Virginia than farther Northward. . . .

Fifth Stage or the Upper Plains. To the North Westward of the Endless Mountains is a Country of vast Extent, and in a manner as high as the Mountains themselves. To look at the abrupt Termination of it, near the Sea Level, as is the Case on the West Side of Hudson's River, below Albany, it looks as a vast high Mountain; for the Kaats Kills, though of more lofty Stature than any other Mountains in these Parts of America, are but the Continuation of the Plains on the Top; and the Cliffs of them, in the Front they present towards Kinderhook. These UPPER PLAINS are of extraordinary rich level Land, and extend from the Mohocks River, through the Country of the Confederates. Their Termination Northward is at a little Distance from Lake Ontario; but what it is Westward is not known, for those most extensive Plains of Ohio are Part of them which continue to widen as they extend further Westward, even far beyond the Missisippi; and its Boundary Southward is a little Chain of broken Hills, about 10 or 15 Miles South of the Ohio River. 'Tis an odd Phaenomenon to observe how near the Tide comes up Hudson's River to the Heads of Delaware and Susquehanna; when these two Rivers are obliged to go so far to meet it in their own Channels. The Reason is, Delaware and Susquehanna have their Heads in the *Plains*, and Hudson's River the Tide at the Foot of them. The English are no where yet in these Plains, but towards the Head of Susquehanna, and on the Mohocks River.

BUFFON

Georges Louis Leclerc, Comte de Buffon (1707–1788), the French naturalist.

THE COLLISION HYPOTHESIS OF EARTH ORIGIN

From *The Natural History of Animals, Vegetables, and Minerals, with the Theory of the Earth in General*, translated by W. Kenrick and Others, Vol. V, pp. 358–373, London, 1776.

The vast extent of the solar system, which amounts to the same sphere of attraction of the sun does not confine itself to the orbs of the planets, but extends to a remote distance, always decreasing, in the same ratio as the square of the distance increases; it is demonstrated that the comets which are lost to our sight in the sky, obey this power, and that their motion, like that of the planets, depends on that of the sun; all these stars whose tracts are so different, describe around the sun, areas proportional to time. The planets in elipsis's more or less approaching a circle, comets in elongated ellipsis's. Comets and planets move therefore by virtue of two forces, the one of attraction, the other of impulsion, which acting at one time, obliges them to describe these courses; but it must be remarked that the planets* pass over the solar system in all manner of directions, and that the inclinations of their orbits are very different from each other, insomuch that, although subject like the planets, to the same force of attraction, the comets have nothing common in their motions of impulsion, they appear in this respect independent of each other: the planets, on the contrary, all turn around the sun in a like direction, and almost in the same plane, having there only 7½ degrees of inclination between the planes, the most distant from their orbit; this conformity of position and direction in the motion of the planets, necessarily supposes somewhat common in this motion of impulsion, and must make us suspect that it has been communicated to them by one and the same cause.

Can it not be imagined with some degree of probability, that a comet falling on the surface of the sun, will displace the planet and separate from it some parts to which it communicates a motion of

* Obviously a misprint for comets.—EDITORS.

impulsion by a like direction and stroke, insomuch that the planets should formerly have belonged to the body of the sun, and been detached therefrom by an impulsive force common to all, and which they still preserve.

This appears to me at least as probable as the opinion of Leibnitz, who pretends that the planets and earth have been suns, and I think that his system, of which we shall find the precise account in the fifth article, would have acquired a great degree of generality, and somewhat more probability if he had raised himself to this idea. The case is here the same as with him, to think that the matter happened in the time when Moses said that God divided light from darkness; for according to Leibnitz, light was divided from darkness when the planets were extinguished. But here the representation is physical and real, since the opaque matter which composes the body of the planets, was really divided from the luminous matter which composes the sun.

This probability will augment prodigiously by the second analogy, which is, that the inclination of the orbits do not exceed $7\frac{1}{3}$ degrees; for by comparing the spaces, we shall find there is twenty-four to one, that two planets are formed in the most distant planes, and consequently $\frac{5}{24}$, or, 7692624 to one, that it is not by chance that all six are thus placed and shut up in the space of $7\frac{1}{2}$ degrees, or what amounts to the same; there is a probability, that they have somewhat in common in the motion which has given them this position. But what can they have in common in the impression of an impulsive motion, if it is not the force and direction of bodies which communicate it? It may therefore be concluded with a very great probability, that the planets received their motion of impulsion by one single stroke. This probability, which is almost equivalent to a certainty, being acquired, I seek after what bodies in motion could make this stroke and produce this effect, and I see only the comets capable of communicating so great a motion to such vast bodies.

Provided we a little examine the course of comets, we shall be easily persuaded that it is almost necessary for them sometimes to fall into the sun. That of 1680 approached it so near, that its perihelium was not distant from it the sixth part of the solar diameter, and if it returns, as there is all appearance it will in 2255, it may then possibly fall into the sun. That depends on the rencounters it will meet with in its road, and of the retardment it suffered in passing through the atmosphere of the sun. Vide Newton, 3 edit. p. 525.

We may therefore presume with this philosopher, that comets sometimes fall into the sun: but this fall may be made in different manners. If they fall directly in, or in a direction not very oblique, they will remain in the sun, and will serve for food to the fire which the planet consumes, and the motion of impulsion which they will have lost, and communicated to the sun, will produce no other effect than that of displacing it more or less, according as the mass of the comet will be more or less considerable; but if the fall of the comet is made in a very oblique direction, which must oftener happen than in any other, then the comet will only graze the surface of the sun, or slightly furrow it; and in this case it may drive out some parts of matter to which it will communicate a common motion of impulsion, and these parts impelled out of the body of the sun and comet, might then become planets which will turn around this planet in the same direction and plane. We might perhaps calculate what mass, what velocity and what direction a comet should have to impel from the sun an equal quantity of matter to that which the six planets and their satellites contain; but this research would be here out of its proper place; it will be sufficient to observe, that all the planets with their satellites, do not make the sixty-fifth* part of the mass of the sun, (Vide Newton, p. 405), because the density of the large planets, Saturn and Jupiter, is less than that of the sun, and although the earth be ten times and the moon near six times denser than the sun, they are nevertheless but as atoms in comparison of the mass of this planet.

I own, that however inconsiderable the 160th* part is, at the first glance it appears to require a very powerful comet to separate this part of the body of the sun; but if we reflect on the prodigious velocity of comets in their perihelium, a velocity so much the greater, as their track is more direct, and as they approach nearer the sun. If besides, we pay attention to the density, fixity and solidity of the matter of which they must be composed, to suffer, without being destroyed, the inconceivable heat they endure near the sun; and if at the same time we reflect, they present to the sight of the observers a bright and solid body, which strongly

---

* In the original, Buffon wrote "650me," which is of course more nearly correct.—EDITORS.

reflects the light of the sun through the immense atmosphere of the comet which surrounds it, and must obscure it: we cannot doubt that the comets are not composed of a very solid and dense matter, and that they do not contain a great quantity of matter under a small volume; that consequently a comet cannot have sufficient mass and velocity to displace the sun, and give a projectile motion to a quantity of matter so considerable as is the 650th part of the mass of this planet. This perfectly agrees with what is known concerning the density of the planets: we think it is so much the less as the planets are farther distant from the sun, and as they have less heat to support, so that Saturn is less dense than Jupiter, and Jupiter much less dense than the earth: and in fact, if the density of the planets was as Newton pretends, proportionable to the quantity of heat which they have to support, Mercury would be seven times denser than the earth, and twenty-eight times denser than the sun; the comet of 1680 would be 28000 times denser than the earth, or 112000 times denser than the sun, and by supposing it as large as the earth, it would contain nearly an equal quantity of matter to the ninth part of the sun, or by giving it only the 100th part of the size of the earth, its mass would be still equal to the 900th part of the sun. From whence it is easy to conclude, that such a mass as a small comet is, might separate and drive from the sun a 900th part, or a 650th part of its mass, particularly if we pay attention to the immense *acquired velocity* with which comets move when they pass in the vicinity of the sun.

Another analogy which deserved some attention, is, the conformity between the density of the matter of the planets and the matter of the sun. We know on the surface of the earth there are some matters 14 or 15000 times denser than others. The densities of gold and air are nearly in this relation. But the internal part of the earth and the body of the planets are composed of more similar parts whose comparative density varies much less, and the conformity of the matter of the planets and of the density of the sun is such, that of 650 parts which compose the whole of the matter of the planets, there is more than 640 which are almost of the same density as the matter of the sun, and not ten parts out of these 650 of a greater density; for Saturn and Jupiter are nearly of the same density as the sun, and the quantity of matter which these planets contain, is at least 64 times greater than the quantity of matter of the four inferior planets, Mars, the Earth, Venus and Mercury. We must therefore say, that the matters of what the planets are generally composed of, is nearly the same as that of the sun, and that consequently this matter may have been separated from it.

But, without dwelling any longer on the objections which might be made, no more than on the proofs which analogies might furnish in favour of my hypothesis; let us pursue the object and deduce inductions from it: let us therefore see what has happened when these planets and particularly the earth received this impulsive motion, and in what state they were found after having been separated from the sun. The comet having by a single stroke communicated a projectile motion to a quantity of matter equal to the 690th part of the sun, the less dense periheliums will be separated from the dense and will have formed by their mutual alteration globes of different densities: Saturn composed of the most gross and light parts, will be the most remote from the sun: afterwards Jupiter who is denser than Saturn will be less distant and so on. The larger and less [dense] planets are the most remote, because they have received an impulsive motion, stronger than the smallest and densest: for, the force of impulsion communicating itself by surfaces, the same stroke will have moved the grosser and lighter parts of the matter of the sun with more velocity than the smallest and most massive; a separation therefore will be made of the dense parts of different degrees, so that the density of the sun being equal to 100, that of Saturn is equal to 67, that of Jupiter to $94\frac{1}{2}$, that of Mars to 200, that of earth to 400, Venus to 800, and Mercury to 2800. But the force of attraction not communicating like that of impulsion, by the surface and acting on the contrary on all parts of the mass it will have retained the densest portions of matter, and it is for this reason that the densest planets are the nighest the sun, and turn round that planet with greater rapidity than the less dense planets, which are also the most remote.

The comet having therefore by its oblique fall furrowed the surface of the sun, will have driven out of its body a part of matter equal to the 650th part of its whole mass; this matter which must be considered in a state of fluidity, or rather of liquefaction, will at first have formed a torrent, the grosser and less dense parts of which will have been driven the farthest, and the densest parts having received only the like impulsion, will not be so very remote, the force of the sun's attraction having retained them. Every part detached by the comet and impelled one by the other will have been constrained to circulate around this planet, and at the same time the mutual attraction of the parts of matter will have formed globes at different distances, the nearest of which to the sun will have necessarily retained more rapidity to turn perpetually afterwards round this planet.

But, it will be said a second time, if the matter which composes the planets has been separated from the sun, the planets should be like the sun, burning and luminous, and not cold and opaque as they are: nothing resembles this globe of fire less than a globe of earth and water, and to judge of it by comparison, the matter of the earth and planets is perfectly different from that of the sun.

To this it may be answered, that in the separation which was made of the more or less dense particles, the matter has changed form and the light or fire is extinguished by this separation caused by the motion of impulsion. Besides, may it not be suspected that if the sun, or a burning or luminous star moves of itself with so much velocity as the planets move, the fire would be extinguished perhaps, that is the reason why that all luminous stars are fixed, and that those stars which are called new, and which have probably changed place, are extinguished from sight? this is confirmed by what has been observed on comets, they must burn to the center when they pass to their perihelium. Nevertheless they do not become luminous of themselves, we see only that they exhale burning vapours of which they leave a considerable part by the way.

The earth and planets therefore at the time of their quitting the sun, were burning and in a state of total liquefaction; this state remained only as long as the violence of the heat which had produced it; by degrees the planets cooled, and it was in this state of fluidity, caused by the fire, that they took their form, and that their motion of rotation raised the parts of the equator by lowering the poles. This figure, which agrees so perfectly with the laws of hydrostatics, necessarily supposes that the earth and planets have been in a state of fluidity, and I here follow Leibnitz's opinion, that though fluidity was a liquefaction caused by the violence of the heat, the internal part of the earth must be a vitrifiable matter, of which sand, gravel, etc. are the fragments and scoria.

It may therefore with some probability be thought, that the planets appertained to the sun, that they were separated by a single stroke which gave to them a motion of impulsion in the same direction and plane, and that their position at different distances from the sun, proceeds only from their different densities. It now remains to explain by the like theory, the motion of the rotation of the planets, and the formation of the satellites; but this, far from adding difficulties or impossibilities to our hypothesis, seems on the contrary, to confirm it.

For the motion of rotation depends solely on the obliquity of the stroke, and it is necessary that an impulsion when it is oblique to the surface of a body, gives it a rotative motion: this motion will be equal and always the same, if the body which receives it is homogenous, and it will be unequal if the body is composed of heterogenous parts, or of different densities; and hence we must conclude, that in every planet the matter is homogenous, since the motions are equal. Another proof of the separation of the dense and less dense parts when they are formed.

But the obliquity of the stroke might be such, as to separate from the body of the principal planet a small part of matter, which will preserve the same direction of motion as the principal planet itself; these parts will be united according to their densities, at different distances from the planet by the force of their mutual attraction, and at the same time, necessarily follow the planet in its course around the sun, by turning themselves around the planet, nearly in the plane of its orbit. We see plainly, that those small parts which the great obliquity of the stroke will have separated, are the satellites: thus the formation, position, and direction of the motions of the satellites perfectly agree with theory; for they have all the same motion in concentrical circles round their principal planet; their motion is in the same plane, and this plane is that of the orbit of the planet. All these effects which are common to them, and which depend on their motion of impulsion, can proceed only from one common cause, i.e. from a common impulsion of motion, which has been communicated to them by one and the same stroke, given under a certain obliquity.

Epochs of the History of the Earth

Translated from *Époques de la nature*, Paris, 1807.

First Epoch. *When the earth and the planets took their form.* In the first days when the earth, molten and turning on herself, took her form and was raised at the equator on being lowered under the poles, the other planets, being in the same state of liquefaction and turning on themselves, took, like the earth, a form swollen on their equators and flattened under their poles. This swelling and this depression were proportional to the rapidity of their rotation. The globe Jupiter furnishes us proof of this. As it turns much more quickly than the earth, it is consequently more elevated at its equator, more lowered under its poles; observations show us that the two diameters of this planet differ more than a thirteenth while those of the earth differ but a 230th part. . . .

It has been seen by earlier experiments, that to warm a body to the point of fusion at least a fifteenth part of the time taken to cool it is necessary. Taking into consideration the great size of the earth and other planets, it would be necessary for them to have been stationed near the sun during many thousands of years to receive the degree of heat necessary for their liquefaction. But there is no evidence in the universe that any body, any planet, any comet remains stationary near the sun, even for an instant. On the contrary, the more the comets approach it (the sun) the more rapid is their movement; the time of their perihelium is extremely short and the fire of this star, while burning on the surface, has not time to penetrate the mass of the comets which approach it most closely.

Thus, all concurs to prove that it would not suffice that the earth and planets, like certain comets, have passed in the vicinity of the sun for their liquefaction to have taken place there. We ought, then, to assume that the matter of the planets has previously belonged to the body of the sun itself and been separated from it, as we have said, by a single and same impulse. . . .

Let us consider the state and aspect of our universe in its first youth: all the planets, newly consolidated at the surface, were still liquid in the interior, and shot out a very bright light. They were like little suns detached from the great one, which excelled them in nothing but volume, and they diffused light and heat in the same way. This time of incandescence lasted until the planet had been consolidated clear to the center, that is to say, about 2936 years for the earth, 644 years for the moon, 2127 years for Mercury, 1130 years for Mars, 3596 years for Venus, 5140 years for Saturn, and 9433 years for Jupiter.

The satellites of these two great planets as well as the rings which surround Saturn, which are all in the plane of the equator of their principal planet, were projected during the time of liquefaction by the centrifugal force of these great planets which turn on themselves with a prodigious rapidity. . . .

I ought to reply to a criticism that has already been made concerning the overlong duration of time. "Why throw us," I have been asked, "into a space as vague as a duration of 168,000 years? According to your picture, the earth is 75,000 years old, and the existing nature ought to continue another 93,000 years. Is it easy, is it even possible, to form an idea of all or of the parts of a series of centuries as long as that?" My only reply is to present the evidence and my deductions from the works of nature. I shall therefore give the details and the dates in the epochs which are going to follow this. It will be seen that, far from having increased the duration of time unnecessarily, I may have shortened it far too much.

To render this opinion more comprehensible, let us give an example. Let us calculate how much time is necessary for the construction of a hill of clay six thousand feet high. The successive deposits of the waters have formed all the beds of the hill from bottom to top. But, we can judge the successive and daily deposits of the waters by the laminae of the shales: they are so thin that one can count a dozen in a *ligne* of thickness. Suppose then that each tide deposits sediment of a twelfth of a *ligne* of thickness—that is to say, of a sixth of a *ligne* every day—the deposit will grow one *ligne* in six days . . . and consequently about five inches in a year. Even this time appears too short if it is compared with what takes place before our eyes on certain shores of the sea where muds and clays are being deposited, as on the coast of Normandy. There the increase of the deposit is not discernible and is much less than five inches in a year. . . .

Second Epoch. *When matter, being consolidated, formed the interior rock of the globe as well as the great vitreous masses at its surface.* During this epoch, and even for a long time afterward, as long as the heat was excessive, there was a separation and forcing out of all the volatile substances, such as water, air, and other substances that the great heat drove off. These could only exist in a region more temperate than the surface of the earth was then. All the volatile materials gathered about the globe in the form of atmosphere, while the molten matter, on being consolidated, formed the interior rock of the globe and the nucleus of those great mountains which are composed of vitreous materials. Thus, the first disposition of the great chains of mountains belongs to this second epoch, whereas the formation of the limestone mountains which have existed only since the establishment of the waters, inasmuch as their composition presupposes the existence of shells and other substances that the sea raises and nourishes, were formed many centuries later. As long as the surface of the globe had not been cooled to the point of allowing the water to remain there without being driven off in vapor, all our seas were in the atmosphere. They were not able to fall and establish themselves upon the earth before the moment when the surface was sufficiently cooled to reject the water no longer by too strong a bubbling. This time of the establishment of the waters on the face of the globe preceded by only a few centuries the moment when one could touch this surface without burning oneself. Thus, counting 75,000 years since the formation of the earth and half that time for its cooling to the point of being able to touch it, perhaps 25,000 of the first years passed before the water, hitherto rejected into the atmosphere, was able to establish itself permanently on the surface of the globe. . . .

During these first 37,000 years, all the great veins and the great lodes of mines where minerals occur, were formed by sublimation. The metallic substances were separated from the other vitreous matter by the long and constant heat which sublimated them and pushed them from the interior of the mass of the globe into all the eminences of its surface, where the contraction of material caused by rapid cooling left fissures and cavities which were encrusted and sometimes filled by the metallic substances we find there today. The same distinction must be made in the origin of mines that we have shown for the origin of vitreous and calcareous matter; the former have been produced by the action of fire, the latter by the medium of water. In the metal mines, the principal vein, or, if one wishes, the primordial mass, has been produced by fusion and sublimation, that is to say, by the action of fire. The other mines that are regarded as secondary and parasite veins, have been produced later, by means of water. . . .

But let us return to our principal topic, the topography of the globe before the descent of the waters. We have but few indices still remaining as to the first form of its surface. The highest mountains composed of vitreous matter are the sole vestiges of this ancient state. They were then even higher than they are today, for, since that time and after the establishment of the waters, the movement of the sea, the rains, the winds, the frosts, the currents of water, and the fall of torrents, in short, all the destructive activity of the elements of air and water and the vibrations due to subterranean movement, have not ceased bringing them down, trenching through them, and even overturning the less solid parts. We cannot doubt that the valleys which are at the foot of the mountains were once much deeper than they are today.

Third Epoch. *When the waters covered our continents.* At a time thirty or thirty-five thousand years after the formation of the planet, the earth found itself sufficiently cooled to receive the waters without rejecting them as vapor. The chaos of the atmosphere was slowly dissipated. Not only the water, but all the volatile matter that the excessive heat had held suspended there, fell in its proper order. It filled all the depths, it covered all the plains, all the spaces it found between the eminences on the surface of the globe, and even surmounted all of those that were not excessively high. There is definite proof that the seas have covered the continent of Europe to 1500 fathoms above the level of the present sea, because shells and other marine products are found in the Alps and in the Pyrenees to that height. There are similar proofs for the continents of Asia, Africa, and even America, where the mountains are higher than in Europe. Marine shells have been found at more than 2000 fathoms above the level of the southern sea.

It is certain that in these early days the diameter of the earth was two leagues greater, as it was enveloped in water to an elevation of 2000 fathoms. The surface of the earth in general was much higher than it is today; and during a long period of time, the seas covered it entirely, except perhaps for some very high lands and the summits of tall mountains which may have projected above this universal sea, which reached at least to that height where shells cease to appear. From this we should infer that the animals to which these remains belonged could be regarded as the first inhabitants of the globe. Judging by the immense quantity of their remains and detritus, this population was innumerable, because these same remains and detritus have formed all the beds of limestone, marble, chalk, and tufa that compose our hills and stretch over great areas in all parts of the earth.

But in the beginning of this sojourn of the waters on the surface of the globe, did they not have a degree of heat which our fish and our shell fish existing at present would be unable to support? And ought we not to presume that the first inhabitants of a still boiling sea would be different from those which it offers us today? Such great heat could agree only with shell fish and fish of another sort; and consequently it is in the earlier part of this epoch, that is to say, from thirty to forty thousand years after the formation of the earth, that one might expect the existence of lost species, living analogues of which are not to be found. These first species, now annihilated, subsisted during the ten or fifteen thousand years that followed the time when the waters were first established.

The time of the formation of the clays followed immediately that of the establishment of the waters. The time of the formation of the first shell fish ought to be placed some centuries later; and the time of transport of their remains followed almost immediately. There is no more interval than that which Nature has put between the birth and the death of these shell animals. The action of the water would constantly convert the vitreous sands into clays. As its movement transported them from place to place, it would carry along at the same time the shells and other debris of marine creatures. Depositing it all as sediments, it formed the beds of clay where today we find these relics, the most ancient mementos of organized Nature.

The formation of shales, slates, coals, and bituminous material date from almost the same time. These are ordinarily found at rather great depth in the clays. They seem to have preceded the deposition of the last clay beds, for below 130 feet of clay with beds containing belemnites, horns of Ammon, and other debris of the most ancient shells, I have found carbonaceous and inflammable material. And it is known that most coal mines are more or less covered by beds of argillaceous earth.

Moreover, it is certain that the two northern continents were not yet separated and even that their separation did not take place until a long time after life had appeared in our northern climates, because elephants have existed at the same time in Siberia and in Canada. This proves incontrovertibly the continuity of Asia or of Europe with America. On the other hand, it seems equally certain that Africa was separated from southern America from the first, because in this part of the new world not a single animal of the old continent has been found nor any remains which could indicate that they previously existed there.

During the long space of time that the sea sojourned over our lands, the sediments and the deposits of the waters formed the horizontal beds of the earth, the lower ones of clay and the upper of limestone. It was in the sea itself that the petrification of marble and of stones took place. These materials, being soft at first, were deposited successively, one on the other, as the waters brought them and let them fall in the form of sediments. Then they were hardened gradually by the force of the affinity of their constituent parts, and finally they formed all the mass of calcareous rocks which are composed of horizontal or evenly inclined beds, like all other materials deposited by the waters.

Fourth Epoch. *When the waters withdrew and the volcanos became active.* . . . In this time when the lands were raised above the waters, they were covered with great trees and vegetation of all kinds. The world-wide sea was populated throughout with fish and shell fish. It was also the universal receptacle for all that which broke away from the lands which rose above it. The fragments of primitive vitreous and vegetable material were carried down from the heights of the land into the depths of the sea, on the bottom of which they formed the first beds of vitreous sand, clay, schist, and slate as well as the ores of coal, of salt, and of bitumens which later were spread throughout all the seas. The quantity of vegetable products and remains in these first lands is too great to be described. When we reduce the surface of all the elevated lands then above the sea to a hundredth or even a two hundredth part of the surface of the globe, that is to say, 130,000 square leagues, it is easy to imagine how much this vast terrain could produce in the way of trees and plants during several thousands of years, how their remains accumulated, and in what enormous quantity they were transported and deposited under the waters, where they formed the great volume of coal in all the mines found in so many places. It is the same for the salt mines, those of granular iron, pyrites, and all other substances in the composition of which acid enters and which could not be formed until after the fall of the waters. These materials have been carried down and deposited in the low places and in the fissures of the rock of the globe where, finding substances already sublimated by the great heat of the earth, they formed the essential material for the alimentation of the volcanos to come. I say "to come," for there was no volcano in activity before the establishment of the waters. They could not begin activity, or at least a permanent activity, until after a lowering of the waters. One ought to distinguish terrestrial volcanos from marine volcanos; the latter only make explosions, momentary so to speak, because at the instant the fire is started by the effervescence of the combustible and pyrite-bearing stones, it is immediately extinguished by the water which covers it. The land volcanos have, on the contrary, a lasting activity and one proportionate to the quantity of material they contain. These materials need a certain quantity of water to enter into effervescence, and it is only then, by the shock of a great volume of fire against a great volume of water, that they can produce their violent eruptions. It is for this reason that all the volcanos at present in activity are in the islands or near the seacoast and that one can count a hundred times more extinguished ones than active ones; for, as the waters withdraw too far from the foot of the volcanos, their eruptions gradually diminish and finally cease. . . .

Up to the time of the activity of the volcanos there existed on the globe only three sorts of materials: first, the vitreous products from the primitive fire; second, the limestones formed by the medium of water; third, all substances produced by the remains of animals and plants. But the fire of the volcanos has given birth to materials of a fourth sort, which often resemble the other three. The fourth class is that of materials raised and thrown out by the volcanos. Some of this appears to be a mixture of the first with other materials, and other portions appear to have undergone a second action of the fire which has given them a new character.

Fifth Epoch. *When the elephants and other animals of the south inhabited the lands of the north.* All the living creatures which exist today have been able to exist in the same way since the temperature of the earth became similar to that of the present. But for a long time the northern countries of the globe enjoyed the same degree of heat that the southern countries enjoy today and in the time when these countries of the north enjoyed this temperature, the lands toward the south were still burning and remained as deserts. . . .

Elephants lived, reproduced, and multiplied during many centuries in Siberia and in the north of Russia.   Then they occupied the lands between the fortieth and fiftieth parallel, where they subsisted even longer than in their native country.   They lived for a still longer time in the country of the fortieth and thirtieth parallels, and so forth, because the gradual cooling of the globe has always been slower in the regions approaching the equator, due as much to the greater thickness of the globe as to the greater heat of the sun.

But this regular migration that the first and greatest animals of our continent have undergone seems to have encountered obstacles in the other continents.   It is certain that armament and skeletons of elephants have been found, and it is probable that more will be found in Canada, in the country of the Illinois, in Mexico, and in other places of northern America, but we have as yet no observation that would show the same fact for the countries of southern America.

Sixth Epoch.   *When the separation of the continents was effected.* The time of the separation of the continents is certainly subsequent to the time when the elephants inhabited the lands of the north, since their species existed equally in America, in Europe, and in Asia. . . .

If Europe is today separated from Greenland, it is probably because there was a considerable subsidence between the lands of Greenland and those of Norway and the point of Scotland, in which region the Orkneys, the islands of Shetland, those of Faroé, of Iceland, and of Hola appear to be only the higher portions of submerged land.   If the continent of Asia is no longer contiguous to that of America toward the north, it is without doubt a consequence of a very similar change. . . .

We further imagine that not only has Greenland been joined to Norway and Scotland, but also that Canada could have been joined to Spain by the banks of Newfoundland, the Azores, and other islands and shallows which are found in this portion of the sea. . . . The submergence of this land is perhaps more modern than that of the continent of Iceland, since tradition appears to record it.   The history of Atlantis, reported by Diodorus and

Plato, could only be applied to a very great land which stretched far to the west of Spain. . . .

Seventh Epoch. *When the power of man was added to that of nature.* The first men, witnesses of the convulsive movements of the earth, still occurring frequently, having only the mountains as asylums against floods and often driven from these same asylums by the fire of volcanos, trembling on an earth which trembled under their feet, naked of spirit and of body, exposed to the injuries of all the elements, victims of the fury of ferocious animals whose prey they were, all equally impressed with the sentiment of melancholy terror, did they not seek promptly to unite, first to defend themselves by their number, then to work together to make dwellings and gain protection? They began by sharpening those hard stones, those jades and "fire flints" in the form of hatchets that have been considered as fallen from the clouds and formed by thunder, whereas they really are the first relics of the art of man in the state of pure Nature. . . .

TARGIONI-TOZZETTI

Giovanni Targioni-Tozzetti (1712–1784), a Tuscan physician and naturalist, was one of the first to recognize the erosive power of running water.

VALLEYS FORMED BY STREAM EROSION

Translated from *Relazioni d'alcuni viaggi fatti in diverse parti della Toscana,* Vol. III, pp. 407 *et seq.,* Florence, 1752.

From the Ponte a Moriano to the mouth of the Valley of Anchiano, the Serchio runs through a narrow and winding channel and by the impact of its waters is shaping, eroding, and dividing the vast mountain masses. I shall furnish evidence of this as I proceed, but if anyone should not care to accept this physical truth, I shall not be annoyed. I shall only ask him to deign to make this short journey. I am satisfied that he will be convinced that the facts are as I deem them, and am sure that he will not consider the hypothesis of the Taglio della Golfolina and of the Foce di Ripafratta too daring. Visual inspection of this channel of the Serchio also brings realization that the opinion of the illustrious Signor Buffon is not generally adequate. It is his opinion that the winding channels which are found on the surface of the earth and through which flowing waters today make their way have been formed by the rushing currents of the sea when our terraqueous globe was covered by its waters. Surely the deep channel of the Serchio from the Ponte a Moriano clear to Anchiano, that of the same river to Ripafratta, those of the Arno through the Valle dell'Infierno and from the Incisa to the Ponte a Rignana, and all the other channels of rivers that I have not yet seen were not thus excavated when these regions were covered with the waters of the sea. On the contrary, they were not formed until after the lowering of the level of the sea, when the fresh water, flowing toward the lowered sea, began to descend and acquired velocity. Then it was that the mountain streams, arranged in a network, began to erode and have continued to deepen their channels until, on approaching the level of the modern sea, they lose their velocity and the force of their impact. It might seem to some that, according to the laws of hydrostatics, erosion and channeling in the primitive mountains ought to have been accomplished in straight lines, rather than twisted and at angles as we find it. However, that is dependent on the diverse resistances of the materials composing the primitive mountains which it must cut and channel. . . . Other important causes of the variation in direction, of course, have been the torrents which, with different directions and different amounts of water, were discharging into the main river, particularly when, on account of flood, they have more impetus than it. Finally, another cause of this winding is due to the fact that the looseness of the material of one mountain is greater than that of another adjoining. . . .

ARDUINO

Giovanni Arduino (1713–1795), director of mines of Tuscany and professor of mineralogy at Padua, first distinguished the tertiary division of strata and made important observations concerning metamorphism.

ORIGIN OF MARBLE AND DOLOMITE

Translated from the French translation of his *Osservazioni* (Venice, 1779), *Bulletin de la Société Géologique de France*, Vol. IV, pp. 112–114, 1883–1884.

. . . Speaking of marbles (or altered rocks) like those of Ena or other masses of the same nature, as far as I have observed I have never found them save among the great rifts in the limestone beds of our mountains. These rifts show that these masses are sunk deeply and vertically or in a nearly vertical direction, and *that they were produced long ago by a subterranean igneous force.*

The materials which thus fill the crevasses, sometimes very long and very often wide, in our Alps, have the characters of having undergone the action of fire or of fusion. In some places this marble is jumbled confusedly in fragments of various sizes, and traversed by veins so that breccia results.

On meditating over the probable origin of these rocks, I could imagine nothing other than that they were derived from the same limestones whose crevices enclose them. The limestone rocks would have been broken and ground by the volcanic fire, they would have been calcined, and their dark color would have given place to the most glaring white; finally, they would have been mixed strangely with other burnt materials and penetrated by water; repetrified, they would have taken this new form, passing from the state of ordinary reddish limestone to that of marble.

I imagine that is explained by my idea that *magnesia is nothing but lime endowed with peculiar properties as a result of pyrogenic subterranean action.* I believe it all the more because the limestones forming the beds of the mountains and enclosing the strips of rock in question show no trace of this foreign constituent part.

GUETTARD

Jean Étienne Guettard (1715–1786), a French physician and naturalist. Guettard's excellent publications, based on an amount and type of field work previously unknown, were of great importance to geology.

MAP SHOWING THE ROCKS THAT TRAVERSE FRANCE AND ENGLAND

Translated from *Mémoires de l'Académie royale française*, pp. 363–392, 1746.

Nothing can contribute more toward providing us with a general physical theory of the earth than numerous observations made on the different terrains and the fossils which they contain; there is also nothing which can render that aid more perceptible than to assemble the different observations and present them at a glance by mineralogic maps. I have traveled with the view of instructing myself on the first point, and in accord with the sentiment of the Academy, which, when I had the honor of rendering it an account of a part of my work, seemed to wish to see a map of it, nothing has pleased me more than to fulfill its desires. This map will be the summary of all the account which I propose to give of what I have observed in my travels. Although I have traversed a rather large portion of France, nevertheless I have not seen many of the provinces of this great kingdom. Also it is readily understood that I could not go in all the localities of those provinces which I could traverse. In order to remedy this defect, I have made use of my reading, where I was always careful to mark that which might concern my project. And, what has been still more useful to me, I have written to many of the places where I have not been, for enlightenment concerning the rocks and other fossils which could be found there. I have usually received the information I desired and it was nearly always accompanied by samples of the fossils. . . .

In this map, it is my intention to show that there is a certain regularity in the distribution of stones, metals, and the majority of other fossils. One does not find such and such stones, such and such metals, indiscriminately in all sorts of regions. There are certain regions where it is absolutely impossible to find quarries or mines of certain stones or certain metals, whereas they are very frequent in others where, had they not been found, there would have been no more reason to expect to find them than elsewhere. I was struck by this type of uniformity in some journeys that I made some years ago in lower Poitou. I was amazed to see that one passed through regions one after another, where, almost suddenly, the stones and terrain became definitely different. . . .

One of the first ideas which came to me after all this work was to assure myself whether England was like this kingdom, entirely or in part. I was led to this by the general and confused information which I already had. I knew that Cornwall was famous for its tin mines, that many places of that province and some others furnished much coal. This led me to the thought that, Cornwall being in alignment with lower Normandy, it could well be that there was a similarity between those two provinces. It was even possible that such might be found between the rest of France and of England. I sought, therefore, to test this idea by the perusal of pamphlets which treated of the matter. The perusal which I made of the works of Childrey and Gerard Boate on the natural history of England and Ireland proved the conjecture to my satisfaction. I recognized that if there was a difference, it was not great; and that the greatest would come from the difference in width between the two kingdoms. Therefore I believe that it is proper that I speak of England and France at the same time, and that the map include both kingdoms. In this way, one will be in a better position to judge this similarity, and it can have no other result than a more certain and complete proof of my thought.

The Fluviatile Origin of the Puddingstones*
[Conglomerates] of the Paris Basin

Translated from *Mémoires de l'Académie royale des sciences*, 1753, pp. 63–96, 1757.

It seems to me that the first inference that can be drawn from the description is that the original cause of the beds formed of all the materials mentioned is the Seine, which made these deposits in remote times and has buried trees at different depths by successive

---

* This name (*Poudingues*) is English. The naturalists of England have assigned to a species of stone the name of a ragout composed of different ingredients, to which mixture the stone in question has some resemblance, being a mass of pebbles assembled and fastened together by any matter whatsoever.

accretions. The various clays, the stones which seem to be only these clays more or less hardened, those which are composed of gravel, and especially the wood seem rather good proof. . . .

Whatever there may be in these observations, I do not think that it can be denied that they concur to give a rather apt explanation of the bed of pebbles in the environs of Paris and to prove, as I have postulated above, that it is formed solely of materials that the Seine has transported. . . .

One might say that, for complete proof, it should be shown that the Seine still rolls stones of the same type. If it should be otherwise, it nevertheless does not follow that earlier it did not bring those which form the bed in question. . . . The mountains which border it are encumbered with houses and castles, many with villages, towns, and even cities; thus many of the mountains no longer furnish the stones formerly torn out by the rainstorms, which now have no purchase on the stones there. Besides that, the mountains without buildings are cultivated and covered with crops at the times when the rains are greatest.

One finds however that the river still carries them. All the difference that I find is that, aside from their greater rarity, they are ordinarily smaller and the gravel is finer. . . . The streams have probably changed greatly in their courses and the work done on these streams often obliged them to take different circuits, to slow their current and thus give the stones they roll more facility to grind themselves. It is not that rather large pieces of stone are not found in our day, but that they are much more rare than those in the ancient basin. . . .

Here is the art which I imagine Nature to have employed in the composition of this stone:

The little lenticular pebbles that I have said form part of it, being stones from which lime can be made, fall little by little into dissolution from the water or the humidity which is always found underground. Thus these pebbles are dissolved on their surfaces and stick to each other and to the grains of sand which surround them. The humidity charged with dissolved parts becomes more active and can act more efficiently on the pebbles from which its dissolved substance was gained, on those of flint, and on the sand itself; then the cement is still better qualified to make a strong, tight binding. This cement is composed of sand, clayey or earth particles, saline parts, and ferruginous parts. . . .

MICHELL

John Michell (1724–1793), English astronomer and professor of geology at Cambridge, made important contributions to the sciences of astronomy and physics as well as to geology. His responsibility for the first torsion balance is described by Cavendish, *q.v.* p. 103.

THE NATURE AND ORIGIN OF EARTHQUAKES

From *Philosophical Transactions of the Royal Society of London*, Vol. LI, pp. 566–74, 1760.

It has been the general opinion of philosophers, that earthquakes owe their origin to some sudden explosion in the internal parts of the earth. This opinion is very agreeable to the phaenomena, which seem plainly to point out something of that kind. The conjectures, however, concerning the cause of such an explosion, have not been yet, I think, sufficiently supported by facts; nor have the more particular effects, which will arise from it, been traced out; and the connexion of them with the phaenomena explained. To do this, is the intent of the following pages; and this we are now the better enabled to do, as the late dreadful earthquake of the 1st of November 1755 [the Lisbon earthquake] supplies us with more facts, and those better related, than any other earthquake of which we have an account.

That these concussions should owe their origin to something in the air, as it has sometimes been imagined, seems very ill to correspond with the phaenomena. This, I apprehend, will sufficiently appear, as those phaenomena are hereafter recounted; nor does there appear to be any such certain and regular connexion between earthquakes and the state of the air, when they happen, as is supposed by those who hold this opinion. It is said, for instance, that earthquakes always happen in calm still weather: but that this is not always so, may be seen in an account of the earthquakes in Sicily of 1693, where we are told, "the south winds have blown very much, which still have been impetuous in the most sensible earthquakes, and the like has happened at other times."

Other examples to the same purpose we have in an account of the earthquakes that happened in New England in 1727 and 1728:

the author of which says, that he could neither observe any connexion between the weather and the earthquakes, nor any
prognostic of them; for that they happened alike in all kinds of
weather, at all times of the tides, and at all times of the moon.

If, however, it should still be supposed, notwithstanding these
instances to the contrary, that there is some general connexion
between earthquakes and the weather, at the time when they
happen, yet, surely, it is far more probable, that the air should be
affected by the causes of earthquakes, than that the earth should
be affected in so extraordinary a manner, and to so great a depth;
and that this, and all other circumstances attending these motions,
should be owing to some cause residing in the air.

Let us then, rejecting this hypothesis, suppose, that earthquakes
have their origin under ground, and we need not go far in search of
a cause, whose real existence in nature we have certain evidence
of, and which is capable of producing all the appearances of these
extraordinary motions.   The cause I mean is subterraneous fires.
These fires, if a large quantity of water should be let out upon them
suddenly, may produce a vapour, whose quantity and elastic force
may be fully sufficient for that purpose.   The principal facts, from
which I would prove, that these fires are the real cause of earthquakes, are as follow.

First, the same places are subject to returns of earthquakes,
not only at small intervals for some time after any considerable
one has happened, but also at greater intervals of some ages.

Both these facts sufficiently appear, from the accounts we have
of earthquakes.   The tremblings and shocks of the earth at
Jamaica in 1692, at Sicily in 1693, and at Lisbon in 1755, were
repeated sometimes at larger, and sometimes at smaller intervals,
for several months.   The same thing has been observed in all other
very violent earthquakes.   At Lima, from the 28th October 1746,
to the 24th February 1747 (the time when the account of them was
sent from thence), there had been numbered no less than 451
shocks, many of them little inferior to the first great one, which
destroyed that city.

The returns of earthquakes also, in the same places, at larger
distances of time, are confirmed by all history.   Constantinople,
and many parts of Asia Minor, have suffered by them, in many
different ages: Sicily has been subject to them, as far back as the

remains even of fabulous history can inform us of: Lisbon did not feel the effects of them for the first time in 1755: Jamaica has frequently been troubled with them, since the English first settled there; and the Spaniards, who were there before, used to build their houses of wood, and only one story high, for fear of them: Lima, Callao, and the parts adjacent, were almost totally destroyed by them twice, within the compass of about sixty years, scarce any building being left standing, and the latter being both times over-flowed by the sea: nor were these the only instances of the like kind, which have happened there; for, from the year 1582 to 1746, they have had no less than sixteen very violent earthquakes, besides an infinity of less considerable ones; and the Spaniards, at their first settling there, were told by the old inhabitants, when they saw them building high houses, that they were building their own sepulchres.

Secondly, those places that are in the neighbourhood of burning mountains, are always subject to frequent earth-quakes; and the eruptions of those mountains, when violent, are generally attended with them.

Asia Minor and Constantinople may be looked upon as in the neighbourhood of Santerini. The countries also about AEtna, Vesuvius, mount Haecla, &c. afford us sufficient proofs to the same purpose. But, of all the places in the known world, I sup-pose, no countries are so subject to earthquakes, as Peru, Chili, and all the western parts of South America; nor is there any country in the known world so full of volcanoes: for, throughout all that long range of mountains, known by the name of the Andes, from 45 degrees south latitude, to several degrees north of the line, as also throughout all Mexico, being about 5000 miles in extent, there is a continued chain of them.

Thirdly, the motion of the earth in earthquakes is partly tremulous, and partly propagated by waves, which succeed one another sometimes at larger and sometimes at smaller distances; and this latter motion is generally propagated much farther than the former.

The former part of this proposition wants no confirmation; for the proof of the latter, *viz.* the wave-like motion of the earth, we may appeal to many accounts of earthquakes: it was very remarkable in the two, which happened at Jamaica in 1687–8 and 1692. In an account of the former, it is said, that a gentleman there saw the ground rise like the sea in a wave, as the earthquake passed along, and that he could distinguish the effects of it, to some miles distance, by the motion of the tops of the trees on the hills. Again, in an account of the latter, it is said, "the ground heaved and swelled, like a rolling swelling sea," insomuch, that people could hardly stand upon their legs by reason of it.

The same has been observed in the earthquakes of New England, where it has been very remarkable. A gentleman giving an account of one, that happened there the 18th November 1755, says, the earth rose in a wave, which made the tops of the trees vibrate ten feet, and that he was forced to support himself, to avoid falling, whilst it was passing.

The same also was observed at Lisbon, in the earthquake of the 1st November 1755, as may be plainly collected from many of the accounts that have been published concerning it, some of which affirm it expresly: and this wave-like motion was propagated to far greater distances than the other tremulous one, being perceived by the motion of waters, and the hanging branches in churches, through all Germany, amongst the Alps, in Denmark, Sweden, Norway and all over the British isles.

Fourthly, it is observed in many places, which are subject to frequent earthquakes, that they generally come to one and the same place, from the same point of the Compass. I may add, also, that the velocity, with which they proceed, (as far as one can collect it from the accounts of them) is the same; but the velocity of the earthquakes of different countries is very different.

Thus all the shocks, that succeeded the first great one at Lisbon in 1755, as well as the first itself, came from the north-west. This is asserted by the person, who says, he was about writing a history of the earthquakes there: all the other accounts also confirm the same thing: for what some say, that they came from the north, and others, that they came from the west, cannot be looked on as any reasonable objection to this, but rather the contrary. The velocity also, with which they were all propagated, was the same, being at least equal to that of sound; for they all followed immediately after the noise that preceded them, or rather the noise and the earthquake came together: and this velocity agrees very well with the intervals between the time when the first shock was felt at Lisbon, and the time when it was felt at other places, from the comparison of which, it seems to have travelled at the rate of more than twenty miles *per* minute.

An historical account of the earthquakes, which have happened in New England, says, that, of five considerable ones, three are known to have come from the same point of the compass, *viz.* the north-west: it is uncertain from what point the other two came, but it is supposed that they came from the same with the former. The velocity of these has been much less than that of the Lisbon earthquakes: this appears from the interval between the preceding noise, and the shock, as well as from the wave-like motion before-mentioned.

The Earth Composed of Regular and Uniform Strata

From *Philosophical Transactions of the Royal Society of London*, Vol. LI, Pt. 2, pp. 582 *et seq.*, 1760.

The earth then (as far as one can judge from the appearances), is not composed of heaps of matter casually thrown together, but of regular and uniform strata. These strata, though they frequently do not exceed a few feet, or perhaps a few inches, in thickness, yet often extend in length and breadth for many miles, and this without varying their thickness considerably. The same stratum also preserves a uniform character throughout, though the strata immediately next to each other are very often totally different. Thus, for instance, we shall have, perhaps, a stratum of potters clay; above that, a stratum of coal; then another stratum of some other kind of clay; next, a sharp grit sand stone; then clay again; next, perhaps, sand stone again; and coal again above that; and it frequently happens, that none of these exceed a few yards in thickness. There are, however, many instances in which the same kind of matter is extended to the depth of some hundreds of yards; but in all these, a very few only excepted, the whole of each is not one continued mass, but is again subdivided into a great number of thin laminae, that seldom are more than one, two, or three feet thick, and frequently not so much.

Beside the horizontal division of the earth into strata, these strata are again divided and shattered by many perpendicular fissures, which are in some places few and narrow, but oftentimes many, and of considerable width.   There are also many instances, where a particular stratum shall have almost no fissures at all, though the strata both above and below it are considerably broken: this happens frequently in clay, probably on account of the softness of it, which may have made it yield to the pressure of the superincumbent matter, and fill up those fissures which it originally had; for we sometimes meet with instances in mines, where the correspondent fissures in an upper and lower stratum are interrupted in an intermediate stratum composed of clay, or some soft matter.

Though these fissures do sometimes correspond to one another in the upper and lower strata, yet this is not generally the case, at least not to any great distance; those clefts, however, in which the larger veins of the ores of metals are found, are an exception to this observation; for they sometimes pass through many strata, and those of different kinds, to unknown depths.

From this constitution of the earth; *viz.* the want of correspondence in the fissures of the upper and lower strata, as well as on account of those strata which are little or not at all shattered, it will come to pass, that the earth cannot easily be separated in a direction perpendicular to the horizon, if we take any considerable portion of it together; but in the horizontal direction, as there is little or no adhesion between one stratum and another, it may be separated without difficulty.

Those fissures which are at some depth below the surface of the earth, are generally found full of water; but all those that are below the level of the sea, must always be so, either from the oozing of the sea, or rather of the land waters between the strata.

The strata of the earth are frequently very much bent, being raised in some places, and depressed in others, and this sometimes with a very quick ascent or descent; but as these ascents and descents, in a great measure, compensate one another, if we take a large extent of country together, we may look upon the whole set of strata, as lying nearly horizontally.   What is very remarkable, however, in their situation, is, that from most, if not all, large tracts of high and mountainous countries, the strata lie in a situation more inclined to the horizon, than the country itself, the mountainous countries being generally, if not always, formed out of the lower strata of earth.   This situation of the strata may be not unaptly represented in the following manner.   Let a number of leaves of paper, of several different sorts or colours, be pasted upon one another; then bending them up together into a ridge in the middle, conceive them to be reduced again to a level surface, by a plane so passing through them, as to cut off all the part that had been raised; let the middle now be raised a little, and this will be a good general representation of most, if not all, large tracts of mountainous countries, together with the parts adjacent, throughout the whole world.

From this formation of the earth, it will follow, that we ought to meet with the same kinds of earths, stones, and minerals, appearing at the surface, in long narrow slips, and lying parallel to the greatest rise of any long ridges of mountains; and so in fact, we find them. The Andes of South America, as it has been said before, have a chain of volcanos, that extend in length above 5000 miles: these volcanos, in all probability, are all derived from the same stratum. Parallel to the Andes, is the Sierra, another long ridge of mountains, that run between the Andes and the sea; and "these two ridges of mountains run within sight of one another, and almost equally, for above a thousand leagues together," being each, at a medium, about twenty leagues wide. The gold and silver mines wrought by the Spaniards, are found in a tract of country parallel to the direction of these, and extending through a great part of the length of them.

The same thing is found to obtain in North America also. The great lakes, which give rise to the river St. Laurence, are kept up by a long ridge of mountains, that run parallel to the eastern coast. In descending from these towards the sea, the same sets of strata, and in the same order are generally met with throughout the greatest part of their length.

In Great Britain, we have another instance to the same purpose, where the direction of the ridge varies about a point from due north and south, lying nearly N. by E. to S. by W. There are many more instances of this to be met with in the world, if we may judge from circumstances, which make it highly probable, that it obtains in a great number of places, and in several they seem to put it almost out of doubt.

The reader is not to suppose, however, that, in any instances, the highest rise of the ridge, and the inclination of the strata from thence to the countries on each side, is perfectly uniform; for they have frequently very considerable inequalities, and these inequalities are sometimes so great, that the strata are bent for some small distance, even the contrary way from the general inclination of them. This often makes it difficult to trace the appearance I have been relating, which, without a general knowledge of the fossil bodies of a large tract of country, it is hardly possible to do.

At considerable distances from large ridges of mountains, the strata, for the most part, assume a situation nearly level; and as the mountainous countries are generally formed out of the lower strata, so the more level countries are generally formed out of the upper strata of the earth.

Hence it comes to pass, that, in the countries of this kind, the same strata are found to extend themselves a great way, as well in breadth as in length: we have an instance of this in the chalky and flinty countries of England and France, which (excepting the interruption of the Channel, and the clays, sands, &c. of a few counties) compose a tract of about three hundred miles each way.

Besides the raising of the strata in a ridge, there is another very remarkable appearance in the structure of the earth, though a very common one; and this is what is usually called by miners, the trapping down of the strata; that is, the whole set of strata on one side a cleft are sunk down below the level of the corresponding strata of the other side. If, in some cases, this difference in the level of the strata, on the different sides of the cleft, should be very considerable, it may have a great effect in producing some of the singularities of particular earthquakes.

KANT

Immanuel Kant (1724–1804), German philosopher and professor in the University of Königsberg, was a pioneer in the interpretation of the sidereal universe. Zittel points out that Kant's cosmogony was neglected for ninety years, until unearthed by Humboldt.

A Naturalistic Hypothesis of Earth Origin

Translated from *Allgemeine Naturgeschichte und Theorie des Himmels*, pp. 27–34, 129–130, Königsberg and Leipzig, 1755.

I postulate that all the matter from which were formed the spheres pertaining to our solar system, all the planets and comets, dispersed in the beginning into their primary elements, filled the whole area of the cosmos where these bodies now circle. . . . Nature, which came into being at the same time as the Creation, was as crude and as unformed as it could be. . . . But *the variety in the kinds of elements* contributed above all to an initial movement and to the organization of the chaos, whereby the repose which would prevail in the case of general uniformity among scattered elements was lifted and the chaos began to take shape about the points of the more strongly attracting particles.

In an area thus filled, the general repose would last only an instant. The elements have essential forces to set one another in motion and are sources of life for themselves. Matter is immediately in travail to take form. Owing to the attraction, the scattered elements of the denser type would attract from the region around them all matter of less specific gravity.

The cause of the formation of planets is not to be sought in the Newtonian attraction alone. This would be much too slow and weak among particles of such exceeding fineness. It would be much better to say that in this space the first forming was occasioned by the converging of several elements which united in response to the ordinary laws of cohesion, until the clumps thus formed gradually grew so large that the Newtonian force of attraction, acting over a distance, was able constantly to increase them.

The forming of the planets according to this system has this advantage over all other possible hypotheses—that the origin of the masses accounts for the origin of the motion and the location of the orbits at the same instant; yes, that even the deviation from the strictest accuracy in these determinations, as well as the conformity itself, becomes clear at a glance. The planets are formed from the little parts, which, as they float in the heavens, move in a circular orbit: *therefore the masses collected from these will continue exactly the same motion, to just the same degree, and in the identical direction.*

———————

Why is the mid-point of such a system taken by an incandescent body? Our planetary world structure has the sun as a central body, and the fixed stars which we see are all apparently midpoints of similar systems. In order to understand why, in the formation of a world structure, the body which serves as mid-point of attraction had to become a fiery one, whereas the remaining globes in its sphere of attraction remain dark and cold world bodies, one has only to recall the way a world structure is produced. In the widely distended space wherein the diffused primary elements get ready for form and systematic motion, the planets and comets are built only of those parts of the attracted elements which, through gravity and the reciprocal effects of all the particles, were forced to the exact limit of direction and velocity requisite for rotation. This part is, as demonstrated above, the least of the entire amount of downward-sinking matter and indeed only the chaff of the denser types which could have come to this degree of closeness through the resistance of the others. There are, in this mixture, the suspended, exceptionally light varieties which, slowed by the resistance of space, do not have the gravity to reach the necessary rapidity of periodic rotation. As a result of the weakness of their motion, they were plunged all together into the central body. As it is just these lighter and volatile parts which are most effective to maintain the fire, we see that, by their addition, the body at the mid-point of the system receives the wherewithal to become an incandescent body—in a word, a sun. On the other hand, the heavier and uncrusted stuff, deficient in these fire-nourishing parts, will make only the cold, dead lumps of the planets, which are deprived of such [incandescent] properties.

This addition of lighter matter is also the reason for the fact that the sun has a lesser specific density.

DESMAREST

Nicolas Desmarest (1725–1805), a French physician, geologist, and miner-alogist, whose writings show exceptional clarity of thought. His appeal to observable facts was well calculated to project the light of reason into the heat of unrestrained argument concerning the real nature of extruded igneous rocks at a time when many leading scientists were inclined to the theory of their aqueous origin.

THE VOLCANIC ORIGIN OF BASALT

Translated from *Mémoires de l'Académie royale des sciences*, 1771, pp. 705–775, 1774.

I have seen many of these isolated masses of basalt, and I admit that if I had been limited to these masses in my observations, I would not have been able to decide that basalt was a compact lava. It is only in going from the simple to the composite that I put myself in a position to establish this truth and to generalize on its application. The infinitely varied results of the operations of fire which are exposed in Auvergne presented the most favorable circumstances in some places and elsewhere displayed the greatest changes in the conditions. I paid attention first to the flows in which the prismatic basalt occupied the center and the edges and in which I have recognized an uninterrupted continuity from their most distant extremities to the open mouth of a volcano. They appeared before me accompanied at the same time by all the phenomena of porous lavas, scoriae, pumices, and baked earth, and lying for the most part on unchanged bases over which the melted materials progressed. Such are the circumstances which guided me in the beginning of my observations. Once enlightened on the primitive state of the phenomena, I have reasoned that the alterations undergone in the disposition of the masses of basalt in certain districts could not invalidate that which has been recognized and well proven in others.

Thus where I have found isolated buttes, composed of black stone, which had the same grain, the same brilliant and glassy points, the same prismatic or rounded form as basalt, I could not persuade myself that if the first that I had observed in the recog-nized and continuous flows were lava, the latter would not be a similar product of fire.

Age Determinations in a Volcanic Region

Translated from *Mémoires de la Classe des sciences mathématiques et physiques, Institut national de France*, Vol. VI, pp. 219–289, 1806.

I believed it necessary, elsewhere, to describe the first alterations that wiped away the craters and the scoriae, and finally the changes which have taken place in the sites which these flows occupied in the bottom of the valleys. All these circumstances seemed to me to bespeak variations which have practically the same sequence. As soon as craters have their edges blunted or widened, or even have begun to fill up; as soon as the scoriae begin to be reduced to a pulverized earthy substance, then the flows which have emerged from these centers of eruption and which are stripped of their scoriae, no longer occupy the bottom of the valleys. They are situated halfway up the hill, the valley being deepened since the flows were established on its former floor. The torrents which skirt them or traverse them, moreover, have lowered part of their mass by jagged cuts which are strewn with falls and cascades.

As it is only by a very long series of centuries that all these forms and all these circumstances have changed, it is easy to show the causes as well as the progress of the variations. Observation first informed me that the scoriae and spongy substances experience a very noticeable comminution and are finally reduced, in a rather short space of time, to pulverized earths. In addition it has shown me the water of rains and melted snows continually displacing these mobile materials. As a result of this double work of water, the edges of the craters, formed mainly of scoriae, must be blunted; then these mouths fill by imperceptible gradations, and finally disappear entirely.

It is in this epoch that the horizontal beds formed in the first and most ancient epoch were cut by valleys; that the craters open during the middle epoch were destroyed or filled up; . . . that the different parts of the flows themselves, established on the surface of the horizontal beds, have been separated by cuts which have gradually become valleys of the first order. It would appear that during the most recent epoch the valleys which separate the portions of the same flow must have increased and become deeper at the same time that the destruction of the craters and the comminution of the scoriae took place.

HUTTON

James Hutton (1726–1797), "a private gentleman" of Edinburgh, was trained as a physician. His followers formed the Plutonist school, opposing Werner. Hutton also announced the principle of Uniformitarianism.

Theory of the Earth

From *Theory of the Earth*, Vol. I, pp. 280–281, Vol. II, pp. 540–564, Edinburgh, 1795.

In examining things which actually exist, and which have proceeded in a certain order, it is natural to look for that which had been first; man desires to know what had been the beginning of those things which now appear. But when, in forming a theory of the earth, a geologist shall indulge his fancy in framing, without evidence, that which had preceded the present order of things, he then either misleads himself, or writes a fable for the amusement of his reader. A theory of the earth, which has for object truth, can have no retrospect to that which had preceded the present order of this world; for, this order alone is what we have to reason upon; and to reason without data is nothing but delusion. A theory, therefore, which is limited to the actual constitution of this earth, cannot be allowed to proceed one step beyond the present order of things.

The system of this earth appears to comprehend many different operations; and it exhibits various powers co-operating for the production of those effects which we perceive. Of this we are informed by studying natural appearances; and in this manner we are led to understand the nature of things, in knowing causes.

That our land, which is now above the level of the sea, had been formerly under water, is a fact for which there is every where the testimony of a multitude of observations. This indeed is a fact which is admitted upon all hands; it is a fact upon which the speculations of philosophers have been already much employed; but it is a fact still more important, in my opinion, than it has been ever yet considered. It is not, however, as a solitary fact that any rational system may be founded upon this truth, That the earth had been formerly at the bottom of the sea; we must also see the nature and constitution of this earth as necessarily subsisting in continual change; and we must see the means employed by nature for constructing a continent of solid land in the fluid bosom of the deep. It is then that we may judge of that design, by finding ends and means contrived in wisdom, that is to say, properly adapted to each other. . . .

If it should be admitted, that this earth had been formed by the collection of materials deposited within the sea, there will then appear to be certain things which ought to be explained by a theory, before that theory be received as belonging to this earth. These are as follows:

First, We ought to show how it came about that this whole earth, or by far the greatest part in all the quarters of the globe, had been formed of transported materials collected together in the sea. It must be here remembered, that the highest of our mountainous countries are equally formed of those travelled materials as are the lowest of our plains; we are not therefore to have recourse to any thing that we see at present for the origin of those materials which actually compose the earth; and we must show from whence had come those travelled materials, manufactured by water, which were employed in composing the highest places of our land.

Secondly, We must explain how those loose and incoherent materials had been consolidated, as we find they are at present. We are not here to allow ourselves the liberty, which naturalists have assumed without the least foundation, of explaining every thing of this sort by *infiltration*, a term in this case expressing nothing but our ignorance.

Thirdly, The strata are not always equally consolidated. We often find contiguous strata in very different states with respect to solidity; and sometimes the most solid masses are found involved in the most porous substance. Some explanation surely would be expected for this appearance, which is of a nature so conclusive as ought to attract the attention of a theorist.

Fourthly, It is not sufficient to show how the earth in general had been consolidated; we must also explain, how it comes to pass that the consolidated bodies are always broken and intersected by veins and fissures. In this case, the reason commonly given, that the earth exposed to the atmosphere had shrunk like moist clay, or contracted by the operation of drying, can only show that such naturalists have thought but little upon the subject. The effect in no shape or degree corresponds to that cause; and veins and fissures, in the solid bodies, are no less frequent under the level of the sea, than on the summits of our mountains.

Fifthly, Having found a cause for the fracture and separation of the solid masses, we must also tell from whence the matter with which those chasms are filled, matter which is foreign both to the earth and to the sea, had been introduced into the veins that intersect the strata. If we fail in this particular, What credit could be given to such hypotheses as are contrived for the explanation of more ambiguous appearances, even when those suppositions should appear most probable?

Sixthly, Supposing that hitherto every thing had been explained in the most satisfactory manner, the most important appearances of our earth still remain to be considered. We find those strata that were originally formed continuous in their substance, and horizontal in their position, now broken, bended, and inclined, in every manner and degree; we must give some reason in our theory for such a general state and disposition of things; and we must tell by what power this event, whether accidental or intended, had been brought about.

Lastly, Whatever powers had been employed in preparing land, while situated under water, or at the bottom of the sea, the most powerful operation yet remains to be explained; this is the means by which the lowest surface of the solid globe was made to be the highest upon the earth. Unless we can show a power of sufficient force, and placed in a proper situation for that purpose, our theory would go for nothing, among people who investigate the nature of things, and who, founding on experience, reason by induction from effect to cause.

Nothing can be admitted as a theory of the earth which does not, in a satisfactory manner, give the efficient causes for all these effects already enumerated. For, as things are universally to be acknowledged in the earth, it is essential in a theory to explain those natural appearances.

But this is not all. We live in a world where order every where prevails; and where final causes are as well known, at least, as those which are efficient. The muscles, for example, by which I move my fingers when I write, are no more the efficient cause of that motion, than this motion is the final cause for which the muscles had been made. Thus, the circulation of the blood is the efficient cause of life; but, life is the final cause, not only for the circulation of the blood, but for the revolution of the globe: Without a central luminary, and a revolution of the planetary body, there could not have been a living creature upon the face of this earth; and, while we see a living system on this earth, we must acknowledge, that in the solar system, we see a final cause.

Now, in a theory which considers this earth as placed in a system of things where ends are at least attained, if not contrived in wisdom, final causes must appear to be an object of consideration, as well as those which are efficient. A living world is evidently an object in the design of things, by whatever Being those things had been designed, and however either wisdom or folly may appear in that design. Therefore the explanation which must be given of the different phenomena of the earth, must be consistent with the actual constitution of this earth as a living world, that is, a world maintaining a system of living animals and plants.

Not only are no powers to be employed that are not natural to the globe, no action to be admitted of except those of which we know the principle, and no extraordinary events to be alledged in order to explain a common appearance, the powers of nature are not to be employed in order to destroy the very object of those powers; we are not to make nature act in violation to that order which we actually observe, and in subversion of that end which is to be perceived in the system of created things. In whatever manner, therefore, we are to employ the great agents, fire and water, for producing those things which appear, it ought to be in such a way as is consistent with the propagation of plants and the life of animals upon the surface of the earth. Chaos and confusion are not to be introduced into the order of nature, because certain things appear to our partial views as being in some disorder. Nor are we to proceed in feigning causes, when those seem insufficient which occur in our experience.

In examining the structure of our earth, we find it no less evidently formed of loose and incoherent materials, than that those materials had been collected from different parts, and gathered together at the bottom of the sea. Consequently, if this continent of land, first collected in the sea, and then raised above its surface, is to remain a habitable earth, and to resist the moving waters of the globe, certain degrees of solidity or consolidation must be given to that collection of loose materials; and certain degrees of hardness must be given to bodies which were soft or incoherent, and consequently so extremely perishable in the situation where they now are placed.

But, at the same time that this earth must have solidity and. hardness to resist the sudden changes which its moving fluids would occasion, it must be made subject to decay and waste upon the surface exposed to the atmosphere; for, such an earth as were made incapable of change, or not subject to decay, could not afford that fertile soil which is required in the system of this world, a soil on which depends the growth of plants and life of animals,— the end of its intention.

Now, we find this earth endued precisely with that degree of hardness and consolidation as qualifies it at the same time to be a fruitful earth, and to maintain its station with all the permanency compatible with the nature of things, which are not formed to remain unchangeable. . . .

We are thus led to inquire into the efficient causes of this constitution of things, by which solidity and stability had been bestowed upon a mass of loose materials, and by which this solid earth, formed first at the bottom of the sea, had been placed in the atmosphere, where plants and animals find the necessary conditions of their life.

Now, we have shown, that subterraneous fire and heat had been employed in the consolidation of our earth, and in the erection of that consolidated body into the place of land. The prejudices of mankind, who cannot see the steps by which we come at this conclusion, are against the doctrine; but, prejudice must give way to evidence. No other Theory will in any degree explain appearances, while almost every appearance is easily explained by this Theory.

We do not dispute the chymical action and efficacy of water, or any other substance which is found among the materials collected at the bottom of the sea; we only mean to affirm, that every action of this kind is incapable of producing perfect solidity in the body of the earth in that situation of things, whatever time should be allowed for that operation, and that whatever may have been the operations of water, aided by fire, and evaporated by heat, the various appearances of mineralization, (every where presented to us in the solid earth, and the most perfect objects of examination), are plainly inexplicable upon the principle of aqueous solution. On the other hand, the operation of heat, melting incoherent bodies, and introducing softness into rigid substances which are to be united, is not only a cause which is proper to explain the effects in question, but also appears, from a multitude of different circumstances, to have been actually exerted among the consolidated bodies of our earth, and in the mineral veins with which the solid bodies of the earth abound.

The doctrine, therefore, of our Theory is briefly this, That, whatever may have been the operation of dissolving water, and the chymical action of it upon the materials accumulated at the bottom of the sea, the general solidity of that mass of earth, and the placing of it in the atmosphere above the surface of the sea, has been the immediate operation of fire or heat melting and expanding bodies. Here is a proposition which may be tried, in applying it to all the phenomena of the mineral region; so far as I have seen, it is perfectly verified in that application.

We have another proposition in our Theory; one which is still more interesting to consider. It is this, That as, in the mineral regions, the loose or incoherent materials of our land had been consolidated by the action of heat; so, upon the surface of this earth exposed to the fluid elements of air and water, there is a necessary principle of dissolution and decay, for that consolidated earth which from the mineral region is exposed to the day. The solid body being thus gradually impaired, there are moving powers continually employed, by which the summits of our land are constantly degraded, and the materials of this decaying surface travelled towards the coast. There are other powers which act upon the shore, by which the coast is necessarily impaired, and our land subjected to the perpetual encroachment of the ocean. . . .

We have now seen, that in every quarter of the globe, and in every climate of the earth, there is formed, by means of the decay of solid rocks, and by the transportation of those movable materials, that beautiful system of mountains and valleys, of hills and plains, covered with growing plants, and inhabited by animals. We have seen, that, with this system of animal and vegetable economy, which depends on soil and climate, there is also a system of moving water, poured upon the surface of the earth, in the most beneficial manner possible for the use of vegetation, and the preservation of our soil; and that this water is gathered together again by running to the lowest place, in order to avoid accumulation of water upon the surface, which would be noxious.

It is in this manner that we first have streams or torrents, which only run in times of rain. But the rain-water absorbed onto the earth is made to issue out in springs, which run perpetually, and which, gathering together as they run, form rivulets, watering valleys, and delighting the various inhabitants of this earth. The rivulets again are united in their turn, and form those rivers which overflow our plains, and which alternately bring permanent fertility and casual devastation to our land. Those rivers, augmenting in their volume as they unite, pour at last their mighty waters into the ocean; and thus is completed that circulation of wholesome fluids, which the earth requires in order to be a habitable world.

Our Theory farther shows, that in the ocean there is a system of animals which have contributed so materially to the formation of our land. These animals are necessarily maintained by the vegetable provision, which is returned in the rivers to the sea, and which the land alone or principally produces. Thus we may perceive the mutual dependence upon each other of those two habitable worlds,—the fluid ocean and the fertile earth.

The land is formed in the sea, and in great part by inhabitants of that fluid world. But those animals, which form with their *exuviae* such a portion of the land, are maintained, like those upon the surface of the earth, by the produce of that land to which they formerly had contributed. Thus the vegetable matter, which is produced upon the surface of the earth in such abundance for the use of animals, and which, in such various shapes, is carried by the rivers into the sea, there sustains that living system which is daily employed to make materials for a future land.

Our solid earth is every where wasted, where exposed to the day. The summits of the mountains are necessarily degraded. The solid weighty materials of those mountains are every where urged through the valleys, by the force of running water. The soil, which is produced in the destruction of the solid earth, is gradually travelled by the moving water, but is constantly supplying vegetation with its necessary aid. This travelled soil is at last deposited upon the coast, where it forms most fertile countries. But the billows of the ocean agitate the loose materials upon the shore, and wear away the coast, with the endless repetitions of this act of power, or this imparted force. Thus the continent of our earth, sapped in its foundation, is carried away into the deep, and sunk again at the bottom of the sea, from whence it had originated.

We are thus led to see a circulation in the matter of this globe, and a system of beautiful economy in the works of nature. This earth, like the body of an animal, is wasted at the same time that it is repaired. It has a state of growth and augmentation; it has another state, which is that of diminution and decay. This world is thus destroyed in one part, but it is renewed in another; and the operations by which this world is thus constantly renewed, are as evident to the scientific eye, as are those in which it is necessarily destroyed. The marks of the internal fire, by which the rocks beneath the sea are hardened, and by which the land is produced above the surface of the sea, have nothing in them which is doubtful or ambiguous. The destroying operations again, though placed within the reach of our examination, and evident almost to every observer, are no more acknowledged by mankind, than is that system of renovation which philosophy alone discovers.

It is only in science that any question concerning the origin and end of things is formed; and it is in science only that the resolution of those questions is to be attained. The natural operations of this globe, by which the size and shape of our land are changed, are so slow as to be altogether imperceptible to men who are employed in pursuing the various occupations of life and literature. We must not ask the industrious inhabitant, for the end or origin of this earth: he sees the present, and he looks no farther into the works of time than his experience can supply his reason. We must not ask the statesman, who looks into the history of time past, for the rise and fall of empires; he proceeds upon the idea of a stationary earth, and most justly has respect to nothing but the influence of moral causes. It is in the philosophy of nature that the natural history of this earth is to be studied; and we must not allow ourselves ever to reason without proper data, or to fabricate a system of apparent wisdom in the folly of a hypothetical delusion.

When, to a scientific view of the subject, we join the proof which has been given, that in all the quarters of the globe, in every place upon the surface of the earth, there are the most undoubted marks of the continued progress of those operations which wear away and waste the land, both in its heighth and wideth, its elevation and extention, and that for a space of duration in which our measures of time are lost, we must sit down contented with this limitation of our retrospect, as well as prospect, and acknowledge, that it is in vain to seek for any computation of the time, during which the materials of this earth had been prepared in a preceding world, and collected at the bottom of a former sea.

SPALLANZANI

Lazarro Spallanzani (1729–1799), Italian abbot, Professor of Natural History at Padua, was one of the first to use experimental methods in the study of rocks.

ON BASALTIFORM LAVAS AND THE ROLE OF GASES IN VOLCANIC ERUPTIONS

From *Travels in the Two Sicilies*, translated anonymously, London, 1798.

The first time I ventured to explore the bottom of the crater of Vulcano, I only found some fragments of this prismatic lava: but when I repeated my visits, and had divested myself of the fear I at first felt, and more carefully examined this dreary bottom, I was enabled to complete my discovery by ascertaining the origin of these prismatic, or, as some may choose to call them, these basaltiform lavas.   For raising my eyes to that part of the crater which was over my head, and facing the northeast; I perceived a large stratum of lava, almost perpendicular, divided lengthwise into complete prisms, some of which were continued with the lava and made one body with it. . . .

The production of these basaltiform lavas, which, from their situation, and their forming a whole with the lava, no one can doubt derive their origin from fire, may, I conceive, be thus explained.   In former times an effervescence took place in the melted lava in the crater, which, after having swelled, and perhaps overflowed its edges, slowly sunk in the cavity of the crater, from the diminution of the fire, and the impellent elastic substances, while a portion of the lava attaching itself to the internal sides and hastily cooled by the atmospheric air, contracted, and divided into regular parts, such as are the forms of the hexagon prisms above mentioned.

In various parts of this work mention has been made of the gases of volcanos.   It has been shown that, by their elasticity, stony substances fused in the fire are rarefied, inflated, and become cellular; as is proved by the great number of lavas, glasses and enamels.   We have seen that, by the violence of these gases, the

101

liquefied matters are hastily raised from the bottom of the craters to the top, filling their whole internal capacity, and flowing over their sides; since, by the action of the same gases, we frequently observe similar phenomena in the furnace.

I shall now proceed to enquire what part this aeriform vapour acts in the eruptions of volcanos. Where it exists in the depths of a volcanic crater, abundantly mixed with a liquid lava violently urged by subterranean conflagrations, I can easily conceive that by its energetic force it may raise the lava to the top of the crater, and compel it to flow over the sides, and form a current.

It is likewise probable that this elastic vapour, when collected in a large quantity, if it finds under the earth any impenetrable obstacle, produces local earthquakes, and subterraneous thunders and roarings; bursting open the sides of the lava and forcing out the lava. We have an example of this, if I may so speak, in miniature, in the two matrasses* broken by this fluid from its exuberance and the resistance it met with.

---

* Containers to hold crushed material melted by Spallanzani in his experiments.—EDITORS.

CAVENDISH

Henry Cavendish (1731–1810), English chemist and physicist.

Experiments to Determine the Density of the Earth

From *Philosophical Transactions of The Royal Society of London*, Vol. LXXXVIII, pp. 469–526, 1798.

Many years ago, the late Rev. JOHN MICHELL, of this Society, contrived a method of determining the density of the earth, by rendering sensible the attraction of small quantities of matter; but, as he was engaged in other pursuits, he did not complete the apparatus till a short time before his death, and did not live to make any experiments with it. After his death, the apparatus came to the Rev. FRANCIS JOHN HYDE WOLLAS-TON, Jacksonian Professor at Cambridge, who, not having conveniences for making experiments with it, in the manner he could wish, was so good as to give it to me.

The apparatus is very simple; it consists of a wooden arm, 6 feet long, made so as to unite great strength with little weight. This arm is suspended in an horizontal position, by a slender wire 40 inches long, and to each extremity is hung a leaden ball, about 2 inches in diameter; and the whole is inclosed in a narrow wooden case, to defend it from the wind.

As no more force is required to make this arm turn round on its centre, than what is necessary to twist the suspending wire, it is plain, that if the wire is sufficiently slender, the most minute force, such as the attraction of a leaden weight a few inches in diameter, will be sufficient to draw the arm sensibly aside. The weights which Mr. MICHELL intended to use were 8 inches diameter. One of these was to be placed on one side the case, opposite to one of the balls, and as near it as could conveniently be done, and the other on the other side, opposite to the other ball, so that the attraction of both these weights would conspire in drawing the arm aside; and, when its position, as affected by these weights, was ascertained, the weights were to be removed to the other side of the case, so as to draw the arm the contrary way, and the position of the arm was to be again determined; and, consequently, half the difference of these positions would shew how much the arm was drawn aside by the attraction of the weights.

In order to determine from hence the density of the earth, it is necessary to ascertain what force is required to draw the arm aside through a given space. This Mr. MICHELL intended to do, by

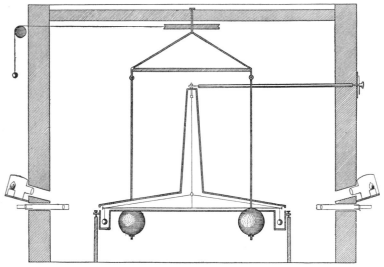

Fig. 7.—The apparatus used by Cavendish to determine the density of the earth, 1798.

putting the arm in motion, and observing the time of its vibrations, from which it may easily be computed.*

As I was convinced of the necessity of guarding against this source of error [*changes of temperature*], I resolved to place the apparatus in a room which should remain constantly shut, and to observe the motion of the arm from without, by means of a telescope; and to suspend the leaden weights in such a manner, that I could move them without entering into the room.   This difference in the manner of observing, rendered it necessary to make some alteration in Mr. MICHELL'S apparatus; and, as there were some

---

* Mr. Coulomb has, in a variety of cases, used a contrivance of this kind for trying small attractions; but Mr. Michell informed me of his intention of making this experiment, and of the method he intended to use, before the publication of any of Mr. Coulomb's experiments.

parts of it which I thought not so convenient as could be wished, I chose to make the greatest part of it afresh. . . .

Before I proceed to the account of the experiments, it will be proper to say something of the manner of observing. Suppose the arm to be at rest, and its position to be observed, let the weights then be moved, the arm will not only be drawn aside thereby, but it will be made to vibrate, and its vibrations will continue a great while; so that, in order to determine how much the arm is drawn aside, it is necessary to observe the extreme points of the vibrations, and from thence to determine the point which it would rest at if its motion was destroyed, or the point of rest, as I shall call it. To do this, I observe three successive extreme points of vibration, and take the mean between the first and third of these points, as the extreme point of vibration in one direction, and then assume the mean between this and the second extreme, as the point of rest; for, as the vibrations are continually diminishing, it is evident, that the mean between two extreme points will not give the true point of rest. . . .

It appears, therefore, that on account of the resistance of the air, the time at which the arm comes to the middle point of the vibration, is not exactly the mean between the times of its coming to the extreme points, which causes some inaccuracy in my method of finding the time of vibration. It must be observed, however, that as the time of coming to the middle point is before the middle of the vibration, both in the first and last vibration, and in general is nearly equally so, the error produced from this cause must be inconsiderable; and, on the whole, I see no method of finding the time of a vibration which is liable to less objection. . . .

In my first experiments, the wire by which the arm was suspended was $39\frac{1}{4}$ inches long, and was of copper silvered, one foot of which weighed $2\frac{4}{10}$ grains; its stiffness was such, as to make the arm perform a vibration in about 15 minutes. I immediately found, indeed, that it was not stiff enough, as the attraction of the weights drew the balls so much aside, as to make them touch the sides of the case; I, however, chose to make some experiments with it, before I changed it. . . .

Conclusion

From this table it appears, that though the experiments agree pretty well together, yet the difference between them, both in the quantity of motion of the arm and in the time of vibration, is greater than can proceed merely from the error of observation. As

The following Table contains the Result of the Experiments.

Exper.	Mot. weight	Mot. arm	Do. corr.	Time vib.	Do. corr.	Density.
1	m. to +	14,32	13,42	′ ″	-	5,5
	+ to m.	14,1	13,17	14,55	-	5,61
2	m. to +	15,87	14,69	-	-	4,88
	+ to m.	15,45	14,14	14,42	-	5,07
3	+ to m.	15,22	13,56	14,39	-	5,26
	m. to +	14,5	13,28	14,54	-	5,55
4	m. to +	3,1	2,95		6,54	5,36
	+ to −	6,18	-	7,1	-	5,29
	− to +	5,92	-	7,3	-	5,58
5	+ to −	5,9	-	7,5	-	5,65
	− to +	5,98	-	7,5	-	5,57
6	m. to −	3,03	2,9	-	-	5,53
	− to +	5,9	5,71	-	-	5,62
7	m. to −	3,15	3,03	7,4 by mean.	6,57	5,29
	− to +	6,1	5,9			5,44
8	m. to −	3,13	3,00	-	-	5,34
	− to +	5,72	5,54	-	-	5,79
9	+ to −	6,32	-	6,58	-	5,1
10	+ to −	6,15	-	6,59	-	5,27
11	+ to −	6,07	-	7,1	-	5,39
12	− to +	6,09	-	7,3	-	5,42
13	− to +	6,12	-	7,6	-	5,47
	+ to −	5,97	-	7,7	-	5,63
14	− to +	6,27	-	7,6	-	5.34
	+ to −	6,13	-	7,6	-	5,46
15	− to +	6.34	-	7,7	-	5,3
16	− to +	6,1	-	7,16	-	5,75
17	− to +	5,78	·	7,2	-	5,68
	+ to −	5,64	-	7,3	-	5,85

FIG. 8.—Table published by Cavendish to show the results of his experiments to determine the density of the earth, 1798.

to the difference in the motion of the arm, it may very well be accounted for, from the current of air produced by the difference of temperature; but, whether this can account for the difference in the time of vibration, is doubtful. If the current of air was regular and of the same swiftness in all parts of the vibration of the ball, I think it could not; but, as there will most likely be much irregularity in the current. it may very likely be sufficient to account for the difference.

———————

By a mean of the experiments made with the wire first used, the density of the earth comes out 5,48 times greater than that of water; and by a mean of those made with the latter wire, it comes out the same; and the extreme difference of the results of the 23 observations made with this wire, is only ,75; so that the extreme results do not differ from the mean by more than ,38, or $\frac{1}{14}$ of the whole, and therefore the density should seem to be determined hereby, to great exactness. It, indeed, may be objected, that as the result appears to be influenced by the current of air, or some other cause, the laws of which we are not well acquainted with, this cause may perhaps act always, or commonly, in the same direction, and thereby make a considerable error in the result. But yet, as the experiments were tried in various weathers, and with considerable variety in the difference of temperature of the weights and air, and with the arm resting at different distances from the sides of the case, it seems very unlikely that this cause should act so uniformly in the same way, as to make the error of the mean result nearly equal to the difference between this and the extreme; and, therefore, it seems very unlikely that the density of the earth should differ from 5,48 by so much as $\frac{1}{14}$ of the whole. . . .

According to the experiments made by Dr. MASKELYNE, on the attraction of the hill Schahillien, the density of the earth is $4\frac{1}{2}$ times that of water; which differs rather more from the preceding determination than I should have expected. But I forbear entering into any consideration of which determination is most to be depended on, till I have examined more carefully how much the preceding determination is affected by irregularities whose quantity I cannot measure.

DE L'ISLE

Jean Baptiste Louis Rome de l'Isle (1736–1790), French mineralogist, is credited with many of the ideas on which Hauy later built the science of crystallography.

LAWS OF CRYSTALLIZATION

Translated from *Essai de cristallographie,* Paris, 1772.

When I say that the theory of crystals, relative to their geometric figures, can throw a great light on this part of natural history, I am far from wishing to insinuate that a geometrician could ever give an account of the formation of different compounds by purely geometric speculations. Neither do I pretend that the figures which these compounds offer us ought to be taken precisely, or that they ever have the regularity and precision of those which mathematicians trace for us. I wish only to say that these figures, in spite of their variety without end, proving to be the same or about the same in the diverse saline, stony, and metallic substances, seem to indicate a hidden affinity in these substances, which some day perhaps we shall discover.

Germs being inadmissible to explain the formation of crystals, it must necessarily be assumed that the integrant molecules of the body have, each according to its own nature, a constant and determinate shape; and that those of the molecules which have some analogy between themselves tend to draw together reciprocally and unite, sometimes indefinitely by all their faces, sometimes by those faces which can have the most absolute and closest contact between themselves. But as the *first elements* of bodies are, and probably always will be, unknown to us because of the littleness of their parts, which escape the best microscopes, we can determine only the figure of the *secondary elements*.

Among these last, the salts, without doubt, have first place. No one today is unaware that, except for a very small number of salts that remain always in fluid form, all take, by the joining of their integrant molecules, a determinate figure, essentially the same in each species of salt. Thus the *marine salt* is always cubical, the *nitre* prismatical, the *alum* pyramidal, the *vitriol* rhomboidal, etc. It is true that these figures are more or less perfect, more or less regular, according to the circumstances which have favored or disturbed the joining of their constituent parts and conformable with the greater or lesser purity of the fluid which held them in dissolution.

There is cause to believe that the slight progress which has been made so far in the knowledge of the proper and essential forms of each species of salt must be attributed to the variety which one finds in the crystallization of the same salt. It has been believed that such variable forms were but little worthy of attention and exerted no influence over the nature of these salts. The primitive forms have not been distinguished with sufficient care from those resulting from the medley or confusion of such forms. Thus although in our laboratories the crystals of marine salt often present hollow and inverted quadrangular pyramids, the cubical figure is nevertheless the primitive and essential form of this salt, since these pyramids are all formed accidentally by the assembly of many quadrangular prisms which are themselves composed of cubes adhering successively to the sides of the first cube.*

That which takes place in the crystallization of salts can instruct us in the course of nature in the forming of other crystallized substances. We see there:

1. That the immediate effect of crystallization is the assembly of many saline molecules into polyhedral and determinate masses.

2. That these molecules have the admirable property of uniting in great numbers while keeping a symmetrical order among themselves, so that they form regular bodies, differently shaped in accordance with the nature of each salt.

3. That this assembly cannot take place if these molecules have not first been dissolved and separated from each other by the interposition of a fluid.

4. That it is by the evaporation, chilling, or subtraction of a part of this fluid that the molecules approach each other and manage to touch and unite.

---

* M. Rouelle has shown that the crystals of marine salt take a cubical form only in gradual evaporation, because then the crystals do not float to the surface of the fluid which holds them in dissolution but are precipitated as they form. As a result, the first saline unions or the primitive crystal being cubical and the new molecules which unite with them being equally cubical, more or less regular cubical crystals ought to result from these unions.

5. That consequently the concurrence of the surrounding air, of heat, and of cold is equally necessary to crystallization.

6. That the assembly of integrant molecules can also take place when they have reached such a degree of proximity that they can easily cross the space that separates them, by virtue of their attraction for each other.

7. That these molecules form masses with a constant and regular figure when they have the time and freedom to unite with each other by the faces that are most disposed toward this union.

8. That these same molecules form irregular and infinitely varied masses when the subtraction of interposed liquid is made so quickly that the parts which it separates find themselves brought together and at the point of contact before being able to take the position toward which they naturally tend.

9. That a like effect can arise from the agitation which the fluid may have undergone at the time of crystallization; for then the molecules are joined haphazardly by the faces which chance placed together in forced contact.

10. Finally, water enters the saline crystals as a constituent part without being essential to the nature of these salts, since one can take it out without destroying their properties, although the crystal form cannot exist without it.

RASPE

Rudolph Eric Raspe (1737–1794), a German mineralogist and archaeologist who escaped to England. Raspe was a competent geologist but a notorious rogue. He is supposed to have written the *Adventures of Baron Munchausen.*

Columnar Jointing in Basaltic Lava

From *An Account of Some German Volcanos*, London, 1776.

The large extensive strata of the black rock under consideration are, for the most part, in a nearly horizontal position, and may thus be seen at different altitudes in the Habichwald. Moreover, at different places they are found without any respect to specific gravity, now under, now above, other lighter strata of ashes, tufo, and flags. I beg leave to consider these solid rock-strata as lavas, cooled in great extensive melted masses under ground, or cooled after their eruption as large lava flows. . . . The great stratum of this kind, or rather the most extensive lava, which I had an opportunity to observe in these parts, is at the top of the Carlsberg in the Habichwald, under the foundations of the octogon. Behind the grotto of Polyphemus and on both its sides, especially to the right, it appears bare and open on the slopes of the steep mountain.

Its length and extent may be far greater; and perhaps all the solid vitreous black rocks on both sides of the octogon, and on the whole summit of the Carlsberg, together with the romantic heaps of its dreadful ruins, scattered down its steep descent into the lower valleys, might be correctly considered as belonging to the same lava.

The black vitreous rocks in solid isolated points below the level of the sea, considered as remains of former, or as nuclei of ancient, volcanos

Such solid ridges and isolated points, standing free and bare on all sides, are very common in Lower-Hesse.

Having ventured to designate the solid strata of the black rock as cooled lavas, in erupted flows or unerupted melted subterranean masses, a denomination which their situation and many other

111

qualities will justify; I venture, in the same hope, to consider these ridges and isolated points as remains of old lava flows, or as kernels and nuclei of old volcanos, deprived by rain, snow, overflowing water, and a long series of destroying centuries, of their external covering of ashes, cinders and sands. . . .

Different constant forms of the black rocks. The prismatical ones, called basaltes, discovered to be lavas

The *form* and *figure*, which they spontaneously affect when broken, is very different. I have in this regard observed the following varieties:

1. Their form is *irregular* in the great lava-stratum on the Carlsberg, as clearly appears in the subterranean gallery and in many fragments, dispersed every where in the valleys and on the ground immediately below its steep summit.
2. They appear in irregular slates or tables, and in vertical position, near the old Winterkasten on the Carlsberg.
3. In *large globular masses*, composed of concentric and less compact coats, and containing in the middle a remarkably hard and compact globular nodulus or nucleus; these are obvious in a large stratum near the Moriz-grotto behind the Sneckenberg and in various other places at the Carlsberg.
4. They are irregular polygons, tending to a prismatic columnar form, in a stratum behind the great basin up to the Moriz-grotto at Wissenstine, at the level with the Sneckenberg.
5. In a *columnar prismatic regular form* they appear in large masses or strata of close-lying columns in the castle-hill at Felsberg and Aldenburg, in the Maderstine near Gudensberg, in the Widelsberg near Wolfhagen, in the Holestine under the Dornberg, and, as I have been told, but never observed myself, in the Dornberg.

These prisms or columns are of a quadrangular, pentagonal, hexagonal, and heptagonal form; flat at the top; and in the same stratum, or rather the same mass, of the same length; but, the thickness of these different masses being different, they differ likewise in length from six feet to twenty and more.

Till very lately, the origin and nature of these singular stones was considered as inexplicable; but a great many correspondent observations, and a closer examination of their substance and situation, has, since the year 1768, convinced Mr. Desmarest, myself, and Mr. Ferber:

a. That the substance of these figured prismatical and columnar vitreous black stones is to be considered almost everywhere as a ferrugineous mass, impregnated with much iron and smelted by heat or fire.

b. That, appearing in prismatical forms, it is named, by the Ancients and the Mineralogists, basaltes.

c. That, in an irregular form, it goes in Italy under the name of lava, or selce and pietre dure.

d. That by refrigeration it gets a prismatical form, just as most parts of smelted ores and metals get by cooling a regular crystallized form in their substance.

But why have not all our black rocks and all the Italian lavas a similar, constantly regular prismatical form? Even the prismatical columnar lavas or basaltes show varieties of forms. The Irish basaltes in the Giants-causeway in the county of Antrim, appear in polygonal articulate columns or prisms. There ought to be some natural reason for that, whatever it be. Is it owing to the difference of their substance and mixture? to the different sloping of the ground on which these fiery melted masses run forward? to the varying quickness of their motion? to their different fluidity? or is it owing rather to the manner of their refrigeration? All these circumstances may, without any doubt, have influenced their different form; but our black rocks and prismatical basaltes being of the same mixture and substance, seem to indicate that their different fluidity, fusion, motion, and refrigeration, have been the proximate causes of their before-described regular and irregular forms and figures.

SAUSSURE

Horace Benedicte de Saussure (1740–1799), professor of philosophy at Geneva, is remembered as one of the first great students of the Alps.

ALPINE GEOLOGY

Translated from *Voyages dans les Alpes*, Geneva, Vol. 1, pp. 167–172, 196–205, 210–212, Vol. 2, pp. 44–47, 267–274, 1787.

Experiments on Granite

I sought a granite in which the constituent elements, quartz, mica, and feldspar, were all well characterized and distinct. The niton stone,* that great rolled rock in the lake at the entrance to the port of our city, possesses these qualities to an eminent degree. Its feldspar is in great white and opaque crystals; its quartz is in morsels of indefinite form, but transparent and of a color bordering on violet; and its mica is in little blackish flakes.

I reduced a fragment of this granite to fine powder. I exposed it to the most violent fire of my furnace. It was transformed into a glass of greenish gray color, semitransparent, thoroughly fused, brilliant on the surface but filled with extremely small bubbles. The lens shows white grains of quartz there, which, being less fine than the others, have resisted vitrification.

Under the same muffle, and at the side of the crucible that contained this pulverized granite, another crucible held fragments of the same rock. The experiments made in this way on the unbroken morsels are much more instructive, because the diverse changes that the different substances of the mixed body undergo can be recognized. After having endured the action of fire, these fragments were united, fused. They filled the bottom of the crucible. The surface of the melted matter was concave and brilliant. On breaking this vitreous material, one recognized distinctly the three elements of granite: the mica, melted to a black glass, tending

---

* Niton stone—"an old stone near Geneva which the inhabitants have regarded since time immemorial as an altar of Neptune."—*Grand Dict. Larousse.*

toward brown and green, strewn with bubbles the size of a millet seed; the feldspar, reduced to a transparent, colorless glass, filled with bubbles visible only under the lens, hard to the point of cutting window glass and of sparking against steel; and finally, the quartz, preserved intact even to its smallest parts, having lost only its transparency, which was taken away by the innumerable cracks contracted in the fire, rendering it a beautiful, dull white.

The vitrification of this granite has, therefore, no resemblance to a homogeneous basalt. Greater intensity of fire, if it were capable of attacking and finally dissolving the quartz, would reduce the granite to a much harder and still more transparent glass, which would resemble basalt still less. And a lesser intensity, as I have proven, would give friable and incoherent masses at first, then cavernous frits without bond and without homogeneity, so that it seemed to me impossible that such a granite would ever give a material which would resemble a homogeneous lava.

After all these experiments, it does not seem possible that any stone of the class of granites, a mixture of quartz and feldspar, could serve as material for basalts or homogeneous lavas. The fires known to us do not render them at all homogeneous. A fire capable of rendering them so would change them to a transparent glass, extremely hard and absolutely different from basalts.

The Origin of Rolled Pebbles and Boulders

Everyone knows the so-called *galets* or *rolled pebbles*, stones of rounded form, or at least with blunted angles, that are ordinarily found in the bed of streams and in the neighboring plains, especially near mountains where these streams have their source. The name given these pebbles comes, without doubt, from the presumption that they have been rolled and rounded by the waters. . . .

It will be seen that most of these pebbles and rocks are of granite, schistose rock, or other alpine and primitive stones, whereas the floor on which they have been deposited is limestone or sandstone and, consequently, of an absolutely different nature. . . .

It cannot longer be doubted that water was the agent, because these pebbles, large and small, are found deposited in horizontal ledges, mixed with sand and gravel, as the waters drifted them. If any of these fragments are seen exposed on a rock, inspection of that place shows clearly that rain water or melted snow has swept away the lightest parts which formerly surrounded these great masses.

Fire is the sole agent that could dispute the transport of these stones with water; but has anyone seen an instance of an explosion that could throw blocks of many cubic *toises** for twelve or fifteen leagues, as we often find them in our region? If one wished to admit this hypothesis, it would be necessary to suppose fire of an extreme size and violence in order to explain such great consequences; but such fires would have melted or calcined these rocks, or at least have thrown out lavas or vitreous matter with them. But no trace of the action of fire is found on the blocks nor in the material that surrounds them, and on the contrary the sand and gravel that accompany them are indubitable relics of the passage of waters. . . .

But, one will say, what was the origin of these waters? What gave them so violent an impulse? How could these masses of rock have been transported onto heights separated from the primitive Alps by wide, deep valleys? . . .

The waters of the ocean in which our mountains were formed still covered a part of these mountains when a violent earthquake suddenly opened great cavities that had previously been empty and caused the rupture of a great number of rocks.

The waters rushed toward these abysses with extreme violence, proportionate to the elevation that they then had, cutting deep valleys and sweeping along immense quantities of earth, sand, and fragments of all sorts of rock. This semi-liquid heap, pushed by the weight of the waters, piled up to the heights where we still see many of these scattered fragments.

Then the waters, continuing to flow but with a rapidity which diminished gradually in proportion to the diminution of their elevation, swept away the lightest parts, little by little. They purged the valleys of this heap of mud and debris, leaving behind only the heavier masses and those whose position or more solid seat protected them from this action.

The Origin of the Jura Valley

The mountain of Vouache seems to be a continuation of the main line of the Jura. This main line, with a general direction

---

* A *toise* is about six feet.—EDITORS.

trending northeast to southwest, changes its direction on approaching Eculfe; there it swings toward the south, which is the trend of the Vouache as well. The beds of the Jura at this end are nearly perpendicular to the horizon. For the greater part they do not diverge more than fifteen degrees from the vertical, and this slope is directed downwards toward the east. This arrangement of the Jura beds is visible near the summit of the mountain above the Fort. Lower, near the Fort itself, their form is not so clearly distinguishable. This position of the beds is also recognized in the slope that descends from the Fort to the banks of the Rhone, and even more distinctly behind the little chapel that lies two or three hundred paces from the Fort on the Geneva side. The beds of Vouache have exactly the same arrangement. They are seen to cut the course of the Rhone transversely, a little above the Eculfe Fort. Their plates are like those of the Jura beds, nearly perpendicular to the horizon; and they diverge, like the Jura beds, about fifteen degrees from the vertical, to descend also to the eastern side.

The position of these beds is so remarkable, so unusually and precisely fixed, that it proves to my satisfaction—as much as a thing of this kind can be proven—that the Vouache and the Jura were formerly united, forming one and the same mountain, leaving, therefore, no egress to the waters enclosed in our basin.

But how was this opening formed? The shaking of an earthquake is a convenient explanation; but it is almost the *deus ex machina*. It must not be employed, save when indubitable indications are seen or when no other explanation is left. I believe that we can avoid it here. It would suffice that the height of the mountain had been a little less in that locality; that it had formed a sort of gorge there. The waters would have taken this route and, little by little, gnawed and excavated their bed to the point where we see it now.

I have sought traces of these erosions. I have skirted the bed of the Rhone, descending from the point where it begins to narrow near the rocks of Jura to below the Fort. It elated me to find the wide and deep furrows that it has engraved on the limestone rocks. On a rock above the Rhone, between Cologne and the Eculfe Fort, there is an ancient ruin that the people of the country call the Château of Folly. The Rhone wets the foot of the rock on which this ruin rests, and it is there particularly that one can observe traces of a part of the higher course which the Rhone formerly followed.   The most remarkable of these traces is a furrow, sunk almost horizontally into the rock.   This furrow has a width of four or five feet and is excavated at least two feet deep in the rock.   Its edges and all its contours are rounded as they always are in the excavations produced by the waters.   It is situated more than twenty feet above the point reached by the Rhone today at the time of highest water.

The Inorganic Nature of Oölites

I have observed this type of calcareous rock composed of rounded grains, in divers localities.   The yellow marble found in Burgundy and known at Dijon under the name of *corgoloin* is composed of these little grains.   I myself have found rocks composed of similar grains, not only on the Dôle and on Mount Salève, but also near Bath in England, near Verona, at the fountain of Vaucluse, at Lieftal in the canton of Basle, and in various other places.

Many naturalists have regarded these little grains as fish eggs and have called these rocks oöliths—in German *Rogenstein*.   Others, believing them millet seed, have called them *cenchrites* (from the Greek κέγχρος, which means millet) and in German *Hirsenstein*.

On examining these little grains with a strong lens, I find that some, those of Verona for example, are composed of concentric layers, smooth on their surface and presenting no indication of organic structure.   Others appear to be a single, entirely homogeneous, mass.   Others seem to have a nucleus of different material, or at least of different color.   Some are exactly spherical; others, elongate.   All these varieties are found together in the stone of Corgoloin.   In that of the Dôle, most of the grains are homogeneous and rounded, although others are less regular.   And there are some in which one or two concentric layers can be clearly recognized.

I could no longer admit that these grains were seeds of millet or any other plant.   They did not seem to be bodies that had ever been organic.   I think rather that they are deposits or crystallizations, rounded by the motion of water at the time of their formation.

The stony concretions known under the name of *Tivoli pills* have a similar origin.

The most beautiful concretions of this sort that I know are those I have seen take form at San Filipo between Siena and Rome. Thermal waters, heated to the thirty-sixth degree on the Réaumur thermometer, saturated with calcareous alabaster, on cooling precipitate the alabaster they hold in solution. The movement of the waters rounds this alabaster as it crystallizes and fashions it into grains which, when broken, appear composed of concentric coats. These are the same waters that are forced to fall over concave sulphur [plaques] modeled from the antique bas-reliefs. The alabaster is deposited on the sulphur, fills its concavities, and forms bas-reliefs of a perfectly white stone which reproduce with the greatest exactitude the figures on which the plaques have been molded.

This explanation of the formation of *cenchrites*, confirmed by similar operations which take place before our eyes, frees us then from having recourse to chemical dissolution, as was done in a journal of physics for the year 1778.

Moreover the calcareous, and in no way neutral, properties of marbles and other rocks composed of these bodies proves that no acid, unless it be the fixed air, has intervened in their formation.

The Action of Glaciers

The sliding of snow masses in the form of avalanches is a well-known phenomenon and one to which we shall have occasion to refer elsewhere. That of glaciers, taking place more slowly and ordinarily with less noise, is not so well known.

Nearly all the glaciers, of the first as well as of the second type, repose on inclined floors, and all those of any considerable size have currents of water below them even in winter, which flow between the ice and the floor on which it rests.

One understands then that these frozen masses, drawn by the slope of the floor on which they rest, loosened by the waters from any binding they might have formed with this same floor, sometimes even raised by these waters, must slide little by little and descend along the slope of the valleys or hills that they cover.

It is this slow but continual sliding of the ice masses on their inclined bases that carries them as far as the low valleys and maintains the heaps of ice in lowlands warm enough to produce great trees and even abundant harvests.   In the bottom of the valley of Chamonix, for example, no glacier forms.   Even the snow disappears in the months of May and June.   Nevertheless the glaciers of Buisson, Bois, and Argentière descend to the bottom of this valley.   But the lower ice of these glaciers has not been formed in this place.   It carries, so to speak, the evidence of the place of its birth, for it is full of debris from the rocks that border the highest extremity of the glacial valley, and these rocks differ from those found in the mountains that border the lower part of this same valley.

All the great glaciers have, at their lower end and along their edges, great heaps of sand and debris, products of landslides from the mountains that tower above them.   Often the glaciers are even encased for their entire length by a species of parapet or earthworks composed of the same debris, which the lateral ice of these glaciers has deposited on their ridges.   In glaciers that were once larger than they are today, the parapets rise above the ice itself.   On the other hand, in those which are greater than they ever have been before, these parapets are lower than the ice. Finally, some are seen where they are at the same level.   The peasants of Chamonix call these piles of debris the moraine of the glacier.

The stones which heap up to form these parapets are mostly rounded, either because their angles have been dulled in rolling down from the mountain tops or because the ice has broken them by rubbing and holding them against its floor and edges.   But those which have remained at the surface of the ice, without having undergone any great amount of rubbing, have kept their jagged corners intact and sharp.   As for their composition, those found at the upper end of the glaciers are the same kinds of stone as the mountains that rise above them; but as the ice carries them down toward the valleys, they lie among mountains of a nature entirely different from theirs.

It seems somewhat more difficult to account for the heaps of rock and sand found piled in the middle of the *vallées de glace*, at such a great distance from the bordering mountains as to make it appear impossible that they came from them.

The stones are ordinarily arranged in lines parallel to the edge of the glacier, and one often sees many of these lines separated by bands of live, pure ice. On crossing the great glacier, two leagues above Montanvert, one has to pass over four or five earthworks of this kind, some of which are raised thirty or forty feet above the surface of the glacier. This is due in part to the quantity of stones gathered there, in part to the ice itself, which, sheltered from the sun and rain by the debris, stands higher than those parts that were originally higher but are bare and exposed to all the destructive action of the atmosphere.

I have known inhabitants of the Alps who, not knowing how to explain the origin of these ridges, said that the ice pushed up and thrust to the surface all foreign bodies found in its interior, and even the loose rocks and sand which were below it. But aside from the fact that such a force would be absolutely inconceivable, there is a still greater difficulty; namely, that the ice, as I have just said, is much higher under these benches of debris than in the rest of the glacier, so that the debris covers only the peaks of the ice, which are sometimes fifteen or twenty feet higher than the bare ice which separates them. It would be necessary then to suppose that the ice thrust itself up, and did that solely and precisely in the places where it is laden with the greatest weight. This is entirely absurd; all the more so because a perfect continuity is observed between this covered ice and that which is not covered—the same openings, the same fractures are seen continuing from one to the other so that it cannot be maintained that one originated at the bottom and the other at the surface. I believe the following is the true explanation of the phenomenon.

In the high Alps, as in the plains, mountains are found in such a state of decay that they continually detach fragments, either whole or crumbled to the form of earth and sand. And this takes place either because the mountains break naturally into fragments of different form or because the destructive action of the atmosphere wears them down and decomposes them. Especially in springtime—at the time of thaw, warm rain, and melting of snow—the particles of rock, sand, and earth that the frost has raised and moved fall on the ice in the high valleys. These stones, piled up on the edges of the glaciers, then follow the motion of the ice that carries them. But we have already seen that all this ice has a progressive movement, that it slides on its inclined floor, that it descends gradually to the low valleys, that there it is melted by the summer heat, and that what is thus destroyed is continually replaced by the progressive movement of the glacier. But the lower part of the glacial valleys is not the only place where ice melts. On fine summer days, especially when south winds prevail or warm rains fall, it melts throughout all the extent of the glaciers. The waters produced by this melting unite and form wide, deep ravines on the ice itself. The glaciers are divided by great crevasses, and as the valleys all have more or less the form of a cradle with their floor carved deeper than their sides, the ice compresses and contracts toward the middle of the valleys. The ice at the edge withdraws from the slopes, sliding toward the lowest point and carrying with it the earth and stones with which it is covered.

The proof of this truth is that toward the end of summer one sees in many localities, especially in the widest valleys, considerable spaces between the foot of the mountain and the edge of the glacier. These spaces arise, not only from the melting of the lateral ice but also because it is diverted from the margin in descending toward the center of the valley. During the next winter, these spaces are filled with snow. The snow absorbs water and is converted into ice. The edges of this new ice nearest to the mountain are covered anew with debris. These covered lines, in their turn, advance toward the center of the glacier. And thus are formed these parallel benches that move obliquely in a composite motion arising from the slope of the ground toward the center of the valley and from the slope of this same valley toward the foot of the mountain.

Finally, the proof of the origin of these benches is completed by the fact that none of them form in localities where glaciers are bordered by rocks of indestructible granite or when the slopes of the mountains are covered with snow or ice.

It seems, at first sight, that these parallel lines of sand and debris ought to mark the years and so serve to determine the age of different parts of the glaciers. But when the benches come from the two sides of a glacier, they mingle near the middle; also, the irregular slope of their bed often disturbs their order and parallelism.

There are localities where there are mountains which are breaking up on only one side of the glacier, and there this calculation could be made with less uncertainty.

PALLAS

Peter Simon Pallas (1741–1811), Prussian scientist, was professor of natural history at St. Petersburg. His explorations in Russia were unusually fruitful.

Mountains of Various Ages

Translated from *Observations sur la formation des montagnes*, St. Petersburg, 1771.

According to our knowledge of the Swedish, Swiss, and Tyrolese Alps, the Apennines, the mountains that surround Bohemia, the Caucasus, the mountains of Siberia, and even the Andes, one can grant as an axiom, that the highest mountains of the globe, which form the continous chains, are made of the rock known as granite, of which the base is always quartz, more or less mixed with feldspar, mica, and small basalts scattered without any order and as irregular fragments in different portions.

Both general and detailed observations reveal that this ancient rock, which we call granite and which is never found in beds but in blocks and crags or at least in masses heaped somewhat on the others, never contains the slightest trace of petrifaction or organic imprint, so that it seems to have been anterior to all organic Nature. . . .

I have already said that *the band of heterogeneous, primitive, schistose mountains*, which accompany the granitic chain throughout the earth and include the mixture of quartzose, spathic, and horn rocks, the pure sandstones, porphyry, and jasper, all rocks in beds that are either nearly perpendicular or at least very steeply inclined (the most favorable to the filtration of the waters), seem, as well as the granite, anterior to organic creation. . . .

We can speak more decisively of the *secondary* and *tertiary* mountains of the Empire.

Throughout the extent of the vast Russian domains, as well as in all Europe, attentive observers have noted that generally the schistose band of the great chains is found directly covered or flanked by the *limestone band*. The latter forms two orders of mountains, very different in height, in the attitude of their beds, and in the composition of the limestone that forms them. . . . My explanation of the two orders of limestone mountains is based principally on those to the west of the Ural chain.

This side of the said chain consists, over a width of fifty to a hundred versts, of solid limestone, very compact, which either does not contain a trace of marine products or only preserves imprints as slight as they are rare. This rock rises in mountains of very considerable height, irregular, steep, and cut by small cliffed valleys. Its generally thick beds are not level, but incline greatly from the horizontal, and are mainly parallel to the direction of the chain, which is also ordinarily that of the schistose band. . . .

Farther away from the chain, the limestone beds flatten rather rapidly, assume a horizontal position and contain an abundance of shells, madrepores, and other marine remains.

. . . I should speak next of an order of mountains that are certainly younger than the marine beds, since the latter serve them as base. Until the present no suite of these *tertiary mountains*—the result of the most recent catastrophe of our globe, so marked and so powerful—has been observed except that which borders the Ural chain on the western side for its entire length. This suite of mountains, mainly composed of sandstone and reddish marls, intermingled with diversely mixed beds, forms a chain separated everywhere by a wide valley from the band of limestone rock of which we have spoken.

In these same sandy and often loamy deposits lie the remains of great animals of India, bones of elephants, rhinoceroses, monstrous buffaloes. . . .

These great bones, sometimes scattered, sometimes piled in skeletons and even in hecatombs, studied where they lie, have definitely convinced me of the reality of a deluge over our land—a catastrophe the probability of which I admit I had not conceived before having traversed these shores and seen for myself all that can serve there as proof of this memorable event. An infinite number of these bones, embedded with a mixture of slightly calcined Tellines, bones of fish, glossoptera, ocher-impregnated wood, etc., proves beyond doubt that they have been transported by inundations. But the carcass of a rhinoceros, found with his skin entire and with remnants of tendons, ligaments, and cartilages, in the frozen lands on the banks of the Vilyuy, the best preserved parts of which I have deposited in the Cabinet of the Academy, formed convincing further proof that it must have been a most violent and rapid movement of inundation which long ago transported these cadavers to our frozen climates before corruption had time to destroy their soft parts.

Most natural philosophers who have treated of the physical geography of the world agree in considering all the isles of the South Seas as elevated on immense vaults of a common furnace. The first eruption of these fires, which raised the floor of the very deep sea there and which perhaps in a single stroke or by rapidly succeeding throes gave birth to the Sunda Isles, the Moluccas, and a part of the Philippines and austral lands, must have expelled from all parts a mass of water that surpasses the imagination. This, hurtling against the barrier opposed it on the north by the continuous chains of Asia and Europe, must have caused enormous overturnings and breaches in the lowlands of these continents, . . . and surmounting the lower parts of the chains which form the middle of the continents, . . . have entombed haphazardly the remains of many great animals which were enveloped in the ruin, and formed by successive depositions the tertiary mountains and the alluvions of Siberia.

LAVOISIER

Antoine Laurent Lavoisier (1743–1794), French chemist, largely responsible for the modern concept of the elements, worked with Guettard on the geologic map of France.

LITTORAL AND PELAGIC BEDS

Translated from *Mémoires de l'Académie des sciences*, 1789, pp. 351–371, 1793.

A part of the materials which are present at the surface in the lower parts of the earth and as far down as it is permitted us to penetrate are placed in horizontal beds. There one finds immense masses of marine bodies of all species, so that it cannot be doubted that, in very distant time, the sea covered a great part of the earth that is now inhabited.

But if this first glance is followed by a more profound examination of the arrangement of the beds and the materials that compose them, one is astonished to see all that is characteristic of order, uniformity, and tranquillity as well as all that marks disorder and movement.

———

These first reflections lead us to a natural conclusion, which is that there should exist in the mineral kingdom two very distinct types of beds, one formed in open sea at great depth, which, following the example of M. Rouelle, I shall call *pelagic* beds; the others formed along the coast, which I shall call *littoral* beds. These two types of beds ought to have distinctive characters which would not permit confusion. The first should present masses of calcareous material, debris of animals, shells, and marine bodies accumulated slowly and peacefully through an immense sequence of years and centuries. The other, on the contrary, should present throughout the mark of movement, destruction, and tumult.

. . . The beds formed in the open sea, or *pelagic*, should be and are in fact composed of nearly pure calcareous matter, that is to say, simply the material of the shells accumulated without mixture. The beds formed at the coast, the *littoral* beds, on the contrary, can

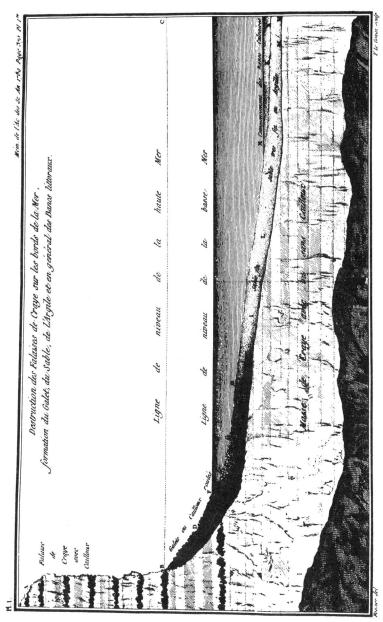

FIG. 9.—Diagram accompanying Lavoisier's study of littoral deposits, 1793.

be composed of materials of an infinity of types, depending on the nature of the coasts. . . .

But that which might escape the first glance—although it would be realized after minutes of reflection—is that the materials of which the littoral beds are formed ought not to be indiscriminately mixed.   On the contrary, they should be arranged and disposed in accordance with certain laws.   The movement of the waters of the sea, decreasing steadily from the surface to the bottom, at least to a depth of forty to fifty feet, ought to carry on a definite washing process analogous to that in the treatment of ores, along the edges of the sea.   This should even increase in extent as the steepness of the slope decreases.   The coarsest materials, like the cobbles, should occupy the highest part and mark the limit of high tide.   Lower, the coarse sands, which are nothing but attenuated pebbles, should range themselves.   Below, in the parts where the sea is less tumultuous and the movements less violent, the fine sands should be deposited.   Finally, the lightest materials, the most finely divided, like clay or silicious earth in a pulverized state, should remain suspended for a long time.   These materials ought to be deposited only at a rather great distance from the coast and at such a depth that the movement of the sea is almost nil.

Without entering the calculations which the most learned analysis would necessitate, it is seen in general that the curve of the bottom, from the coast to open sea, should fairly approximate a segment of parabola, the axis of which would be parallel to the horizon.   That is to say, the inclination of the coast to the horizon at the limit of the open sea should approach forty-five degrees. Thereafter it should go, diminishing, to the place where the water of the sea is in absolute repose, and from there its base should tend to become absolutely horizontal.

The first plate is intended to give an idea of what thus takes place at the edges of the sea, in localities where the coast is chalk. It is that which one observes in upper Normandy and on the corresponding coasts of England.

HAUY

Abbé René Just Hauy (1743–1822), French mineralogist, was the founder of the science of crystallography and thus of systematic mineralogy.

A Basis for a System of Mineralogy

Translated from *Traité de minéralogie*, Paris, Vol. 1, pp. xiii–xvii, 1801.

Geometry has direct and inescapable relations with mineralogy in the description of crystalline forms and still more by its numerous applications to the structure of crystals. This structure is solely the result of a natural geometry submitted to special rules, with each solid having its figure determined by the combination of an infinite number of other little solids, which are, as it were, the elements of the first. A casual glance at crystals may lead to the idea that they were *pure sports of nature*, but this is simply an elegant way of declaring one's ignorance. With a thoughtful examination of them, we discover laws of arrangement. With the help of these, calculation portrays and links up the observed results. How variable, and at the same time how precise and regular are these laws! How simple they are ordinarily, without losing anything of their significance!

The theory which has served to develop these laws is based entirely on a fact, the existence of which has hitherto been vaguely discerned, rather than demonstrated. This fact is that in all minerals which belong to the same species these little solids, which are the crystal elements and which I call their *integrant molecules*, have an invariable form, in which the faces lie in the direction of the natural joints indicated by the mechanical division of the crystals. Their angles and dimensions are derived from calculations combined with observation. Moreover, the integrant molecules of the different species also have more or less marked differences from each other; except in a small number of cases where their forms have similar characteristics that serve as points of contact between those species. It follows that the determination of the integrant molecules ought to have a great influence on the classification of the species. This consideration led me, more than once, either to subdivide a group into many species which in the former systems were considered as but one or to assemble and reunite the scattered members of a single species from which many distinct species had been made. Some of these separations and regroupings, made at a time when analysis had not yet unveiled the true nature of substances which it sought, are found to be confirmed today by the results of chemistry. I even dare say that, if no mineral substance had ever been chemically analyzed, one would be able, with the aid of sustained work on the integrant molecules, to form groups that would have been justifiably regarded as belonging to as many, clearly circumscribed, species;* so that to distribute them later in a well-ordered system, the analysis of a single sample from each of them would have sufficed.

From that, one understands the sense in which it is necessary to take my statement above, that the determination of the species belongs to chemistry. Perhaps it would be nearer the truth to say that chemistry completes that determination by acquainting us with the principal molecules, the components of the integrant molecules. At this point it is easy to perceive (and the rest of the book will offer many examples) how important it is that research with regard to these two sorts of molecules should strive toward a common end; that the chemist and the mineralogist should enlighten each other in their work; and that the goniometer, which furnishes data for the calculation of crystalline forms, should be associated with the scales that weigh the products of analysis.

---

* These groups should not have been limited to the crystals proper; lamellar masses could be placed in them, or even those which show no mechanical division, because these last frequently are so related to their crystalline analogues in position and appearance as to permit their recognition as members of the same species. Thus these masses, featureless by themselves, can be determined, at least approximately, by the crystals which serve in a way as their interpreters.

PLAYFAIR

John Playfair (1748–1819) was professor of mathematics and philosophy at Edinburgh. Playfair's explanation and development of Hutton's ideas were largely responsible for their acceptance.

PROOFS OF THE HUTTONIAN THEORIES

From *Illustrations of the Huttonian Theory of the Earth*, pp. 100–105, 312–315, 384–389, Edinburgh, 1802.

The Action of the Sea

If the coast is bold and rocky, it speaks a language easy to be interpreted. Its broken and abrupt contour, the deep gulphs and salient promontories by which it is indented, and the proportion which these irregularities bear to the force of the waves, combined with the inequality of hardness in the rocks, prove, that the present line of the shore has been determined by the action of the sea. The naked and precipitous cliffs which overhang the deep, the rocks hollowed, perforated, as they are farther advanced in the sea, and at last insulated, lead to the same conclusion, and mark very clearly so many different stages of decay. It is true, we do not see the successive steps of this progress exemplified in the states of the same individual rock, but we see them clearly in different individuals; and the conviction thus produced, when the phenomena are sufficiently multiplied and varied, is as irresistible, as if we saw the changes actually effected in the moment of observation. . . .

Again, where the sea-coast is flat, we have abundant evidence of the degradation of the land in the beaches of sand and small gravel; the sand banks and shoals that are continually changing; the alluvial land at the mouths of the rivers; the bars that seem to oppose their discharge into the sea, and the shallowness of the sea itself. On such coasts, the land usually seems to gain upon the sea, whereas, on shores of a bolder aspect, it is the sea that generally appears to gain upon the land. What the land acquires in extent, however, it loses in elevation; and, whether its surface increase or diminish, the depredations made on it are in both cases evinced with equal certainty.

Stream Activity and Uniformitarianism

If we proceed in our survey from the shores, inland, we meet at every step with the fullest evidence of the same truths, and particularly in the nature and economy of rivers.   Every river appears to consist of a main trunk, fed from a variety of branches, each running in a valley proportioned to its size, and all of them together forming a system of vallies, communicating with one another, and having such a nice adjustment of their declivities, that none of them join the principal valley, either on too high or too low a level; a circumstance which would be infinitely improbable, if each of these vallies were not the work of the stream that flows in it.

If indeed a river consisted of a single stream, without branches, running in a straight valley, it might be supposed that some great concussion, or some powerful torrent, had opened at once the channel by which its waters are conducted to the ocean; but, when the usual form of a river is considered, the trunk divided into many branches, which rise at a great distance from one another, and these again subdivided into an infinity of smaller ramifications, it becomes strongly impressed upon the mind, that all these channels have been cut by the waters themselves; that they have been slowly dug out by the washing and erosion of the land; and that it is by the repeated touches of the same instrument, that this curious assemblage of lines has been engraved so deeply on the surface of the globe.

The changes which have taken place in the courses of rivers, are also to be traced, in many instances, by successive platforms of flat alluvial land, rising one above another, and marking the different levels on which the river has run at different periods of time.   Of these, the number to be distinguished, in some instances, is not less than four or even five; and this necessarily carries us back, like all the operations we are now treating of, to an antiquity extremely remote: for, if it be considered, that each change which the river makes in its bed, obliterates at least a part of the monuments of former changes, we shall be convinced, that only a small part of the progression can leave any distinct memorial behind it, and that there is no reason to think, that in the part which we see, the beginning is included.

In the same manner, when a river undermines its banks, it often discovers deposites of sand and gravel, that have been made when it ran on a higher level than it does at present. In other instances, the same strata are seen on both the banks, though the bed of the river is now sunk deep between them, and perhaps holds as winding a course through the solid rock, as if it flowed along the surface; a proof that it must have begun to sink its bed, when it ran through such loose materials as opposed but a very inconsiderable resistance to its stream. A river, of which the course is both serpentine and deeply excavated in the rock, is among the phenomena, by which the slow waste of the land, and also the cause of that waste, are most directly pointed out.

It is, however, where rivers issue through narrow defiles among mountains, that the identity of the strata on both sides is most easily recognized, and remarked at the same time with the greatest wonder. On observing the Patowmack, where it penetrates the ridge of the Allegany mountains, or the Irtish, as it issues from the defiles of Altai, there is no man, however little addicted to geological speculations, who does not immediately acknowledge, that the mountain was once continued quite across the space in which the river now flows; and, if he ventures to reason concerning the cause of so wonderful a change, he ascribes it to some great convulsion of nature, which has torn the mountain asunder, and opened a passage for the waters. It is only the philosopher, who has deeply meditated on the effects which action long continued is able to produce, and on the simplicity of the means which nature employs in all her operations, who sees in this nothing but the gradual working of a stream, that once flowed as high as the top of the ridge which it now so deeply intersects, and has cut its course through the rock, in the same way, and almost with the same instrument, by which the lapidary divides a block of marble or granite.

It is highly interesting to trace up, in this manner, the action of causes with which we are familiar, to the production of effects, which at first seem to require the introduction of unknown and extraordinary powers; and it is no less interesting to observe, how skilfully nature has balanced the action of all the minute causes of waste, and rendered them conducive to the general good. . . .

The Origin of Granitic Veins

It must, however, be admitted, that a difference of character is often to be observed between the granite mass and the veins proceeding from it; sometimes the substances in the latter are more highly crystallized than in the former; sometimes, but more rarely, they are less crystallized, and, in some instances, an ingredient that enters into the mass seems entirely wanting in the vein.   These varieties, for what we yet know, are not subject to any general rule; but they have been held out as a proof, that the masses and the veins are not of the same formation.   It may be answered, that a perfect similarity between substances that, on every hypothesis, must have crystallized in very different circumstances, is not always to be looked for; but the most direct answer is, that this perfect similarity does sometimes occur, insomuch that, in certain instances, no difference whatsoever can be discovered between the mass and the vein, but they consist of the same ingredients, and have the same degree of crystallization.   Some instances of this are just about to be remarked.

A strong objection to the supposed origin of granitic veins from infiltration, and indeed to their formation in any way but by igneous fusion, arises from the number of fragments of schistus, often contained, and completely insulated in those veins.   How these fragments were introduced into the fissures of the schistus, and sustained till they were surrounded by the matter deposited by water, is very hard to be conceived; but if they were carried in by melted granite, nothing is more easily understood.

The following are some of the places where the phenomena of granite veins may be distinctly seen.

The island of Arran, remarkable for collecting into a very small compass a great number of the most interesting facts of geology, exhibits many instances of the penetration of schistus by veins of granite.   A group of granite mountains occupies the northern extremity of the island, the highest of which, Goatfield, rises nearly to the height of 3000 feet, and on the south side is covered with schistus to the height of 1100.   From thence, the line of junction, or that at which the granite emerges from under the schistus, winds, so far as I was able to observe, round the whole group of mountains, with many wavings and irregularities, rising sometimes to a greater, and descending sometimes to a much lower level, than that just mentioned.   Along this line, particularly on the south, wherever the rock is laid bare, and cut into by the torrents, innumerable veins of granite are to be seen entering into the schistus, growing narrower as they advance into it; and being directed, in very many cases, from below upwards, they are precisely of the kind which the infiltration of water could not produce, even were that fluid capable of dissolving the substances which the vein consists of. From this south face of the mountain, and from the bed of a torrent that intersects it very deeply, Dr. Hutton brought a block of schistus, of several hundredweight, curiously penetrated by granite veins, including in them many insulated fragments of the schistus.

The Problem of "Loose Stones"

The loose stones found on the sides of hills, and the bottoms of valleys, when traced back to their original place, point out with demonstrative evidence the great changes which have happened since the commencement of their journey; and in particular serve to show, that many valleys which now deeply intersect the surface, had not begun to be cut out when these stones were first detached from their native rocks. We know, for instance, that stones under the influence of such forces as we are now considering, cannot have first descended from one ridge, and then ascended on the side of an opposite ridge. But the granite of Mont Blanc has been found, as mentioned above, on the sides of Jura, and even on the side of it farthest from the Alps. Now, in the present state of the earth's surface, between the central chain of the Alps, from which these pieces of granite must have come, and the ridge of Mont Jura, besides many smaller valleys, there is the great valley of the Rhone, from the bottom of which, to the place where they now lie, is a height of not less than 3000 feet. Stones could not by any force, that we know of, be made to ascend over this height. We must therefore suppose, that when they travelled from Mont Blanc to Jura, this deep valley did not exist, but that such a uniform declivity, as water can run on with rapidity, extended from the one summit to the other. This supposition accords well with what has been already said concerning the recent formation of the Leman Lake, and of the present valley of the Rhone.

We can derive, in a matter of this sort, but little aid from calculation; yet we may discover by it, whether our hypothesis transgresses materially against the laws of probability, and is inconsistent with physical principles already established. The horizontal distance from Mont Jura to the granite mountains, at the head of the Arve, may be accounted fifty geographic miles. Though we suppose Mont Blanc, and the rest of those mountains, to have been originally much higher than they are at present, the ridge of Jura must have been so likewise; and though probably not by an equal quantity, yet it is the fairest way to suppose the difference of their height to have been nearly the same in former ages as it is at present, and it may therefore be taken at 10,000 feet. The declivity of a plane from the top of Mont Jura to the top of Mont Blanc, would therefore be about one mile and three-quarters in fifty, or one foot in thirty; an inclination much greater than is necessary for water to run on even with extreme rapidity, and more than sufficient to enable a river or a torrent to carry with it stones or fragments of rock, almost to any distance.

Saussure, in relating the fact that pieces of granite are found among the high passes near the summits of Mont Jura, alleges, that they are ónly found in spots from which the central chain of the Alps may be seen. But it should seem that this coincidence is accidental, because, from whatever cause the transportation of these blocks has proceeded, the form of the mountains, especially of Mont Jura, must be too much changed to admit of the supposition, that the places on it from which Mont Blanc is now visible, are the same from which that mountain was visible when these stones were transported hither. It may be, however, that the passes which now exist in Mont Jura are the remains of valleys or beds of torrents, which once flowed westward from the Alps; and it is natural, that the fragments from the latter mountains should be found in the neighbourhood of those ancient water-tracks.

Glacial Erosion and Transportation

Saussure observed in another part of the Alps, that where the Drance descends from the sides of Mont Velan and the Great St. Bernard, to join the Rhone in the Vallais, the valley it runs in lies between mountains of primary schistus, in which no granite appears, and yet that the bottom of this valley, toward its lower extremity, is for a considerable way covered with loose blocks of granite.* His familiar acquaintance with all the rocks of those mountains, led him immediately to suspect, that these stones came from the granite chain of Mont Blanc, which is westward of the Drance, and considerably higher than the intervening mountains. This conjecture was verified by the observations of one of his

---

* *Voyages aux Alpes*, tom. ii.   §1022.

friends, who found the stones in question to agree exactly with a rock at the point of Ornex, the nearest part of the granite chain.

In the present state of the surface, however, the valley of Orsières lies between the rocks of Ornex and the valley of the Drance, and would certainly have intercepted the granite blocks in their way from the one of these points to the other, if it had existed at the time when they were passing over that tract. The valley of Orsières, therefore, was not formed, when the torrents, or the glaciers transported these fragments from their native place.

Mountainous countries, when carefully examined, afford so many facts similar to the preceding, that we should never have done were we to enumerate all the instances in which they occur. They lead to conclusions of great use, if we would compare the machinery which nature actually employs in the transportation of rocks, with the largest fragments of rock which appear to have been removed, at some former period, from their native place.

For the moving of large masses of rock, the most powerful engines without doubt which nature employs are the glaciers, those lakes or rivers of ice which are formed in the highest valleys of the Alps, and other mountains of the first order. These great masses are in perpetual motion, undermined by the influx of heat from the earth, and impelled down the declivities on which they rest by their own enormous weight, together with that of the innumerable fragments of rock with which they are loaded. These fragments they gradually transport to their utmost boundaries, where a formidable wall ascertains the magnitude, and attests the force, of the great engine by which it was erected. The immense quantity and size of the rocks thus transported, have been remarked with astonishment by every observer,* and explain sufficiently how fragments of rock may be put in motion, even where there is but little declivity, and where the actual surface of the ground is considerably uneven. In this manner, before the valleys were cut out in the form they now are, and when the mountains were still more elevated, huge fragments of rock may have been carried to a great distance; and it is not wonderful, if these same masses, greatly diminished in size, and reduced to gravel or sand, have reached the shores, or even the bottom, of the ocean.

---

* The stones collected on the *Glacier de Miage*, when Saussure visited it, were in such quantity as to conceal the ice entirely. *Voyages aux Alpes*, tom. iv. § 854.

WERNER

Abraham Gottlob Werner (1749–1817) was for many years the most famous of the professors in the Freiberg School of Mines. His Neptunist school may have retarded geological thought, but his inspiring teaching and his earnest effort to classify all data did even more to advance it.

The Aqueous Origin of Basalt

Translated from *Journal de physique* (de Rogier), Vol. XXXVIII, pp. 409–420, 1791—a translation by "J. P. B. W. B." of *Neue Entdeckung, Intelligenz-Blattes des Allgemeinen Literatur Zeitung*, No. 57, 1788.

The unexpected observation which I made last summer at the hill of Scheibenberg, well known as basaltic, ought to be infinitely important because of the relations of the basalt with the rocks on which it rests.* It should be considered impartially by every geognost observer, especially at a time when the nature and origin of basalt provoke the inquiries of savants and hold their attention.

Earlier I had noticed from a distance, a great, white mine dump near the summit of this basaltic mountain, which is situated a scant quarter league and almost due south from the little town of Scheibenberg. On inquiry, I was told that it was the dump of a sand pit that had served the needs of the town since it was founded. A mine of sand at the summit of a basaltic mountain seemed a very singular thing to me. So it was my first care, on climbing this mountain to make a mineralogic examination, to direct my attention to this sand pit.

I had already seen from afar that the hill, or rather its summit, was cut in one place so that I would find a nearly perpendicular section there. Thus I would be able to reconnoiter the interior of the basaltic mountain a little. It will be seen that I was not mistaken in my opinion. Nevertheless I thought that it was only a bank of sand which surrounded the foot of the basaltic summit, in

---

* That is, its relations with the *gneiss* on which *the basalt and the beds which form its base* are found here, not as the product of an *eruption* and of a *volcanic heaping*, but always as a precipitation by the humid way.

the way it generally was believed then that the sand and clay were deposited at Pohlberg, near Annaberg, where, as is known, these fossils [fragments of basalt] were extracted in great quantity.

But how surprised I was to see, at the first glance on arriving, first, at the base, a thick bank of quartzose sand, then above, a bed of clay, finally a bed of argillaceous stone called wacke, and resting on this last, the basalt! I saw the first three beds bury themselves nearly horizontally under the basalt and thus form its base, the sand becoming finer above, more argillaceous, and finally changing into true clay, as the clay was converted into wacke in its upper part, and finally the wacke into basalt. In a word, I found here a perfect transition from pure sand to argillaceous sand, from the latter to sandy clay and from the sandy clay by many gradations to the fat clay, to the wacke, and finally to the basalt.

At this sight, I was led at once and irresistibly to think (as would, without doubt, any impartial connoisseur, struck by the consequences of this phenomenon)—I was, I say, irresistibly led to the following ideas. This basalt, this wacke, this clay, and this sand are of one and the same formation. They are all the result of a precipitation by the humid way in one and the same submersion of this country. The waters which covered it then transported first the sand, then deposited the clay, and gradually changed their precipitation into wacke and finally into true basalt.

Space does not permit me to enter into more detail on this great and important observation, but I shall certainly give a more ample description of it soon in one of our journals. Now what will the large party among our mineralogists who are very much biased in favor of the volcanic origin of the basalts say?

I shall add some further short remarks to this observation. The basalt presents a considerable section here, but it is nearly perpendicular and it is divided into columns. The gaps which separate these basaltic columns descend into the wacke and penetrate in some places across the bed at the base. The wacke has almost a schistose structure. One cannot see the base of the bed of sand, as it is covered by the dump, but one observes that it becomes coarser towards the bottom and changes into true gravel or pebbly sand. The gneiss that constitutes the country rock of all this region is found immediately below the sand dump.

The New Theory of Veins, and of the Mode of Their Formation

From *New Theory of the Formation of Veins*, translated by Charles Anderson, Edinburgh, 1795.

All *true veins* were originally, and of necessity, rents open in their upper part, which have afterwards *filled up from above.*

Rents may be produced by many different causes.   Mountains have been formed by a successive accumulation of different beds or layers placed or heaped upon one another.   The mass of these beds was at first wet, and possessed little solidity, so that when the accumulation of matter had attained a certain height, the mass of the mountain yielded to its weight, and must consequently have sunk and cracked.   As the waters which formerly assisted in supporting the mass of the mountain began to lower their level; these masses then lost their former support, yielded to the action of their weight, and began to separate and be detached from the rest of the mountain, falling to the free side, or that where the least resistance was opposed.   The shrinking of the mass of the mountain, produced by desiccation, and still more by earthquakes, and other similar causes may also have contributed to the formation of rents.

The same *precipitation*, which in the humid way formed the *strata* and *beds* of rocks, (also the minerals contained in these rocks), furnished and produced the *substance* of *veins;* this took place during the time, when the solution from which the precipitate was formed, covered the already existing rents, which were as yet wholly or in part empty, and open in their upper part.

Veins (whether considered as rents, or as the substance constituting the vein) have been produced at very different times, and the antiquity or *relative age* of each can be easily assigned.

The *distinguishing characteristics* for the *relative age* of veins, and their substances, are the following:

1. Every vein which *intersects* another, is *newer* than the one traversed, and is of *later formation* than all those which *it traverses;* of course, the *oldest vein is traversed* by all those that are of a *posterior formation*, and the newer veins always cross those that are older.

When two veins cross, one of them without suffering any derangement or interruption traverses the other; this last is interrupted and cut across through its whole thickness by the former. . . . This crossing of veins is of great importance, . . . yet, till now, it has always escaped the observation of mineralogists.

2. The *middle part of veins* is commonly of later formation than that portion which is nearest their walls; and what we find in the *upper part* of a vein is *newer* than what we meet with in the lower part.

3. In a specimen composed of different minerals, the *superimposed* portion is always of newer formation than that on which it rests, which is of course older.

In recapitulating the state of our present knowledge, it is obvious that we know with certainty, that the flötz and primitive mountains have been produced by a series of precipitations and depositions formed in succession; that they took place from water which covered the globe, existing always more or less generally, and containing the different substances which have been produced from them. We are also certain that the fossils [minerals] which constitute the beds and strata of mountains were dissolved in this universal water and were precipitated from it: consequently the metals and minerals found in the primitive rocks, and in the beds of flëtz mountains, were also contained in this universal solvent, that they also were formed from it by precipitation. We are still farther certain, that, at different periods, different fossils have been formed from it, at one time earth, at another metallic minerals, at a third time other fossils. We know too, from the position of these fossils, one above another, to determine with the utmost precision, which are the oldest, and which the newest precipitates. We are also convinced, that the solid mass of our globe has been produced by a series of precipitations formed in succession, (in the humid way). . . . The precipitates which formed the beds of mountains, have, of necessity, deposited on the bottom of the general reservoir, solid and compact materials; whilst the matter which composed the greater part of the mass of veins, being deposited by degrees on their walls, has there formed druses: Afterwards, minerals of different natures have been successively deposited upon one another.

. . . The geognost, who is possessed of the necessary knowledge of chemistry, and consequently of the impossibility of one elementary substance being transmuted into another, will see that there are only two ways in which the following question can be answered. At what time the metallic, earthy and other substances, which were, and still are in part, contained in the general solution; at what time, I say, have these substances entered into the general solution? It may be answered, either that these substances have altogether, and from the beginning, been contained in the universal solvent, or that they may have been introduced from time to time . . . , and if we admit the first answer, it is not possible to understand . . . why successive depositions should have been formed of so different a nature. Thus, it is not possible to conceive, why, in a mountain of gneiss, the strata of this rock should alternate many times with beds, in some instances, of limestone, sometimes of hornblende, lead-glance, and other metallic minerals: sometimes of magnetic ironstone, quartz, feldspar, etc.; all of which are essentially different from gneiss: sometimes also of limestone, clay, marl, lead-glance with calamine, chalk and flint; and this perhaps for more than a hundred times. . . . It is therefore most probable, that at different periods the universal solvent contained mixtures as various as the different precipitates; and that the universal waters held in solution at one time one substance, at another, another.

From what has been said in this section, it must be obvious that the natural history of veins cannot be thoroughly understood without a knowledge of the primitive and flötz rocks, as well as their mode of formation. . . . In studying more particularly the different rock formations, we must begin with the newest, which are the alluvial; and from these, ascend successively to the most ancient. From the alluvial we pass to the newest flötz mountains, and so on through the transition to the oldest primitive mountains.

LAPLACE

Pierre Simon, Marquis de Laplace (1749–1847), the French mathematician, author of two epoch-making astronomical works, the *Méchanique céleste* and the *Exposition du système du monde*. It was in the latter, published first in 1796, that he proposed the hypothesis for the origin of the planetary system that was generally accepted throughout the nineteenth century.

THE NEBULAR HYPOTHESIS

From *The System of the World*, translated by the Rev. Henry H. Harte, Vol. II, 1830.

However arbitrary the elements of the system of the planets may be, there exists between them some very remarkable relations, which may throw light on their origin. Considering it with attention, we are astonished to see all the planets move around the Sun from west to east, and nearly in the same plane, all the satellites moving round their respective planets in the same direction, and nearly in the same plane with the planets. Lastly, the Sun, the planets, and those satellites in which a motion of rotation have been observed, turn on their own axes, in the same direction, and nearly in the same plane as their motion of projection.

The satellites exhibit in this respect a remarkable peculiarity. Their motion of rotation is exactly equal to their motion of revolution; so that they always present the same hemisphere to their primary. At least, this has been observed for the Moon, for the four satellites of Jupiter, and for the last satellite of Saturn, the only satellites whose rotation has been hitherto recognized.

Phenomena so extraordinary, are not the effect of irregular causes. By subjecting their probability to computation, it is found that there is more than two thousand to one against the hypothesis that they are the effect of chance, which is a probability much greater than that on which most of the events of history, respecting which there does not exist a doubt, depends. We ought, therefore, to be assured with the same confidence, that a primitive cause has directed the planetary motions.

Another phenomenon of the solar system, equally remarkable, is the small eccentricity of the orbits of the planets and their satellites, while those of comets are very much extended.   The orbits of this system present no intermediate shades between a great and small eccentricity.   We are here compelled to acknowledge the effect of a regular cause; chance alone could not have given a form nearly circular to the orbits of all the planets.   It is, therefore, necessary that the cause which determined the motions of these bodies, rendered them also nearly circular.   This cause then must also have influenced the great eccentricity of the orbits of comets, and their motion in every direction; for, considering the orbits of retrograde comets, as being inclined more than one hundred degrees to the ecliptic, we find that the mean inclination of the orbits of all the observed comets, approaches near to one hundred degrees, which would be the case if the bodies had been projected at random.

What is this primitive cause?   In the concluding note of this work I will suggest an hypothesis which appears to me to result with a great degree of probability, from the preceding phenomena, which, however, I present with that diffidence, which ought always to attach to whatever is not the result of observation and computation.

Whatever be the true cause, it is certain that the elements of the planetary system are so arranged as to enjoy the greatest possible stability, unless it is deranged by the intervention of foreign causes.   From the sole circumstance that the motions of the planets and satellites are performed in orbits nearly circular, in the same direction, and in planes which are inconsiderably inclined to each other, the system will always oscillate about a mean state, from which it will deviate but by very small quantities. The mean motions of rotation and of revolution of these different bodies are uniform, and their mean distances from the foci of the principal forces which actuate them are constant; all the secular inequalities are periodic. . . .

From the preceding chapter it appears, that we have the five following phenomena to assist us in investigating the cause of the primitive motions of the planetary system.   The motions of the planets in the same direction, and very nearly in the same plane; the motions of the satellites in the same direction as those of the planets; the motions of rotation of these different bodies and also of the Sun, in the same direction as their motions of projection, and in planes very little inclined to each other; the small eccentricity of the orbits of the planets and satellites; finally, the great eccentricity of the orbits of the comets, their inclinations being at the same time entirely indeterminate.

Buffon is the only individual that I know of, who, since the discovery of the true system of the world, endeavoured to investigate the origin of the planets and satellites. He supposed that a comet, by impinging on the Sun, carried away a torrent of matter, which was reunited far off, into globes of different magnitudes, and at different distances from this star. These globes, when they cool and become hardened, are the planets and their satellites. This hypothesis satisfied the first of the five preceding phenomena; for it is evident that all bodies thus formed should move very nearly in the plane which passes through the centre of the Sun, and through the direction of the torrent of matter which has produced them: but the four remaining phenomena appear to me inexplicable on this supposition. Indeed the absolute motion of the molecules of a planet ought to be in the same direction as the motion of its centre of gravity; but it by no means follows from this, that the motion of rotation of a planet should be also in the same direction. Thus the Earth may revolve from east to west, and yet the absolute motion of each of its molecules may be directed from west to east. This observation applies also to the revolution of the satellites, of which the direction, in the same hypothesis, is not necessarily the same as that of the motion of projection of the planets.

The small eccentricity of the planetary orbits is a phenomenon, not only difficult to explain on this hypothesis, but altogether inconsistent with it. We know from the theory of central forces, that if a body which moves in a re-entrant orbit about the Sun. passes very near the body of the Sun, it will return constantly to it, at the end of each revolution. Hence it follows that if the planets were originally detached from the Sun, they would touch it, at each return to this star; and their orbits, instead of being nearly circular, would be very eccentric. Indeed it must be admitted that a torrent of matter detached from the Sun, cannot be compared to a globe which just skims by its surface: from the impulsions which the parts of this torrent receive from each other, combined with their mutual attraction, they may, by changing the direction of their motions, increase the distances of their perihelions from the

Sun. But their orbits should be extremely eccentric, or at least all the orbits would not be circular, except by the most extraordinary chance. Finally, no reason can be assigned on the hypothesis of Buffon, why the orbits of more than one hundred comets, which have been already observed, should be all very eccentric. This hypothesis, therefore, is far from satisfying the preceding phenomena. Let us consider whether we can assign the true cause.

Whatever may be its nature, since it has produced or influenced the direction of the planetary motions, it must have embraced them all within the sphere of its action; and considering the immense distance which intervenes between them, nothing could have effected this but a fluid of almost indefinite extent. In order to have impressed on them all a motion circular and in the same direction about the Sun, this fluid must environ this star, like an atmosphere. From a consideration of the planetary motions, we are, therefore, brought to the conclusion, that in consequence of an excessive heat, the solar atmosphere originally extended beyond the orbits of all the planets, and that it has successively contracted itself within its present limits.

In the primitive state in which we have supposed the Sun to be, it resembles those substances which are termed nebulae, which, when seen through telescopes, appear to be composed of a nucleus, more or less brilliant, surrounded by a nebulosity, which, by condensing on its surface, transforms it into a star. If all the stars are conceived to be similarly formed, we can suppose their anterior state of nebulosity to be preceded by other states, in which the nebulous matter was more or less diffuse, the nucleus being at the same time more or less brilliant. By going back in this manner, we shall arrive at a state of nebulosity so diffuse, that its existence can with difficulty be conceived.

For a considerable time back, the particular arrangement of some stars visible to the naked eye, has engaged the attention of philosophers. Mitchel remarked long since how extremely improbable it was that the stars composing the constellation called the Pleiades, for example, should be confined within the narrow space which contains them, by the sole chance of hazard; from which he inferred that this group of stars, and the similar groups which the heavens present to us, are the effects of a primitive cause, or of a primitive law of nature. These groups are a general result of the condensation of nebulae of several nuclei; for it is evident that the nebulous matter being perpetually attracted by these different nuclei, ought at length to form a group of stars, like to that of the Pleiades. The condensation of nebulae consisting of two nuclei, will in like manner form stars very near to each other, revolving the one about the other like to the double stars, whose respective motions have been already recognized.

But in what manner has the solar atmosphere determined the motions of rotation and revolution of the planets and satellites? If these bodies had penetrated deeply into this atmosphere, its resistance would cause them to fall on the Sun. We may, therefore, suppose that the planets were formed at its successive limits, by the condensation of zones of vapours, which it must, while it was cooling, have abandoned in the plane of its equator.

Let us resume the results which we have given in the tenth chapter of the preceding book. The Sun's atmosphere cannot extend indefinitely; its limit is the point where the centrifugal force arising from the motion of rotation balances the gravity; but according as the cooling contracts the atmosphere, and condenses the molecules which are near to it, on the surface of the star, the motion of rotation increases; for in virtue of the principle of areas, the sum of the areas described by the radius vector of each particle of the Sun and of its atmosphere, and projected on the plane of its equator, is always the same. Consequently, the rotation ought to be quicker, when these particles approach to the centre of the Sun. The centrifugal force arising from this motion becoming thus greater, the point where the gravity is equal to it, is nearer to the centre of the Sun. Supposing, therefore, what is natural to admit, that the atmosphere extended at any epoch as far as this limit, it ought, according as it cooled, to abandon the molecules, which are situated at this limit, and at the successive limits produced by the increased rotation of the Sun. These particles, after being abandoned, have continued to circulate about this star, because their centrifugal force was balanced by their gravity. But as this equality does not obtain for those molecules of the atmosphere which are situated on the parallels to the Sun's equator, these have come nearer by their gravity to the atmosphere according as it condensed, and they have not ceased to belong to it, inasmuch as by this motion, they have approached to the plane of this equator.

Let us now consider the zones of vapours, which have been successively abandoned. These zones, ought, according to all probability, to form by their condensation, and by the mutual attraction of their particles, several concentrical rings of vapours circulating about the Sun. The mutual friction of the molecules of each ring ought to accelerate some and retard others, until they all had acquired the same angular motion. Consequently, the real velocities of the molecules which are farther from the Sun, ought to be greatest. The following cause ought, likewise, to contribute to this difference of velocities: The most distant particles of the Sun, which, by the effects of cooling and of condensation, have collected so as to constitute the superior part of the ring, have always described areas proportional to the times, because the central force by which they are actuated has been constantly directed to this star; but this constancy of areas requires an increase of velocity, according as they approach more to each other. It appears that the same cause ought to diminish the velocity of the particles, which, situated near the ring, constitute its inferior part.

If all the particles of a ring of vapours continued to condense without separating, they would at length constitute a solid or a liquid ring. But the regularity which this formation requires in all the parts of the ring, and in their cooling, ought to make this phenomenon very rare. Thus the solar system presents but one example of it; that of the rings of Saturn. Almost always each ring of vapours ought to be divided into several masses, which, being moved with velocities which differ little from each other, should continue to revolve at the same distance about the Sun. These masses should assume a spheroidical form, with a rotatory motion in the direction of that of their revolution, because their inferior particles have a less real velocity than the superior; they have, therefore constituted so many planets in a state of vapour. But if one of them was sufficiently powerful to unite successively by its attraction all the others about its centre, the ring of vapours would be changed into one sole spheroidical mass, circulating about the Sun, with a motion of rotation in the same direction with that of revolution. This last case has been the most common; however, the solar system presents to us the first case, in the four small planets which revolve between Mars and Jupiter, at least unless we suppose with Olbers, that they originally formed one planet only, which was divided by an explosion into several parts, and actuated by different velocities. Now if we trace the changes which a farther cooling ought to produce in the planets formed of vapours, and of which we have suggested the formation, we shall see to arise in the centre of each of them, a nucleus increasing continually, by the condensation of the atmosphere which environs it. In this state, the planet resembles the Sun in the nebulous state, in which we have first supposed it to be; the cooling should, therefore, produce at the different limits of its atmosphere, phenomena similar to those which have been described, namely, rings and satellites circulating about its centre in the direction of its motion of rotation, and revolving in the same direction on their axes. The regular distribution of the mass of rings of Saturn about its centre and in the plane of its equator, results naturally from this hypothesis, and, without it, is inexplicable. Those rings appear to me to be existing proofs of the primitive extension of the atmosphere of Saturn, and of its successive condensations. Thus the singular phenomena of the small eccentricities of the orbits of the planets and satellites, of the small inclination of these orbits to the solar equator, and of the identity in the direction of the motions of rotation and revolution of all those bodies with that of the rotation of the Sun, follow from the hypothesis which has been suggested, and render it extremely probable. If the solar system was formed with perfect regularity, the orbits of the bodies which compose it would be circles, of which the planes, as well as those of the various equators and rings, would coincide with the plane of the solar equator. But we may suppose that the innumerable varieties which must necessarily exist in the temperature and density of different parts of these great masses, ought to produce the eccentricities of their orbits, and the deviations of their motions, from the plane of this equator. . . .

If in the zones abandoned by the atmosphere of the Sun, there are any molecules too volatile to be united to each other, or to the planets, they ought, in their circulation about this star, to exhibit all the appearances of the zodiacal light, without opposing any sensible resistance to the different bodies of the planetary system, both on account of their great rarity, and also because their motion is very nearly the same as that of the planets which they meet.

An attentive examination of all the circumstances of this system renders our hypothesis still more probable. The primitive fluidity of the planets is clearly indicated by the compression of their figure, conformably to the laws of the mutual attraction of their molecules; it is, moreover, demonstrated by the regular diminution of gravity, as we proceed from the equator to the poles. This state of primitive fluidity to which we are conducted by astronomical phenomena, is also apparent from those which natural history points out. But in order fully to estimate them, we should take into account the immense variety of combinations formed by all the terrestrial substances which were mixed together in a state of vapour, when the depression of their temperature enabled their elements to unite; it is necessary, likewise, to consider the wonderful changes which this depression ought to cause in the interior and at the surface of the earth, in all its productions, in the constitution and pressure of the atmosphere, in the ocean, and in all substances which it held in a state of solution. Finally, we should take into account the sudden changes, such as great volcanic eruptions, which must at different epochs have deranged the regularity of these changes. Geology, thus studied under the point of view which connects it with astronomy, may, with respect to several objects, acquire both precision and certainty.

DOLOMIEU

Guy S. Tancrède de Dolomieu (1750–1801), French army officer, was a professor in the École des mines at Paris. His work was mainly with the igneous rocks and with the region later known as the Dolomites.

Rhyolitic Lavas

Translated from *Mémoire sur les Îles Ponces*, Paris, 1785.

The island of Leponte is formed principally of two materials, definitely distinct in color and in other characters. The first is this black lava of which I have spoken in the description of the island and which is crystallized in massive basalts. It would seem that the single crater, whose site is occupied by the port, has produced them both. The black lava must have emerged from it in the last period of its eruption, as it dominates all the other materials thrown from the same mouth, which are of an entirely different nature.

The black lavas are present in very small quantity, in comparison with the white and whitish lavas. This last color seems essential to nearly all the materials which form the island. The interior of all the mountains, all the escarpments, nearly all the crags are white. Their appearance and their color could lead to doubt of their volcanic origin, if one did not know that black is not a color essential to all the products of volcanos.

If nearly all the black lavas seem to have hornstone as a base, or massive *schoerl* in which are found crystals of *schoerl*, garnet, and feldspar, the greater part of the white lavas, at least those of the island of Leponte, seem to belong more particularly to the granite and to the foliate granitic rocks. The substances which ordinarily constitute this type of composite rock are recognized in nearly all the white materials of this island—that is, granular quartz, black scaly mica, and more or less pure feldspar. . . .

There are no naturalists who are not acquainted with the type of rocks, placed between the granite, the gneiss, and the porphyries, which has a little in common with all three, and of which the base or groundmass is ordinarily an impure feldspar of a slightly scaly

151

texture.   It is principally among these that the primary materials of the whitish lavas of the island of Leponte must be sought.

THE RELATIONS BETWEEN THE AUVERGNE VOLCANICS AND THE GRANITE ON WHICH THEY REST

Translated from "Rapport fait à l'Institut national, par le Citoyen Dolomieu," *Journal des mines*, Vol. XLI, pp. 385–402, 1789.

The mountains of which I speak [in Auvergne] have reached the surface through these granitic masses.   They have pierced these to place on top of them materials which lay below and which, without the activity of volcanic agents, would have escaped our observation forever.   Most of these volcanic products are entirely different in their composition from the granites on which they have come to rest.   Even those which approach it most closely have such dissimilarities that it would be impossible to confuse them. The granitic rock, at whatever depth it is trenched by the valley openings, does not contain any essential or foreign substance which could be attributed to the effects of the volcanos.

The first conclusions to be drawn from this relationship—the most simple results of this type of equation—are: (1) that the volcanic products belong here to a body of materials different from the granites and which reposes under them; (2) that the volcanic agents have lain below the granite here, and worked in depths far below it, as a mole works below the turf, putting earth from a bed underneath on top of the meadow; (3) that the granite is not the primordial rock here, as it is necessarily posterior to the material which supports its weight, although it has priority of position over all that has since come to cover it; (4) that the substances which directly produce, or in some fashion contribute to, the volcanic phenomena should be found in this body of materials anterior to the granite; (5) that these substances which we have not yet reached can resemble some of those we know, but can also be different, and that their nature will probably remain conjectural for a long time to come, although their existence is proven by their effects, which are still for the most part inexplicable to us; (6) finally, the base of the lavas here belongs to the most ancient bodies of which we have any knowledge and will retain the dignity that priority gives until we have an opportunity to learn what reposes below them, and as long as we admit that there is a solid nucleus on which the beds of rock are successively placed.

To be as precise as possible in announcing this conclusion, I have consistently used the adverb "here," in order to restrict the conclusions I have drawn solely to the localities which furnished me the observations. But I have reason to believe that conditions are similar in all other volcanos, whatever be the nature of the ground which surrounds them. I think that the volcanic agents, as well as the sources of all ejections, lie at great depths within or below the consolidated crust of the earth everywhere, and that the causes which contribute to the combustion accompanying the eruptions and those which produce the fluidity of the lavas remain hidden there.

This also seems to me to validate completely an opinion that I have long sustained, to wit: that the volcanic hearths are not located in the secondary beds as various writers have supposed, that they do not lie in the beds of coal or other combustible materials of vegetable or animal origin, and that if there truly exists a subterranean fire, it is not by that sort of substance that it is fed.

In advancing the hypothesis of the fluidity of the center of the globe—or rather in believing in its possibility—and in deducing its likelihood from the phenomena which it would serve to explain, I do not undertake to show or even to indicate the agent, whatever it may be, that prevents the complete solidification of the materials of which it is composed. . . .

But I can say that in deducing this type of pasty fluidity from that of the lavas which I suppose to belong to this central fluid, I do not believe that it can be compared with that produced by the heat of the fire in our furnaces on materials analogous to those which serve as bases of the lavas. It should not be a vitreous fluidity, as Buffon supposed. If heat aids in its production, as I think, and serves to maintain it, it is not by its direct action on the earthy molecules but with the aid of some vehicle which holds the integrant molecules apart when they do not exert a very great affinity of aggregation against it. Those of the molecules whose affinity of aggregation has more energy can unite and form crystals; such are abundant in nearly all lavas.

For I shall repeat, perhaps for the hundredth time, compact lavas are not vitrifications, and their fluidity on emerging from volcanos—which is retained much longer than their cooling should permit—is a singular effect of a cause which is not yet determined.

On the Distinctions between Dolomite and Limestone

Translated from *Journal de physique*, Vol. XXXIX, pp. 3–10, 1791.

For a long time, my excellent friend, I have recognized that effervescence with acids was not an essential characteristic of calcareous stones, although this property was indicated by all naturalists as the most certain sign by which one could recognize stones of this type. I have observed that many stones of this composition undergo attack by acids without producing the great liberation of air which effervescence occasions. . . . It has often happened that I have spread acid on the surface of stones which seemed calcareous by all the external characters, without producing the effervescence which I awaited. I would always have doubted that the stone was entirely calcareous; I would have believed that a very small quantity of earth of that composition was combined with other earths without losing the aerial acid which belonged to it, if I had not used other proofs to establish its genus undeniably.

Finally, eighteen months ago, while making mineralogical excursions in the mountains of Tyrol with M. Fleuriau de Bellevue, . . . I found an immense quantity of these same limestones which do not produce sudden effervescence in the test with acids. . . .

When I made this observation I was in the midst of primitive mountains, surrounded by granites, porphyries, and other composite rocks; . . . I believed then that this sort of limestone belonged particularly to this type of mountain, . . . but on returning to Italy I found that the rocks of the calcareous mountains, which succeed those of porphyry between Bolsano and Trente, had this same peculiarity although they were in horizontal beds and I found some imprints of shells. The sole difference that I recognized then between the rocks of the younger formation and those of the much more remote epoch consisted in the grain and the texture.

Having had occasion to re-examine these slightly effervescent rocks, I have just discovered another more singular property. This is their phosphorescence on impact, but this faculty of giving light belongs only to the limestones of the primitive mountains of the Tyrol and not to those of the secondary or tertiary mountains.

SOULAVIE

Jean Louis Giraud, Abbé Soulavie (1752–1813), French churchman, author, and natural philosopher.

METHODS OF DETERMINING THE CHRONOLOGICAL SEQUENCE OF ROCKS

Translated from *La Chronologie physique des volcans*, Paris, 1781.

In the most brilliant age of the French Monarchy, at that time in the reign of Louis XIV when all the laws of the nation were brightened by new lights, legislation established the Control of Acts. Public registries were opened where the notaries were obliged to inscribe the date of contracts.

Since this institution the notaries have not been able to fabricate clandestine acts. This public and general registry assures the fortune of the citizens. As long as the Registry of the Control exists one can verify the authenticity of originals, acts, and extracts. They present forever the precise date of these contracts; they serve always as affirmatory documents for all private acts of society; and they conserve perpetually the evidence of the date and the comparative succession of the contracts of citizens.

Registries of control exist likewise in nature. These place the successive facts of the physical world in nature and demonstrate the verity of the various periods and ages of nature. The volcanoes which have broken through all sorts of beds of ancient and modern land, which they have riddled with fire-vomiting mouths, have pierced the bottom of the sea basins and the continents. All these volcanoes and their products may be placed today in a certain order by reason of these curious "registries of control" of nature. The study of these monuments is, in the mineral kingdom, the veritable art of verifying the dates of nature, as the study of herbiferous or shell-bearing rocks is the art of verifying the dates and eras of the ancient history of organized beings in the kingdom of the living world. We shall give in the following the principles of this new branch of natural history. . . .

155

Thus the chronology of the extinguished volcanoes of southern France is proven (1) by the superposition of substances, (2) by their state of relative preservation, and (3) by their comparative elevation above the present level of the seas. I give here the recapitulation of these proofs according to their degrees of probability.

IT IS EVIDENT (I) that a volcanic vein inserted in a rock of granite—the oldest of all—is more ancient than a bed of conglomerate of limestone, granite, and basalt; that this bed, the result of an aqueous deposit, is more ancient than the superimposed stream of lava; that this horizontal flow is anterior to the valley that it contains; that this valley has been formed before the little volcanoes located on its floor; and that among these volcanoes the lower beds basal to the fire-vomiting mountain have been made before the upper ones. Thus the proofs are established by the superposition of the masses.

IT IS PROBABLE (II), if one recognizes the destruction of the volcanic features during the progress of time, that the volcanoes of which the traces have been most nearly removed are the most ancient. Thus the volcanoes represented only by veins in the granite—the oldest of rocks—are older than volcanoes which appear merely as mounds. The latter antedate those in which the flows are preserved but are cut by valleys. These in turn precede the uninterrupted basaltic lavas on the valley floors and the volcanoes with clearly defined craters. In the first case, all extruded volcanic material has been removed. In the second, only the mounds remain. In the third, only a few basaltic remnants may still be observed. In the fourth, only the flows on the valley floor are found. In the fifth, the flows are related to a crater, and in the sixth, practically nothing has been destroyed. Thus the proofs established by the degree of preservation may be summarized. They agree with the preceding, based on superposition.

IT APPEARS REASONABLE, FINALLY (if it is true that the seas aroused the volcanic fires and if they have uncovered our regions little by little after having inundated them) that the volcanoes of which the granitic base is the highest are the most ancient; thus those of the high Mézin, those of the upper plateau of Mézillac, Lachamp-Raphael, those of Coiron, are earlier than those of the Dornas, Privas, Aps, Antraigues, Jaujac, Craux, than those of the bottom of the valleys and than all those, finally, of which the base is lower than the level of the seas, as Vesuvius, Etna, and the submarine volcano described by M. le Chevalier Hamilton. But this third type of proof, drawn from the elevation of volcanoes, is subject to exceptions, for one knows of volcanoes with craters of the latest date which are very high and some volcanoes on the continents are now active.

HALL

Sir James Hall (1761–1832), Scottish geologist and chemist, in addition to his other experimental work, made the first known attempts to simulate earth folds in the laboratory.

Marble from Limestone in the Laboratory

From "Account of a Series of Experiments," read in the Royal Society of Edinburgh, June 3, 1805, *Transactions of the Royal Society of Edinburgh*, Vol. VI, pp. 73–75, 80–88, 151–155, 1812.

Of all mineral substances, the *Carbonate of Lime* is unquestionably the most important in a general view. As limestone or marble, it constitutes a very considerable part of the solid mass of many countries; and, in the form of veins and nodules of spar, pervades every species of stone. Its history is thus interwoven in such a manner with that of the mineral kingdom at large, that the fate of any geological theory must very much depend upon its successful application to the various conditions of this substance. But, till Dr Black, by his discovery of Carbonic Acid, explained the chemical nature of the carbonate, no rational theory could be formed, of the chemical revolutions which it has undoubtedly undergone.

This discovery was, in the first instance, hostile to the supposed action of fire; for the decomposition of limestone by fire in every common kiln being thus proved, it seemed absurd to ascribe to that same agent the formation of limestone, or of any mass containing it.

The contemplation of this difficulty led Dr Hutton to view the action of fire in a manner peculiar to himself, and thus to form a geological theory, by which, in my opinion, he has furnished the world with the true solution of one of the most interesting problems that has ever engaged the attention of men of science.

He supposed,

1. That Heat has acted, at some remote period, on all rocks.
2. That during the action of heat, all these rocks (even such as now appear at the surface) lay covered by a superincumbent mass, of great weight and strength.

3. THAT in consequence of the combined action of Heat and Pressure, effects were produced different from those of heat on common occasions; in particular, that the carbonate of lime was reduced to a state of fusion, more or less complete, without any calcination.

THE essential and characteristic principle of this theory is thus comprised in the word *Compression;* and by one bold hypothesis, founded on this principle, he undertook to meet all the objections to the action of fire, and to account for those circumstances in which minerals are found to differ from the usual products of our furnaces. . . .

AFTER three years of almost daily warfare with Dr HUTTON, on the subject of his theory, I began to view his fundamental principles with less and less repugnance.

. . . If we take a hollow tube or barrel closed at one end, and open at the other, of one foot or more in length; it is evident, that by introducing one end into a furnace, we can supply to it as great heat as art can produce, while the other end is kept cool, or, if necessary, exposed to extreme cold. If, then, the substance which we mean to subject to the combined action of heat and pressure, be introduced into the breech or closed end of the barrel, and if the middle part be filled with some refractory substance, leaving a small empty space at the muzzle, we can apply heat to the muzzle, while the breech containing the subject of experiment, is kept cool, and thus close the barrel by any of the numerous modes which heat affords, from the welding of iron to the melting of sealing wax. Things being then reversed, and the breech put into the furnace a heat of any required intensity may be applied to the subject of experiment, now in a state of constraint.

MY first application of this scheme was carried on with a common gun-barrel, cut off at the touch-hole, and welded very strongly at the breech by means of a plug of iron. Into it I introduced the carbonate, previously rammed into a cartridge of paper or pasteboard, in order to protect it from the iron, by which, in some former trials, the subject of experiment had been contaminated throughout during the action of heat. I then rammed the rest of the barrel full of pounded clay, previously baked in a strong heat, and I had the muzzle closed like the breech, by a plug of iron welded upon it in a common forge; the rest of the barrel being kept cold during this operation, by means of wet cloths. The breech of the barrel was then introduced horizontally into a common muffle heated to about 25° of Wedgwood. To the muzzle a rope was fixed, in such a manner, that the barrel could be withdrawn without dangers from an explosion. I likewise, about this time, closed the muzzle of the barrel, by means of a plug, fixed by solder only; which method has this peculiar advantage, that I could shut and open the barrel, without having recourse to a workman. In these trials, though many barrels yielded to the expansive force, others resisted it, and afforded some results that were in the highest degree encouraging, and even satisfactory, could they have been obtained with certainty on repetition of the process. In many of them, chalk, or common limestone previously pulverised, was agglutinated into a stony mass, which required a smart blow of a hammer to break it, and felt under the knife like a common limestone; at the same time, the substance, when thrown into nitric acid, dissolved entirely with violent effervescence.

On the third of March of the same year (1801), I made a similar experiment, in which a pyrometer-piece was placed within the barrel, and another in the muffle; they agreed in indicating 23 degrees. The inner tube which was of Reaumur's porcelain, contained 80 grains of pounded chalk. The carbonate was found, after the experiment, to have lost $3\frac{1}{2}$ grains. A thin rim, less than the 20th of an inch in thickness, of whitish matter, appeared on the outside of the mass. In other respects, the carbonate was in a very perfect state; it was of a yellowish color, and had a decided semitransparency and saline fracture. But what renders this result of the greatest value, is, that on breaking the mass, a space of more than the tenth of an inch square, was found to be completely crystallized, having acquired the rhomboidal fracture of calcareous spar. It was white and opaque, and presented to the view three sets of parallel plates which are seen under three different angles.

I HAVE likewise made some experiments with coal, treated in the same manner as the carbonate of lime: but I have found it much less tractable; for the bitumen, when heat is applied to it, tends to escape by its simple elasticity, whereas the carbonic acid in marble, is in part retained by the chemical force of quicklime. I succeeded, however, in constraining the bituminous matter of the coal, to a certain degree, in red heats, so as to bring the substance into a complete fusion, and to retain its faculty of burning with flame. But, I could not accomplish this in heats capable of agglutinating the carbonate; for I have found, where I rammed them successively into the same tube, and where the vessel has withstood the expansive force, that the carbonate has been agglutinated into a good limestone, but that the coal has lost about half its weight, together with its power of giving flame when burnt, remaining in a very compact state, with a shining fracture. Although this experiment has not afforded the desired result, it answers another purpose admirably well. It is known, that where a bed of coal is crossed by a dike of whinstone, the coal is found in a peculiar state in the immediate neighbourhood of the whin: the substance in such places being incapable of giving flame, it is distinguished by the name of *blind coal.*

I FOUND that the organization of animal substance was entirely obliterated by a slight action of heat, but that a stronger heat was required to perform the entire fusion of vegetable matter. This, however, was accomplished; and in several experiments, pieces of wood were changed to a jet-black and inflammable substance, generally very porous, in which no trace could be discovered of the original organization. In others, the vegetable fibres were still visible, and are forced asunder by large and shining air-bubbles.

RESULTS OF THE SLOW COOLING OF MELTED ROCK

From "Experiments on Whinstone and Lava," *Transactions of the Royal Society of Edinburgh*, Vol. V, pp. 43–48, 61, 66, 1805.

The experiments described in this paper were suggested to me many years ago, when employed in studying the *Geological System* of the late Dr HUTTON, by the following plausible objection, to which it seems liable.

Granite, porphyry, and basaltes, are supposed by Dr HUTTON to have flowed in a state of perfect fusion into their present position; but their internal structure, being universally rough and stony, appears to contradict this hypothesis; for the result of the fusion of earthy substances, hitherto observed in our experiments, either is glass, or possesses, in some degree, the vitreous character.

This objection, however, loses much of its force, when we attend to the peculiar circumstances under which, according to this theory, the action of heat was exerted. These substances, when in fusion, and long after their congelation, are supposed to have occupied a subterraneous position far below what was then the surface of the earth; and Dr HUTTON has ascribed to the modification of heat, occasioned by the pressure of the superincumbent mass, many important phenomena of the mineral kingdom, which he has thus reconciled to his system.

One necessary consequence of the position of these bodies, seems, however, to have been overlooked by Dr HUTTON himself; I mean, that, after their fusion, they must have cooled very slowly; and it appeared to me probable, on that account, that, during their congelation, a crystallization had taken place, with more or less regularity, producing the stony and crystallized structure, common to all unstratified substances, from the large grained granite, to the fine grained and almost homogeneous basalt. This conjecture derived additional probability from an accident similar to those formerly observed by Mr KEIR, which had just happened at Leith: a large glass-house pot, filled with green bottle glass in fusion, having cooled slowly, its contents had lost every character of glass, and had completely assumed the stony structure. . . .

Encouraged by this reasoning, I began my projected series of experiments in the course of the same year (1790), with very promising appearances of success. I found that I could command the result which had occurred accidentally at the glass-house; for, by means of slow cooling, I converted bottle glass, after fusion, into a stony substance, which again, by the application of strong heat, and subsequent rapid cooling, I restored to the state of perfect glass. This operation I performed repeatedly with the same specimen, so as to ascertain that the character of the result was stony or vitreous, according to the mode of its cooling.

Some peculiar circumstances interrupted the prosecution of these experiments till last winter, when I determined to resume them. Deliberating on the substance most proper to submit to experiment on this occasion, I was decided by the advice of Dr HOPE, well known by his discovery of the Earth of Strontites, to give the preference to whinstone.

The term whinstone, as used in most parts of Scotland, denotes a numerous class of stones, distinguished in other countries by the names of basaltes, trap, wacken, grünstein and porphyry. As they are, in my opinion, mere varieties of the same class, I conceive that they ought to be connected by some common name, and have made use of this, already familiar to us, and which seems liable to no objection, since it is not confined to any particular species.* . . .

The whinstone first employed was taken from a quarry† near the Dean, on the Water of Leith, in the neighbourhood of Edinburgh. This stone is an aggregate of black and greenish-black hornblend, intimately mixed with a pale reddish-brown matter, which has some resemblance to felspar, but is far more fusible. Both substances are imperfectly and confusedly crystallized in minute grains. The hornblend is in the greatest proportion; and its fracture appears to be striated, though in some parts foliated; that of the reddish-brown matter is foliated. The fracture of the stone *en masse* is uneven, and it abounds in small facettes, which have some degree of lustre. It may be scratched, though with difficulty, by a knife, and gives an earthy smell when breathed on. It frequently contains small specks of pyrites.

On the 17th of January 1798, I introduced a black lead [graphite] crucible, filled with fragments of this stone, into the great reverberating furnace at Mr BARKER's iron foundery. In about a quarter of an hour, I found that the substance had entered into fusion, and was agitated by a strong ebullition. I removed the crucible, and allowed it to cool rapidly. The result was a black glass, with a tolerably clean fracture, interrupted however by some specks. . . .

At last, on the 27th of January, I succeeded completely in the object I had in view. A crucible, containing a quantity of whinstone, melted in the manner above described, being removed from the reverberatory, and conveyed rapidly to a large open fire, was immediately surrounded with burning coals, and the fire, after being maintained several hours, was allowed to go out. The

---

* In characterising the particular specimens, I have adopted, with scarcely any variation, descriptions drawn up by Dr KENNEDY, whose name I shall have occasion frequently to mention in the course of this paper. In the employment of terms, we have profited by the advice of Mr DERIABIN, a gentleman well versed in the language of the Wernerian School.

† Called Bell's Mills Quarry.

crucible, when cold, was broken, and was found to contain a substance, differing in all respects from glass, and in texture completely resembling whinstone. Its fracture was rough, stony and crystalline; and a number of shining facettes were interspersed through the whole mass. The crystallization was still more apparent in cavities produced by air bubbles, the internal surface of which was lined with distinct crystals.*

Lava of Sta Venere

This current has flowed in the neighbourhood of a little chapel, called Sta Venere, above the village of Piedimonte, on the north side of Mount Aetna. Owing to the strong resemblance which it bears to stones supposed not volcanic, we took care that our specimens should be broken from the actual current; and to one of them, though mostly compact, is attached a scorified maşs, which had made part of the external surface. The solid part is of a black, or rather dark blue, colour, very fine grained and homogeneous, having a multitude of minute and shining facettes visible in the sun; in this, and in other circumstances, it greatly resembles the rock of Edinburgh Castle. This lava is the second in M. Dolomieu's *Catalogue*, and is well described, p. 185.†

The pure black glass formed from this lava yielded, in the regulated heat, the most highly crystallized mass we have obtained from any lava or whin.

These experiments seem to establish, in a direct manner what I had deduced, analogically, from the properties of whinstone, namely, that the stony character of a lava is fully accounted for by slow cooling after the most perfect fusion; and, consequently, that no argument against the intensity of volcanic fire can be founded upon that character. We are therefore justified in believing, as numberless facts indicate, that volcanic heat has often been of excessive intensity.

---

* I showed this result at a meeting of the Society on 5th of February.

† "Lave homogène noire: son grain est fin et ferré, il est un peu brillant, comme micacé lorsqu'on le présente au soleil; sa cassure nette et sèche est conchéide comme elle du silex."

In the comparison instituted between whin and lava, the two classes are found to agree so exactly in all their properties which we have examined, as to lead to a belief of their absolute identity.

THE ROLE OF HEAT IN THE CONSOLIDATION OF STRATA

From "On the Consolidation of the Strata," *Transactions of the Royal Society of Edinburgh*, Vol. X, pp. 314–329, 1826.

It had often been urged, and apparently with good reason, against this branch of the Huttonian Theory, that no amount of heat applied to loose sand, gravel, or shingle, would occasion the parts to consolidate into a compact stone. And as all my experience led to the same conclusion, I saw that, unless, along with heat some flux were introduced amongst the materials, no agglutination of the particles would take place. The striking circumstance above alluded to, as occurring near Dunglass, and which will be particularly described presently, having suggested to me the idea that the salt of the ocean might possibly have been the agent in causing the requisite degree of fusion, I instituted a series of experiments, the details of which I am about to bring before the Society. By these, I conceive it will be shown, that this material, under various modifications, is fully adequate to explain the consolidation of the strata, and many other effects which we see on the surface of the Earth. . . .

Dry salt was placed along with sand, sometimes in a separate layer, at the bottom of the crucible, and sometimes mixed throughout the experiment: the whole was then exposed to heat from below. I found that the salt was invariably sent in fumes through the loose mass, and by its action produced solid stone in a manner completely satisfactory, as illustrative of the facts in Aikengaw; and so as to give a good explanation of the production of sandstone in general.

These artificial stones are of various degrees of durability and hardness;—some of them do not stand exposure to the elements, and crumble when immersed in water;—some resist exposure for years;—others are so soft as not to preserve their form for any length of time;—while some bear to be dressed by the chisel; and, it may be remarked generally, that, as far as the results of my experiments have been compared with natural sandstone, the same boundless variety exists in both cases. A striking instance of this resemblance occurs in the case of the Salt-Heugh, the sandstone of which, when immersed in water, crumbles down, exactly in the same manner as those results of my experiments which taste much of salt. . . .

So far the results were satisfactory. But it next occurred, that it might be plausibly objected, that the presence of the super-incumbent cool ocean, would interfere with the process, on the principles of latent heat. To put this to the test, I proceeded to expose a quantity of sand, covered to the depth of several inches with common salt-water, to the heat of a furnace, and, as the liquid boiled away, replenished it from time to time by additions from the sea. Of course it gradually approached to a state of brine. But this proved a very tedious operation, requiring a continued ebullition, during three weeks without discharge of the fresh-water; and I thought it much easier, and no less satisfactory, to employ brine from the first, formed at once by loading the water with as much salt as it could dissolve, amounting to about one-third of its weight.

The vessels employed in these early experiments, were the large black-lead crucibles used by the brass-founders. I filled the vessel, which was 18 inches high and 10 broad, nearly to the brim with brine of full saturation, the lower portion being occupied, to the depth of about 15 inches, with loose sand from the sea-shore, and thoroughly drenched with the brine. In order to have a view of the progress of the experiment, I placed an earthen-ware tube, about the size and shape of a gun-barrel, closed at bottom, and open at the top, in a vertical position, having its lower extremity immersed in the sand, and reaching to within about an inch of the bottom of the pot, while the other end rose a foot above the surface of the brine, and could be looked into without inconvenience.

After a great number of experiments, furnishing an unbounded variety of results, I at length obtained a confirmation of the main object in view. I observed that the bottom of the porcelain barrel, and of course the sand in which it rested, became red-hot, whilst the brine, which, during the experiment, had been constantly replenished from a separate vessel, continued merely in a state of ebullition: the upper portion of the sand, drenched with the liquid, remained permanently quite loose, but the lower portion of the sand had formed itself into a solid cake.

On allowing the whole to cool, after it had been exposed to a high heat for many hours, and breaking up the mass, I was delighted to find the result, occupying the lower part of the pot, possessed of all the qualities of a perfect sandstone, as may be seen in the specimens now presented to the Society. Whenever the heat was not maintained so long, the sandstone which resulted was less perfect in its structure, tasted strongly of salt, and sometimes crumbled to sand when placed in water.

MACLURE

William Maclure (1763–1840), a Scottish geologist who settled in Virginia, made the first geological survey of what was then the United States.

Observations on the Geology of the United States

From *Observations on the Geology of the United States of America*, pp. iv–vi, 14–21, Philadelphia, 1817.

In all speculations on the origin, or agents that have produced the changes on this globe, it is probable that we ought to keep within the boundaries of the probable effects resulting from the regular operations of the great laws of nature which our experience and observation has brought within the sphere of our knowledge. When we overleap those limits, and suppose a total change in nature's laws, we embark on the sea of uncertainty, where one conjecture is perhaps as probable as another; for none of them can have any support, or derive any authority from the practical facts wherewith our experience has brought us acquainted. The equator has been supposed to have been once where the poles are now, to account for the bones of the animals now living near the tropics being found in the higher latitudes; yet without any change either in the poles or equator, it is certainly not impossible but even probable, that these animals, before their tyrant man obstructed their passage, might migrate to the north during nearly three months of the summer; and might have a sufficient quantity of heat, and a much greater abundance of nourishing vegetable food, than the torrid zone could afford them at that season.

There does not appear to be any thing either in the climate or food that could prevent the elephants, rhinoceroses, etc. from following the spring into the north, and arriving in the summer even to the latitude of 50 or 60 degrees, and retiring to the warmer climates on the approach of the winter; on the contrary, it would appear to be the natural course of things, and what I believe our buffaloes in the uninhabited parts of our continent still continue to do; that is, to migrate in vast droves from south to north, and from north to south, in search of their food, according to the season.

The birds and the fish continue their migrations, passing by roads out of the reach of man; the natural change of place which their wants require, has not been barred and obstructed by the united power and industry of the lords of the creation.*

The short period of time that mankind seem to have been capable of correct observation, and the minute segment of the immense circle of nature's operations, that has revolved during the comparatively short period, renders all speculations on the origin of the crust of the earth mere conjectures, founded on distant and obscure analogy. Were it possible to separate this metaphysical part from the collection and classification of facts, the truth and accuracy of observation would be much augmented, and the progress of knowledge much more certain and uniform; but the pleasure of indulging the imagination is so superior to that derived from the labour and drudgery of observation—the self-love of mankind is so flattered by the intoxicating idea of acting a part in the creation—that we can scarcely expect to find any great collection of facts, untinged by the false colouring of systems.

The peculiar structure of the continent of North America, by the extended continuity of the immense masses of rocks of the same formation or class, with the uniform structure and regularity of their uninterrupted stratification, forces the observer's attention to the limits which separate the great and principal classes; on the tracing of which, he finds so much order and regularity, that the bare collection of the *facts* partake somewhat of the delusion of theory.

The prominent feature of the eastern side of the continent of North America, is an extended range of mountains, running nearly north-east and south-west from the St. Lawrence to the Mississippi, the most elevated parts as well as the greatest mass of which

---

* Until lately we have restricted nature to two modes of acting; by fire, and by water: now, it is found, that she can change and metallize rocks in the dry way, without any solution or fluidity; and the galvanic pile may be formed in the stratifications of a mountain, as well as in a chemist's laboratory. These are two other modes wherein we must now allow her to change and modify the surface of this earth; and who can say how many more means yet unknown, she may possess? each of which, when found out by accurate and impartial observation, must make a change in former theories.

consists of *primitive* as far south as the Hudson river, decreasing in height and breadth as it traverses the state of New Jersey. The primitive occupies but a small part of the lower country, where it passes through the states of Pennsylvania and Maryland, where the highest part of the range of mountains to the west consists of transition, with some intervening vallies of secondary. In Virginia, the primitive increases in breadth, and proportionally in height, occupying the greatest mass, as well as the most elevated points of the range of mountains in the states of North Carolina and Georgia, where it takes a more westerly direction.

Though this primitive formation contains all the variety of primitive rocks found in the mountains of Europe, yet neither their relative situation in the order of succession, or their relative heights in the range of mountains, correspond with what has been observed in Europe. The order of succession from the clay state to the granite, as well as the gradual diminishing height of the strata, from the granite through the gneiss, mica slate, hornblende rocks, down to the clay slate, is so often inverted and mixed, as to render the arrangement of any regular series impracticable.

No secondary limestone has been found on the south-east side of the primitive, nor any series of other secondary rocks, except some partial beds of the old red sandstone formation, which partly cover its lower edge; in this, it seems to resemble some of the European chains, such as the Carpathian, Bohemian, Saxon, Tyrolian and Alpine or Swiss mountains, all of which, though covered with very extensive secondary limestone formations on their north and west flanks, have little secondary limestone on their southern and eastern sides.

The old red sandstone above mentioned, covers partially the lower levels of the primitive, from twelve miles south of Connecticut river to near the Rappahannock, a range of nearly four hundred miles; and though often interrupted, yet retains through the whole distance that uniform feature of resemblance so remarkable in the other formations of this continent. The same nature of sandstone strata is observable, running in nearly the same direction, partially covered with wacke and greenstone-trap, and containing the same metallic substances. The above uniformity is equally observable in the great alluvial formation which covers the southeast edge of the primitive, from Long island to the gulf of Mexico, consisting of sand, gravel, etc. with marsh and sea mud or clay, containing both vegetable and animal remains, found from thirty to forty feet below the surface.

Along the north-west edge of the primitive, commences the *transition* formation, occupying, after the primitive, some of the highest mountains in the range, and appears to be both higher and wider to the west in the states of Pennsylvania, Maryland, and part of Virginia, where the primitive is least extended, and lowest in height. It contains all the varieties of rocks found in the same formation in Europe, as the mountains in the Crimea, etc. and resembles in this the chain of the Carpathian, Bohemian and Saxon mountains, which have all a very considerable transition formation, succeeding the secondary limestone on their northern sides. Anthracite has been found in different places of this formation, and has not yet been discovered in any of the other formations in North America.

The necessity of such a class or division of rocks as the *transition*, has been doubted by some, nor is it now generally used in the south of Europe; but such rocks are found, and in very considerable quantities, in almost every country that has been examined. There are only two classes, the primitive or secondary, in which they can be placed. They are excluded from the primitive, by containing pebbles, evidently rounded by attrition when in an insulated state, and by the remains of organic substances being found, though rarely, in them; and yet many of the variety of transition rocks, such as the grey wacke slate, and quartzose aggregates, are hardly distinguishable from primitive slate and quartz when fresh; it is only in a state of decomposition, that the grain of the transition rocks appears, and facilitates the discrimination.

If they are placed with the secondary, they would form another division in the class, already rather confusedly divided; as their hardness, the glossy, slaty, and almost chrystalline structure of the cement of a great proportion of the transition aggregates, would exclude them from any division, as yet defined, of the other secondary rocks. Besides the objections arising out of their individual structure, the nature of their stratification removes them still further from the secondary, and makes them approach still nearer to the primitive. They are found regularly stratified, generally dipping at an angle above twenty and not exceeding forty-five degrees from the horizon; whereas, the secondary rocks are either horizontal or undulating with the inequalities of the surface. A bed of grey wacke, or grey wacke slate and transition limestone, runs south-west from the Potomac to near the Yadkin river, a distance of two hundred miles, from one to five miles in breadth, having the primitive formation on each side, dipping the same as the primitive, though at a less angle, the strata running in the same direction; and from its relative situation, dip, and stratification, bearing no characters of the secondary, not having been yet found alternating with secondary rocks, it cannot be classed with them, without destroying all order and introducing confusion. To class it with the primitive, would be making the primitive include not only aggregates composed of pieces of different kinds of rocks rounded by attrition, but also limestone with a dull fracture, coloured by organic or other combustible matter, which it loses by being burnt. It would perhaps add to the precision of the classification, if this class was augmented by placing some of the porphyritic and other rocks in it, which are more of an earthy than chrystalline fracture, but which at present are considered as primitive.

It might have been as well if, when giving names to the different classes of rocks, all reference to the relative period of their origin or formation had been avoided; and in place of *primitive* and *second-ary*, some other names had been adopted, taken from the most prominent feature or general property of the class of rocks intended to be designated, such as perhaps chrystalline in place of primitive —deposition or horizontal in place of secondary, etc.; but as those old names are in general use, and consecrated by time and long habit, it is more than probable that the present state of our knowledge does not authorise us to change them. The adoption of new names, on account of some new property discovered in the substance is the cause of much complication and inconvenience already; and if adopted as a precedent in future, will create a confused accumulation of terms calculated to retard the progress of the science. When we change the names given to defined substances, by those who went before us, what right have we to suppose, that posterity will respect our own nomenclature?

On the north-west side of the transition formation, along the whole range of mountains, lays the great *secondary* formation, which, for the extent of the surface it covers and the uniformity of its deposition, is equal in magnitude and importance, if not superior, to any yet known: there is no doubt of its extending to the borders of the great lakes to the north, and some hundred miles beyond the Mississippi to the west. We have indeed every reason to believe, from what is already known, that the limits of this great basin to the west, is not far distant from the foot of the Stony mountains; and to the north, that it reaches beyond Lake Superior, giving an area extending from east to west from Fort Ann, near Lake Champlain, to near the foot of the Stony mountains, of about fifteen hundred miles, and from south to north from the Natchez to the upper side of the great lakes, about twelve hundred miles.

This extensive basin is filled with most of the species of rocks, attending the secondary formation elsewhere, nor is their continuity interrupted on the east side of the Mississippi by the interposition of any other formation except the alluvial deposits on the banks of the large rivers. The foundation of most of the level countries is generally limestone, and the hills or ridges in some places consist of sandstone: a kind of dark coloured slaty clay, containing vegetable impressions, with a little mixture of carbon, frequently alternates with all the strata of this formation, the whole of which is nearly horizontal. The highest mountains are on the external borders of the basin, gradually diminishing in height towards its centre.

SCHLOTHEIM

Ernst Friedrich, Baron von Schlotheim (1764–1832) was the first German paleontologist to recognize the stratigraphic significance of the occurrence of fossils.

On the Use of Fossils in Geological Investigations

Translated from *Taschenbuch für die Gesammte Mineralogie* (Leonhard), Vol. VII, pp. 2–134, 1813.

Obviously the occurrence of fossils can furnish us with the most important aid in the closer determination of the relative ages of many kinds of rock, and enables us to distinguish the synchronous or asynchronous origins of these and their subordinate beds. It can furnish us with deep insight into the nature of those great earth revolutions which formed the upper earth crust that we see and which repeatedly changed it. Perhaps it can even help determine the epochs of certain of those revolutions in terms of the span of years in which they took place. Moreover, it can inform us whether such revolutions were spread generally over the earth's surface or were merely local. And finally, it can reveal to us the remarkable facts of organic creation: plant and animal types so different in the different epochs, developing so wonderfully from the earliest times. Where could the naturalist seek for more telling documents of the history of creation than in the fossils themselves? The most difficult question, whether a greater part of these belongs to one or to several entirely distinct creations, following each other at great time intervals—creations entirely distinct from the present one—cannot be answered definitely for a long time. For present purposes, its solution is hardly of any essential importance. . . .

It is true that in recent years our mineralogical writers have paid much more attention to the determination of the formations in which the fossils occur; but, for the most part, we learn simply that Ammonites, Terebratulae, Lenticulites, Turbinites, etc., occur in some Transition formations and in the overlying Floetz formations, without being informed as to the different species of these. This is done regardless of the fact that the main goal to which we must direct our scientific efforts, if we would gain the desired results, is the accurate determination of the individual species of fossils which occur in the different formations and which appear to belong exclusively to certain of these. There is an endless variety of Ammonites, Terebratulae, etc. Of these, certain species may belong only in the Transition Series, certain ones in Alpine limestone, others in the Jura limestone, and still others in our so-called Kupferschiefer, Upper Muschelkalk, and Sandstone formations, etc. Should this supposition be affirmed, we would thus be practically in a position to determine the characteristic fossils for several series of formations and then could reach many unexpected conclusions concerning the rock layers and stratum members belonging to a given formation. Linked with other geognostic research, one can expect highly interesting information concerning the history of organic creation and the relative age of earth revolutions.

NICOL

William Nicol (c. 1768–1851), Scottish natural philosopher, is best known among geologists for his contributions to the technique of microscopy.

The Nicol Prism

From *Edinburgh New Philosophical Journal*, Vol. VI, pp. 83–84, 1829.

The following simple method of constructing a prism of calcareous-spar, so that only one image may be seen at a time, will, perhaps, prove interesting to those who are in the habit of examining the optical properties of crystallised bodies by polarised light.

Let a rhomboid of calcareous-spar one inch long be reduced in breadth and thickness to three-tenths of an inch; let the obliquity of its terminal planes be increased about three degrees; or, in other words, let the angles formed by the terminal planes, and the adjoining obtuse lateral edges, be made equal to 68°, by operating on the terminal planes: These planes may now be polished. The rhomboid is then to be divided into two equal portions, by a plane passing through the acute lateral edges, and nearly touching the two obtuse solid angles. The sectional plane of each of the two halves must now be made to form exactly an angle of 90° with the terminal plane, and then carefully polished. The two portions are now to be firmly cemented together by means of Canada balsam, so as to form a rhomboid similar to what it was before its division.

If a ray of common light fall on the end of such a rhomboid in a direction parallel to the lateral edges, the two rays into which it is divided, in passing through the spar, will deviate so far from each other, that only the ordinary image will be seen. That image, too, will appear exactly in its true position, and free from colour. The range of the ordinary ray will be found considerably greater than the whole field of vision, as may easily be seen by making the rhomboid revolve on an axis parallel to the longer diagonal of the terminal planes. There is a tinge of blue where the ordinary ray vanishes on one side, and a tinge of orange, accompanied by a number of extremely minute obscurely coloured fringes, where it terminates on the other. If the rhomboid revolve beyond the fringes, the ordinary image will disappear, and the extraordinary image come into view. The latter, however, from the great obliquity of the incident light, occupies a smaller range, and is less distinct than the other. The ordinary ray passing out of the rhomboid in a direction parallel to its lateral edges, is therefore the best adapted for analytical purposes; and as calcareous-spar, when pure, and free from flaws, is not only transparent, but perfectly colourless, a rhomboid of that substance, of the above construction, developes the coloured rings of crystallized bodies with a degree of brilliancy not to be equalled by a plate of tourmaline, or perhaps by any other substance.

With the view of rendering the structure of the analyzing rhomboid more easily understood, I have supposed a piece of the spar to be divided into two equal portions. Such a division, however, would be a difficult task; but if two similar pieces of spar be taken, it will be found a very easy matter to remove one-half of each of them, either by grinding, or by the action of a file. The pieces of spar should not be much less than an inch long, and they need not be longer than 1.4 inch. If the latter dimension be adopted, the breadth and thickness will require to be about .48 of an inch.

In cementing the two pieces together, it will be proper to let the pointed end of the one project a little over the terminal plane of the other. By so doing, a more firm contact is obtained at the edges; and when the cement is sufficiently indurated, the whole of the projecting parts may easily be removed, according to their cleavages. The lateral planes should be left quite rough, to prevent the reflection of extraneous light.

THE PREPARATION OF THIN SECTIONS

From "Process of Preparing Fossil Plants for the Microscope," p. 45 in *Observations on Fossil Vegetables*, by Henry Witham, London, 1831.

Let a thin slice be cut off from the fossil wood, in a direction perpendicular to the length of its fibres. The slice thus obtained must be ground perfectly flat, and then polished. The polished surface is to be cemented to a piece of plate or mirror glass, a little larger than itself, and this may be done by means of Canada balsam. A thin layer of the substance must be applied to the polished surface of the slice, and also to one side of the glass. The slice and the glass are now to be laid on any thin plate of metal, as a common fire-shovel, and gradually heated over a slow fire, with a view to concentrate the balsam. . . . When the balsam is thought to be sufficiently concentrated, and all air-bubbles completely removed, the slice and the glass may be taken from the shovel, and applied to each other. . . . When the whole is cooled down to the temperature of the air, and the balsam becomes solid, that part of the balsam adhering to the surface of the glass surrounding the slice should be removed by the point of a penknife; and it may be right to remark, that, in this operation, it will at once be seen whether the balsam has undergone the requisite concentration. . . .

The slice must now be ground down to that degree of thinness which will permit its structure to be seen by the help of a microscope. To facilitate this part of the grinding, the lapidary will find it advantageous to fix the glass in a groove made in a small piece of wood, of which half inch deal will answer the purpose. . . .

A lapidary by attending to the above directions, will find no difficulty in reducing any piece of petrified wood to that degree of thinness sufficient to render its structure visible; and anyone, even without the aid of the mechanism employed by the lapidary, may accomplish that object.

HUMBOLDT

Friedrich Heinrich Alexander, Baron von Humboldt (1769–1859), Prussian natural scientist and world traveler.

Volcanic Activity in Mexico

From "Essai sur la Nouvelle Espagne," translated in *Journal of Natural Philosophy, Chemistry, and the Arts*, (ed. by William Nicholson), Vol. XXVI, pp. 81–86, 1810.

Large hill thrown up by a volcano in 1759. The grand catastrophe in which this volcanic mountain [Jorullo] issued from the earth, and by which the face of a considerable extent of ground was totally altered, was perhaps one of the most extensive physical changes, that the history of our globe exhibits. Geology points out spots in the ocean, where, within the last two thousand years, volcanic islets have arisen above the surface of the sea, as near the Azores, in the Archipelago, and on the south of Iceland: but it records no instance of a mountain of scoriae and ashes, 517 meters (563 yards) above the old level of the neighbouring plains, suddenly formed in the centre of a thousand small burning cones, thirty-six leagues from the seashore, and forty-two leagues from any other volcano. This phenomenon remained unknown to the mineralogists and natural philosophers of Europe, though it took place but fifty years ago, and within six days journey of the capital of Mexico.

Country described. Descending from the central flat toward the coasts of the Pacific ocean, a vast plain extends from the hills of Aguasarco to the villages of Toipa, and Patatlan, equally celebrated for their fine cotton plantations. Between the picachos del Mortero and the cerras de las Cuevas and de Cuiche, this plain is only from 750 to 800 met. (820 to 880 yards) above the level of the sea. Basaltic hills rise in the midst of a country, in which porphyry with a base of greenstone predominates. Their summits are crowned with oaks always in verdure, and the foliage of laurels and olives intermingled with dwarf fan palms. This beautiful vegetation forms a singular contrast with the arid plain, which has been laid waste by volcanic fire.

179

A fertile plain shaken by an earthquake, and a hill raised on it.
To the middle of the eighteenth century fields of sugarcanes and
indigo extended between two rivulets, called Cuitimba and San
Pedro. They were skirted by basaltic mountains, the structure
of which seems to indicate, that all the country, in remote periods,
has several times experienced the violent action of volcanoes.
Those fields, irrigated by art, belonged to the estate of San Pedro
de Jorullo (Xorullo, or Juvriso), one of the largest and most
valuable in the country. In the month of June, 1759, fearful
rumbling noises were accompanied with frequent shocks of an
earthquake, which succeeded each other at intervals for fifty
or sixty days, and threw the inhabitants of the estate into the
greatest consternation. From the beginning of the month
of September, every thing seemed perfectly quiet, when in the
night of the 28th of that month a terrible subterranean noise
was heard anew. The frightened Indians fled to the mountains of Aguasarco. A space of three or four square miles, known
by the name of Malpays, rose in the shape of a bladder. The
boundaries of this rising are still distinguishable in the ruptured
strata. The Malpays toward the edge is only 12 met. (13 yards)
above the former level of the plain, called las playas de Jorullo;
but the convexity of the ground increases progressively toward the
centre, till it reaches the height of 160 met. (175 yards).

The event described. They who witnessed this grand catastrophe
from the top of Aguasarco assert, that they saw flames issue out of
the ground for the space of more than half a league square; that
fragments of red hot rocks were thrown to a prodigious height;
and that through a thick cloud of ashes, illumined by the volcanic
fire, and resembling a stormy sea, the softened crust of the earth
was seen to swell up. The rivers of Cuitimba and San Pedro then
precipitated themselves into the burning crevices. The decomposition of the water contributed to reanimate the flames, which
were perceptible at the city of Pascuoro, though standing on a
very wide plain 1400 met. (1530 yards) above the level of the
playas de Jorullo. Eruptions of mud, particularly of the strata of
clay including decomposed nodules of basaltes with concentric
layers, seem to prove, that subterranean waters had no small part
in this extraordinary revolution. Thousands of small cones, only
two or three yards high, which the Indians call ovens, issued from
the raised dome of the Malpays. Though the heat of these volcanic ovens has diminished greatly within these fifteen years,

according to the testimony of the Indians, I found the thermometer rise to 95° (if centig. 203°F.) in the crevices that emitted an aqueous vapour. Each little cone is a chimney [*fumarole*, in the original], from which a thick smoke rises to the height of ten or fifteen met. (11 or 16 yards). In several a subterranean noise is heard like that of some fluid boiling at no great depth.

Six large hills in one line. Amid these ovens, in a fissure, the direction of which is from N.N.E. to S.S.E., six large hummocks rise 400 or 500 met. (440 or 550 yards) above the old level of the plain. This is the phenomenon of Monte Novo at Naples repeated several times in a row of volcanic hills. The loftiest of these huge hummocks, which reminded me of the country of Auvergne, is the large volcano of Jorullo. It is constantly burning, and has thrown out on the north side an immense quantity of scorified and basaltic lava, including fragments of primitive rocks. These grand eruptions of the central volcano continued till February, 1760. In the succeeding years they became gradually less frequent. The Indians, alarmed by the horrible noise of the new volcano, at first deserted the villages for seven or eight leagues round the plain of Jorullo. In a few months they became familiar with the alarming sight, returned to their huts, and went down to the mountains of Aguasarco and Santa Ines, to admire the sheaves of fire thrown out by an infinite number of large and small volcanic openings. The ashes then covered the houses of Queretoro, more than 48 leagues (120 miles) in a right line from the place of the explosion. Though the subterranean fire appears to be in no great activity* at present, and the Malpays and the great volcano begin to be covered with vegetables, we found the air so heated by the little ovens, that in the shade, and at a considerable distance from the ground, the thermometer rose to 43° (109.4°F.). This fact evinces, that there is no exaggeration in the report of some of the old Indians, who say, that the plains of Jorullo were uninhabitable for several years, and even to a considerable distance from the ground raised up, on account of the excessive heat.

Line of volcanoes in Mexico crossing the chain of hills. The situation of the new volcano of Jorullo leads to a very curious geological

----

* In the bottom of the crater we found the heat of the air 47° (116.6°F.), and in some places 58° and 60° (136.4° and 140°). We had to pass over cracks exhaling sulphurous vapours, in which the thermometer rose to 85° (185°). From these cracks, and the heaps of scoriae that cover considerable hollows, the descent into the craters is not without danger.

observation. It has already been observed in the 3d chapter, that there is in New Spain a line of great heights, or a narrow zone included between the latitudes of 18°59′ and 19°12′, in which are all the summits of Anahuac that rise above the region of perpetual snow. These summits are either volcanoes still actually burning; or mountains, the form of which, as well as the nature of their rocks, renders it extremely probable, that they formerly contained subterranean fire. Setting out from the coast of the Gulf of Mexico, and proceeding westward, we find the peak of Oribaza, the two volcanoes of la Puebla, the Nevado de Toluca, the peak of Tancitaro, and the volcano of Colima. These great heights, instead of forming the ridge of the cordillera of Anahuac, and following its direction, which is from S.E. to N.W., are on the contrary in a line perpendicular to the axis of the great chain of mountains. It is certainly worthy remark, that in the year 1759 the new volcano of Jorullo was formed in the continuation of this line, and on the same parallel as the ancient Mexican volcanoes.

Indicate a long interior fissure in the earth. A view of my plan of the environs of Jorullo will show, that the six large hummocks have risen out of the earth on a vein, that crosses the plain from the cerro of Las Cuevas to the picacho del Montero. The new mouths of Vesuvius too are found ranged along a fissure. Do not these analogies give us reason to suppose, that there exists in this part of Mexico, at a great depth within the Earth, a fissure stretching from east to west through a space of 137 leagues (343 miles), and through which the volcanic fire has made its way at different times, bursting the outer crust of porphyritic rocks, from the coasts of the Gulf of Mexico to the South Sea? Is this fissure prolonged to that little groupe of islands, called by Colluet the Archipelago of Regigedo, and round which, in the same parallel with the Mexican volcanoes, pumice stone has been seen floating?

Earthquakes and Volcanoes in the Americas

From "Account of the Earthquake Which Destroyed the Town of Caraccas on the 26th March 1812," *Edinburgh Philosophical Journal,* Vol. I, pp. 272–280, 1819.

There are few events in the physical world which are calculated to excite so deep and permanent an interest as the earthquake which destroyed the town of Caraccas, and by which more than 20,000 persons perished, almost at the same instant, in the province of Venezuela. . . .

The 26th of March was a remarkably hot day. The air was calm, and the sky unclouded. It was Holy Thursday, and a great part of the population was assembled in the churches. Nothing seemed to presage the calamities of the day. At seven minutes after four in the afternoon the first shock was felt; it was sufficiently powerful, to make the bells of the churches toll; it lasted five or six seconds, during which time, the ground was in a continual undulating movement, and seemed to heave up like a boiling liquid. The danger was thought to be past, when a tremendous subterraneous noise was heard, resembling the rolling of thunder, but louder, and of longer continuance, than that heard within the tropics in time of storms. This noise preceded a perpendicular motion of three or four seconds, followed by an undulatory movement somewhat longer. The shocks were in opposite directions, from north to south, and from east to west. Nothing could resist the movement from beneath upward, and undulations crossing each other. The town of Caraccas was entirely overthrown. Between nine and ten thousand of the inhabitants were buried under the ruins of the houses and churches. . . . Nine-tenths of the fine town of Caraccas were entirely destroyed. The walls of the houses that were not thrown down, as those of the street San Juan, near the Capuchin Hospital, were cracked in such a manner, that it was impossible to run the risk of inhabiting them.

Shocks as violent as those which, in the space of one minute,* overthrew the city of Caraccas, could not be confined to a small portion of the continent. Their fatal effects extended as far as the provinces of Venezuela, Varinas, and Maracaybo, along the coast; and still more to the inland mountains. La Guayra, Mayquetia, Antimano, Baruta, La Vega, San Felipe, and Merida, were almost entirely destroyed. The number of the dead exceeded four or five thousand at La Guayra, and at the town of San Felipe, near the copper-mines of Aroa. It appears that it was on a line running east north-east, and west south-west, from La Guayra and Caraccas to the lofty mountains of Niquitao and Merida, that the violence of the earthquake was principally directed. It was felt in the kingdom of New Granada from the branches of the high Sierra de Santa Marta as far as Santa Fé de Bogota and Honda,

---

* The duration of the earthquake, that is to say the whole of the movements of undulation and rising which occasioned the horrible catastrophe of the 26th of March 1812, was estimated by some at 50″, by others at 1′12″.

on the banks of the Magdalena, 180 leagues from Caraccas. It was every where more violent in the Cordilleras of gneiss and mica-slate, or immediately at their foot, than in the plains: and this difference was particularly striking in the savannahs of Varinas and Casanara. In the valleys of Aragua, situate between Caraccas and the town of San Felipe, the commotions were very weak: and La Victoria, Maracay, and Valentia, scarcely suffered at all, not-withstanding their proximity to the capital. At Valecillo, a few leagues from Valencia, the earth, opening, threw out such an immense quantity of water, that it formed a new torrent. The same phenomenon took place near Porto-Cabello. On the other hand, the lake of Maracaybo diminished sensibly. At Coro no commotion was felt, though the town is situated upon the coast, between other towns which suffered from the earthquake.

Fifteen or eighteen hours after the great catastrophe, the ground remained tranquil . . . the commotions did not recommence till after the 27th. They were then attended with a very loud and long continued subterranean noise. The inhabitants of Caraccas wandered into the country; but the villages and farms having suffered as much as the town, they could find no shelter till they were beyond the mountains of Los Teques, in the valleys of Aragua, and in the Llanos or Savannahs. No less than fifteen oscillations were often felt in one day. On the 5th of April there was almost as violent an earthquake, as that which overthrew the capital. During several hours the ground was in a state of perpetual undulation. Large masses of earth fell in the mountains; and enormous rocks were detached from the Silla of Caraccas. It was even asserted and believed that the two domes of the Silla sunk fifty or sixty toises;* but this assertion is founded on no measurement whatever.

While violent commotions were felt at the same time in the valley of the Mississippi, in the island of St. Vincent, and in the province of Venezuela, the inhabitants of Caraccas, of Calabozo, situated in the midst of the steppes, and on the borders of the Rio Apura, in a space of 4000 square leagues, were terrified on the 30th of April 1812, by a subterraneous noise, which resembled frequent discharges of the largest cannon. This noise began at two in the morning. It was accompanied by no shock; and, what is very remarkable, it was as loud on the coast as at eighty leagues

---

* A *toise* is about six feet.—Editors.

distance inland. It was every where believed to be transmitted through the air; and was so far from being thought a subterraneous noise, that at Caraccas, as well as at Calabozo, preparations were made to put the place into a state of defence against an enemy, who seemed to be advancing with heavy artillery. Mr. Palacio, crossing the Rio Apura near the junction of the Rio Nula, was told by the inhabitants that the "*firing of cannon*" had been heard as distinctly at the western extremity of the province of Varinas, as at the port of La Guayra to the north of the chain of the coast.

The day on which the inhabitants of Terra Firma were alarmed by a subterraneous noise, was that on which happened the eruption of the volcano in the island of St. Vincent. This mountain, near five hundred toises high, had not thrown out any lava since the year 1718. Scarcely was any smoke perceived to issue from its top, when, in the month of May 1811, frequent shocks announced, that the volcanic fire was either rekindled, or directed anew toward that part of the West Indies. The first eruption did not take place till the 27th of April 1812, at noon. It was only an ejection of ashes, but attended with a tremendous noise. On the 30th, the lava passed the brink of the crater, and, after a course of four hours, reached the sea. The noise of the explosion "resembled that of alternate discharges of very large cannon and of musketry; and, what is well worthy of remark, it seemed much louder at sea, at a great distance from the island, than in sight of land, and near the burning volcano."

The distance in a straight line from the volcano of St. Vincent to the Rio Apura, near the mouth of the Nula, is 210 nautical leagues. The explosions were consequently heard at a distance equal to that between Vesuvius and Paris. This phenomenon, connected with a great number of facts observed in the Cordilleras of the Andes, shows how much more extensive the subterranean sphere of activity of a volcano is, than we are disposed to admit from the small changes effected at the surface of the globe. The detonations heard during whole days together in the New World, 80, 100, or even 200 leagues distant from a crater, do not reach us by the propagation of sound through the air; they are transmitted to us by the ground. The little town of Honda, on the banks of the Magdalena, is not less than 145 leagues from Cotopaxi; and yet in the great explosions of this volcano, in 1744, a subterraneous noise was heard at Honda, and supposed to be discharges of heavy artillery. The monks of St. Francis spread the news, that the town of Carthagena was bombarded by the English; and the intelligence was believed.   Now the volcano of Cotopaxi is a cone, more than 1800 toises above the basin of Honda, and rises from a table-land, the elevation of which is more than 1500 toises above the valley of the Magdalena.   In all the colossal mountains of Quito, of the provinces of Los Pastos, and of Popayan, crevices and valleys without number are interposed.   It cannot be admitted, under these circumstances, that the noise could be transmitted through the air, or by the superior surface of the globe, and that it came from that point, where the cone and crater of Cotopaxi are placed.   It appears probable, that the higher part of the kingdom of Quito and the neighbouring Cordilleras, far from being a group of distinct volcanoes, constitute a single swollen mass, an enormous volcanic wall, stretching from south to north, and the crest of which exhibits a surface of more than six hundred square leagues. Cotopaxi, Tunguragua, Antisana, and Pichincha, are placed on this same vault, on this raised ground.   The fire issues sometimes from one, sometimes from another of these summits.   The obstructed craters appear to be extinguished volcanoes; but we may presume, that, while Cotopaxi or Tunguragua have only one or two eruptions in the course of a century, the fire is not less continually active under the town of Quito, under Pichincha and Imbaburu.

Advancing toward the north, we find, between the volcano of Cotopaxi and the town of Honda, two other *systems of volcanic mountains,* those of Los Pastos and of Popayan.   The connection of these systems was manifested in the Andes in an incontestible manner by a phenomenon, which I have already had occasion to notice.   Since the month of November 1796, a thick column of smoke had issued from the volcano of Pasto, west of the town of that name, and near the valley of Rio Guaytara.   The mouths of the volcano are lateral, and placed on its western declivity, yet during three successive months the column rose so much higher than the ridge of the mountain, that it was constantly visible to the inhabitants of the town of Pasto.   They related to us their astonishment, when, on the 4th of February 1797, they observed the smoke disappear in an instant, without feeling any shock whatever.   At that very moment, sixty-five leagues to the south, between Chimborazo, Tunguragua, and the Altar (Capac Urcu), the town of Riobamba was overthrown by the most dreadful earthquake of which tradition has transmitted the history.   Is it possible to doubt from this coincidence of phenomena, that the vapours issuing from the small apertures or *ventanillas* of the volcano of Pasto, had an influence on the pressure of those elastic fluids, which shook the ground of the kingdom of Quito, and destroyed in a few minutes thirty or forty thousand inhabitants?

In order to explain these great effects of *volcanic reactions*, and to prove, that the group or system of the volcanoes of the West India Islands may sometimes shake the continent, it was necessary to cite the Cordillera of the Andes. Geological reasoning can be supported only on the analogy of facts that are recent, and consequently well authenticated: and in what other region of the globe could we find greater, and at the same time more varied volcanic phenomena, than in that double chain of mountains heaved up by fire? in that land, where Nature has covered every summit and every valley with her wonders? If we consider a burning crater only as an insulated phenomenon, if we satisfy ourselves with examining the mass of stony substances which it has thrown up, the volcanic action at the surface of the globe will appear neither very powerful nor very extensive. But the image of this action swells in the mind, when we study the relations that link together volcanoes of the same group; for instance, those of Naples and Sicily, of the Canary Islands, of the Azores, of the Caribbee Islands, of Mexico, of Guatimala, and of the table-land of Quito; when we examine either the reactions of these different systems of volcanoes on one another, or the distance to which, by subterranean communications, they at the same moment shake the Earth."*

---

* The following is the series of phenomena which M. Humboldt supposes to have had the same origin:

27th September 1796. Eruption in the West India Islands. Volcano of Guadaloupe.—November 1796. The volcano of Pasto begins to emit smoke.—14th of December 1796. Destruction of Cumana.—4th of February 1797. Destruction of Riobamba.—30th of January 1811. Appearance of Sabrina Island, in the Azores. It increases particularly on the 15th of June 1811.—May 1811. Beginning of the earthquakes in the Island St. Vincent, which lasted till May 1812.—16th of December 1811. Beginning of the commotions in the Valley of the Mississippi and the Ohio, which lasted till 1813.—December 1811. Earthquake at Caraccas.—26th of March 1812. Destruction of Caraccas. Earthquakes which continued till 1813.—30th April 1812. Eruption of the volcano at St. Vincent's; and the same day subterranean noises at Caraccas, and on the banks of the Apura.

CUVIER

Léopold Chrétien Frédéric Dagobert (pseudonym "Georges"), Baron Cuvier (1769–1832), the great comparative anatomist who brought a rational basis to paleontology. Of Swiss parentage, Cuvier was born in Montbéliard, which at that time was a part of Germany, and completed his formal education in Stuttgart. He later became a professor in the Jardin des Plantes and eventually a peer of France.

REVOLUTIONS AND CATASTROPHES IN THE HISTORY OF THE EARTH

From *Essay on the Theory of the Earth*, translated by Robert Jameson, pp. 7–17, 106–108, Edinburgh, 1817.

First Proofs of Revolutions on the Surface of the Globe

The lowest and most level parts of the earth, when penetrated to a very great depth, exhibit nothing but horizontal strata composed of various substances, and containing almost all of them innumerable marine productions. Similar strata, with the same kind of productions, compose the hills even to a great height. Sometimes the shells are so numerous as to constitute the entire body of the stratum. They are almost everywhere in such a perfect state of preservation, that even the smallest of them retain their most delicate parts, their sharpest ridges, and their finest and tenderest processes. They are found in elevations far above the level of every part of the ocean, and in places to which the sea could not be conveyed by any existing cause. They are not only inclosed in loose sand, but are often incrusted and penetrated on all sides by the hardest stones. Every part of the earth, every hemisphere, every continent, every island of any size, exhibits the same phenomenon. We are therefore forcibly led to believe, not only that the sea has at one period or another covered all our plains, but that it must have remained there for a long time, and in a state of tranquillity; which circumstance was necessary for the formation of deposits so extensive, so thick, in part so solid, and containing exuviae so perfectly preserved.

The time is past for ignorance to assert that these remains of organized bodies are mere *lusus naturae*,—productions generated in the womb of the earth by its own creative powers. A nice and scrupulous comparison of their forms, of their contexture, and frequently even of their composition, cannot detect the slightest difference between these shells and the shells which still inhabit the sea. They have therefore once lived in the sea, and been deposited by it: the sea consequently must have rested in the places where the deposition has taken place. Hence it is evident that the basin or reservoir containing the sea has undergone some change at least, either in extent, or in situation, or in both. Such is the result of the very first search, and of the most superficial examination.

The traces of revolutions become still more apparent and decisive when we ascend a little higher, and approach nearer to the foot of the great chains of mountains. There are still found many beds of shells; some of these are even larger and more solid; the shells are quite as numerous and as entirely preserved; but they are not of the same species with those which were found in the less elevated regions. The strata which contain them are not so generally horizontal; they have various degrees of inclination, and are sometimes situated vertically. While in the plains and low hills it was necessary to dig deep in order to detect the succession of the strata, here we perceive them by means of the vallies which time or violence has produced, and which disclose their edges to the eye of the observer. At the bottom of these declivities, huge masses of their *debris* are collected, and form round hills, the height of which is augmented by the operation of every thaw and of every storm.

These inclined or vertical strata, which form the ridges of the secondary mountains, do not rest on the horizontal strata of the hills which are situated at their base, and serve as their first steps; but, on the contrary, are situated underneath them. The latter are placed upon the declivities of the former. When we dig through the horizontal strata in the neighbourhood of the inclined strata, the inclined strata are invariably found below. Nay, sometimes, when the inclined strata are not too much elevated, their summit is surmounted by horizontal strata. The inclined strata are therefore more ancient than the horizontal strata. And as they must necessarily have been formed in a horizontal position, they have been subsequently shifted into their inclined or vertical position, and that too before the horizontal strata were placed above them.

Thus the sea, previous to the formation of the horizontal strata, had formed others, which, by some means, have been broken, lifted up, and overturned in a thousand ways. There had therefore been also at least one change in the basin of that sea which preceded ours; it had also experienced at least one revolution; and as several of these inclined strata which it had formed first, are elevated above the level of the horizontal strata which have succeeded and which surround them, this revolution, while it gave them their present inclination, had also caused them to project above the level of the sea, so as to form islands, or at least rocks and inequalities; and this must have happened whether one of their edges was lifted up above the water, or the depression of the opposite edge caused the water to subside. This is the second result, not less obvious, nor less clearly demonstrated, than the first, to every one who will take the trouble of studying carefully the remains by which it is illustrated and proved.

Proofs That Such Revolutions Have Been Numerous

If we institute a more detailed comparison between the various strata and those remains of animals which they contain, we shall soon discover still more numerous differences among them, indicating a proportional number of changes in their condition. The sea has not always deposited stony substances of the same kind. It has observed a regular succession as to the nature of its deposits; the more ancient the strata are, so much the more uniform and extensive are they; and the more recent they are, the more limited are they, and the more variation is observed in them at small distances. Thus the great catastrophes which have produced revolutions in the basin of the sea, were preceded, accompanied, and followed by changes in the nature of the fluid and of the substances which it held in solution; and when the surface of the seas came to be divided by islands and projecting ridges, different changes took place in every separate basin.

Amidst these changes of the general fluid, it must have been almost impossible for the same kind of animals to continue to live:—nor did they do so in fact. Their species, and even their genera, change with the strata; and although the same species occasionally recur at small distances, it is generally the case that the shells of the ancient strata have forms peculiar to themselves; that they gradually disappear, till they are not to be seen at all in the recent strata, still less in the existing seas, in which, indeed, we never discover their corresponding species, and where several even of their genera are not to be found; that, on the contrary, the shells of the recent strata resemble, as it respects the genus, those which still exist in the sea; and that in the last-formed and loosest of these strata there are some species which the eye of the most expert naturalist cannot distinguish from those which at present inhabit the ocean.

In animal nature, therefore, there has been a succession of changes corresponding to those which have taken place in the chemical nature of the fluid; and when the sea last receded from our continent, its inhabitants were not very different from those which it still continues to support.

Finally, if we examine with greater care these remains of organized bodies, we shall discover, in the midst even of the most ancient secondary strata, other strata that are crowded with animal or vegetable productions, which belong to the land and to fresh water; and amongst the more recent strata, that is, the strata which are nearest the surface, there are some of them in which land animals are buried under heaps of marine productions. Thus the various catastrophes of our planet have not only caused the different parts of our continent to rise by degrees from the basin of the sea, but it has also frequently happened, that lands which had been laid dry have been again covered by the water, in consequence either of these lands sinking down below the level of the sea, or of the sea being raised above the level of the lands. The particular portions of the earth also which the sea has abandoned by its last retreat, had been laid dry once before, and had at that time produced quadrupeds, birds, plants, and all kinds of terrestrial productions; it had then been inundated by the sea, which has since retired from it and left it to be occupied by its own proper inhabitants.

The changes which have taken place in the productions of the shelly strata have not, therefore, been entirely owing to a gradual and general retreat of the waters, but to successive irruptions and retreats, the final result of which, however, has been an universal depression of the level of the sea.

Proofs That the Revolutions Have Been Sudden

These repeated irruptions and retreats of the sea have neither been slow nor gradual; most of the catastrophes which have occasioned them have been sudden; and this is easily proved, especially with regard to the last of them, the traces of which are most conspicuous. In the northern regions it has left the carcases of some large quadrupeds which the ice had arrested, and which are preserved even to the present day with their skin, their hair, and their flesh. If they had not been frozen as soon as killed they must quickly have been decomposed by putrefaction. But this eternal frost could not have taken possession of the regions which these animals inhabited except by the same cause which destroyed them;* this cause, therefore, must have been as sudden as its effect. The breaking to pieces and overturnings of the strata, which happened in former catastrophes, shew plainly enough that they were sudden and violent like the last; and the heaps of *debris* and rounded pebbles which are found in various places among the solid strata, demonstrate the vast force of the motions excited in the mass of waters by these overturnings. Life, therefore, has been often disturbed on this earth by terrible events—calamities which, at their commencement, have perhaps moved and overturned to a great depth the entire outer crust of the globe, but which, since these first commotions, have uniformly acted at a less depth and less generally. Numberless living beings have been the victims of these catastrophes; some have been destroyed by sudden inundations, others have been laid dry in consequence of the bottom of the seas being instantaneously elevated. Their races even have become extinct, and have left no memorial of them except some small fragments which the naturalist can scarcely recognise.

Such are the conclusions which necessarily result from the objects that we meet with at every step of our enquiry, and which we can always verify by examples drawn from almost every country. Every part of the globe bears the impress of these great and terrible events so distinctly, that they must be visible to all who are qualified to read their history in the remains which they have left behind.

Relations of the Species of Fossil Bones, with the Strata in Which They Are Found

The most important consideration, that which has been the chief object of my researches, and which constitutes their legitimate connection with the theory of the earth, is to ascertain the particu-

---

* The two most remarkable phenomena of this kind, and which must for ever banish all idea of a slow and gradual revolution, are the rhinoceros discovered in 1771 in the banks of the *Vilhoui*, and the elephant recently found by M. Adams near the mouth of the *Lena*. This last retained its flesh and skin, on which was hair of two kinds; one short, fine, and crisped, resembling wool, and the other like long bristles. The flesh was still in such high preservation, that it was eaten by dogs.

lar strata in which each of the species was found, and to enquire if any of the general laws could be ascertained, relative either to the zoological subdivisions, or to the greater or less resemblance between these fossil species and those which still exist upon the earth.

The laws already recognised with respect to these relations are very distinct and satisfactory.

It is, in the first place, clearly ascertained, that the oviparous quadrupeds are found considerably earlier, or in more ancient strata, than those of the viviparous class. Thus the crocodiles of Honfleur and of England are found underneath the chalk. The *monitors* of Thuringia would be still more ancient, if, according to the Wernerian school, the copper-slate in which they are contained, along with a great number of fishes supposed to have belonged to fresh water, is to be placed among the most ancient strata of the secondary or flætz formations. The great alligators, or crocodiles, and the tortoises of Maestricht, are found in the chalk formation; but these are both marine animals.

This earliest appearance of fossil bones seems to indicate, that dry lands and fresh waters must have existed before the formation of the chalk strata. Yet neither at that early epoch, nor during the formation of the chalk strata, nor even for a long period afterwards, do we find any fossil remains of mammiferous land-quadrupeds.

We begin to find the bones of mammiferous sea-animals, namely, of the lamantin and of seals, in the coarse shell limestone which immediately covers the chalk strata in the neighbourhood of Paris. But no bones of mammiferous land-quadrupeds are to be found in that formation; and notwithstanding the most careful investigations, I have never been able to discover the slightest traces of this class, except in the formations which lie over the coarse limestone strata; but immediately on reaching these more recent formations, the bones of land-quadrupeds are discovered in great abundance.

As it is reasonable to believe that shells and fish did not exist at the period of the formation of the primitive rocks, we are also led to conclude that the oviparous quadrupeds began to exist along with the fishes, and at the commencement of the period which produced the secondary formations; while the land-quadrupeds did not appear upon the earth till long afterwards, and until the coarse shell limestone had been already deposited, which contains the greater part of our genera of shells, although of quite different species from those that are now found in a natural state.

CUVIER AND BRONGNIART

Alexandre Brongniart (1770–1847), French mineralogist and zoologist, was associated with Cuvier in the epoch-making studies which established new standards and new methods in stratigraphic geology.

STRATIGRAPHY OF THE PARIS BASIN

Translated from *Essai sur la géographie minéralogique des environs de Paris*, Paris, 1811.

It appears that the materials which compose the Paris basin, as we have delimited it, were deposited in a vast, down-warped place—a kind of gulf with shores of chalk. Possibly this gulf was completely enclosed, a sort of great lake, but we cannot be certain of this, inasmuch as its southwest borders are covered by the extensive sandy bed of which we spoke earlier   Moreover, this great bed is not the only one which has covered the chalk.   There are many in Champagne and Picardy which, although smaller, are of the same nature and could have been formed at the same time.   They are placed, like it, immediately above the chalk in the places where the latter was too high to be covered by the beds of the Paris basin.

First we shall describe the chalk, the oldest rock in our region. We shall terminate with the sandy bed, the youngest of our geological formations.   Between these two extremes we shall treat of less extensive but more varied materials which filled the great depression in the chalk before the bed of sand was deposited over all.

These materials can be divided into two stages.   The first, which covers the chalk everywhere that it was not too high and which has filled all the bottom of the gulf, is itself divided into two parts which are at the same level and are placed, not one on top of the other, but end to end.   These are:

The bed of silicious limestone without shells;

The bed of coarse, shell-bearing limestone.

We know the limits of the chalk coast during this stage rather well because they were not covered then, but those boundaries are masked in many places by the second stage and by the great sandy bed that forms the third and that covers a great part of the two others.

The second stage is formed of gypsum and marl. It is not widely spread but only occurs in places and, as it were, in patches. Also these deposits differ greatly from each other in their thickness and in the details of their composition.

These two intermediate stages, as well as the two extreme stages, are covered and all the spaces they have left are partially filled by another type of terrain, also mixed with marl and silica, which we call the fresh-water terrain because it swarms with fresh-water shells.

Such are the great masses of which our district is composed and which form its different stages. But, by subdividing each stage, still greater precision can be attained. One obtains it from closer mineralogical determinations which give us as many as eleven distinct types of beds, which we shall first enumerate and then describe as to their distinctive characteristics.

Enumeration of the various sorts of terrain or of formations* which constitute the floor of the Paris region.

1. Formation of the chalk.
2. ——— Of the plastic clay.
3. ——— Of the coarse limestone and its marine sandstone.
4. ——— Of the silicious limestone.
5. ——— Of the gypsum with bones and of the first fresh-water terrain.
6. ——— Of the marine marls.
7. ——— Of the sandstone without shells and of the sand.
8. ——— Of the upper marine sandstone.

---

* To name the various sorts of terrain, we shall make frequent use of the word *formation* adopted by the school of Freiberg to designate an assembly of beds of the same or of different nature but formed at the same time.

The greater part of these formations have been unknown so far to the geologists of the famous school of Freiberg. At least we have been able to recognize almost none of them in the works they have published which we have had occasion to consult. Meanwhile, as it is possible that these various formations exist elsewhere than in the environs of Paris, it seemed advisable to give them definite names, so that geologists could designate them clearly if they recognize them elsewhere.

9 ——— Of the burrstones without shells and of the argillaceous sand.

10. ——— Of the second fresh-water terrain, including the
marls and the burrstones with fresh-water shells.

11. ——— Of the alluvial clay, both ancient and modern,
including the rounded pebbles, the pudding-stones,
the black argillaceous marls, and the peats.

To avoid repetition we shall not follow the order of the above
table exactly, in the descriptions that we are about to make of
these last formations. We shall sometimes unite under the same
heading both the terrains that are absolutely alike in their mineralogic nature and those that follow them and are, so to speak,
dependent on the others although differing in their mode of
formation and in their mineral composition.

Concerning the Chalk

By many geologists, the chalk has been regarded as a very
recent formation, of little distinction and of little importance.
As a result of this false idea, it has been poorly described. We are
going to try to rectify and complete the description of its characteristics in accordance with the observations that we have made
on the abundant chalk in the Paris basin and on that which we
have seen in England and in various parts of France. . . .

Its characteristics in general are (1) that it presents masses in
which the bedding is often very indistinct. These layers are
horizontal but are not readily separable horizontally like those
of the coarse limestone. (2) These masses nearly always include
discontinuous beds; or irregularly shaped flint of which the surfaces
adhering to the chalk blend, so to speak, these two substances
into each other; or nodules harder than the rest of the mass that
have the form of flints and are scattered like them. . . .

But this formation is essentially characterized by the fossils it
encloses; fossils entirely different, not only in species but often in
genera, from all those enclosed by the coarse limestone. . . .

Not one of these species is found in the coarse limestone. The
genus Belemnite is the characteristic fossil of the chalk. This
formation is then perfectly distinct from the marine limestone

which covers it. It does not seem that there was a gradual transition between them, at least in the area that we have studied.*

Concerning the Plastic Clay

Nearly all the surface of the chalk mass is covered by a bed of plastic clay, which has certain very remarkable characteristics throughout, although it presents noticeable differences at various points. . . .

If we compare the description we have just given of the chalk bed with that of the plastic-clay bed, we notice (1) that none of the fossils met in chalk are found in the clay; (2) that nowhere is there an imperceptible passage from the chalk to the clay, as those parts of the clay bed nearest the chalk do not include any more lime than the other parts.

It seems to us that one can conclude from these observations; first, that the liquid which deposited the bed of plastic clay differed greatly from that which deposited the chalk, as it did not contain appreciable amounts of calcium carbonate and as none of the animals that inhabited the waters which deposited the chalk lived in it.

Secondly, that there must have been a clean-cut separation, and perhaps even a long lapse of time, between the deposition of the chalk and that of the clay, since there is no transition between these two types of terrain. The type of breccia, with fragments of chalk and paste of clay, that we have noticed at Meudon, seems even to prove that the chalk was already solid when the clay was deposited. Earth was insinuated between the fragments of chalk produced at the surface of the chalky terrain by the motion of the waters or by some entirely different cause.

The Coarse Limestone and Its Shell-bearing Marine Sandstone

The coarse limestone does not always lie immediately above the clay; it is often separated from it by a bed of sand of variable thickness. We cannot say whether this sand belongs to the limestone formation or to that of the clay. The fact that we have not found shells in the localities where we have observed this sand (it is true

---

* All these characteristics, which are likewise found in the limestone of the mountain of Maestricht, make us believe that that terrain belongs to the chalk formation. M. Defrance has recognized absolutely the same species of belemnite as in the chalk of Meudon.

these places were not numerous) would connect the sand with the clay formation. But as the lowest limestone bed usually has some sand and is always filled with shells, we do not know yet whether this sand is different from the first or whether it is the same deposit. The fact that the sand associated with the clay is generally rather pure, although colored red or bluish gray, makes us suspect that it is different. It is refractory and often very coarse grained.

The limestone formation, aside from this sand, is composed of alternating beds of more or less pure, coarse limestone, of argillaceous marl, even of clay foliated into very thin beds, and of calcareous marl. But it must not be understood from this that these various beds are placed there at random and without system. They always follow the same order of superposition in the considerable stretch of terrain that we have traversed. Sometimes there are many beds that are lacking or are very thin; but what is in the lower part in one district never is in the upper in another.

This constancy in the order of superposition of the thinnest beds throughout a distance of at least twelve *myriametres* (about seventy-five miles) is, we believe, one of the most remarkable facts which we have definitely established in the course of our researches. For the arts and for geology, consequences as interesting as they are certain ought to result from this.

The means we have employed to recognize a bed observed before in a very distant locality, among so great a number of limestone beds, is the determination of the fossils enclosed in each bed. These fossils are nearly always the same in corresponding beds and present rather notable difference in species from one system of beds to another. It is a badge for recognition that, to the present, has not misled us.

One should not believe, however, that the difference from one bed to another is as clear cut as that from the chalk to the clay. If it were so, one would have that many individual formations. But the characteristic fossils of a bed become less numerous in the bed above and disappear absolutely in the others, or are replaced gradually by new fossils that have not appeared before.

Following this course, we are going to indicate the principal system of beds that can be observed in the coarse limestone. In the following chapters will be found the complete description, bed by bed, of the numerous layers that we have examined and the list of the fossil species that we have recognized there. It is from these observations that we have drawn the results that we present here in a general way.

The Gypsum, the First Fresh-water Formation, and the Marine Marls

The terrain of which we are about to trace the history is one of the clearest examples of the proper conception of *formation*. In it may be seen beds differing greatly from each other in their chemical composition but evidently formed at the same time.

The terrain that we call gypseous is composed not only of gypsum; it consists of alternating beds of gypsum, argillaceous marl, and limestone. The beds have followed an order of superposition that has always been the same in the great gypseous band that we have studied from Meaux to Triel and Grisy. Some beds are lacking in certain districts; but those that remain are always in the same respective position.

The gypsum is placed immediately above the marine limestone and it is not possible to doubt this superposition. . . .

The gypsum hills or buttes have a distinctive aspect that makes them recognizable from afar, as they are always placed on the limestone. They appear as second, very distinct, elongated or conical hills on top of the highest hills.

The fossils enclosed by this mass and those in the marl above it present features of quite different nature.

It is in this first mass that the discovery of skeletons and scattered bones of unknown birds and quadrupeds is an everyday occurrence. To the north of Paris, they are in the gypseous mass itself. There they have preserved their solidity and are surrounded only by a very thin bed of calcareous marl. But in the layers to the south, they are often in the marl that divides the gypseous bands and are very friable. We shall not speak of the manner in which they are situated in the mass, of their state of preservation, of their species, etc.; these data have been sufficiently elucidated in earlier memoirs. Turtle bones and fish skeletons have also been found in this mass.

But what is truly more remarkable and much more important in its resulting consequences is that fresh-water shells are occasionally found there. For that matter, one alone would suffice to demonstrate the truth of Lamanon's opinion and that of other naturalists who think that the gypsum of Montmartre and other hills of the Paris basin was crystallized in fresh-water lakes. . . .

It is in the same system of beds that shells of the genera Limnæus and Planorbus, which hardly differ from the species living in our swamps, are found in nearly all the layers of the butte of Chaumont and even those of eastern Montmartre. The fossils prove that these marls are of fresh-water origin as are the gypsum beds below.

The gypsum beds, the bands of marl which separate them, and those that cover them as high as, and including, the white marl constitute the first or the oldest fresh-water formation in the Paris basin. It is seen that the fresh-water shells which characterize this formation are found principally in the white calcareous marl. Moreover, neither burrstone nor other silicious rock is known in this first fresh-water formation, except for the menilites and horny flints of the upper beds of the main mass of gypsum.

Numerous and often massive bands of argillaceous or calcareous marls appear again above the white marls. No fossil has been discovered in them as yet. We therefore cannot say to which formation they belong.

Succeeding these one finds a band of foliate yellowish marl that encloses kidney-shaped masses of earthy strontium sulphate in the lower part and, a little above, a thin bed of small bivalve shells that lie flat and crowded together. We ascribe these shells to the genus Cytherea. This bed, which seems to have very little importance, is remarkable, first, for its great extent. We have observed it over an area more than ten leagues long and more than four wide, always in the same place and of the same thickness. It is so thin that it is necessary to know exactly where one ought to look in order to find it. In the second place, it serves as a limit to the fresh-water formation and indicates the beginning of a new marine formation.

SMITH

William Smith (1769–1839), an English civil engineer, is credited with the first adequate concept of the relations of strata to each other and of the significance of fossils in stratigraphic geology. Smith's ideas and maps, rather than his scanty writings, justify his preeminent place in the history of the science.

THE STRATA OF ENGLAND

From *Memoir to Map of Strata of England*, London, 1815.

After twenty-four years of intense application to such an abstruse subject as the discovery and delineation of the British Strata, the reader may easily conceive the great satisfaction I feel in bringing it to its present state of perfection. The chances were thought much against my ever completing it on a map, of the greater part of our island, large enough to show the general course and width of each stratum of the soil and minerals, with a section of their proportions, dip, and direction, in the colours most proper to make them striking and just representations of nature; and which is the first general mineralogical survey of the island.

The map also contains the relative altitude of the hills, which seem proportioned to the nature of the rocks of which they are formed; the highways, the streams, rivers, canals, and railways, upon a larger scale, and more correctly, than any map before published; also the situation of collieries and mines.

The wealth of a country primarily consists in the industry of its inhabitants, and in its vegetable and mineral productions; the application of the latter of which to the purposes of manufacture, within memory, has principally enabled our happy island to attain her present pre-eminence among the nations of the earth. Whatever, therefore, tends to facilitate the discoveries and improvements of the one or the other, may with just propriety be considered a national concern, and cannot more properly be laid before the public than at a period when the wisdom and vigour of His Majesty's councils have given peace to a distracted world, and may render this happy event the commencement of a new era in the history of natural science.

The immense sums of money imprudently expended in searching for coal and other minerals, out of the regular course of the strata which constantly attend such productions; and in forming canals, where no bulky materials were afterwards found to be carried upon them; prove the necessity of better general information on this extensive subject. And I presume to think, that the accurate surveys and examinations of the strata, as well near the surface of the earth as in its interior, to the greatest depths to which art has hitherto penetrated, by the sinking of wells, mines, and other excavations, to which I have devoted the whole period of my life, have enabled me to prove that there is a great degree of regularity in the position and thickness of all these strata; and although considerable dislocations are found in collieries and mines, and some vacancies in the superficial courses of them, yet that the general order is preserved; and that each stratum is also possessed of properties peculiar to itself, has the same exterior characters and chemical qualities, and the same extraneous or organized fossils throughout its course. I have, with immense labour and expense, collected specimens of each stratum, and of the peculiar extraneous fossils, organic remains and vegetable impressions, and compared them with others from very distant parts of the island, with reference to the exact habitation of each, and have arranged them in the same order as they lay in the earth; which arrangement must readily convince every scientific or discerning person, that the earth is formed as well as governed, like the other works of its great Creator, according to regular and immutable laws, which are discoverable by human industry and observation, and which form a legitimate and most important object of science. The discoveries and improvements, both in mining and agriculture, which are now confined to a few parts of the kingdom, may be fully extended to many more, and in some degree to all, by a better knowledge of geology; and a faithful general view of the soil and substrata of our island (in which no beds are omitted that can well be described in such a map) will be found a work of great convenience, in considering the various applications which are made to the legislature for canals, roads, and railways alone. I am prepared also to give more minute and detailed delineations on a new impression of county or other maps, of the largest scale, and to illustrate, by lectures and by specimens, the particular sites of the numerous animal remains and vegetable impressions found in each stratum, with an accurate detail of every characteristic mark which has led to these discoveries, and to publish a complete illustration of my geological system.

By a knowledge of the alluvial deposits in low marshy grounds around the coast, which I have had frequent opportunities of investigating, and by more correct information concerning the shoals and sandbanks adjacent, great benefits may accrue both to the landed and commercial interests of the country in the draining of such low lands, and in the improvement of sandy and bar harbours. The higher lands of the interior contain a wonderful admixture of soils, stones, shells, marls, minerals, and fossils, very regularly deposited in strata, which rise successively to the surface of the earth. The purposes to which many of them may be applied, are doubtless still unknown. It has been the chief object of my research to simplify and extend this kind of knowledge, whence practical applications the most important may result; proofs of which might be given in the many works which I have executed on these principles, in different parts of the island; and by their more general diffusion, various works of art, and agricultural experiments in particular, will be generally conducted with more skill and certainty of effect. By a classification of soils, according to the substrata, good practical farmers may choose such as are best suited to their accustomed mode of management, and they may thus be tempted to transfer useful and well-established practices in husbandry to many parts of the same stratum, which are still highly susceptible to improvement; and beneficial results will be recorded with more regularity for the advantage of others, desirous of trying experiments upon the same strata.

On these principles also, the most proper soil will be known for plantations of timber: miners and colliers, in searching for metals and coal; builders, for freestone, limestone, and brick-earth; the inhabitants of dry countries, for water; the farmer, for fossil manures; will all be directed to proper situations, in search of the various articles they require; and will be prevented from expensive trials, where there can be no prospect of success.

Finding good materials for roads, in the nearest places, will reduce that heavy public charge.

Tracing the courses of springs beneath the surface, will show the best methods of draining and improving land; and the collecting of water from those natural subterraneous reservoirs, the caverns of hills and joints of rocks, for the supply of canals.

Much of the art of constructing those public works, their value and utility, and the products of collieries and mines, depends upon this science, as also the perfection and extent of potteries, glass, salt, alum, vitriol, and saltpetre works; the procurement of fuller's earth, founders' and glassmakers' sand, of materials for chemists and colourmen, and the various substances used in grinding and polishing metals and marble: in fact, there are few arts or employments which may not derive some useful hint or improvement from a better knowledge of the products of our soil and substrata.

It will appear as unnecessary, as it would be difficult to enumerate all the advantages, when it is considered what numerous coincidences and indisputable facts have occurred (in the course of so many years constant observation and experiment on strata, in different parts of the country) to found this extensive investigation, which must lead to accurate ideas of all the surface of the earth, if not to a complete knowledge of its internal structure, and the progress and periods of formation; for nothing can be more strongly and distinctly marked than the line which separates the animal from the vegetable fossils, and the courses of numerous strata, which are designated by these and other characters, the most intelligible and useful.

MACCULLOCH

John Macculloch (1773–1835), a Scottish physician and geologist, was the first geologist to be appointed to an official survey, the Trigonometrical Survey of Great Britain. His recognition of the true nature of the "Basement Complex" exposed in the Western Islands of Scotland is one of the important milestones in the progress of geological science.

METAMORPHIC AND IGNEOUS ROCKS OF THE WESTERN ISLANDS OF SCOTLAND

From *A Description of the Western Islands of Scotland*, Vol. I, pp. 48–49, 218–221, 394, London, 1819.

The Granite and Gneiss of Tirey

The gneiss of Tirey is more remarkable for containing masses of limestone. One of these has long been known by the flesh-coloured marble which it affords; of which a quantity has been exported for the purpose of ornamental architecture since the time when it was first pointed out by Raspe. It is improperly called a bed, as it is only an irregular rock, lying among the gneiss without stratification or continuity. In this respect it resembles the greater number of the primary limestones found in gneiss and in mica slate, and may be considered a large nodule. There is considerable obscurity attending these detached masses of limestone. Of all the rocks which occur in extended masses, granite, trap, and porphyry, only, are unstratified; while the others possess characters of stratification almost always very unequivocal, although in many cases attended with marks of posterior derangement. It is possible that the masses of limestone thus found in gneiss have once been stratified, and that they have suffered some posterior changes by which the appearances of this disposition have been obliterated. In illustration of this opinion I may point out the state of the white marble of Sky, hereafter described; which, though at present as shapeless as the limestones in question, has been once undoubtedly stratified, since it forms portions of a series of parallel strata containing organic remains. . . .

All the varieties of gneiss are occasionally intersected by granite veins, and they are indeed almost characteristic of this rock; being rarely absent for any considerable space, and seldom traversing micaceous schist unless under circumstances where they can be traced to some neighbouring mass of granite.   They are however most abundant in the granitic division. . . . In some varieties of gneiss they are so abundant as nearly to exclude the original rock, so that the mass presents little else than a congeries of veins. An instance of this nature occurs in the Flannan isles, but the most striking are to be seen on the north west coast between Loch Laxford and Cape Wrath. . . . The hornblende schist and the gneiss are broken into pieces and entangled among the veins in the same manner as the stratified rocks are in the trap of Sky;* but with infinitely greater intricacy, so as rather to resemble a red and white veined marble with imbedded fragments of black. . . .

Graphic granite is much more frequent in the veins that traverse gneiss than in the others, although not absolutely limited to these. The felspar is generally the predominant substance in this class of veins and often presents a common polarity throughout the whole mass, as already noticed in Coll.   The mutual disposition of the felspar and quartz is various.   Occasionally it is partially laminar, as in Rona; more commonly the quartz is in prismatic forms, triangular and hexagonal, or occasionally, even hollow and filled with felspar.   In a very few instances the summits of the quartz crystals are perfect, and protrude into a vacant space.   From considering the relative forms of the quartz and felspar it will sometimes appear that the one and sometimes that the other has first crystallized, and thus determined the shape of its associate.

The Trap Veins of Sky

. . . I have here, as on other occasions, applied to these veins the general term trap, for the reasons assigned in speaking of the rocks of this class, namely, because they vary in composition; although basalt is perhaps the prevalent substance in them.   The order allotted for them in this description is also that which they

* It is not easy to admit the arguments derived from these appearances in favour of the igneous origin of trap and refuse them in the instance of granite. There is in truth no difference in the cases but that which arises from the difference of the materials engaged.

hold in nature, since they traverse every rock that lies in their way, from the most ancient to the most recent; seldom suffering any change either of direction or composition in this varying course. As the same vein is therefore found to pass indiscriminately through rocks of all ages, it is plain that its association with these can afford no register of the period of its formation. . . . It is only where they interfere with each other that a register more extensive can be found. I have always sought for such examples wherever these veins abound, and, among other places, in Sky, but I have never yet traced more than two distinct sets. . . . We have no means of knowing what distance of time has intervened between these veins. The angle of the courses of both kinds with the horizon is various, but in a very considerable proportion it is vertical or nearly so.

VON BUCH

Leopold von Buch (1774–1852), Pomeranian geologist, whose *Ehrebungs* (upheaval) hypothesis for the elevation of volcanoes persisted many years. He edited the first geological map of Germany.

THE IGNEOUS ORIGIN OF BASALT

Translated from *Mineralogische Briefe aus Auvergne*, 1802, in *Gesammelte Schriften*, Vol. I, pp. 486–487, 494, 1867.

The lava, in its characteristics, still resembles the scoria on the edge or in the interior of craters. In the dense, blackish-gray groundmass we see remnants of glassy feldspar and very small hornblende crystals. We have the composition of domite (trachyte) on the slope of mountains, but always less recognizable and in a blacker groundmass. However, in the upper part the lava is porous like all lava streams, and no crystals are developed as fossils there. Instead, there is so great a number of flakes of specular hematite that they fill the inner surface of the cavities as definite druses, and because of them, the whole mass of lava gleams metallically in the sunlight. The lava is relatively darker the more of this hematite it contains (lighter if this is lacking), so that specimens from these prove to be in no way contradictory. The black color of this lava is, after all, only a result of the iron that is mixed with it.

So the phenomena of these mountains lead us to the following, unexpected, conclusion: the lavas of Volvic are domite that flowed. The transition from gray-white domite to black lava is unbroken in the flows, so much so that we would never look on the end members of the series as having flowed if we did not find them in the middle of the lava streams. The specular hematite penetrates the domite as well as the granite of the Puy de Chopine; accumulation of this pushes away feldspar and hornblende; and, finally, the mass it has colored, flows. Domite is formed from the granite. Therefore the granite is the primary mass out of which the lava of Volvic arose. Through a series of modifying operations, the granite is changed to lava! And the seat of this volcano is therefore in the granite, itself.

The mass of lava of the Puy de l'Enfer beyond Aydat is true basalt. There is a little augite and olivine, and scattered and indistinct tablets of gray feldspar.

Might we not expect to see the flows of Come and of Louchadière, and those which flow down to Nechers on the sides of Mondor de Murol, maintain an individual character? But does not this variation of streams mean a difference in the causes which brought them about? Conversely, might we not conjecture an identity of the lava or close geological relationship? In this case, would they not prove, thereby, their contemporaneity and their common origin from one volcano?

The Raised Coasts of Sweden

Translated from *Reise durch Norwegen und Lappland*, Vol. II, 1810, in *Gesammelte Schriften*, Vol. II, pp. 503–504, 1867.

A mile farther along I came to Innerviken on a small bay. A few years before, one crossed it in boats, but now it is so dried up that the road goes across it. The people of the region, who note the withdrawal daily, believe they will live to see the bottom of the bay changed to fields and meadows. There is hardly a spot here that does not confirm this withdrawal. To suggest doubts about it to the people anywhere around the gulf here means to make oneself truly ridiculous in their eyes. It is an extremely strange, remarkable, and astonishing phenomenon. How many questions it suggests! What a field of research for Swedish physicists! Is the decrease the same in like periods of time? Is it of equal amount in all places, or perhaps greater and faster in the interior of the Gulf of Bothnyia?

In front of Gefle and near Calmar, marks were hewn exactly at sea level by Celsius sixty years ago so that the recession can be determined with the greatest accuracy. The well-known engineers, Robsahm and Hällstrom, examined these marks at Gefle as well as at Calmar and found confirmation of the recent recession. But their calculations have not been made known, and remain in the hands of Baron Hermmelin. May the physicists not withhold them much longer! Linnaeus mentions in his travels in Skåne, that he also made an exact mark, a quarter mile from Trälleborg, on a block which would not be moved, and describes the immediate surroundings with the accuracy of a botanist. Would not the investigation of this place and what has taken place there be worth a short journey from Lund or Copenhagen?   It is certain that sea level cannot sink: the equilibrium of the sea absolutely does not permit it.   But since the phenomenon of the recession may not be doubted now, our present understanding shows us at least that there is no other solution than the conviction that all Sweden is being slowly raised up, from Frederickshald as far as Åbo and perhaps even to St. Petersburg.   Near Bergen on the coast of Norway in Söndmör and Nördmör, as Magistrate Wibe of Bergen has assured me, something of this recession has also been perceived with the aid of excellent marine charts of the west coast of Norway. Cliffs which were formerly covered with water now project above it.   However, the belief in the recession of the sea is evidently not so widespread on the west coast; neither as universal nor as certain as in the Gulf of Bothnyia.   Also the variable and excessive tides of the North Sea hinder precise observation.   It is also possible that Sweden has risen more than Norway; the northern part more than the southern.

THE UPHEAVAL OF VOLCANOES

Translated from *Physikalische Beschreibung der canarischen Inseln*, 1825, in *Gesammelte Schriften*, Vol. III, pp. 510–513, 1867.

Therefore the whole Canary Island group cannot be considered as other than a collection of islands which were raised from the bottom of the sea, one after the other.   The force which was able to accomplish so great a result must have been stored up for a long time in the interior and increased in strength before it could over- come the resistance of the downpressing mass.   From there it split and pushed the basaltic and conglomeratic beds of the bottom of the sea, as well as those below them, above the surface.   It then escaped through the great upheaval crater.   But so great a raised mass will fall back again and close the opening as soon as the force which lifted it is released.   There is no volcano formed.   The Peak [of Teneriffe] was raised, however, as a higher dome of trachyte in the middle of such an upheaval crater.   Thus the permanent connection of the interior with the outer air is disclosed.   Vapors break out continuously and, if there is an obstacle to their release, they can, at the foot of volcanoes or at some distance, push out as individual lava streams and need not raise whole islands to over- come it.   The volcano remains the focal point of these manifesta- tions.   It is plugged by chilling and slumping of the melted mass only at the top, never at depth. Hence there is only one volcano on the Canary Islands, the Peak of Teyde. It is a central volcano.

All the volcanoes of the earth's surface fall into two classes, essentially distinct from each other: central and line volcanoes. In the former, active eruptions always form a mid-point of a great number of eruptions nearly equal in intensity in all directions. The latter, the line volcanoes, lie in a series along a line, often but a slight distance apart like chimneys on a great crack—which they well might be. At such places one sometimes counts twenty, thirty, or even more volcanoes. Thus they stretch out over an important part of the earth's surface. As regards their location, volcanoes are, again, of a twofold kind. Either they lift themselves up as separate cone islands from the bottom of the sea, in which case a primitive mountain, whose foot they seem to indicate, ordinarily trends in the same direction to one side of them; or these volcanoes stand on the highest ridge of this mountain range and form the very peak. In their composition and their derived products, the two types of volcano are not different from each other. They are nearly always, with but few exceptions, trachyte mountains; and their solid products can be derived from such trachyte.

If the mountain ranges themselves are considered as masses which have ascended in great fissures through the activity of black (augite-) porphyries, the location of these volcanoes can be understood to a certain extent. Either that which is active in the volcanoes finds much more facility to force its way to the surface in this major fracture, in which case the volcanoes push up to the surface of the mountain itself; or the primitive mountain masses over the fissure are still too great an obstacle to them; in which case they would break out on the edge of the fissure, as the black porphyry in fact ordinarily does, at the foot of the mountains.

But if the material below the surface, which wishes to break out, finds no such fissure to furnish a route for its active force, or if the obstacle on the fissure is entirely too large; then the force will grow below the surface until it has the power to overcome the obstacle, and even to shatter the mountain masses which overlie it. It will even make a new fissure for itself and hold open a constant connection with this if it is strong enough. Then central volcanoes are engendered. However, these would seldom push out before they have cleared a way in the form of upheaval islands with upheaval craters.

These last structures seem to need no extraordinary conjunction of especially favorable conditions nor any unusual state of the earth's crust such as that involved in a mountain chain. Therefore they can continue forever; and this also seems to be a fact. Islands are raised up out of the sea before our eyes. If one follows the continual new discoveries of voyagers in the South Seas or studies carefully the sagacious and informative description of the South Sea islands by Herr von Chamisso, one cannot fail to believe that an imposing number of new islands is continually coming into existence, either just below the surface of the ocean or above it. Indeed, the history of the flora would demonstrate this.

Various systems can be detected on the surface of the earth in accordance with these various types of volcanoes. The accurate notation and explanation of these systems is all the more important to physical geography because the whole form, possibly the structure of the continent, appears to bear a definite relation to them.

THE IMPORTANCE OF FOSSILS

Translated from "Über die Ammoniten in den älteren Gebirgs-Schichten," *Abhandlungen der königlichen Akademie der Wissenschaften*, 1830, pp. 135–138, 1832.

In its present state, geognosy can hardly proceed farther without the accurate determination of the organic forms found in the rocks. It will reach its results all the more steadily and truly, the more carefully the nature of the fossils is determined. Many formations can be distinguished only by this. The bare examination of the beds would accomplish it only with the greatest difficulty. Geognosy urgently needs the instruction of zoology.

This instruction is necessary, at least for all formations that are customarily grouped under the name Tertiary—all those that lie over the Cretaceous. It was fortunate that Lamarck, when he prepared his brilliant review of the *Conchylidae*, found, in his neighborhood near Grignon and Cortagnon, deposits of mussel shells which were admirably preserved yet not similar to those still living in the sea. It brought to his attention the fact that these forms also must be studied, because the only hope of holding the threads which tie all organic creatures together is the study of all forms that ever appeared and not simply of those still living that are brought to us in slight numbers through some lucky dash of the waves or fortunate cast of a trawl. Since then, the study of fossils has ceased to be regarded as a part of mineralogy. Since then, one actually finds some fossils in zoological museums, but as yet as mere individual examples in the above-mentioned mineralogical collections, very rarely as essential constituents of the museum itself.

The zoological information on the fossils of the older formations is also somewhat overlooked. One of the most remarkable, and in many respects most important, classes of lost creatures, that of the ammonites, has hardly been investigated at all. This lack is all the more painful to geognosts as there are almost none of the older formations which do not have their definitely characteristic ammonite. There is cause to wonder over this neglect as every one knows how the strange structure of the ammonites has held the attention of the natural philosophers, and with what diligence they have been collected in nearly every country of Europe since the time of Conrad Gesner. . . .

Cuvier, in about 1802, was the first who dared to connect the inhabitants of the ammonite shells with other known animal forms. He was the first to emphasize that they must be squidlike animals—cephalopods. And this opinion was brilliantly confirmed shortly thereafter by the famous spirula that Peron brought back from his voyage around the world. Since then ammonites have been regarded, without exception, as one of the outer branches of a series which begins or ends with the shell-less Octopus or Loligo. That was truly a great advance. One is now in a position to postulate what was necessary for the life of an ammonite, how the animal grew, how he built his shell.

We are indebted to M. Élie de Beaumont for the observation that even in the Muschelkalk, no ammonites with toothed lobes occur. Herr Bronn has gone further and observed that in still older rocks, only ammonites with angular lobes are found—especially in the so-called Transition limestone. From this, a type of ammonite gradually changing in successive formations became plausible. Now it became important to learn what the forms might be in the beds separating the Muschelkalk from the Shale limestone. That applies especially to the generally widespread Coal Series.

CORDIER

Pierre Louis Antoine Cordier (1777–1862), French mining engineer and professor of geology at the Jardin des plantes.

THE CRYSTALLINE NATURE OF VOLCANIC ROCKS

Translated from *Journal de physique*, Vol. LXXXIII, pp. 135–163, 285–307, 352–386, 1816.

The scope of my work embraces all of the substances, whether vitreous or scorified, whether compact, earthy, friable, or pulverulent *lithoids*, which act as paste or groundmass in the volcanic rocks of all ages. In the following, speaking frequently in a collective fashion, I shall designate such substances indifferently under the generic denominations of *indeterminate bases, indeterminate pastes*. In any event, unless I warn to the contrary, I shall constantly consider them as that which they essentially are— abstractions made from the crystals or amorphous crystalline grains which they ordinarily enclose, without reference to their usual porosity or to the odd fragments that nearly always appear in some of them.

My first attempts to force the different substances under consideration to disclose their intimate texture had little success. I drew no other fruit from them than encouraging indications and the discovery of titaniferous iron. After having uselessly tried different methods of calcination and different chemical agents, I put the substance under the microscope and tested the particles in the flame of the blowpipe by Saussure's method; from then on, satisfying results appeared.

Some of the substances which I tried were solid, others soft, friable, or pulverulent. To observe the first, I found it sufficed to detach very thin splinters and expose them on a glass disk used as a slide, either whole or slightly broken, or reduced by slight pressure into a rather fine powder. As regards the second, I could not obtain exact results, save by thinning them out in water and classifying the particles according to size, by washing, decanting, and drying. This process, which was suggested to me by the operations of large-scale washing practiced in mines treating poor ores, and which is very time-consuming, can also be usefully applied to solid volcanic materials after they have been properly pulverized.

I have found most useful, and I cannot overpraise, the method of Saussure for determining the conditions of fusion of minerals with the aid of the blowpipe. . . . His method consists essentially of fixing the minute fragments of the minerals to be tested to the end of a very thin wire of cyanite (or sappare) moistened with gummy water, warming briskly to solder them to the support, and melting them without the addition of any foreign substance. An ordinary blowpipe and the flame of a large candle are used. . . . The phenomena of the fusion are examined under the microscope. The diameter of the greatest globule that one can get on melting . . . serves with the aid of rather delicate refinements to determine approximately the ratio of fusibility expressed in degrees of the Wedgwood pyrometer. . . .

Examination of the Lithoid Pastes Which Enter into the Composition of Lava Flows of All Ages

The general term of lithoid paste which I use here embraces the varieties of every type hitherto designated under the names of uniform basaltic lava, basalt, the base of *graustein*, the base of leucitic lavas, the base of petrosiliceous lavas, volcanic *bornstein*, *klingstein*, phonolite, sonorous compact feldspar, domite, and the base of porphyritic feldspathic lavas.

All the lithoid pastes whatsoever, without distinction of epoch, are found to be composed of perfectly discernible, heterogeneous parts, very distinct from each other and appearing under the form of grains with crystallized structure, variously colored and interlaced as in ordinary granite. . . .

The mineralogical notion, previously conventional for basalt, will change to the following: compact pyroxene, mixed with many microscopic particles of feldspar and titaniferous iron, with which are sometimes associated particles of olivine, white garnet, and iron oxide. . . .

I shall not enter into this subject [of earlier ideas], but I shall reproduce the results of my observations under another form; I shall say that it is shown that the materials of the interior lava flows (those of obsidian excepted) crystallize entirely by cooling, and change into an infinity of very small crystals or grains, belonging to well-determined species of minerals, solidly interlaced, leaving between them occasional thin vacuoles.

The Temperature Gradient of the Earth's Interior

From Essai sur la température de l'intérieur de la terre, *Mémoires de l'Académie des sciences pour l'année* 1827, translated from the French by the Junior Class (1827) in Amherst College, Amherst, 1828.

The experiments on subterranean temperature, which have been heretofore published, are of two kinds.

The first has had for its object, the investigation of the temperature of ordinary fountains; that of rivers, . . . and that of waters flowing either from caverns, or galleries, which serve to drain the water from large mines.

The design of the other has been to determine the temperature of those natural or artificial cavities, by means of which, we are able to penetrate into the bosom of the earth. . . .

Nearly two thirds of these observations upon the temperature have been made upon the air contained in subterranean cavities, and most of the remainder upon the water which presents itself in various ways in these cavities. . . .

By means of the precautions to which I resorted, I hope that my own particular experiments will be regarded as sufficiently exact. The most of them have been made in the coal mines in France, situated at considerable distances from one another, and which I have chosen as presenting the most favourable circumstances.

It will be sufficient at this time to add, that my experiments were made, in the first instance, in August, 1823; in the second, in September, 1825; and in the third, in November 1822, and September 1825.

I made use of mercurial thermometers which I had carefully verified and compared with one another; and which, in all cases, where I do not state to the contrary, have been employed in the experiment with the bulb naked.

. . . The works in coal, a substance easily excavated, advance with rapidity, so that the fore part of them has not time to undergo a sensible change of its peculiar and original temperature. Further, one can in a few minutes, pierce the coal with deep holes, in which the thermometer, placed with the proper precautions, will unquestionably take the temperature of the rock.   Now this is the method I have adopted.

The thermometers which I employed were covered in such a manner as to retain, for a sufficient length of time, the temperature acquired in the earth.   For this purpose, each instrument was rolled up closely in a leaf of silk paper, forming seven entire folds.   This roll, accurately fitted around the bulb, was closed up by a thread a little below the other extremity, in order that the part of the tube, necessary to be got at for observing the scale might be drawn out at pleasure, without fear from the contact of air.   The whole was contained in a case of sheet tin.

———————

. . . I found, that at the works of Ravin, for a difference of level amounting to 558 feet, there was a difference of 7°,1* in the temperature; and at the excavation of Castillan, for 591 feet the difference was 11°,4: in other words, in the first of these mines the heat increased one degree for every 79 feet, and in the second, one degree for every 52 feet.

I confess that this so great difference between the two results, obtained at two places so little distant from each other, astonished me.   I do not doubt but that it was owing to a circumstance altogether local, depending upon the slight thickness of the coal formation and upon the unequal conducting power of the vertical beds of the primary rock beneath.   In fact, the works of Ravin are situated in the direction of an extensive vein of copper, which shows itself two miles from thence, on the side of Rosaire.

———————

Before closing the second part of my work, I shall present the results in the two following Tables [on pp. 218 and 219].

We see definitely from these tables that the depth, corresponding to the increase of one degree of subterranean heat, should be fixed (in round numbers) at 64 feet for Carmeaux, 35 for Littry, and 27 for Decise.

Such are the last elements which we have to exhibit.   It now only remains to make a recapitulation: and it seems to us, that we have a foundation for drawing the following conclusions from all which precedes.

———————

* The Amherst translators changed all the readings, originally given in centigrade, to Fahrenheit.   They are to be so read.—EDITORS.

1st.   Our experiments fully prove the existence of an internal heat, which is natural to the terrestrial globe, which depends not on the influence of the sun's rays, and which increases rapidly with the depth.

Table of the data furnished by the direct experiments made on the temperature of the earth at Carmeaux, Littry and Decise.			
Places of the experiments.	No. of Exps.	Depth of the stations.	Observed Temp
		feet.	
Carmeaux. Water of the Verias well. .	1	20.34	55°,2 a
Water of the well at Bigorre.	2	37,72	55.6 b
Rock at the bot of the mine of Ravin.	3	596.81	62,7 c
Rock at the bot. of the M. of Castillan.	4	629,95	67,1 c
Littry. Exterior surface of the mines.	1	0	51,8 d
Rock at the bot } Station a.	2	324.82	60,8
tom of the mine } Station b.	3	324,82	61,28
of St. Charles. } Mean for the 2 sta.	4	324,82	61,04
Decise. Water of the well at Pelisson.	1	28,87	52.5 e
Water of the well at Pavillons.	2	57.45	53,1 e
Rock at the bottom } Upper station.	3	351.07	63,1
of the mine Jacobe. } Lower station.	4	561,05	71,78

   a This well is but a short distance from the mine called Ravin.
   b This well is immediately above the station taken in the mine at Castillan.
   c The distance between these two mines is about one and a half miles.
   d A temperature which must be considered equal to the mean temperature of the place.
   e These wells are situated almost immediately above the stations taken in the mine.

Fig. 10.—Cordier's table of data concerning the temperature of the interior of the earth, 1828.

2nd.   The increase of the subterranean heat in proportion to the depth, does not follow the same law throughout the whole earth.   It may be twice, or even thrice as great, in one country as in another.

3rd.   These differences are not in a constant ratio to the latitude or longitude.

4th. Finally, the increase is certainly much more rapid than has heretofore been supposed; it may be as great as 27, or even 24 feet, for a degree, in some countries. Provisionally, however, the mean must not be put lower than 46 feet.

Table of results obtained by calculation from the preceding data.			
Places of the experiments,	*Nos. of the observations in the preceding table which are compared.*	*Depth cor. to increase of 1° of heat.*	*Remarks.*
Carmeaux.	No. 1 & 3. No. 1 & 4. No. 2 & 3. No. 2 & 4.	feet. 76.25 51,31 78.63 51,8	These results are set down from memory. The cause of the great difference between these two results being unknown, the mean should be taken, which is 65,2.
Littry.	No. 1 & 4.	35,14	
Decise.	No. 1 & 3. No. 2 & 3. No. 1 & 4. No. 2 & 4. No. 3 & 4.	28.29 27.34 27.63 27,19 27,	The coincidence of these results is remarkable.

Fig. 11.—Cordier's table of results of his calculations of the temperature gradient in the interior of the earth, 1828.

These important conclusions give us with certainty, in their various modifications, the principles, according to which the mathematical theory of the diffusion of heat in large bodies, may be applied to the terrestrial globe. They harmonize with the consequences drawn from very different phenomena of nature, which have so long indicated the intense heat of the earth's interior.

OMALIUS D'HALLOY

Jean Baptiste Julien d'Omalius d'Halloy (1783–1875), Belgian geologist, prepared one of the earliest geologic maps of France and adjoining regions. The article from which this excerpt is taken was written in 1813.

THE SYSTEMATIC CLASSIFICATION OF GEOLOGIC FORMATIONS

Translated from *Annales des mines*, Vol. VII, pp. 353–376, 1822.

Two main points of view seem to lead equally to the division of a country into physical regions determined by the nature of the ground: the one considers it geologically, that is to say, by the epoch of formation; the other looks upon it solely from its mineralogic or rather chemical nature. . . . It is readily realized, moreover, that the geological consideration is much more useful for the progress of science, . . . and that the ability of uniting many systems into one group allows doing without detailed observations which are necessary in the other case because of the frequent changes in the nature of the dominant substances in a terrain formed at one time.

The old division of "primitive and secondary terrains," that is, anterior and posterior to the existence of organic beings, no longer agrees with the intimate connection that has been noted between the primitive terrains and certain beds filled with the debris of living beings. The celebrated school of Freiberg has introduced an intermediate class in which to place these last beds. Since then, new observations have proven that these intermediate terrains, instead of being constantly posterior to all the rocks which have the general character assigned to primitive terrains, are found intercalated between crystalline rocks which have no fossils.

. . . Therefore one ought not to be astonished if I propose to reunite these terrains into one great class, that I shall designate as "primordial terrains," a term which has been used before in a less definite sense than the word "primitive." . . .

The secondary terrains do not show the same uncertainties as the primordial; the superpositions are evident there, and although part of them have undergone inclination, it is not in as violent nor as irregular a manner.

The rocks classed as of the first secondary group do not ordinarily, by themselves, cover a great extent of country, but are often found in countries where the primordial terrains, especially those of granite, predominate. . . .

To form the second group, I unite many systems, of which the more important have been designated by the names Zechstein, or old limestone of the Alps, parti-colored sandstone [*grès bigarré*], Muschelkalk, Quadersandstein, and Jurassic limestone.

The chalk formation as I have defined it in a preceding memoir, that is, including the tuffs, sands, and marls which are found below the chalk proper, constitutes the third group.

I unite in the fourth group all the terrains later than the chalk, of which the aqueous origin is not contested. . . .

Here, moreover, are the names which I propose to give to the five groups which I have believed ought to be established in the secondary terrains.

I shall call the first *terrains pénéens,* which is only, so to speak, a translation of *todteliegende* and which reminds one of the fact that the most characteristic beds are ordinarily *poor* in animal debris.

The second group will be designated by the name of *terrains ammonéens,* words which recall that all the systems of which it is composed have been formed at an epoch when these remarkable animals called ammonites existed.

The third, which corresponds to that which has already been called the chalk formation, will be designated by the name of cretaceous terrain.

The name mastozootic, applied to the fourth group, will recall that it is within these terrains we find the mammal skeletons the study of which inaugurated the science of geology among us.

Finally, the fifth group will be designated by the name pyroid, which, without affirming anything as to the manner in which these terrains were formed, will state that they resemble all those which have a proven igneous origin.

SEDGWICK

Adam Sedgwick (1785–1873), English stratigrapher, was throughout the greater part of his life professor of geology at the University of Cambridge. His pioneer work in Wales led to the recognition of the Cambrian series, the validity of which he stoutly defended in his long and bitter controversy with Murchison.

The Metamorphism of Sedimentary Rocks*

From *Transactions of the Geological Society of London*, Vol. III, pp. 74–105, 1847.

ALL solid mineral masses must have undergone some change since the time of their first production. Beds of secondary limestone and sandstone did not drop to the bottom of the sea, layer upon layer, in a solid form; and it is equally certain (though not equally obvious) that large unstratified crystalline masses were not created as we now find them. No one supposes that columnar basalt was originally built up of solid parallel jointed pillars, or that the structure of a granitoid rock was effected by a mere fortuitous concourse of the crystalline parts. We believe that these phaenomena are the necessary consequences of a certain anterior condition of the materials we examine. Sometimes, indeed, we can imitate these conditions, and then (as the laws of nature are unchangeable) we can do over again that which has been done a thousand times before in the laboratory of nature.

Many large mineral masses appear to have been once in a state of igneous fusion. Such masses, in passing from a fluid (or semifluid) to a solid state, necessarily put on a form more or less crystalline. The crystalline form is therefore the first and inevitable change. But there is another effect, arising out of such changes, of great geological importance. The mass which has changed its temperature, and become solid, has also changed its dimensions. Contraction must produce tension on the whole mass; and this tension, acting mechanically, will in many instances produce joints and fissures, and sometimes contortions: these

---

* A paper read before the Geological Society of London on March 11, 1835.

effects will be of greater or less regularity according to the conditions of each particular case.

The original modifications in the structure of an igneous rock *may* have been produced in a comparatively short period of time; and the same remark applies to some *metamorphic* rocks. The saccharoid texture (for example) of limestone, when in contact with trap, *may* have been produced during a very short period; for we know that this effect has been beautifully imitated in a chemical laboratory. In general, when *metamorphic* rocks appear to have been in a state of igneous fusion, it is obvious that all questions, respecting the length of time during which their crystalline structure was elaborated, must come very nearly under the rule that affects igneous rocks.

There is, however, a large class of *metamorphic* rocks, the structure of which can only have been produced by causes acting during long periods of time. I am not now speaking of gneiss, mica slate, and other old formations of crystalline strata. To assume that *all* such rocks are *metamorphic* is nothing better than to beg some of the greatest fundamental questions in geology. But in cases where a new mineral structure appears certainly to have been superinduced by direct igneous action, we sometimes meet with phaenomena utterly at variance with the hypothesis of a chemical action continued only through a short period. Rocks, it is well known, are bad conductors of heat; yet among stratified rocks the manifestations of igneous action are sometimes propagated to great distances. Such phaenomena may be readily explained. Masses of granite and porphyry were not necessarily protruded instantaneously. They may have been many years in assuming their present relative position among the stratified formations. Again, the effect produced by such protruded masses might be modified, almost indefinitely, by the conducting powers of the materials among which they rose. One mass may have been pushed out into the sea or the open air; another, after its first elevation, may still have been covered up by a vast thickness of badly conducting strata. Nor is this all. It is by no means necessary to suppose that all changes produced by igneous agents on stratified rocks took place only during periods of eruption. We may suppose, for example, that the lower slate formations of Cornwall and Cumberland formed a dome, overhanging the great subterranean fires, for many years, or even for many centuries, before a contraction of the upper surface, or a mechanical action from below, pushed up the great bosses of granite among the altered and half-molten beds. Hence, although it be certain that the structure of the altered rocks has in many cases been produced by a sudden action, we are by no means limited in our hypothesis, but may fairly suppose such periods of duration as in each case, are necessary to the elaboration of our phaenomena.

My chief object is, however, to describe some of the changes produced on mechanical, stratified rocks by causes acting under a comparatively low temperature, and often during indefinite periods of time. Very few of these changes can be imitated in a laboratory, because it is impossible to imitate the conditions under which they have been brought about. They admit not, therefore, of synthetic proof; but they are unquestionably subordinate to chemical and mechanical laws, which we can study experimentally, and establish on appropriate evidence. By assuming the existence of these laws, and by studying the conditions under which they have acted (not as matters of experiment but of observation), we may gradually ascend towards an explanation of some of the perplexing phaenomena presented by sedimentary rocks.

That these rocks are greatly changed since their first origin is too obvious to require any formal proof. Take, for example, a mass of Hertfordshire pudding-stone. There can be no doubt that it once formed a portion of a shingle bank of rolled chalk flints and finely comminuted sand. The materials are now so closely agglutinated by siliceous cement, that a fracture passes indifferently through the sandstone and the imbedded flint pebbles. Again, rocks of this kind are sometimes divided into prismatic masses by cross joints; and these joints pass without any deviation through the imbedded flints, so as to produce a series of smooth surfaces. I merely mention this as an example of a great change produced naturally (and unquestionably without any very high temperature) upon a coarse, mechanical deposit.

The most striking modifications of structure enumerated in the preceding portion of this paper extend to comparatively short distances from given centres of chemical action: those I am about to notice (especially the transverse cleavage of the various slate rocks), are of a very different character. As the finest examples of slaty cleavage are derived from the great Cumbrian cluster of mountains, and from the chains of North Wales, it may be well, in the first place, briefly to compare the physical structure of the two regions.

The zone of the Cumbrian green slate alternates with an indefinite number of tabular or stratified masses of felspathic and porphyritic rocks. The slates are crystalline, and have been so firmly packed in among the alternating porphyries, as to undergo very few contortions or undulations during the period of their elevation. Moreover, they contain no organic remains: such remains being, perhaps, obliterated; or, more probably, organic beings not having propagated in an ocean exposed to continual incursions of felspathic rocks. All the masses alternating with the porphyritic system exhibit in greater or less perfection a *cleavage*, which is in no instance parallel to the true beds. These facts were stated in a former paper; and I expressed my belief that the felspathic tabular masses were of Plutonic origin, and that even the great alternating beds of slate (especially the more crystalline and chloritic varieties), might, in part, have derived their materials from Plutonic sediment. I also imagined (at the time the paper was written) that each great alternating mass of slate had an independent cleavage, produced probably by crystal-line forces acting under a high temperature; and this high tempera-ture seemed to be naturally accounted for by the presence of the porphyries.

In the great chain of North Wales we have the same indefinite alternations: but the porphyries are less abundant in proportion to the other masses, and have produced a less impress on the slate system Some of the slates are crystalline and some earthy; and in *both varieties* we find (though rarely) traces of organic remains. The whole system of slates and tabular porphyries has been thrown into a number of great undulations, producing through the chain a series of longitudinal anticlinal and synclinal lines. Lastly, parallel lines of cleavage not merely affect given beds; but some-times run, without deviation, even through coarse mechanical subordinate strata, affecting whole ranges of mountains, and preserving their parallelism in spite of undulations and anticlinal lines.   These facts have led me to give up the opinion, that the cleavage planes have been materially modified by any action of the alternating porphyries.

Leaving all further comparison between the structure of Cumberland and Wales, I return to the description of the most general facts exhibited in slaty cleavage   If we examine a quarry where this structure is well developed, we find a nearly homogeneous mass, easily separable into thin parallel laminae.   But the thickness of these laminae is not defined by *joints* (*i.e.* by fissures at definite distances); for the cleavage of each part may be carried on indefinitely, or at least so far as the operation is not interrupted by a mere mechanical difficulty.   That this arrangement is crystalline it is impossible to doubt, when we examine the planes of cleavage, and see them coated over with flakes of chlorite and semicrystalline matter, which not merely define the planes in question, but strike in parallel flakes through the whole mass of the rock.   Were there any doubt of this conclusion, it is further confirmed by the fact, that these planes of cleavage are inclined at various angles to the planes of stratification, and are, perhaps, in no instance coincident with them.   This last fact is of great importance, and is now generally admitted by English geologists. But it requires to be made still more prominent; for it is not considered in its proper extent by Continental geologists, and by some of them is actually denied.

I think it obvious that the contortions of slate rocks are phaenomena quite distinct from cleavage, and that the curves presented by such formations are the true lines of disturbed strata.   In many cases a cleavage seems to have been the last change superinduced on rocks before they became entirely solid; and after that time it is not conceivable that any mere mechanical force, however violent, should have thrown them into such contortions as we often see passing through them.   Again, the contorted laminae, so often seen in formations of argillaceous schist, seem to be removed from all analogy with known modes of crystalline action; whereas the great parallel plates of slaty cleavage (however enormous may be their scale), are quite compatible with it.

On the Classification and Nomenclature
of the Lower Paleozoic Rocks
of England and Wales

From *Quarterly Journal of the Geological Society of London*, Vol. VIII, pp. 136–168, 1852.

I will first enumerate (*in ascending order*) the several groups into which the whole Welsh series (Cambrian and Silurian) may, I think, be conveniently subdivided; and I may premise, that I consider all the palaeozoic rocks, from the lowest Cambrian to the highest Permian, as one system—the primary or palaeozoic system. This primary system admits of three great subdivisions; viz. a lower subdivision, including the Cambrian and Silurian series; a middle, including the Devonian series; and an upper, including the Carboniferous and Permian series. These three subdivisions belong to one great *systema naturae*, the subordinate parts of which often pass one into another, by almost insensible gradations; although the species in the several subdivisions and subordinate groups often entirely, or almost entirely, change.* But the *primary system*, thus defined, differs entirely from the *systema naturae* of the *secondary system;* and, in like manner, the *systema naturae* of the secondary system differs almost entirely from the *systema naturae* of the tertiary system. Lastly, we have the actual *systema naturae* of the living world; but between the *tertiary system* and that of living nature no one has yet drawn any intelligible line of demarcation.

I do not pretend to answer a question, whether the primary, secondary, and tertiary systems may not, in progress of discovery, be at length brought in a similar intimate relation; neither do I discuss a question respecting the expediency of any further subdivisions of the secondary system. A good classification only represents the actual condition of our knowledge; and the following remarks relate only to the classification of the subordinate groups of the lower palaeozoic system, as above defined. To avoid all verbal ambiguity, or wrangling about the use or abuse of the word "system," I will provisionally separate the whole *lower palaeozoic series* into two great natural subdivisions—Cambrian and Silurian;

---

* This view of regarding all the Palaeozoic rocks as of *one system* is not new. It has often been discussed in this Society; and it was formally advanced by myself in 1843.—*Proceed. Geol. Soc.*, vol. lv. p. 223.

each of which may be again subdivided into a series of stages or groups, which, collectively, I here designate by the names CAMBRIAN SERIES and SILURIAN SERIES. The Cambrian and Silurian collective groups, thus defined, have a well-marked physical separation; and the Silurian groups are, not unusually, unconformable to the Cambrian: and although several fossils are common to the two collective groups, especially near the planes of junction, yet the fossils of the well-defined lower stages of the Cambrian series are very widely distinct from the fossils of the upper stages of the Silurian series. I believe that this is the case in Wales and Siluria; and I am certain that it is the case in the Cumbrian cluster of mountains. The fossils of the Coniston calcareous slates hardly reappear at all, and certainly not as a group, among the very numerous fossils of the rocks between Kendal and Kirkby Lonsdale (Upper Ludlow). Hence, on mere palaeontological grounds, it would produce nothing but confusion were we to designate the Ludlow rocks south of Kendal, and the calcareous slates of Coniston, &c. as one system, while we adopt the restricted use of the word "system" now in common use.

The whole Cambrian series is exhibited, in vast undulations, from the Menai to the Berwyns; and a part of it, again, in a system of what might be called short independent waves, on the east side of the Berwyns, until the last beds of the series become buried under the carboniferous limestone. But, if we extend our views to the north end of the great undulating series, we find (not, however, without continual breaks and dislocations) the prevailing strike and dip so changed, that the successive beds are seen to plunge, with a northern dip, under the rocks forming the base of the great deposits of Denbigh flagstone which compose the true Silurian series of North Wales. In this view, the physical separation of the Cambrian and Silurian series is not hypothetical, but perfectly natural; and the zoological separation, taken on the whole, is, perhaps, as complete as the physical.

Our whole scheme of nomenclature of the lower Palaeozoic rocks is geographical. This scheme was followed out, from first to last, in the "Silurian System." The system, and all the subordinate groups, were defined by geographical names. Now it is surely an axiom in geological nomenclature, that if we give a new geographical name to any group of strata, that name must refer us to a spot near which we find the group well-developed. In Cambria the whole series of the oldest palaeozoic division is more nobly developed than in any other part of Britain (on this point I can speak from my own experience); while in Siluria we find only the highest group of the whole series. This would have been a sufficient reason for changing the name Silurian into Cambrian, had, by any caprice or accident, the name Silurian been first given to the older Cambrian rocks; but it seems to me a very strange reason for changing the name Cambrian (a right name for a great series of rocks well-developed in Cambria, and a name which had the undoubted priority) into Silurian. If indeed we had a good and perfect series of the older palaeozoic groups in Siluria, then the words "Silurian System" might be stereotyped as a general designation of all the lower palaeozoic rocks of Britain. But Siluria shows us no such typical series, while Cambria does. On the ground, therefore, of geographical propriety, as well as of priority, I vindicate the claims of the Cambrian series for a place in our nomenclature.

PRÉVOST

Louis Constant Prévost (1787–1867), influential French geologist, led the opposition to von Buch's upheaval hypothesis and made important contributions to the correlation of strata.

A New Volcanic Island Not Upheaved

Translated from "Notes sur l'île Julia, pour servir à l'histoire de la formation des montagnes volcaniques," *Mémoires de la Société géologique de France*, Vol. II, pp. 91–124, 1835.

In the month of July, 1831, an island [l'île Julia] appeared in the Mediterranean, between Sicily and Africa, following violent volcanic eruptions that reached the surface through the waters of the sea.

This event excited general attention, and Rear Admiral de Rigny, then *ministre de la marine*, having offered to put at the disposal of the Academy of Sciences the government brig *Flèche*, which he sent, under the command of Captain Lapierre, to determine the exact position of this new island, the Academy confided to me the honorable mission of gathering data and making observations that would be of geological interest.

———

Many days before the first eruptions, the surface of the sea appeared to be boiling, and the waters were troubled. The sea was covered with dead or bloated fish, a great number of which were gathered on the shores of Sicily, more than eight or ten leagues from the point where the eruptions later appeared.

These eruptions began with light vapors, which gradually increasing gave place to a constant column, white and flocculent, 1,500 to 2,000 feet high and 60 to 100 feet wide. At first these vapors alone arose. Later they were mixed with cinders and stones, and other reddish and fuliginous vapors. The column of cinders and stones, which arose intermittently and appeared black during the day and incandescent at its center during the night, was noticed a long time before any solid mass appeared at its base. . . .

The apparition of the island was gradual. Many peaks appeared separately, and united to form around the center of eruption a collar of loose material, the form of which changed continually, and which, originally at the level of the waters, rose gradually to at least two hundred feet. At first the crater was in communication with the sea, sometimes on the north, sometimes on the southeast, depending on the effect of the winds or on that of the waves which aided in the moving and dispersing of the ejected matter. . . .

Finally, the disappearance was slow and gradual; as the appearance had been. Evidently it, as well as the lowering of the ground already sunk below sea level, was produced mainly by the action of the waves, which, after having caused landslides of the cinders, scoriae, and incoherent fragments of which the island was composed, carried away these loose materials. It is probable also that the quakes which have been felt since the eruptions ceased, have contributed to the transformation of the island of Julia into a shoal, covered in some parts with only nine to ten feet of water, and with a form which no longer indicates its origin.

The island of Julia was not formed by an uplift of the ground

If, after what I have just said, I call to mind what I announced in my first report, that the entire island of Julia showed only a heap of cinders, sands, and volcanic scoriae without a single lava flow or any strata of hard and continuous rocks which might have formed the original floor of the sea, and that these various materials present a stratification following two lines of slope inclined in opposite directions, the one toward the center of the crater, and the other toward the outer base of the cone, it becomes unquestionable, I think, that it was truly nothing but the summit of a cone of eruption precisely like the summits of Etna and Vesuvius in form, nature, arrangement of material, and origin. . . .

Since then I have entirely opposed the supposition to which many geologists adhere even today, namely, that the island of Julia was produced by the violent upheaval of the sea floor where the strata were suddenly tilted by violent action. The disappearance of the island in accordance with my prediction and the information concerning its gradual formation that I have since procured have fully confirmed my assertion. Nevertheless, being in favor of the ingenious ideas introduced into the science by

Herr von Buch, I ventured in my first report the conjecture that around the cone of eruption forming the island of Julia there ought to exist, under water, a girdle of rocks formed by the edges of the elevation crater, if the upheaval of the ground had preceded the establishment of the new volcanic center. The knowledge that I have since acquired of the circumstances which accompanied the apparition of the island and, in addition, the observations that I have been able to make in Sicily and Italy have fully convinced me that my conjectures along this line were entirely baseless. Forced to yield to the evidence of facts, I have been led not only to abandon the idea of a circle of elevation about the island of Julia and all the centers of eruption that I have visited, but even to doubt that this fascinating theory of elevation craters can even be applied to any of the volcanoes for which it has been imagined by its author.

GILMER

Francis Walker Gilmer (1790–1826), a Virginian lawyer, aided Jefferson in starting the University of Virginia.

THE NATURAL BRIDGE OF VIRGINIA A RESULT OF EROSION

From *Transactions of the American Philosophical Society*, N.S., Vol. I, pp. 187–192, 1818.

Theories have already been attempted for explaining the formation of the Natural Bridge. As they were generally formed in the age of the mechanical philosophy, before geology or even chemistry had become sciences, we need not wonder to find the solutions they offered, very insufficient. Mr. Jefferson's hypothesis rested entirely upon the supposition, that some sudden and violent convulsion of nature, tore away one part of the hill from the other, and left the bridge remaining over the chasm. . . . It contradicts, also, that beautiful and valuable rule laid down by Newton, "that it is unphilosophical to assign more causes for the natural appearance of things than are both true and sufficient to account for the phenomena." . . .

In the present state of geology, the phenomenon does not require us to resort to the operation of the unknown, or even of doubtful agents. And instead of its being the effect of a sudden convulsion, or an extraordinary deviation from the ordinary laws of nature, it will be found to have been produced by the very slow operation of causes which have always, and must ever continue, to act in the same manner.

To make this manifest, let us consider the situation of the bridge. That the place at which it stands is the highest point of a transverse ridge of hills, with a narrow base, which crosses the ravine at that spot. The country about the bridge, like all that which is west of the mountains, from the Atlantic to the Pacific ocean, is calcareous. The strata of rock, which at different places make different angles with the horizon, are here parallel to it. . . . Here, as in calcareous countries generally, there are frequent and large fissures in the earth, which are sometimes conduits for subterraneous streams, called "sinking rivers," "sinking creeks," &c. . . . It is probable, then, that the water of Cedar Creek originally found a subterraneous passage beneath the arch of the present bridge, then only the continuation of the transverse ridge of hills. The stream has gradually widened, and deepened this ravine to its present situation. Fragments of its sides also yielding to the expansion and contraction of heat and cold, tumbled down even above the height of the water. Or, if there was no subterraneous outlet, the waters opposed by the hill flowed back, and formed a lake, whose contact dissolved the resistance where it was least, wore away the channel through which it now flows, and left the earth standing above its surface. I incline, however, rather to the first hypothesis, because the ravine has the appearance, from its narrow banks, of having been the channel of a stream in all time, and had it been the bed of a lake, the continued action of the water would have widened it into a basin. The stone and earth composing the arch of the bridge, remained there and no where else; because, the hill being of rock, the depth of rock was greatest above the surface of the water where the hill was highest, and this part being very thick, and the strata horizontal, the arch was strong enough to rest on such a base.

QUOY AND GAIMARD

Jean René Constant Quoy (1790–1869), French naval surgeon and naturalist.
Joseph Paul Gaimard (c. 1790–1858), French naturalist.

THE FORMATION OF CORAL ISLANDS

Translated from "Mémoire sur l'Accroissement des polypes lithophytes considéré géologiquement," *Annales des sciences naturelles*, Vol. VI, pp. 273–290, 1823.

Among the phenomena of zoology connected with the theory of the earth, those which concern the solid zoophytes are still far from being explained. In calling the attention of naturalists to these animalcules, we hope to show that all that has been said or thought to have been observed hitherto concerning the enormous amount that they are capable of accomplishing is inexact, always excessively exaggerated, and most often erroneous.

Instead of believing that the Society Islands, some parts of New Ireland [New Mecklenburg], the Louisiade, the archipelago of the Solomons, the lower Friendly Islands, the Mariannes [Ladrones], the Pelews, the Navigators' Islands [Samoa], those of Fiji, the Marquesas, etc., are in part or totally the work of zoophytes, we think that all these lands have as a base the same elements, the same minerals which concur to form all the known islands and continents. There, indeed, are found the schists, as at Timor and Vaigiou; sandstones, as on the coast of New Holland [Australia]. Elsewhere, limestone in horizontal beds forms the island of Boni and surrounds the volcanic peaks of the Mariannes [Ladrone Islands]. Granite is also found sometimes, but most often volcanoes have formed the islands scattered in the southern ocean.

We propose in this memoir

1. To consider how the lithophytes raise their dwellings on bases already established, and what circumstances are favorable or unfavorable to their growth.

2. To show that there are no islands of any considerable size, constantly inhabited by man, which are entirely formed by corals: and that far from raising perpendicular walls from the

depths of the ocean, as has been claimed, these animals only form beds or incrustations of a few toises [fathoms] in thickness.

————————

This is the way this addition, this superposition of the madrepores, operates. In the places where the heat is constantly intense, or the cut-in bays enclose shallow and peaceful waters which are not subject to agitation by great waves or the tropical trade winds, the rock-forming polyps multiply. They build their dwellings on the submarine rocks, enveloping them entirely or in part, but properly speaking they do not form them. Thus all these reefs, all these madreporic belts that one meets rather often in the South Seas to the windward of the islands, are, in our opinion, platforms arising from the conformation of the primitive surface.

BABBAGE

Charles Babbage (1790–1871), English mathematician, was Lucasian professor at Trinity College, Cambridge.

THE TEMPLE OF SERAPIS[*]

From *Quarterly Journal of the Geological Society of London*, Vol. III, pp. 186–217, 1847.

Proofs of the Raising and Lowering of the Temple

THE facts and observations which I have thrown together in the following paper were collected during the month of June 1828, in company with Mr. Head. They relate to a monument of ancient art, which is perhaps more interesting than any other to the geologist.

I shall first state the facts which came under my own observation, without assuming that they have not been previously noticed, though not aware of their having yet been collected into one point. I shall then suggest an explanation of the singular phaenomena which the temple presents, and afterwards briefly sketch those more general views to which I have been led by reflecting on the causes that appear to have produced the alternate subsidence and elevation of the Temple of Serapis.

In the year 1749, the upper portions of three marble columns that had been nearly concealed by underwood, were discovered, in the neighbourhood of the town of Pozzuoli. In the following year excavations were made, and ultimately it was found that they formed part of a large temple which was supposed to have been dedicated to the god Serapis.

The temple is situated about a hundred feet from the sea, . . .

The most remarkable circumstance which first attracts the attention of the observer is the state of the remaining three large columns. Throughout a part of their height, commencing at nearly 11 feet above the floor of the temple, and continuing about 8 feet, they are perforated in all directions by a species of

---

[*] A paper read before the Geological Society of London on March 12, 1834.

boring marine animal, the *Modiola lithophaga* of Lamarck,— which still exists in the adjacent parts of the Mediterranean.

About half a mile along the sea-shore towards the west, and standing at some distance from it, in the sea, are the remains of columns and buildings which bear the names of the temples of the Nymphs and of Neptune.

The tops of the broken columns are nearly on a level with the surface of the water, which is about five feet deep.

At the east foot of Monte Nuovo an ancient beach may be seen for about fifty yards, which is two feet higher than the present beach, and which is covered by about seventeen feet of tuff. . . .

There are also the remains of two Roman roads, at present under water; one of these reached from Pozzuoli to the Lucrine lake.

Another vestige of the art of a remote period which exhibits decided evidence of a change of level, is the series of piers placed in the sea, projecting from the town of Pozzuoli, and known by the name of the bridge of Caligula.

The general depth of the sea around these piers is from thirty-five to fifty feet. . . .

At the height of about four feet above the present level of the sea on the sixth pier is a line of perforations, apparently by the *Modiola* or other boring animal.

It seems then that the temple subsided into the sea; but whether this happened slowly or at intervals, by repeated shocks of earthquakes, does not appear. Nearly at its lowest point there are indications of its having been stationary. For about 6 inches below the highest perforation of the Modiolae the columns are corroded, as if that point remained exposed for some time, alternately to the action of wind and water.

The next period in the history of the temple was its gradual elevation. Whether the deposit out of which it was dug covered it up before or after this event, is not perhaps distinctly evident. From the section behind the temple, I am induced to suppose that it preceded the elevation; and the chance of the columns not being overthrown by any sudden rising, would be considerably increased by the support they would derive from having more than one-half their height imbedded in earth.

The Effects of Changes in Temperature in the Earth's Crust

The preceding conclusions involve no hypothesis, and may be considered as inferences fairly resulting from the specimens collected, from the facts observed on the spot, and from the historical evidence of changes which have happened in the neighbourhood of the temple. I shall now proceed to offer some conjectures relative to the causes of the successive changes in the level of the ground on which this temple stands—conjectures which I wish to be considered as entirely distinct from the former part of this communication.

On examining the country round Pozzuoli it is difficult to avoid the conclusion, that the action of heat is in some way or other the cause of the phaenomena of the change of level of the temple. Its own hot spring, its immediate contiguity to the Solfatara, its nearness to the Monte Nuovo, the hot spring at the Baths of Nero on the opposite side of the bay of Baiae, the boiling springs and ancient volcanos of Ischia on one side and Vesuvius on the other, are the most prominent of a multitude of facts which point to that conclusion.

The mode by which this heat operates is a question of greater difficulty, and in the absence of sufficient data, it may be enough to point out shortly some of its possible results.

It may be imagined that at a considerable depth below the surface a vast reservoir of melted lava exists, containing highly elastic matter imprisoned within it by the pressure of the superincumbent strata. The addition of matter supplying this elastic fluid, or the accession of heat, may increase the force, or on the other hand, the expansion or contraction of some portion of the superior strata may cause a fissure through which the melted lava may be forced up by the elastic fluid. In such circumstances, besides the earthquakes which will be caused by the rent, and the stream of lava which issues through it, the whole of the strata resting on the fluid lava will slowly subside. The cooling of the lava may fill up the rent and the strata again rise as before, until a renewal of the same cause reproduces a renewal of the same effect. It may here be remarked, that the expulsion of the immense quantity of gaseous matter, which some volcanos are known to throw out, may lower the temperature of the cauldron below, more effectually than the abstraction of the lava which is ejected from it.

Another view of the subject is, that there exists below the ground in the neighbourhood of Pozzuoli cavities containing water or other condensed gases in a highly heated state—that any accession or diminution of heat, arising from the volcanic causes in operation in the neighbourhood, will increase or diminish the elasticity of these gases, and thus cause an elevation or subsidence in the strata above.

A different view however of the effect of heat may be taken, one which is well known, and which has in some instances been measured. The solid beds below the temple are themselves liable to expand by the action of heat, and to contract by its abstraction; rents and earthquakes, as well as elevations and depressions of the surface, may be the result of the partial application of this cause. It may perhaps be doubted whether sufficient effect can arise without imagining masses of immense thickness to have altered their temperature; a change which might have required longer time for its completion than the phaenomena admit.

From a series of experiments upon the expansion of various stones by the application of heat, made by Mr. H. C. Bartlett, of the U. S. Engineers, under the direction of Col. Totten, and recorded in the American Journal of Science, vol. xxii. p. 136, it appears that for 1° of Fahrenheit's scale—

Granite expands	·000004825
Marble	·000005668
Sandstone	·000009532

From these data the expansion of those substances has been calculated for various degrees of temperature, and for thicknesses varying from 1 to 500 miles. . . . From this it may be inferred, that if the strata below the temple and its immediate neighbourhood are equally expansible with sandstone, then a change of temperature of only 100°F. acting on a thickness of five miles would cause a change of level of above twenty-five feet—an alteration greater than any of the observed facts at the temple of Serapis require.

———————

The difficulties of this theory are, that some part of the surface at the piers of Caligula's Bridge is at present raised above its former level, and other parts, as the temple of the Nymphs and of Neptune, are still below that level; whilst the temple of Serapis appears to have returned nearly to its former state. The answer to this is, that the thickness of the expanding beds may differ in different parts, or may have a different power of conducting heat— or it may be remarked, if the conducting power and the thickness be the same, the distance from the source of heat may be different, and consequently the full effects may have reached the piers of the bridge, and yet not have attained the other points.

On the whole this explanation is the most tenable, because it is founded on facts—viz. that matter expands by heating; that great accessions of heat have at various times taken place in the neighbourhood of the temple; that it is sufficient to account for the phaenomena by supposing a moderate depth of the beds below it heated to a degree which it is not unreasonable to presume must have taken place; that such changes of level would on the whole occur gradually, although they might be accompanied with earthquakes and occasionally by sudden changes of level—facts of which we have historical evidence as having happened on this spot.

In reflecting on the preceding explanation of the causes which produced the changes of level of the ground in the neighbourhood of Pozzuoli, I was led to consider whether they might not be extended to other instances, and whether there are not natural causes, constantly exerting their influence, which, concurring with the known properties of matter, must necessarily produce those alterations of sea and land, those elevations of continents and mountains, and those vast cycles of which geology gives such incontrovertible proofs.

The following explanation of the origin of the changes which have continually taken place in the forms and the levels of large portions of the earth's surface at many distant periods of time, and which appear still to continue their slow but certain progress, arose from the examination of the temple of Serapis, which has been detailed in the former part of this paper.

The theory rests upon the following principles:—

1st.   That as we descend below the surface of the earth at any point, the temperature increases.

2nd.   That solid rocks expand by being heated, but that clay and some other substances contract under the same circumstances.

3rd.   That different rocks and strata conduct heat differently.

4th.   That the earth radiates heat differently from different parts of its surface, according as it is covered with forests, with mountains, with deserts, or with water.

5th. That existing atmospheric agents and other causes are constantly changing the condition of the earth's surface, and that, assisted by the force of gravity, there is a continual transport of matter from a higher to a lower level.

If we imagine at every point of the earth's surface a line drawn to its centre, then if a point be taken in any one line at a given temperature, there will be contiguous points of exactly the same temperature in all the adjacent lines; and if we conceive a surface to pass through all these points, it will constitute a surface of uniform temperature, or an isothermal surface. This therefore will not be parallel to that of the earth, but will be irregular, descending more towards the centre of the earth, where it passes under deep oceans.

An increase of 1° of Fahrenheit's thermometer, for every fifty or sixty feet we penetrate below the earth's surface, seems nearly the average result of observations. If the rate continue, it is obvious that, at a small distance below the surface, we shall arrive at a heat which will keep all the substances with which we are acquainted in a state of fusion. Without however assuming the fluidity of the *central* nucleus—a question yet unsettled, and which rests on very inferior evidence to that by which the principles here employed are supported—we may yet arrive at important conclusions; and these may be applied to the case of central fluidity, according to the opinions of the several inquirers.

The newly-formed strata will be consolidated by the application of heat; they may, perhaps, contract in bulk, and thus give space for new deposits, which will, in their turn, become similarly consolidated. But the surface of uniform temperature below the bed of the ocean, cannot rise towards the earth's surface, without an increase in the temperature of all the beds of various rock on which it rests; and this increase must take place for a considerable depth. The consequence will be a gradual rise of the ancient bed of the ocean, and of all the deposits newly formed upon it. The shallowness of this altered ocean will, by exposing it to greater evaporation from the effect of the sun's heat, give increased force to the atmospheric causes still operating upon the inequalities of the solid surface and tend more rapidly to fill up the depressions.

Possibly the conducting power of the heated rocks may be so slow, that its total effect may not be produced for centuries after the sea has given place to dry land; and we can conceive in such circumstances, the force of the sun's rays from without, and the increasing heat from below, so consolidating the surface, that the land may again descend below the level of the adjacent seas, even though its first bottom is still subject to the elevatory process. . . .

On the other hand, as the high land gradually wears away by the removal of a portion of its thickness, and as the cooling down of the surface takes place, its contraction might give place to enormous rents. If these cracks penetrate to any great reservoirs of melted matter, such as appear to subsist beneath volcanos, then they will be compressed by the contraction, and the melted matter will rise and fill the cracks, which, when cooled down, become dykes. Rents therefore or veins may arise by contraction from cooling, and proceed from the surface downwards; or they may result from expansive force acting from below and proceed upwards.

If these rents do not reach the internal reservoir of melted matter, and if there exist in the neighbourhood any volcanic vents connected with it, the contraction of the upper strata may give rise to volcanic eruptions through those vents, which might be driven by such a force almost to any height. These eruptions may themselves diminish the heat of the beds immediately above the melting cauldron from which they arise; for the conversion of some of the fluid substances into gases, on the removal of the enormous pressure, will rapidly abstract heat from the melted mass.

As the removal of the upper surface of the high land will diminish its resistance to fracture, so the altered pressure arising from the removal of that weight, and its transfer to the bottom of the ocean, may determine the exit of the melted matter at the nearest points of weakest resistance.

It appears, therefore, that from changes continually going on, by the destruction of forests, the filling up of seas, and the wearing down of elevated lands, the heat radiated from the earth's surface varies considerably at different periods. In consequence of this variation, and also in consequence of the covering up of the bottoms of seas, by the detritus of the land, the *surfaces of equal temperature* within the earth are continually changing their form, and exposing thick beds near the exterior to alterations of temperature. The expansion and contraction of these strata, and, in some cases, their becoming fluid, may form rents and veins, produce earthquakes, determine volcanic eruptions, elevate continents, and possibly raise mountain chains.

MURCHISON

Sir Roderick Impey Murchison (1792–1871), eminent British geologist, was for many years director general of the Geological Survey of the United Kingdom.

The Silurian System

From *London and Edinburgh Philosophical Magazine*, Third Series, Vol. VII, pp. 46–52, 1835.

GEOLOGISTS having long felt that the older sedimentary deposits required a systematic examination, I have devoted the last five years to the study of this class of rocks, hoping thereby to fill up certain pages which were wanting in the chronology of the science. A table published last year was the first attempt to convey to the geological student a correct view of the thickness, variety of strata, and fossil organic contents of a vast system, which, though arranged by nature in a most lucid order of succession, had not previously been pointed out. These rocks, rising from beneath the old red sandstone in Herefordshire, Shropshire, Radnorshire, Brecknockshire, Monmouthshire, and Caermarthenshire, and each distinguished by separate and *peculiar organic remains*, were respectively named after those localities where each of them could be best studied, and their places in the series most clearly established. I have no change to announce in the order detailed in that table (see Lond. and Edinb. Phil. Mag., vol. iv. p. 370), but I wish to simplify it by the abandonment of double names, as applied to any one formation, and by the adoption of the names of those places only where the respective rocky masses lie in juxtaposition.

The names finally adopted, and which will be incorporated in a work now in preparation on this subject, are,

1. *Ludlow rocks*, divided into upper and lower Ludlow rocks, with a central zone of limestone: in this formation no change of name is proposed.

2. *Wenlock limestone* and shale (*equivalent, Dudley*).

3. *Caradoc sandstones*. This name, supplying the place of the Horderley and May Hill rocks, has been derived from the striking and well-known ridge of Caer Caradoc, on the eastern flanks of

which, and lying between it and the Wenlock Edge, are exhibited those peculiar strata which are the equivalents of the shelly sand-stones of Tortworth.

4. *Llandeilo flags* (preferred to "Builth and Llandeilo"). When this table is reprinted, there will naturally be found many additions to the organic remains, some identifications of the British with foreign species, and numerous corrections.

Notwithstanding the adoption of these names, there was still required a comprehensive term by which the whole group could be designated, and at once distinguished from the *old red sandstone* above, and the *slaty rocks* below. Without such a collective name for the group, I found it impracticable to proceed with the work which I had engaged to complete, it being essential to the clear exposition of the subject, no longer to speak of these deposits as "transition rocks" or "fossiliferous grauwacké." The term "transition" might indeed, have been retained, if for no other reason than to impress upon foreign geologists, (the Germans particularly,) how vast a difference exists between the geological horizon of the mountain or *carboniferous* limestone and that of the limestones of Ludlow and Wenlock, which are not only separated by many thousand feet of strata from the limestone of the carbonif-erous system, but, further, contain an entirely distinct class of organic remains. It was, however, utterly hopeless to use the word "transition" in any definite sense as applied to these lower deposits, seeing the extent to which it had been abused. By some it was confined to those older rocks in which the earliest traces of organic remains were supposed to be observed, whilst others had more recently so expanded the meaning as to comprehend in it the whole of the carboniferous series! Thus at a period when, from the rapid advances of the science, it had become indispensable to define the boundaries of groups naturally distinct from each other, dissimilar things were still confounded under one common name! and hence every geologist with whom I am acquainted had been for some time agreed upon the expediency of obliterating the term. The name "transition" is, in truth, not applicable to any one class of stratified deposits in preference to another. Thus, for example, within the area of a map now preparing for publication and embracing parts of ten counties only, I shall be able to show *transitions* into every formation, beginning with the inferior oolite and terminating in descending order with the Llandeilo flags, many thousand feet

below the old red sandstone; whilst the latter overlie other fossiliferous masses, the relative ages of which yet remain to be worked
out! In various memoirs read before the Geological Society I have
described these rocks as "fossiliferous grauwacké," but this term is
in reality a misnomer, as the group contains few if any strata of the
true grauwacké of German mineralogists. But whilst this system
contains no such beds, it is underlaid and sometimes in discordant
stratification, by a vast series of slaty rocks, in which much genuine
grauwacké is exhibited. It was therefore manifest that if used at
all in geological nomenclature, the term "grauwacké" must be
rejected as inapplicable to the first great system below the old red
sandstone, and restricted to rocks which were *now* proved to be of
much higher antiquity. My friend Professor Sedgwick will
doubtless soon dispel the obscurity which hangs over these grauwacké rocks, with which his labours in Wales and Cumberland have
so well enabled him successfully to grapple.

To return, however, to the system under review, I was urged by
leading geologists both at home and abroad to propound an entirely
new name for it. In consonance, therefore, with those views which
have rendered the names used by English geologists so current
throughout the world, I venture to suggest, that as the great mass
of rocks in question, trending from south-west to north-east,
traverses the kingdom of our ancestors the Silures, the term
"Silurian system" should be adopted as expressive of the deposits
which lie between the old red sandstone and the slaty rocks of
Wales, including, as above detailed, the Ludlow, Wenlock, Caradoc,
and Llandeilo formations. . . .

. . . In allusion to this term [Silurian] I have only further to
add, that it is to be hoped that no naturalist will, from its sound,
fall into the mistake of an early English writer who is ridiculed by
Camden for having misapplied the line of Juvenal,

> "Magna qui voce solebat
> Vendere municipes fracta de merce Siluros,"

supposing that the British captives were exposed to sale at Rome,
when the poet spoke of *fishes*, and not of men! My geological
readers do not require to be told that there are no fossilized
remains of the "*Silurus*," or bony Pike, in these deposits, since
M. Agassiz will afford us very different names for the ichthyolites
of the Ludlow rocks.

The Permian System

From *Philosophical Magazine*, Third Series, Vol. XIX, 1841.

Letter to M. Fischer de Waldheim, ex-President of the Society of Naturalists of Moscow

Moscow, Oct. 8, 1841.

My dear Sir,

As you have taken a lively interest in the success of the geological expedition which I have just completed, accompanied by my friends M. de Verneuil, Count de Keyserling, and Lieutenant Koksharoff, I hasten to communicate to you some of its chief results; and I do so with real pleasure, because in requesting you to present them to the Society of Naturalists of Moscow, I acquit myself of a duty towards a distinguished body which has done me the honour of placing my name in the list of its foreign members.

The wide extension in the North of Russia of the Silurian, Devonian and Carboniferous Systems, as proceeding from the last year's survey, by the same observers and our friend the Baron A. de Meyendorf, is already known to you from the abstracts of memoirs communicated to the Geological Societies of London and Paris. Our principal objects this year were,—1st. To study the order of superposition, the relations and geographical distribution of the other and superior sedimentary rocks in the central and southern parts of the empire. 2nd. To examine the Ural Mountains, and to observe the manner in which that chain rises from beneath the horizontal formations of Russia. 3rd. To explore the carboniferous region of the Donetz, and the adjacent rocks on the Sea of Azof.

Our last year's survey had pretty nearly determined the limits of the great tract of carboniferous limestone of the North of Russia. On this occasion we have added to its upper part that remarkable mass of rock which forms the peninsula of the Volga near Samara, and which, clearly exposed in lofty, vertical cliffs, and charged with myriads of the curious fossils *Fusilina*, constitutes one of the striking features of Russian geology.

The carboniferous system is surmounted, to the east of the Volga, by a vast series of beds of marls, schists, limestones, sandstones and conglomerates, to which I propose to give the name of "Permian System," because, although this series represents as a whole, the lower new red sandstone (*Rohte todte liegende*) and the magnesian limestone or *Zechstein*, yet it cannot be classed exactly (whether by the succession of the strata or their contents) with either of the German or British subdivisions of this age.   Moreover the British lithological term of lower new red sandstone, is as inapplicable to the great masses of marls, white and yellow limestones, and gray copper grits, as the name of old *red* sandstone was found to be in reference to the schistose black rocks of Devonshire.

To this "Permian System" we refer the chief deposits of gypsum of Arzamas, of Kazan, and of the rivers Piana, Kama and Oufa, and of the environs of Orenbourg; we also place in it the saline sources of Solikamsk and Sergiefsk, and the rock salt of Iletsk and other localities in the government of Orenbourg, as well as all the copper mines and the large accumulations of plants and petrified wood, of which you have given a list in the "Bulletin" of your Society (anno 1840).   Of the fossils of this system, some undescribed species of *Producti* might seem to connect the Permian with the carboniferous æra; and other shells, together with fishes and Saurians, link it on more closely to the period of the Zechstein, whilst its peculiar plants appear to constitute a Flora of a type intermediate between the epochs of the new red sandstone or "trias" and the coal-measures.   Hence it is that I have ventured to consider this series as worthy of being regarded as a "System."

I have not time to enter upon the numerous and interesting phaenomena of the Ural Mountains, the examination of which occupied us nearly three months.   We there studied alternately the wonders of the gold alluvia, the sites of the entombment of your great mammalia, and sought for the causes of the astonishing metamorphism of the sedimentary rocks of that chain.   For an explanation of the last class of phaenomena, the works of Humboldt and Gustaf Rose must always be consulted.   I will on this occasion simply say, that far from being *primitive*, as was supposed, this chain, with the exception of its eruptive masses, is entirely composed of *Silurian, Devonian* and *Carboniferous* rocks, more or less altered and crystallized, but in which nevertheless we have been able to recognize in a great number of localities my own *Pentamerus Knightii*, and many fossils which clearly define the age of the other strata.   These rocks, though much broken up, are arranged in parallel bands, the mean direction of which in the North Ural is from N. and by W. to S. and by E., whilst in the South Ural, trending N. and S., they assume a fan-shaped arrangement, spreading out towards the southern steppe of the Kirghis, where, interlaced with porphyries and other trap-rocks, they are often converted into the far-famed jaspers of this region.

Accept, dear Sir, the assurance of the affection and esteem of your devoted servant,

RODERICK IMPEY MURCHISON
President of the Geol. Society of London

HITCHCOCK

Edward Hitchcock (1793–1864), American geologist, was long associated with Amherst College, first as professor of chemistry and natural history and later as president. He served as state geologist of Massachusetts from 1830 to 1841, and in 1856 he reorganized the geological survey of Vermont. His name is indelibly associated with the study of the fossil footprints and Triassic rocks of the Connecticut Valley; the ideas which he developed concerning the metamorphism of sedimentary rocks were revolutionary at the time.

Footmarks on New Red Sandstone in Massachusetts

From *American Journal of Science*, Vol. XXIX, pp. 307–340, 1836.

The almost entire absence of birds from the organic remains found in the rocks, has been to geologists a matter of some surprise. Up to a very recent date, I am not aware that any certain examples of these animals in a fossil state have been discovered, except the nine or ten specimens found by Cuvier, in the tertiary gypsum beds near Paris. In the third volume (third edition) of his *Ossemens Fossiles*, he has examined all the cases of fossil birds reported by previous writers, and he regards them, nearly all, as deserving little credit.

For this paucity of ornitholites, geologists have, indeed, assigned probable reasons, derived from the structure and habits of birds. These render them less liable, than quadrupeds and other animals, to be submerged beneath the waters, so as to be preserved in aqueous deposites; and even when they chance to perish in the water, they float so long upon the surface, as to be most certainly discovered, and devoured by rapacious animals.

But although these circumstances satisfactorily explain the fact, above referred to, they do not render the geologist less solicitous to discover any relics of the feathered tribe, that may be found in the fossiliferous rocks: and I have, therefore, been much gratified by some unexpected disclosures of this sort, during the past summer, in the new red sandstone formation on the banks of Connecticut river, in Massachusetts.

My attention was first called to the subject by Dr. James Deane of Greenfield; who sent me some casts of impressions, on a red

micaceous sandstone, brought from the south part of Montague, for
flagging stones. . . . They consist of two slabs, about forty inches
square, originally united face to face; but on separation, presenting
four most distinct depressions on one of them, with four correspondent projections on the other; precisely resembling the impressions of the feet of a large bird in mud. . . .

Not long afterwards, Col. John Wilson of Deerfield, pointed out
to me similar impressions on the flagging stones in that village.
Having ascertained that these were brought from the town of
Gill, . . . I visited the spot, and was gratified to find several
distinct kinds of similar impressions; some of them very small, and
others almost incredibly large. . . .

At the quarries above named, these impressions are exhibited on
the rock in place, as depressions, more or less perfect and deep,
made by an animal with two feet, and usually three toes.   In a few
instances, a fourth or hind toe, has made an impression, not directly
in the rear, but inclining somewhat inward; and in one instance, the
four toes all point forward. . . .

. . . In all cases where there are three toes pointing forward, the
middle toe is the longest; sometimes very much so.   In a majority
of cases, the toes gradually taper, more or less to a point: but in
some most remarkable varieties they are thick and somewhat
knobbed, and terminate abruptly.

In the narrow toed impressions, distinct claws are not often seen,
although sometimes discoverable.   But in the thick toed varieties,
they are often very obvious.   Much, however, in respect to this
appendage, depends upon the nature of the rock.   If it be composed
of fine clay, the claws are usually well marked. . . .

I trust I have proceeded far enough in these details, to justify me
in coming to the conclusion, that the impressions are the tracks of
birds, made while the incipient sandstone and shale were in a
plastic state.   This is the conclusion, to which the most common
observer comes, at once, upon inspecting the specimens.   But the
geologist should be the last of all men to trust to first impressions.
I shall, therefore, briefly state the arguments that sustain this
conclusion.

1. These impressions are evidently the tracks of a biped animal.
For I have not been able to find an instance, where more than a
single row of impressions exists.

2. They could not have been made by any other known biped, except birds.   On this point, I am happy to have the opinion of more than one distinguished zoologist.

3. They correspond very well with the tracks of birds.   They have the same ternary division of their anterior part, as the feet of birds.   Frequently, and perhaps always, the toes, like those of birds, are terminated by claws.   If the toes are sometimes slender and sometimes thick and blunt, so are those of birds.

METAMORPHISM OF THE NEWPORT CONGLOMERATE

From *American Journal of Science*, Second Series, Vol. XXXI, pp. 372–392, 1861.

Doubtless many geologists will demur at my conclusions, as I should have done without visiting the localities.   I can ask them only to suspend their judgment till they have seen the rocks which I describe, especially those at Newport and at Plymouth.   If they shall then propose any more rational theory, I hope I shall be willing and thankful to receive it.

With these preliminary remarks I proceed to the details.   We [including his son, Charles H. Hitchcock] give them as proving and illustrating the following statement. . . .

We have found striking examples where the pebbles of conglomerates have been elongated and flattened so as at length to be converted into the siliceous laminae of the schists and gneiss and the cement into mica, talc, and feldspar.

Perhaps the best exposure of the Rhode Island conglomerate is at the well known "Purgatory," two and a half miles east of Newport, and within the limits of Middletown.   According to the paper of C. H. Hitchcock read before the Am. Association in Aug., 1860, the belt of conglomerate commences a little south of Purgatory, is a mile wide with interstratified belts of slates, and extends N. 30°E. probably as far as Sandy Point, in Portsmouth some 5½ miles.   It shows several folds, is underlaid by a gritty schist or sandstone, and itself underlies the coal measures.

"It is a coarse conglomerate, composed of elongated and flattened pebbles, from the smallest size, to bowlders nearly 12 feet long, cemented by a meagre amount of talcose schist, or sandstone," with numerous small disseminated crystals of magnetite. The pebbles are mostly a fine-grained, or compact quartz rock, which when partly decomposed appears like sandstone; not unfrequently the pebbles seem to pass into an imperfect mica schist, and show lamination. A few of them are gneiss, and probably granite, and occasionally hornblende rock. In their shortest diameter they rarely exceed a foot, while in length, one, two, and three feet are very common, and a few may be seen from 4 to 6, and one, at least, is as long as 12 feet. The following facts as to the pebbles, are of the most interest:

1. They are often very much elongated in the direction of the strike; 2. They are flattened, but not so strikingly as they are elongated; 3. They are indented often deeply by one being pressed into another; 4. They are sometimes a good deal bent, occasionally in two directions; 5. They are cut across by parallel joints or fissures, varying in distance from each other from one or two inches to many feet. The most distinct of these joints, which are a rod or two apart; are perpendicular to the horizon, and nearly at right angles to the strike, and make a clean cut from top to bottom of hills 30 or 40 feet high. Abrading agencies have often removed the rock on one side of these joints, or between two of them, so as to leave walls of pebbles smoothly cut in two; the whole appearing like a pile of wood neatly sawed. Acres of such walls may be seen in the vicinity of "Purgatory." Often the surface of the pebbles thus cut through is not only perfectly even, but smooth and seemingly polished. Yet the two parts of the pebbles thus cut off, perfectly correspond, and one part has never been made to slip over the other. In some minor joints single pebbles are not entirely cut off, but are sometimes drawn out of their beds at one end where the rock is separated, and remain projecting above the cleared surface. These joints do not always extend through the whole rock.

From these facts we could hardly avoid drawing the following conclusions:

1. This rock was once a conglomerate of the usual character, except in the great abundance of the pebbles, and it has subsequently experienced great metamorphoses making the cement crystalline and schistose, and elongating and flattening the pebbles.

2. The pebbles must have been in a state more or less plastic, when they were elongated, flattened and bent. If their shape has been thus altered, their plasticity must of course be admitted; for the attempt to change their present form would result only in fracture and comminution. The degree of plasticity, however must

have varied considerably; for some of them are scarcely flattened or elongated at all—and as has been stated, some are not cut off by the joints.

The neat and clean manner in which the pebbles have been generally severed by the joints, implies plasticity.

Fig. 1 will give some idea of an elongated pebble from **Newport,** which is 10 inches long and 3 inches across its broadest part.

1.

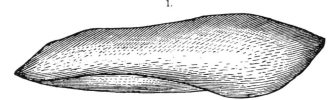

Fig. 2 shows a pebble 8 inches long with a deep indentation.

2.

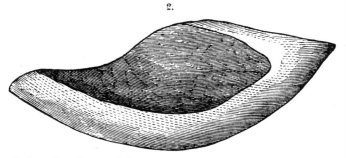

Perhaps I ought to add that sometimes the elongated pebbles **partially** or wholly lose their rounded form at the ends, and begin to assume a foliated or schistose aspect, and to be somewhat blended with the talcose or micaceous cement. This though not general, is frequently the case.

FIG. 12.—Illustrations accompanying Hitchcock's paper on the metamorphism of conglomerates, 1861.

For though occasionally we meet with one that has a somewhat uneven surface, as if mechanically broken, such cases are rare. Whatever may be our theory of the agency that has formed the joints, the conviction is forced upon every observer that the materials must have been in a soft state after their original consolidation. There is no evidence that the opposite walls have slid upon one another at all, as the opposite parts of the pebbles coincide. It seems as if a huge saw or cleaver had done the work.

These proofs of plasticity apply essentially, though less forcibly to the micaceous and talcose cement which has also been cut across by these joints. Though generally small in quantity it sometimes forms layers of considerable thickness interstratified with the pebbles.

Some have imagined that the elongated, flattened, bent, and indented pebbles of this conglomerate may have been worn into their present shape and brought into a parallel arrangement by the mechanical attrition of waves and currents. We feel sure that an extensive and careful examination of the localities, and of beaches where shingle is now being formed, will convince any one that they cannot have had such an origin.

1. We do not believe that any beach can be found with pebbles that have anything more than a slight resemblance to those at Newport. Those somewhat elongated may indeed be found where they are derived from slate rocks. But nowhere does the attrition of pebbles against one another produce deep indentations, and leave the one neatly fitting into the other, nay, one bent partially around the other, as is the case at Newport. If these phenomena were produced by original attrition how strange that they should have such an extraordinary development on Rhode Island, while it is not marked enough in any other conglomerate in our country save in Vermont, to have arrested the attention of geologists.

2. The remarkable joints in this conglomerate prove that the pebbles have been in a plastic state, and since the strata have been much folded, and consequently subjected to strong lateral pressure, how could the pebbles have escaped compression and modification of form? A mass of the conglomerate when broken open along the line of strike, a good deal resembles a plug of tobacco, which has been rolled into lumps and then subjected to strong pressure, so that the lumps are distorted and made to conform to all the irregularities around them.

3. The force by which the pebbles were flattened and indented must have operated laterally, as would result from the plication of the strata; folds in which are frequent. If there were a great superincumbent pressure and less in the direction of the strike, the same lateral force might have elongated the pebbles. But perhaps there may have been also a horizontal curvature in the strata, to aid in the work, as we shall explain when we come to describe the Vermont localities. It may not, however, be easy to show how this

compressing force has operated where rocks have been so folded and disturbed as around Newport, for the conglomerate is in juxtaposition with granite, which has exerted a powerful metamorphic influence on other strata there; but if we can show the results of the agency, our main object will be accomplished.

4. The phenomena of the joints in this rock, conduct us most naturally to some polar force as the chief agent in their production. Mere shrinkage could not have separated the pebbles as smoothly; much less could a strain from beneath have thus fractured them; for sometimes the joints are not more than two or three inches apart, and if we suppose one of them to have been the result of fracture, yet how is the other to be obtained in that manner? A simple inspection of the rock in place will satisfy any one that no mechanical agency is alone sufficient to explain these phenomena. We have been driven to the supposition of some polarizing force acting upon soft materials. If, as Sir John Herschel supposes, cleavage may have resulted from a sort of crystallization in plastic materials, why may not joints come into the same category? Why should the conclusions drawn from the experiments of Mr. Fox upon the lamination of plastic clay, by electric currents, be limited to cleavage?

5. The Newport conglomerate is probably only a special variety of the extensive deposit of highly silicious puddingstone found so abundantly between Boston and Rhode Island. Both have the same geological position, we believe, and were the Roxbury conglomerate to be brought into a plastic state, and the pebbles elongated and flattened by pressure, we think the result would resemble the Newport conglomerate.

STUDER

Bernard Studer (1794–1887), Swiss scientist, was for many years professor of geology at Berne.

METAMORPHIC MESOZOIC ROCKS DISTINGUISHED FROM PALEOZOIC

Translated from "Geognostische Bermerkungen über einige Theile der nordlichen Alpenkette," *Zeitschrift für Mineralogie* (Leonhard), Vol. XXI, pp. 1ff., 1827.

The high mountain ranges between Tödi and Galanda and the northern, limestone mountains of Glärnisch, Mürtschenstock, and Kuhfirsten . . . appear, except for the abnormal alterations which they have undergone in some parts, to be formed of shales and sandstones which it is most difficult, if not impossible, to separate into different formations. . . . Black or gray shales, dull and earthy, or faintly shiny, are predominant. They are often closely associated with bluish gray shaly sandstone or sandy calcareous shales, with the extremely thin beds of which they are interfoliated. Also, where the sandstone is more massive and conspicuous, as in the mountains adjoining Bündten, it usually shows a marked schistose structure. It is further distinguished by a somewhat coarser and more evident grain and by lighter, greenish gray colors. . . . Dark limestones, in great masses rising in the mountains, bedded or irregularly fissured, sometimes lie on these shales and sandstones, sometimes are placed as irregularly defined interbeds or horizontal stocks, and frequently seem to form their base and separate them from the crystalline formations.

Hitherto the greater part of these formations—frequently, all of them—have been listed among the Transition Series; the shales have been considered as true clay schists; the sandstones, as graywackes. The mineralogical character of the rock types, as well as the relations of the beds have seemed to corroborate this hypothesis fully. But important contradictions arise on the paleontologic side, doubts which, in so far as they may be confirmed, must result either in a very different conception of the age of these formations—or in the complete abandonment of the

257

organic-geologic system.   So far as I know, no fossil characteristic of the Transition Series has ever been found in these shales—no orthoceratite, no Productus, no coral.   On the contrary, there are a surprising number of nummulites, *N. Llaevigatus* Lam., frequently more than one inch in diameter, turrilites, agreeing closely with *T. Bergeri*, echinoids, . . . cardium and pectens of ordinary shale, little oysters, etc.   All these are fossils which hitherto have ordinarily been found in the Greensand or Cretaceous.   However, the ammonites characteristic of the Greensand, the inoceramus, the hamites, etc., are lacking. . . . At some places, it was thought that several nummulite layers, one on top of the other and separated by shales, could be distinguished, although this might be an illusion caused solely by the almost incredible tangle of the mountain structure. . . . We found nummulites and green grains in every one of these thick, subordinate limestone masses.   We were convinced that we encountered just as many formation divisions, separated by shales, on the uniform southern dips. Even at the foot of the cliff, the nummulite limestone still appears, broken through vertically by thick veinlike masses of gray quartz rock which still holds faint traces of lime.   How astonished we were when we could have a view of the whole cliff from the other side of the valley!   The limestone masses, more than one hundred feet thick, striking uniformly through half the valley, seem to turn downwards suddenly at one end, upwards at the other, or to bend down at both to unite with lower and upper, and to form one continuous mass which, in the broad winding of the shale mountains, seems to wander from the bottom of the valley to the highest peaks.

. . . But at Rieseten-Pass itself, the character of the rock type alters gradually, and the change seems to extend over the higher parts of the Troskigräte, as well.   The shales become more and more shiny, deceptively resembling the older clay schists.   At the same time, the dark color takes on a reddish cast and becomes violet.   Locally, vivid gray, green, and red colors with distinct silky luster are seen.   Between these schists lay thin beds of granular quartz. . . . In a horizontal direction, just as in depth, no separation of these rocks from the black shales and sandstones is possible.   The transition of the upper, variegated rocks of Riesetengrät into the dark gray, dull rocks of the Fohalp and the Kalfenserberge can be traced distinctly from Seezboden.

On reviewing again the whole extent of the variegated schists and conglomerates, it is remarkable how the direction of their main members, the direction of greatest development, runs almost exactly parallel with that of the Great Valley of Glarner. Diverging greatly from other Alpine formations whose strike-line is also that of the Alpine chain, our line cuts this line with an angle of about 60° northeast. Thus it shows anew that they must not be considered as an important division of the formation series, of the same rank as the gneiss, the shale group, or the Molasse, but only as a special, later occurring modification of one of these features.

DE LA BECHE

Henry Thomas De la Beche (1796–1855), English geologist, was the first director of the Geological Survey of the British Isles, established in 1835. The works issued under his supervision were the most detailed and accurate presentation of geological features that had thus far been prepared in any country. His *Manual of Geology*, published in 1831, is notable "for its ample and clear presentation of the science."

The Drafting of Geological Sections

From *Sections and Views Illustrative of Geological Phenomena*, London, 1830, pp. vii, viii, and description of Pl. 2.

One of the principal objects in the following work is to induce geologists to present us with sections more conformable to nature than is usually done. Sections and views are, or ought to be, miniature representations of nature; and to them we look, perhaps, more than to memoirs, for a right understanding of an author's labours. From a want of attention to this subject, it sometimes happens that the sections and accompanying memoirs are not in strict accordance, particularly after the sections are reduced to their proper proportions. This want of accordance I have had more than one occasion to regret in the following work, as it has compelled me to omit much that appeared valuable. Among the sections here presented, there are doubtless many that are only approximations to the truth. But as approximations they may be valuable, and add to our stock of knowledge.

The advance of geological science has lately been so rapid, that it requires some exertion to keep pace with it. Hasty conclusions can no longer command attention,—there are too many observers in the field to permit errors to remain long uncontradicted,—and it is very desirable for the progress of science, that no deference for a name should cause them to remain uncontradicted. It surely can be no offense to state, that the progress of science has led to new views, and that the consequences that can be deduced from the knowledge of a hundred facts may be very different from those deducible from five. It is also possible that the facts first known may be the exceptions to a rule, and not the rule itself, and

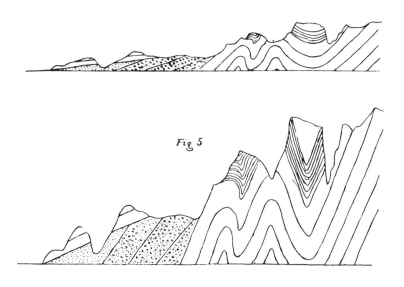

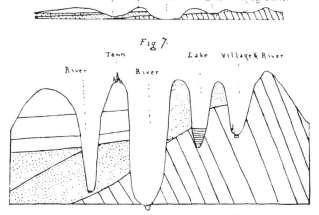

Fig. 13.—Four of the figures used by De la Beche to illustrate his ideas concerning the drafting of geological sections, 1830.

generalizations from these first-known facts, though useful at the time, may be highly mischievous, and impede the progress of the science if retained when it has made some advance.

Explanation of Plate II

This Plate is intended to illustrate the value of proportion in geological sections generally. From a want of attention to this subject, the greater part of such sections are more mischievous than useful, and tend to mislead rather than to instruct the geologist. . . .

Fig. 4. is an imaginary section, supposed proportional, introduced in order to show the mischief of adopting a scale of height differing from that of length.

Fig. 5. is a section, supposed of the same country, the scale of height being three times that of the length; the distortion has produced great differences in the dips, and has destroyed the real outline of the country, but the section still remains in some measure intelligible.

Fig. 6. is another imaginary section, supposed to be proportional.

Fig. 7. is a section of the same country, the scale of height being ten times that of the distance. The confusion that arises in this case is very apparent. The true outline of the country is entirely destroyed, towns are perched on heights, or placed in valleys, on or in which they never would have been built, wide valleys become gorges, and shallow lakes and rivers acquire considerable depth; not only is the outline of the country distorted, but the relative positions of the strata appear altered, and erroneous impressions are produced. In Fig. 6. strata may be imagined to fine off, but in Fig. 7. the same strata appear to abut against each other in considerable thickness.

The object of Fig. 4. 5. 6. and 7. is to induce the geological student to form sections as nearly as possible approaching to the proportions exhibited in nature. Such sections as are represented by Fig. 7. are by no means so uncommon in geological memoirs as might be imagined: it is true that the strata would not be made to abut so suddenly against each other as is here represented; the lines would probably be turned up, but the section would nevertheless be a distorted section, and convey an erroneous idea of nature.

LYELL

Sir Charles Lyell (1797–1875) probably accomplished more in the advancement of geological knowledge than any other one man. The world-wide influence of his treatises and textbooks definitely established the principle of uniformitarianism.

Uniformitarianism

From *Principles of Geology*, 1st ed., Vol. III, pp. 1–5, London, 1833.

All naturalists, who have carefully examined the arrangement of the mineral masses composing the earth's crust, and who have studied their internal structure and fossil contents, have recognized therein the signs of a great succession of former changes; and the causes of these changes have been the object of anxious inquiry. As the first theorists possessed but a scanty acquaintance with the present economy of the animate and inanimate world, and the vicissitudes to which these are subject, we find them in the situation of novices, who attempt to read a history written in a foreign language, doubting about the meaning of the most ordinary terms; disputing, for example, whether a shell was really a shell,— whether sand and pebbles were the result of aqueous trituration, —whether stratification was the effect of successive deposition from water; and a thousand other elementary questions which now appear to us so easy and simple, that we can hardly conceive them to have once afforded matter for warm and tedious controversy.

In the first volume we enumerated many prepossessions which biassed the minds of the earlier inquirers, and checked an impartial desire of arriving at truth. But of all the causes to which we alluded, no one contributed so powerfully to give rise to a false method of philosophizing as the entire unconsciousness of the first geologists of the extent of their own ignorance respecting the operations of the existing agents of change.

They imagined themselves sufficiently acquainted with the mutations now in progress in the animate and inanimate world, to entitle them at once to affirm, whether the solution of certain problems in geology could ever be derived from the observation of the actual economy of nature, and having decided that they could not, they felt themselves at liberty to indulge their imaginations, in guessing at what *might be*, rather than in inquiring *what is;* in other words, they employed themselves in conjecturing what might have been the course of nature at a remote period, rather than in the investigation of what was the course of nature in their own times.

It appeared to them more philosophical to speculate on the possibilities of the past, than patiently to explore the realities of the present, and having invented theories under the influence of such maxims, they were consistently unwilling to test their validity by the criterion of their accordance with the ordinary operations of nature. On the contrary, the claims of each new hypothesis to credibility appeared enhanced by the great contrast of the causes or forces introduced to those now developed in our terrestrial system during a period, as it has been termed, of *repose.*

Never was there a dogma more calculated to foster indolence, and to blunt the keen edge of curiosity, than this assumption of the discordance between the former and the existing causes of change. It produced a state of mind unfavourable in the highest conceivable degree to the candid reception of the evidence of those minute, but incessant mutations, which every part of the earth's surface is undergoing, and by which the condition of its living inhabitants is continually made to vary. The student, instead of being encouraged with the hope of interpreting the enigmas presented to him in the earth's structure,—instead of being prompted to undertake laborious inquiries into the natural history of the organic world, and the complicated effects of the igneous and aqueous causes now in operation, was taught to despond from the first. Geology, it was affirmed, could never rise to the rank of an exact science,—the greater number of phenomena must for ever remain inexplicable, or only be partially elucidated by ingenious conjectures. Even the mystery which invested the subject was said to constitute one of its principal charms, affording, as it did, full scope to the fancy to indulge in a boundless field of speculation.

The course directly opposed to these theoretical views consists in an earnest and patient endeavour to reconcile the former indications of change with the evidence of gradual mutations now in progress; restricting us, in the first instance, to known causes, and then speculating on those which may be in activity in regions inaccessible to us. It seeks an interpretation of geological monuments by comparing the changes of which they give evidence with the vicissitudes now in progress, or *which may be* in progress.

We shall give a few examples in illustration of the practical results already derived from the two distinct methods of theorizing, for we have now the advantage of being enabled to judge by experience of their respective merits, and by the relative value of the fruits which they have produced.

In our historical sketch of the progress of geology, the reader has seen that a controversy was maintained for more than a century, respecting the origin of fossil shells and bones—were they organic or inorganic substances? That the latter opinion should for a long time have prevailed, and that these bodies should have been supposed to be fashioned into their present form by a plastic virtue, or some other mysterious agency, may appear absurd; but it was, perhaps, as reasonable a conjecture as could be expected from those who did not appeal, in the first instance, to the analogy of the living creation, as affording the only source of authentic information. It was only by an accurate examination of living testacea, and by a comparison of the osteology of the existing vertebrated animals with the remains found entombed in ancient strata, that this favourite dogma was exploded, and all were, at length, persuaded that these substances were exclusively of organic origin.

In like manner, when a discussion had arisen as to the nature of basalt and other mineral masses, evidently constituting a particular class of rocks, the popular opinion inclined to a belief that they were of aqueous, not of igneous origin. These rocks, it was said, might have been precipitated from an aqueous solution, from a chaotic fluid, or an ocean which rose over the continents, charged with the requisite mineral ingredients. All are now agreed that it would have been impossible for human ingenuity to invent a theory more distant from the truth; yet we must cease to wonder, on that account, that it gained so many proselytes, when we remember that its claims to probability arose partly from its confirming the assumed want of all analogy between geological causes and those now in action.

By what train of investigation were all theorists brought round at length to an opposite opinion, and induced to assent to the igneous origin of these formations? By an examination of the structure of active volcanos, the mineral composition of their lavas and ejections, and by comparing the undoubted products of fire with the ancient rocks in question.

We are now, for the most part, agreed as to what rocks are of igneous, and what of aqueous origin,—in what manner fossil shells, whether of the sea or of lakes, have been imbedded in strata,—how sand may have been converted into sandstone,—and are unanimous as to other propositions which are not of a complicated nature; but when we ascend to those of a higher order, we find as little disposition, as formerly, to make a strenuous effort, in the first instance, to search out an explanation in the ordinary economy of Nature. If, for example, we seek for the causes why mineral masses are associated together in certain groups; why they are arranged in a certain order which is never inverted; why there are many breaks in the continuity of the series; why different organic remains are found in distinct sets of strata; why there is often an abrupt passage from an assemblage of species contained in one formation to that in another immediately super-imposed,—when these and other topics of an equally extensive kind are discussed, we find the habit of indulging conjectures, respecting irregular and extraordinary causes, to be still in full force.

We hear of sudden and violent revolutions of the globe, of the instantaneous elevation of mountain chains, of paroxysms of volcanic energy, declining according to some, and according to others increasing in violence, from the earliest to the latest ages. We are also told of general catastrophes and a succession of deluges, of the alternation of periods of repose and disorder, of the refrigeration of the globe, of the sudden annihilation of whole races of animals and plants, and other hypotheses, in which we see the ancient spirit of speculation revived, and a desire manifested to cut, rather than patiently to untie, the Gordian knot.

In our attempt to unravel these difficult questions, we shall adopt a different course, restricting ourselves to the known or possible operations of existing causes; feeling assured that we have not yet exhausted the resources which the study of the present course of nature may provide, and therefore that we are not authorized, in the infancy of our science, to recur to extraordinary agents. We shall adhere to this plan, not only on the grounds explained in the first volume, but because, as we have above stated, history informs us that this method has always put geologists on the road that leads to truth,—suggesting views which, although imperfect at first, have been found capable of improvement, until at last adopted by universal consent. On the other hand, the opposite method, that of speculating on a former distinct state of things, has led invariably to a multitude of contradictory systems, which have been overthrown one after the other,— which have been found quite incapable of modification,—and which are often required to be precisely reversed.

In regard to the subjects treated of in our first two volumes, if systematic treatises had been written on these topics, we should willingly have entered at once upon the description of geological monuments properly so called, referring to other authors for the elucidation of elementary and collateral questions, just as we shall appeal to the best authorities in conchology and comparative anatomy, in proof of many positions which, but for the labours of naturalists devoted to these departments, would have demanded long digressions. When we find it asserted, for example, that the bones of a fossil animal at Œningen were those of man, and the fact adduced as a proof of the deluge, we are now able at once to dismiss the arguments as nugatory, and to affirm the skeleton to be that of a reptile, on the authority of an able anatomist; and when we find among ancient writers the opinion of the gigantic stature of the human race in times of old, grounded on the magnitude of certain fossil teeth and bones, we are able to affirm these remains to belong to the elephant and rhinoceros, on the same authority.

But since in our attempt to solve geological problems, we shall be called upon to refer to the operation of aqueous and igneous causes, the geographical distribution of animals and plants, the real existence of species, their successive extinction, and so forth, we were under the necessity of collecting together a variety of facts, and of entering into long trains of reasoning, which could only be accomplished in preliminary treatises.

These topics we regard as constituting the alphabet and grammar of geology; not that we expect from such studies to obtain a key to the interpretation of all geological phenomena, but because they form the groundwork from which we must rise to the contemplation of more general questions relating to the complicated results to which, in an indefinite lapse of ages, the existing causes of change may give rise.

Subdivisions of the Tertiary Epoch

From *Principles of Geology*, 4th ed., Book III, pp. 384–400, London, 1835.

I shall now proceed to consider the subdivisions of tertiary strata which may be founded on the results of a comparison of their respective fossils, and to give names to the periods to which they may be severally referred.   But, first, it will be necessary to explain the difference between the *tertiary* phenomena and those described in the last two books.   In the present work all those geological monuments are called tertiary which are newer than the secondary formations, and which on the other hand cannot be proved to have originated since the earth was inhabited by man.   Part of the changes, whether of the animate or inanimate world, considered in the preceding books, was ascertained by historical testimony to have taken place within the human epoch; as, for example, the accumulation of the newer portions of the deltas of the Po, Rhone, and Nile.   Another part, where history was silent, was proved to belong to the same epoch by the evidence of the fossil remains of man or his works.   All formations, whether igneous or aqueous, which can be shown by any such proofs to be of a date posterior to the introduction of man, will be called *Recent*.   Some authors have applied the term *contemporaneous* in the same sense; but as this word is so frequently in use to express the synchronous origin of distinct rocks of every age, it would be a source of great inconvenience and ambiguity if we were to confine it to a technical meaning.

To resume my classification:—the tertiary strata may be divided into four groups, in the older of which we find an extremely small number of fossils identifiable with species now living; whereas on approaching the superior and newer sets, we find the remains of recent testacea in abundance.   In no instance where we have an opportunity of observing two distinct formations in contact, the one superimposed upon the other, do we meet with an assemblage of organic remains in the uppermost differing more widely from the existing creation than the fossils of the inferior group. If there is occasionally an apparent exception to the rule, it is only where the remains belong to distinct classes of the animal kingdom; as, for example, where a deposit containing the bones of quadrupeds for the most part extinct overlies a stratum in which the imbedded shells are mostly of recent species—such exceptions seem to point to a difference in the comparative duration of species in different classes, but do not invalidate the general proposition before laid down.

Newer Pliocene period. The latest of the four periods before alluded to is that which immediately preceded the Recent era. To this more modern period may be referred a portion of the strata of Sicily, the district around Naples, and several others to be considered in the sequel. They are characterized by a great preponderance of fossil shells referrible to species still living, and may be called the Newer Pliocene strata, the term Pliocene, or "more recent," being derived from πλειων, *major*, and καινος, *recens*, as a large, often by far the largest, part of the fossil shells are of recent species.

Out of 226 fossil species brought from the Sicilian beds above alluded to, M. Deshayes found that no less than 216 were of species still living, and for the most part in the Mediterranean, whereas ten only were of extinct or unknown species. I do not imagine that any of the 'groups referred to this period in the present work contain much more than the proportion of one in ten of extinct species of shells. Nevertheless, the antiquity of some Newer Pliocene strata of Sicily, as contrasted with our most remote historical eras, must be very great, embracing perhaps myriads of years. There are no data for supposing that there is any break, or strong line of demarcation, between the strata and fossils of this and the Recent epoch; but, on the contrary, the monuments of the one seem to pass insensibly into those of the other.

Older Pliocene period. The formations termed Subapennine in the north of Italy, and in Tuscany, contain among their fossil shells a large number which have been identified with living species. The proportion of *recent* shells usually approaches to one half. Out of 569 species examined from these strata in Italy, 238 were found to be still living, and 331 extinct or unknown. Out of 111 from the English crag, M. Deshayes determined forty-five to be recent species, and sixty-six to be extinct or unknown. . . .

Miocene period. This antecedent tertiary epoch I shall name Miocene, or "less recent" from μειων, *minor,* and καινος, *recens,* a small minority only of the fossil shells imbedded in its formations being referrible to living species. After examining 1021 Miocene shells, M. Deshayes found that only 176 were recent, being in proportion of rather more than seventeen in one hundred. As there are a certain number of fossil species which are exclusively confined to the Pliocene period, so also there are many shells equally characteristic of the Miocene. The species which pass from the Miocene into the Pliocene period, or which are common to both, are in number 196, of which 114 are living, and eighty-two extinct. The Miocene strata are largely developed in Touraine, and in the south of France near Bordeaux, in Piedmont, in the basin of Vienna, and other localities. . . .

Eocene period. The period next antecedent may be called Eocene, from ηως, *aurora,* and καινος, *recens,* because the very small proportion of living species contained in these strata indicates what may be considered the first commencement, or *dawn,* of the existing state of the animate creation. To this era the formations first called tertiary, of the Paris and London basins, are referrible. . . .

The total number of fossil shells of this period already known, when the tables of M. Deshayes, before alluded to, were constructed, was 1238, of which number forty-two only are living species, being in the proportion of nearly three and a half in one hundred. Of fossil species, not known as recent, forty-two were found to be common to the Eocene and Miocene epochs.

The present geographical distribution of those recent species which are found fossil in formations of such high antiquity as those of the Paris and London basins, is a subject of the highest interest. In the more modern formations, where so large a proportion of the fossil shells belong to species still living, they also belong, for the most part, to species now inhabiting the seas immediately adjoining the countries where they occur fossil; whereas the recent species, found in the older tertiary strata, are frequently inhabitants of distant latitudes, and usually of warmer climates. Of the forty-two Eocene species, or those found in the earliest tertiary strata, which occur fossil in England, France, and Belgium, and are at the same time still living, about half now inhabit the seas within or near the tropics, and almost all the rest are inhabitants of the more southern and warmer parts of Europe. If some Eocene species still flourish in the same latitudes where they are found fossil, they are species which, like *Lucina divaricata*, are now found in many seas, even those of very distant quarters of the globe; and this wide geographical range indicates a capacity of enduring a variety of external circumstances, which may enable a species to survive considerable changes of climate and other revolutions of the earth's surface. One fluviatile species (*Melania inquinata*), fossil in the Paris basin, is now known only in the Philippine Islands; and, during the lowering of the temperature of certain parts of the earth's surface, may perhaps have escaped destruction by migrating to the south. I have pointed out in the third book how rapidly the eggs of freshwater species might, by the instrumentality of water-fowl, be transported from one region to another. Other Eocene species, which still survive and range from the temperate zone to the equator may formerly have extended from the pole to the temperate zone; and what was once the southern limit of their range may now be the most northern.

Even if geologists had not established several remarkable facts in attestation of the longevity of certain tertiary species, we might still have anticipated that the duration of the living species of aquatic and terrestrial testacea would be very unequal. For it is clear that those which have had a wide range, and inhabit many different regions and climates, may survive the influence of destroying causes, which might extirpate the greater part of species at present their contemporaries. The increasing of existing species, and gradual disappearance of the extinct, as we trace the series of formations from the older to the newer, is somewhat analogous, as was before observed, to the fluctuations of a population such as might be recorded at successive periods, from the time when the oldest of the individuals now living was born to the present moment; and those Eocene testacea which still flourish may be said to have outlived several successive states of the organic world, just as Nestor survived three generations of men.

It appears, then, that the numerical proportion of recent to extinct species of fossil shells in the different tertiary periods may be thus expressed.—In the

Newer Pliocene period about	90	to 95
Older Pliocene period.......	35	to 50
Miocene period...........	17	
Eocene period............	3½	

per cent of *recent* fossils

These numbers, however, must be regarded merely as the results obtained from a careful examination of the first groups which chance has thrown in our way, or which lie in the most accessible parts of Europe.

In thus selecting the proportional number of recent to extinct species of shells as a useful term of comparison for successive tertiary groups, or as one from which a convenient nomenclature may be derived, I have no wish to exalt the mere percentage of living species of fossil shells into the leading characteristic of each group.   The Eocene strata of Paris and London, for example, are marked by the presence of a vast variety of peculiar *extinct* species of testacea, as well as of other animal and vegetable remains, in comparison of which the proportion of living species is a character of subordinate importance.   At the same time it should be observed, that had the geologist collected the fossils of the crag of Norfolk, the blue clay of London, and the coarse white limestone of Paris, and then considered these formations merely with reference to the number of recent shells contained in each, he would have seen, by this character alone, that the Parisian and London strata differed widely from the crag, and agreed very closely with each other.   Afterwards, on extending his examination to the *extinct* species, he would find that those of the Paris and London formations also corresponded, and formed together an assemblage very distinct from the *extinct* species in the crag. In this and many other cases where our zoological investigations are far advanced, a reference to the proportion of recent species would lead to the same general classifications, as the mere consideration of extinct testacea in different tertiary formations.

Many geologists are desirous of connecting divisions such as those above pointed out with sudden and violent interruptions to the ordinary course of events, and they regard them as indicative of successive changes in the organic world, accompanying revolutions equally important in the physical geography of the earth's surface.   But I have already attempted to show, that such apparent breaks in the geological series may be accounted for partly by the mode in which the commemorative processes operate, partly by the removal of strata by denudation, and that they arise, in part, from the small progress which we have hitherto made in the discovery and study of such deposits as are preserved.

From the experience of the last few years, we may anticipate the discovery of many intermediate gradations between the boundary lines first drawn; and if formations are brought to light intervening between the Eocene and Miocene, or between those of the last period and the Pliocene, we may still find an appropriate place for all by forming subdivisions, on the same principle as that which has determined the separation of the lower from the upper Pliocene groups. Thus, for example, we might have three divisions of the Eocene epoch,—the older, middle, and newer; and three similar subdivisions, both of the Miocene and Pliocene epochs. In that case, the formations of the middle period must be considered as the types from which the assemblage of organic remains in the groups on both sides will diverge.

In conclusion, I may observe, that although the lapse of ages comprised within a single period is very much narrowed by the fourfold subdivision above explained, yet when all the Eocene and Miocene deposits are said to be *contemporaneous,* this term must be received with a good deal of latitude. Considerable intervals of time may have elapsed without giving rise to any marked distinction in the imbedded organic remains.

Suppose the growth of the delta of the Nile to cease from this moment, and some new river to begin to transport sediment into the Mediterranean at any other point, and to form a delta in the course of many thousand years, this last formation might contain the same fossils as the marine and fluviatile deposits of the Nile previously accumulated in Lower Egypt; the difference at least might be so trifling that future geologists would regard them as contemporaneous, if they followed the same rules of classification as those laid down in this chapter.

SCROPE

George Poulett Scrope (1797–1876), English geologist and member of
Parliament, made several important contributions to the theory of volcanic
action.

VOLCANOES

From *Considerations on Volcanos*, London, pp. 29–31, 92–94, 241–242, 1825.

The Source of Volcanic Energy

Now let us imagine the existence on any point near the surface
of the globe, of a solid body of lava, of indefinite horizontal and
vertical dimensions, at an elevated temperature, and cut off from
all communication with the external atmosphere, or ocean, by
overlying strata of solid rock.

The pressure sustained by this mass is either wholly null, or
consists of the reaction of the expansive force of the elastic vapours
it contains, against the surfaces by which it is enclosed.

It is obvious that no change can take place in this pressure so
long as the temperature remains fixed, and the confining surfaces
equally stable; nor can we conceive how any variation can be
produced in the stability of those boundaries, while the temperature, and consequently the expansive force of the confined mass,
remains fixed. To effect any change in this state of things, we
must therefore have recourse to the supposition of an accession of
caloric to the lava. This supposition, so far from being gratuitous
or unwarranted, is supported and confirmed with the strongest
evidence by other considerations. Thus we know that a great
quantity of caloric is continually passing off from every active
volcanic vent in combination with the immense volumes of heated
vapour and incandescent lavas, emitted from it. Unless the
loss of caloric thus sustained, were made up by the accession
of an equal, or nearly equal, quantity from below, it would be
impossible for the mass of lava beneath a volcanic vent to continue
in eruption, as it is frequently known to do, for centuries; or to
renew its eruptions, when they had ceased. The phenomena of
thermal springs suggest the same remark.

The observations lately made as to the temperature of mines, which *encreases with their depth*, lead to the conclusion that the interior of the globe, at no great vertical distance, is at an intense temperature. This internal accumulation of caloric must be continually endeavoring to put itself in equilibrio, by passing from the centre towards the circumference, wherever the conducting powers of the substances enveloping the globe, permit the transmission most readily into external space. But the conducting powers of the solid rocks which compose the outer crust of the earth, are necessarily imperfect in the highest degree, from their want of density and frequently porous structure.

What transmission of caloric can take place through the thick formations of secondary limestones, clays, shales, and sandstones?

The crystalline, and particularly the compact granitoidal rocks, seem, however, far better adapted for this purpose; and caloric would no doubt be propagated through them more readily than the former class, or through the schists, and laminar rocks.

From these considerations it is rendered probable that a continual supply of caloric passes off from the interior of the globe towards its circumference, wherever its transmission is facilitated or permitted by the conducting powers of the intervening substances, or by the temporary opening of vents for its more free escape.

If, as I think it will appear, the phenomena of volcanos, under all their various phases of action and quiescence, together with their accompaniments of earthquakes, &c. &c. and perhaps many of the more ambiguous and obscure indications of congenerous causes visible in the constitution of the globe's surface, can be accounted for in the simplest and most satisfactory manner, according to well-known principles of physics, by this single assumption of the exposure of subterranean masses of crystalline rocks, which we know to exist, to a continual accession of caloric from below, which we have the strongest reasons for presuming a priori—in this event we shall be bound by common sense and the simplest rules of induction to accept this hypothesis with the utmost confidence, and it would be the height of irrationality and scepticism to refuse our acquiescence in it.

Relation of Shape of Rock Bodies to Their Composition

In adducing examples of these different modifications of figure assumed by a mass of lava after its emission upon the earth's surface, in obedience to the laws that regulate the motion of all substances in a similar state of semi-liquidity,—it will be difficult not to perceive a remarkable fact, viz. that the lavas which are mineralogically classed as basalts, from the prevalence of the ferruginous minerals—augite, hornblende, or titaniferous iron in their composition,—are almost universally found to have spread into thin *sheets*, or long and shallow *currents*, to a considerable distance from the orifice of protrusion;—while those lava-rocks which consist almost wholly of felspar (trachytes) are as uniformly disposed in massive beds, hummocks, or domes.   Take for examples the extensive and shallow basaltic plateaux of the Mt. Dor and Cantal—of the environs of Le Puy en Velay—of the north of Ireland—of the plains of Iceland—and of the gentle slopes of the isle of Bourbon—take the innumerable long and straggling streams of the Auvergne—of Aetna, Vesuvius, &c.—and compare the conformation of these and other basaltic beds, with the massive trachytic hummocks that are closely grouped around the volcanic centres of the Mt. Dor, Cantal and Mt. Mézen in France, with the bell-shaped masses of the Puy de Dôme, de Sarcouy, and de Cliersou, with the numerous trachytic domes of Hungary, and with the still more stupendous beds of the American Andes.   Such a comparison presents a striking evidence of the truth of the observation. . . .

This so remarkable and constant relation between the mineral nature of a bed of consolidated lava and the proportions of its different dimensions, has been already recognized by Geologists; but many have been unfortunately led by this remark to the adoption of a serious error as to the origin of the trachytic and phonolitic rocks, which are considered by them as in no instance to have flowed on the surface of the earth, but to have been always elevated en masse into the position they now occupy.

Some writers have even gone the length of supposing that they swelled up like a bladder by inflation from below, (De Buch, Humboldt,) and are consequently still *hollow within*—a gratuitous supposition entirely at variance with all that we know for certain concerning the nature and mode of operation of the volcanic energy.

It is on the contrary obvious that the remarkable bulkiness of the felspathic lavas is fully and simply accounted for by their imperfect fluidity, which has been already recognised to diminish, ceteris paribus, with their specific gravity, and by no means induces the necessity of supposing any other mode of volcanic action than that by which the basaltic lavas were also produced. But it is likewise wholly and strikingly untrue that trachytes never occur in sheets or currents (nappes ou coulèes.) On the contrary when reduced to a great degree of comminution and possessing a high liquidity, the felspathic lavas have universally assumed that mode of disposition, as in the instance of the numerous streams of obsidian or glassy trachyte in Lipari, Teneriffe, Bourbon, Iceland, &c.; and under favorable external circumstances—as when the inclination of the ground from the orifice of protrusion was considerable, lavas of this quality, even though extremely coarse-grained, and therefore very imperfectly liquid, are frequently found to have spread laterally to a considerable distance, with this only distinction, that the currents thus formed are far thicker and more bulky than would have been the case with basaltic lavas of the same texture, and under the same outward circumstances.

Three Modes of Rock Formation

It appears to me therefore on the whole, that the formation of the grand mineral masses of every age composing the known crust of the globe, is attributable to *three* primary modes of production, distinct in their nature, but of which the products have been often confused and mingled together, from circumstances of isochronism or collocation.—These are,

I. The chemical precipitation of various mineral substances, but particularly silex and carbonate of lime, from a state of solution in the ocean, or other body of water; as its temperature and solvent powers gradually decreased.

II. The subsidence of particles of mineral matter, of various degrees of coarseness, from a state of suspension in the ocean or other reservoir, into which they had been taken up, either by the violent escape of aqueous vapour from the interior of the globe, by the abrasive force of marine and fluviatile currents, or finally by the decomposition of the shells of molluscus animals, which possessed the faculty of elaborating their coverings from the substances they procured from sea-water.

III. The elevation of crystalline matter through fissures in
the crust of the globe, which had been already formed
in the two former modes; this rise being occasioned either
by the expansion of a lower bed, in which case the rock
was elevated nearly in a solid state; or by its own intumescence, owing to a sudden diminution of compression; in
which case the matter rose in an imperfectly liquid state,
and at a high temperature.

All the characteristic differences observable in the successive
formations of every kind, may, I conceive, be satisfactorily traced
to the gradual diminution in frequency and energy of the productive causes, the varying nature of the original materials acted on,
and the chemical and mechanical changes they have undergone
during the process; and with this consideration in view these three
modes of production are perhaps fully equal to account for the
origin of all the great mineral masses observable on the surface
of the globe.   They have also one immense advantage over most,
perhaps over all, of the hypotheses that have as yet been brought
forward to explain the same appearances; and which speaks
volumes in their favour; and this is, that *they are all still in operation*,—with diminished energy, it is true; but this is the necessary
result of their nature.

THE ORIGIN OF VALLEYS

From *The Geology and Extinct Volcanos of Central France*, pp. 206–208,
London, 1858.

The volcanic district of the Haute Loire presents another chain
of proofs equally conclusive of the same fact.   It is impossible to
doubt that the present valleys of the Loire, and all its tributary
streams within the basin of Le Puy, have been hollowed out since
the flowing of the lava-currents, whose corresponding sections
now fringe the opposite margins of these channels with columnar
ranges of basalt, and which constitute the intervening plains.
Yet these lavas are undeniably of contemporary origin with the
cones of loose scoriae which rise here and there from their surface,
and which would necessarily have been hurried away by any
general and violent rush of waters over this tract of country.
It is indeed obviously impossible that any such flood should have
occurred; and we are therefore driven to conclude that the erosive

force of the streams which still flow in these channels, together with action of direct rains, frost, and other meteoric phenomena, have alone hollowed out this extensive system of deep, and, in some instances (as that of the Loire itself), wide valleys.

The leading idea which is present in all our researches, and which accompanies every fresh observation, the sound which to the ear of the student of Nature seems continually echoed from every part of her works, is—

Time!—Time!—Time!

At least, since by a fortunate concurrence of igneous and aqueous phenomena we are enabled to prove the valleys which intersect the mountainous district of Central France to have been for the most part gradually excavated by the action of such natural causes as are still at work, it is surely incumbent on us to pause before we attribute similar excavations in other lofty tracts of country, in which, from the absence of recent volcanos, evidence of this nature is wanting, to the occurrence of unexampled and unattested catastrophes, of a purely hypothetical nature.

GESNER

Abraham Gesner (1797–1864), Canadian physician and geologist, was provincial geologist of New Brunswick from 1838 to 1843. The geological survey of New Brunswick was the first to be undertaken by any British colony.

The Geology of Grand Manan

From *First Report of the Geological Survey of New Brunswick*, St. John, 1839.*

The Trap Rock Ridge

The west side of the island lies almost upon a straight line, notwithstanding several high headlands that advance into the sea. It is uninhabited on this side, which presents a bold front of overhanging cliffs and lofty mural precipices of majestic grandeur and beauty. Deep caverns are worn out of the solid base of the lofty wall, which tumbles headlong into the sea beneath. Along the straight coast on this side of Grand Manan, there is a lofty ridge of trap† rising most frequently in a perpendicular direction from the sea. The breadth of this ridge is about two and a half miles. The mountain thus skirting the shore is furrowed lengthwise, and is occupied by several small lakes, that fill the deep circular impressions along its summit. It can scarcely be doubted that these basins, now filled with water, were once the craters whence the trap flowed in a liquid state.‡

There is a wide difference between the trap rock and the schistoes formations underlying the eastern side of the mountain. The latter have deep ravines extending from west to east, and those distinguishing grooves and scratches, that point out the course of

---

* The excerpts printed here are from a reprint which appeared in *The Grand Manan Historian*, No. III, published by the Grand Manan Historical Society in 1936. The footnotes were supplied by Buchanan Charles, president of that society.

† The word *trap* comes from the Swedish *trappa*, meaning stair, and became current in geology because this kind of rock sometimes occurs in masses rising like steps.

‡ This, of course, is an erroneous deduction.—K. F. M.

a current of waters once sweeping over them.* The diluvial grooves common in the Province are parallel to the ravines worn out of the slate.

The eastern side of the island is low, and quite level. The different kinds of slate and quartz rock, into which numerous dikes of trap have been injected, compose its base. These slates also have been more or less changed in their characters by the heat attending the filling of the dikes, and the strata are much disordered. It is to be remarked that this island, and almost all those in Passamaquoddy Bay, have their longest diameters in the direction of the course followed by the stratified formations of the Province, and there can be no doubt that the direction of strata in all countries has greatly modified the courses of mountains formed by intrusive rocks.

The western side of the island will average from three to four hundred feet in height. Its lofty mural cliffs stand like rude imitations of masonry, and rival in grandeur those of the celebrated Cape Blomidon in Nova Scotia.

The rock at many places is perfectly basaltic, and appears like large pieces of squared timber placed upright side by side, with a perfection and beauty equal to the basaltic columns of Staffa. These are met by enormous blocks of rhomboidal and amorphous trap, which from their architectural arrangement appear to have been laid by the skill and ingenuity of man.

The amorphous trap is frequently alternated with amygdaloid, which by decomposing more rapidly than the compact variety, hastens the undermining and consequent breaking down of the headlong steep. Whole façades of columns have been broken off and carried away by the sea. The ends of the columns have been polished by the attrition of the waves, constantly moving the sand, and the lofty colonnades stand based upon a natural tesselated pavement.

Whale Cove exhibits a mural precipice that is three hundred feet perpendicular above the level of the sea. This frightful escarpment is composed of alternate layers of amorphous trap and amygdaloid, and resembles a section of the most perfectly stratified rocks. [Because of these layers, which are nearly horizontal, this

---

* These grooves and scratches, called striae, are now identified as being of glacial origin.

cliff is now called Seven Days' Work.—B. C.]   The layers vary
from ten to thirty feet in thickness, and dip to the south-east at
an angle of 15°.   It is remarkable that each alternate layer is
composed of amygdaloid, as there can be no doubt that this rock,
and the amorphous greenstone interstratified with it, are of
volcanic origin.   But, perhaps, this kind of stratification may be
accounted for by referring to the periods of activity, and repose,
common to all volcanoes.   Why a compact trap should be ejected
from a crater at one time, and cellular lava at another, is not
readily explained, unless one be admitted to be the product of
submarine action, and the other to have been cooled by exposure
to the air.*   The amygdaloid abounds in oval cavities, filled with
calcareous spar, zeolite, semi-opal, and heulandite.   Nodules of
these minerals often constitute the greatest portion of the rock.

Submerged Forests and Marshes

But the most remarkable circumstance connected with the
geology of Grand Manan, is the fact, that the whole east side of
the main, and all the small islands in that direction, have, within a
recent period, been submersed to the depth of about eighteen feet.
At the time this submersion took place, the Island was not inhabited, but several persons are still alive who can remember the
tradition, that there once existed between the main, and three
Duck, Nantucket, and other Islands, a kind of marsh, which
occupied several thousand acres, and was only covered by the sea
at high tides.   This kind of marsh has also been seen at Grand
Harbour, the Thoroughfare, and other places along the shore.   It
produced a peculiar kind of grass, which was used for fodder.   All
these marshes have now disappeared, and it was only at a few
places where any parts of them could be found, and wherever any
remnant still remains, it is situated eighteen feet below the mark
of the highest tide, and is covered during every influx of the sea.
Upon examination, I found that not only this marsh, but large
bogs of peat, have been buried beneath the ocean, until its waves,
and the rapid motion of the tides, have almost removed them, and

---

* The layers of "Seven Days' Work" are of volcanic origin, and represent
successive flows of lava.   The differing character of the layers is attributable
to the varying composition and cooling rates of the lavas.   Dr. Gesner's
suggestion of alternating submarine and atmospheric action may be
disregarded.

left their beds to be overflown twice in every twenty-four hours.

The stumps of a great number of trees—the pine, hemlock, and cedar—still remain firmly secured in the sunken earth, by their roots, at the spots where they flourished. This buried forest, with its logs, branches, and leaves, is now covered by each succeeding tide, and the peat taken from the remaining bog, when dry, will burn more rapidly than that taken from the upland. It was by this submersion, that the small Islands became isolated from the main, for the marshes and peat bogs formerly uniting them, were soon removed, when they became exposed to the violence of the sea and its currents.*

---

* The tradition of the existence of a great marsh, connecting the small islands with the main island, within a recent period but before Grand Manan was inhabited, could not have been founded on fact. The small islands were separated much as they are now even before this region of the continent was inhabited by white men.

In actual fact the remains of marsh and forest that Dr. Gesner found appear to have survived a change in level that took place in a very remote age.

There are indications that the coast for several centuries has been nearly, if not perfectly stable. The submergence was [presumably] caused by the return to the sea of the water that had formed the ice sheet during the glacial period, thus indicating that the submergence took place at the close of the ice age.

EMMONS

Ebenezer Emmons (1798–1863), American geologist, professor of natural history at Williams College. The ideas concerning the Taconic System, to which Emmons tenaciously adhered, gave rise to the greatest controversy in the history of American geology.

THE TACONIC SYSTEM

From *Geology of New York, Part II, Comprising the Survey of the Second Geological District*, pp. 4–5, 136–141, Albany, 1842.

In introducing a description of the rocks lying mostly in a belt between the New-York State line and the Hoosic mountain range, I was actuated by a wish to impart a more perfect knowledge of all the lower rocks. . . . In proposing a separation of these rocks from other systems, I was influenced partly by the opinion, more than once expressed, that it would lead to a more thorough knowledge of their characters and relations. These rocks have been termed by some of our ablest geologists, *metamorphic*. If I have interpreted this word rightly, I fully believe that they are by no means entitled to this appellation. It will be observed, that in my account of this system, I have labored to prove that they are not the Loraine shales, or in other words, the Hudson-river group, altered by igneous action. Neither are they the parts of the Primary system, as usually located. They may be *primary rocks* in its true sense, and yet differ from those always placed there, as gneiss, hornblende, mica and talcose slates. Whatever may be the final opinion in relation to these rocks, I have no wish to be supported in the views which I have taken of them, unless they are entitled to support. The position the rocks occupy, and the changes to which they have been subjected, are circumstances which have cast in our path many perplexities and obscurities, such as ought to shield any geologist from censure, though he may fall into some sad mistakes. On these grounds I hope to find refuge, if my well-meant labors have either led me into erroneous doctrines, or into an abortive attempt to establish that which has no substantial foundation.

The Taconic system, as its name is intended to indicate, lies along both sides of the Taconic range of mountains, whose direction is nearly north and south, or for a great distance parallel with the boundary line between the States of New-York, Connecticut, Massachusetts and Vermont. The counties through which the Taconic rocks pass, are Westchester, Columbia, Rensselaer and Washington; and after passing out of the State, they are found stretching through the whole length of Vermont, and into Canada as far north as Quebec. It is, however, in Massachusetts, in the county of Berkshire, that we find the most satisfactory exhibition of these rocks. They form a belt whose width is not far from fifteen miles along the whole western border, and which extends clearly to the western base of the Taconic range. The greatest breadth, therefore, as will be seen by an inspection of any map of this section of country, is wider upon the eastern than upon the western side of this range. In Vermont, they range along the upper members of the Champlain group, and thus become connected with the Second district.

Position and Relation of the Taconic System

The position of this system of rocks deserves an attentive examination; for it is only by a clear understanding of their position, that we shall be able to explain some of the remarkable phenomena found in connection with them. Turning our attention first to the eastern border, we find the primary ranges of New-England, at elevations it is true not very remarkable, but still above the adjacent country upon the west. It is to be noticed, too, that the western slope is rather steep; and it may be considered that it is against this steep slope that the Taconic system reposes. There is one exception, however, to this statement, viz. Saddle mountain rises more than a thousand feet higher than Hoosic mountain. Upon the west is the Taconic range, pursuing its course near the western border of the system, and attaining an elevation of eighteen hundred or two thousand feet. A large portion, then, of its rocks or masses are interlocked between these ranges: the New-England or primary ranges upon the east, the most important of which is the Hoosic mountain; and the Taconic with the more westerly abrupt hills upon the west, or the eastern border of the New-York Transition system. It is this position which is to be taken into view, when we attempt to account for the numerous contortions which exist in the beds lying between these mountains; and there are many facts which favor the view that the rocks lying in this narrow space have been greatly compressed by lateral pressure, and have been forced, as it were, towards the Hoosic mountain range.

The preceding view is favored by the fact, that in the midst of the most mountainous tract, the greatest contortions exist; while in the more level parts, or sections, the contortions and disturbed strata are greatly diminished. In this connection, I may state another result as the consequence of the geographical position of the Taconic system: it is the partial blending of the rocks of the three adjacent systems; the Primary of the Hoosic ranges upon the east, and the New-York Transition system on the west with the Taconic; creating thereby many doubts and perplexities as it regards the true limits of either system; and inasmuch as the whole belt itself of the latter rocks is narrow, doubts are thrown over the whole as it regards the views we are to take of them. It will be more clearly seen in the following pages, how it is that differences of opinions prevail in relation to these rocks. Where they have been crowded together, and especially where the masses are lithologically similar, it is not at all remarkable that the views and opinions of geologists should differ; besides, under the most favorable circumstances, the lines of demarkation between rocks of different eras are often extremely obscured, and cannot be drawn with that exactitude we wish, in consequence of concealment under the soil, or other circumstances equally effective to render their extent and relations indistinct and uncertain.

Taconic System Not Connected with or Related to the Slates and Shales of the Champlain Group

Much difficulty is encountered, as has already been hinted, when we attempt to draw the line of demarkation between the shales and slates east of the Hudson river and Lake Champlain, and the slates of the Taconic system. So nearly do the latter resemble the former in lithological characters, that in specimens of small size, the one might be mistaken for the other. But this is a common difficulty, or one common to all rocks of the same lithological characters, and it is not to be considered as a positive objection to the separation which I now propose.

If then reliance can be placed upon lithological characters, and upon associated minerals, we may raise something more than doubt as it regards the identity of the Taconic rocks with the true Primary system, or certain members of it. In truth, much confidence is felt in the correctness of the principles which have influenced me in proposing their separation, and that they possess characters fully sufficient to give them an independent place in the systems of the day.

If the preceding views are admissible, there is sufficient reason for regarding the rocks which lie between the upper members of the Champlain group and the Hoosic mountain, as a distinct series at least; but I would remark, that by the expression, "lying between," I have reference to geographical position; for considered geologically, they can be regarded in no other light than as inferior to the Potsdam sandstone, or as having been deposited at an era earlier than the lowest member of the New-York Transition system. We have in no instance, however, been able to trace a connection in these masses, and we have never found the Potsdam sandstone resting upon any of the members of the Taconic system. To attempt to explain this remarkable feature, or fact, would be premature. The bare fact that the Potsdam sandstone rests on gneiss or granite, without the interposition of any other rock, we early pointed out; and commencing our series with it, we find it to be unbroken and uninterrupted up to the Old red sandstone. But if we commence an examination at the foot of the Hoosic mountain, which is gneiss, we pass over a series totally different from those of which we have just been speaking, and among which the Potsdam sandstone does not appear, neither a limestone which can be referred to those of the Champlain group, or slate or shale which can be recognized as belonging to the New-York system. If we are correct in this conclusion; if the Taconic rocks differ as much as has been represented from the Primary, and also from the Transition series, then it appears necessary that we should adopt views at least somewhat analogous to those expressed in the preceding pages.

ÉLIE DE BEAUMONT

Jean Baptiste Armand Louis Leonce Élie de Beaumont (1798–1874), professor of geology in the College of France and chief engineer of Mines, attributed mountain ranges to the shrinkage of the earth.

THE NATURE AND CAUSE OF MOUNTAIN BUILDING

Translated from *Notice sur les systèmes de montagnes*, Vol. I, pp. 3–5, Vol. III, pp. 1317–1330, Paris, 1852.

The two great concepts of a series of violent revolutions and of the formation of mountain chains through upheaval having been introduced successively into geology, it is natural to ask whether they are independent of each other; whether mountain chains could rise without producing a veritable revolution on the surface of the globe; and whether the convulsions which necessarily accompany the uprising of masses as extensive and with as complicated a structure as the high mountains were not the equivalent of the revolutions of the surface of the globe which have been established by other reasoning from observation of sedimentary deposits and of extinct groups of animals whose remains they conceal. The question also arises whether the lines of demarcation in the succession of terrains, at each of which the sedimentary deposition appears to have been resumed under new influences, were not merely the result of changes in the extent and physical condition of seas caused by the successive upheavals of mountains. . . .

In nearly all mountain chains, careful observation reveals that the most recent beds extend horizontally to the foot of the mountains, as if laid down in seas or lakes, the shores of which were partly formed by these mountains. In contrast, other beds are upturned and bent along the flanks of the mountains, rising in some points to the very crests. In every range, or at least in every chain, the sedimentary rocks are thus divided into two distinct systems. The variable position in the general series of beds at which the break between the two systems occurs is one of the most distinctive characteristics of each mountain range. And, while the attitude of the old, upturned beds furnishes the best proof of the upheaval of the mountains which they partially compose, the geologic ages of these two systems of beds furnish the most certain means of determining the age of the mountains themselves. Indeed, it is evident that the date of the appearance of the chain is within the period of deposition of the beds which stretch horizontally at the foot of its slopes.

The mountain chains correspond essentially to those parts of the earth's crust which have had their horizontal extent diminished by *transverse compression*. The difference of elevation between the flat terrains situated on either side of the same mountain chain, which is a familiar orographic condition, agrees remarkably with the hypothesis of transverse compression. This hypothesis postulates in general that the portions remaining intact on the sides have ceased to be connected in a fixed relation. They have acted like the two *chops of a vise* between which the intermediate part has been compressed. And this movement has even tended to make them overlap slightly with regard to each other and has produced the unequal elevation at which they remain after the completion of the phenomenon.

The transverse compression has reached all parts of the earth's crust below the mountainous crest. The crushed solid masses, as well as the compressed soft masses, have increased their thickness in proportion as their horizontal extent has been diminished. The parts compressed by the crushing have not been able to make their way to the lower surface of the solid crust of the globe, which is held fixed on the liquid interior by its weight. They have found no other egress than the upper surface, across which they have risen, breaking and uplifting the superficial layers. The greater part of the dislocations, folds, and undulations which the sedimentary beds exhibit are direct or indirect consequences of these upheavals. . . .

In the form under which I have just sketched this theory, all the irregularities which the earth's crust presents—either in its structure, its density, or even in the singular regularity manifested in the disposition of these irregularities—would result theoretically from the dissipation of a part of the heat which the earth held when its crust, now solidified, was in a state of fusion. This heat could not be other than the primitive heat to which, according to the more or less explicit opinion of the most successful interpreters of nature, Descartes, Newton, Leibnitz, Buffon, Laplace, Fourier, etc., the earth owes its spheroidal form and the generally regular arrangement of its beds, from the center out, in the order of their specific weight. This heat could, on the other hand, have a less ancient origin, but one prior, however, to all phenomena geologists can observe, as M. Poisson and other eminent savants have thought. This last could have been the case without essentially altering the nature of the mechanical phenomena that slow disappearance of the heat must have caused and ought yet to produce in the crust.

In either case, the essential point of the theory based on the dispersal of heat is the explanation of the *upheaval* of mountains as a result of the *slow and progressive diminution of the volume of the earth.*

LOGAN AND HUNT

Sir William Edmond Logan (1798–1875), world-famous Canadian geologist, was born and trained in Great Britain, and became first director of the Geological Survey of Canada.

Thomas Sterry Hunt (1826–1892), American chemist and geologist, was associated with the Geological Survey of Canada during the quarter century between 1847 and 1872. For a few years thereafter he lectured on geology at the Massachusetts Institute of Technology. He was largely responsible for the organization of the International Geological Congress, which first met in Paris in 1878.

GEOLOGIC OUTLINE OF CANADA

Translated from *Esquisse géologique de Canada, pour servir à l'intelligence de la carte géologique et de la collection des minéraux économiques envoyées à l'Exposition universelle de Paris,* 1855.

Concerning the Laurentian Mountains

The province of Canada is crossed, in its entire length, by a mountainous terrain which divides it into two basins, which may be called the basin of the North and the basin of the South. These mountains, called the Laurentians, form the northern bank of the St. Lawrence, from the Gulf to Cape Tourmente, near Quebec; starting at this point, they follow the direction of the river, but slip away little by little, and near Montreal, they are ten leagues from the St. Lawrence. Going westward, this mountainous terrain follows the path of the Ontaouais, and crosses it near the Lac des Chats, fifty leagues from Montreal. Then, taking a southern direction, it overtakes the St. Lawrence near the outlet of Lake Ontario, and from here, running toward the northwest, the southern limit of this formation reaches the southeast end of Lake Huron at Matchedash Bay, and forms the eastern bank of the lake at the 47° latitude, where, leaving this lake, the formation reaches Lake Superior and extends in a northwest direction to the Arctic Sea.

South of the St. Lawrence, this same terrain covers an extended area between Lake Ontario and Lake Champlain, and is called the Adirondack Mountains. Excepting for this portion and perhaps a small outcropping in Arkansas and another near the sources of the Mississippi, this formation is not found again south of the St. Lawrence and as it belongs especially to the valley of this river, the Geologic Commission of Canada has called it the Laurentian System.

The Laurentian System

The rocks of this system are almost without exception from ancient sedimentary beds which have become very crystalline; they have been greatly folded and form mountain ranges having a somewhat northeast to southwest direction, rising sometimes to heights of eight hundred or a thousand meters and even above. The rocks of this formation are the oldest known on the American continent and probably correspond to the oldest gneiss of Finland and Scandinavia, and to similar rocks in the north of Scotland.

The rocks of the Laurentian formation are for the most part crystalline schists, principally gneissoid or hornblendite. Associated with these schists are great stratified masses of a crystalline rock which is composed almost entirely of feldspar with a lime and soda base. This rock is sometimes fine-grained, but more commonly is porphyritic and contains cleaving masses of feldspar, which are often several centimeters in diameter. These feldspars belong to the sixteenth system, and ordinarily have a composition of andesine, labradorite, anorthosite, or intermediate varieties. Their colors are often varied, but the striated feldspars are generally bluish or reddish and often present colored reflexes. Hypersthene is very common in these feldspar rocks, but always in small quantity. Titaniferous iron is found in a great number of localities, sometimes in small grains, at other times in considerable masses.

With these schists and feldspars are found strata of quartzite, associated with crystalline limestones which have quite an important place in this formation. The limestones form layers from one to a hundred meters in thickness, and often show a succession of their layers intercalated with layers of gneiss or quartzite. The quartzites are sometimes in the form of conglomerates which have in certain cases a cement of dolomite. Associated with these limestones are layers composed for the most part of wollastonite and pyroxene which evidently owe their origin to the metamorphism that the silicious limestones have undergone. Layers of dolomite, or limestone which is more or less magnesian, are often intercalated with pure limestones.

The limestones of this formation are rarely compact and generally large-grained. They are white or of a reddish, bluish, or grayish color; these colors are sometimes arranged in bands which coincide with the stratification. The principal mineral types which are found in these limestones are: apatite, fluor, serpentine, etc. The chondrodite and graphite are often arranged in bands parallel with the stratification. One sometimes finds layers of a mixture of wollastonite and pyroxene, which are very rich in zircon, sphene, garnet, and idocrase. The most crystalline varieties often exhude a very fetid odor when they are crushed. These limestones do not furnish well-crystallized minerals everywhere; near the Bay of Quinite are beds which have kept their sedimentary character and in which are seen the beginning of metamorphism.

The conditions in which these limestones are sometimes found show that the agents which crystallized them were the type that make carbonate of lime almost liquid and that in this state it underwent great pressure. To uphold this opinion, we find that the limestone often refills fissures in the surrounding silicious layers and envelops detached and often bent fragments of these less-melted strata, in the manner of an igneous rock.

These schists, feldspars, quartzites, and limestones, such as we have described, constitute the stratified part of the Laurentian System; but there are besides granites, syenites, intrusive diorites which form quite important masses; the granites are sometimes albitic, and often contain tourmaline, mica in great sheets, sphene, and sulphide of molybdenum.

Concerning the Cambrian or Huronian System

The shores of Lakes Huron and Superior give us a series of schists, sandstones, and conglomerates intercalated with deep strata of diorite, lying in discordant stratification on the Laurentian System. As these rocks are below the Silurian terrain and as, moreover, they have not yielded any fossil, they may well be added to the Cambrian System (the Lower Cambrian of Mr. Sedgwick). The schists of this system, on Lake Superior, are bluish in color and contain beds of horny silex which has calcareous bands and whose crevices often are full of anthracite.

These rocks are often covered by a considerable thickness of trap, on which are superposed deep layers of white and red sandstone, which sometimes becomes conglomerate, containing orbiculi of quartz and jasper. Beds of reddish, clayey limestone are interposed in these sandstones, which are crossed and covered by a second formation of diorite of great thickness, showing a columnar structure. This formation, which according to Mr. Logan's observations has a total thickness of nearly four thousand meters, is crossed by a great number of dikes of trap.

In the corresponding formation on the northern shore of Lake Huron, there are sandstones having a more vitreous aspect and more abundant conglomerates than on Lake Superior, associated however with schists and schistose conglomerates like those we have just described, the whole showing great masses intercalated with diorite. A bed of limestone sixteen meters thick forms a part of this series, which Mr. Logan estimates is more than three thousand meters deep. Mr. Logan has proved that after the irruption of interstratified diorites there are two systems of diorite dikes and a third of granite of an intermediate period between these last two. The formation of metalliferous veins belongs to a still more recent period.

This Huronian Formation extends along Lake Huron and Lake Superior for nearly one hundred and fifty leagues, and shows everywhere metalliferous veins which up to the present have been little worked. But one cannot doubt that this region contains metallic deposits which will one day become a source of great wealth to Canada. The coal formations in the neighboring state of Michigan will then furnish the fuel necessary for metallurgical operations.

LOGAN

Sir William Edmond Logan (1798–1875), world-famous Canadian geologist, was born and trained in Great Britain, and became the first director of the Geological Survey of Canada.

HURONIAN AND LAURENTIAN DIVISIONS OF THE AZOIC ROCKS

From *Proceedings of the American Association for the Advancement of Science*, 1857, pp. 44–47, 1858.

The sub-Silurian Azoic [Pre-Cambrian] rocks of Canada occupy an area of nearly a quarter of a million of square miles. Independent of their stratification, the parallelism that can be shown to exist, between their lithological character and that of metamorphic rocks of a later age, leaves no doubt on my mind that they are a series of very ancient sedimentary deposits, in an altered condition. The further they are investigated, the greater is the evidence that they must be of very great thickness, and the more strongly is the conviction forced upon me that they are capable of division into stratigraphical groups, the superposition of which will be ultimately demonstrated, while the volume each will be found to possess, and the importance of the economic materials by which some of them are characterized, will render it proper and convenient that they should be recognized by distinct names, and represented by different colors on the geological map.

So early as the year 1845, as will be found by reference to my Report on the Ottawa district (presented to the Canadian government the subsequent year), a division was drawn between that portion which consists of gneiss and its subordinate masses, and that portion consisting of gneiss interstratified with important bands of crystalline limestone. I was then disposed to place the lime-bearing series above the uncalcareous, and although no reason has since been found to contradict this arrangement, nothing has been discovered to especially confirm it; and the complication which subsequent experience has shown to exist in the folds of the whole (apparent dips being from frequent overturns of little value), would induce me to suspend any very positive assertion in respect to their relative superposition, until more extended examination has furnished better evidence.

In the same Report is mentioned, among the Azoic rocks, a formation occurring on Lake Temiscaming, and consisting of silicious slates and slate conglomerates, overlaid by pale sea-green or slightly greenish-white sandstone, with quartzose conglomerates.
. . .

In the Report transmitted to the Canadian government in 1848, on the north shore of Lake Huron, similar rocks are described as constituting the group which is rendered of such economic importance, from its association with copper lodes. This group consists of the same silicious slates and slate conglomerates, holding pebbles of syenite instead of gneiss, similar sandstones, sometimes showing ripple-marks, some of the sandstones pale-red green, and similar quartzose conglomerates, in which blood-red jasper pebbles become largely mingled with those of white quartz, and in great mountain masses predominate over them. But the series is here much intersected and interstratified with greenstone trap, which was not observed on Lake Temiscaming.

The group on Lake Huron we have computed to be about 10,000 feet thick, and from its volume, its distinct lithological character, its clearly marked date posterior to the gneiss, and its economic importance as a copper-bearing formation, it appears to me to require a distinct appellation, and a separate color on the map. Indeed, the investigation of Canadian geology could not be conveniently carried on without it. We have, in consequence, given to the series the title of Huronian.

A distinctive name being given to this portion of the Azoic rock, renders it necessary to apply one to the remaining portion. The only local one that would be appropriate in Canada is that derived from the Laurentide range of mountains, which are composed of it from Lake Huron to Labrador. We have, therefore, designated it as the Laurentian series.

A PRE-CAMBRIAN FOSSIL

From *Geological Survey of Canada, Report of Progress*, pp. 48–49, 1863.

Although the Laurentian rocks have hitherto been considered azoic, certain forms strongly resembling fossils were discovered three years since by Mr. John McMullen, then attached to the

Geological Survey as an explorer, in one of the bands of limestone
belonging to this series at the Grand Calumet. Any organic
remains which may have been entombed in these limestones would,
if they retained their calcareous character, be almost certainly

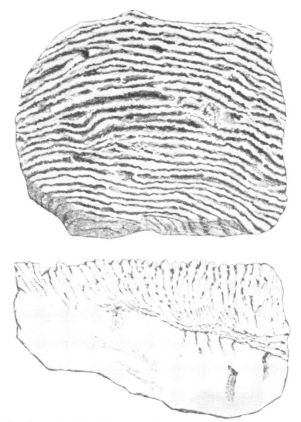

Fig. 14.—Logan's original drawings of *Eozoon canadense*, 1863. The upper
figure represents the weathered surface of a specimen, natural size, and the
lower figure is a vertical transverse section of the same specimen.

obliterated by crystallization, and it would only be through the
replacement of the original carbonate of lime by a different mineral
substance that there would be some chance of the forms being preserved. The specimens obtained from the Grand Calumet present
parallel or apparently concentric layers, resembling those of
Stromatopora rugosa, except that they anastomose at various

points.   The layers are composed of crystalline pyroxene, while the interstices are filled with crystalline carbonate of lime.   These specimens have called to recollection others which were some years ago obtained from Dr. James Wilson of Perth, and then regarded merely as minerals.   They came from loose masses of limestone, in that vicinity, and exhibit similar forms to those of the Calumet composed of dark green concretionary serpentine, while the interstices are filled with crystalline dolomite.   If both are to be regarded as the results of unaided mineral arrangement, it would seem strange that identical forms should be derived from minerals of such different composition.   If the specimens had been obtained from the altered rocks of the Lower Silurian series, there would have been little hesitation in pronouncing them to be fossils. Their resemblance to *Stromatapora rugosa* from the Birdseye and Black River limestone, where this fossil has been replaced by concretionary silica, is very striking.   In the specimens from the Calumet the pyroxene and the carbonate of lime being both white, the forms, although weathered into strong relief on the surface, are not perceptible in fresh fractures until the fragments are subjected to the action of an acid, the application of which shews the peculiar structure throughout the mass.

BRONN

Heinrich Georg Bronn (1800–1862), German paleontologist, professor of zoology at Heidelberg. His *Lethaea geognostica* was the first comprehensive work on fossils in chronological succession.

THE PLAN OF CREATION IN THE SERIES OF GEOLOGICAL AGES

From the translation by W. S. Dallas in *Annals and Magazine of Natural History*, Third Series, Vol. IV, pp. 82–90, London, 1859, of *Untersuchungen über die Entwicklungsgesetze der Organischen Welt*, Stuttgart, 1858.

The Distribution of Fossil Organisms in the Natural Series of Sedimentary Strata

The investigations contained in this work are a confirmation of the laws resulting from the purely geological study of the evolution of the Crust of the earth in relation to the successive appearance of organized beings. They also bring to light certain facts which do not immediately result from those laws, although they are not in contradiction to them—facts which particularly deserve attention.

First fundamental law. 1. Organisms have made their appearance, in the sequence of time and in different localities, in conditions of type and number which were in relation to the external conditions of existence.

2. The appearance of the two organized kingdoms was simultaneous. . . .

3. The population of the surface of the earth was, originally, very uniform in all latitudes. It is only towards the middle of the tertiary period that we see the floras and faunas become essentially differentiated according to zones.

4. Both as regards its constitution and number, the primitive population of the surface of the earth corresponded with a hot climate, of tropical nature, uniform throughout the year. . . .

5. All the successive modifications of the animal and vegetable population of the surface of the globe have been effected by the annihilation of the older species and the continual appearance of new species, without their having ever been any gradual passage from one species to another.

6. The primitive types, whether animal or vegetable, were the most widely different of all from those of existing nature. Some of them differed from the latter so much as to form sub-classes or orders,—most of them at least generically. But in proportion as, in the history of the earth, we approach the present epoch, we observe a constantly increasing concordance of the genera, or even, in certain cases, an identity of species with our existing nature.

7. In all times there have existed faunas and floras topographically distinct, in consequence of differences of conditions in the stations, by reason of the distribution of the seas and the elevation of mountains. But in proportion as the evolution of the surface of the earth multiplied and varied the conditions of stations, in proportion as seas were divided, continents extended, chains of mountains elongated, and summits elevated, we also see a diversification of the organized types and of their mode of grouping and association. Topographical faunas and floras became more clearly defined; and in all cases the number of species living together constantly became more considerable . . .

---

10. A multitude of plants and animals, especially more than three-fourths of the terrestrial insects, birds, and mammals, which, either with regard to their food or habitation, are necessarily connected with certain genera, or even certain species of plants, could, of course, only appear after the latter. The inferior plants and animals are often less intimately connected with other organisms than others which are higher in the series.

11. The principal modifications which the external conditions of existence of organisms had to undergo, consisted, undoubtedly, in the division of the universal ocean into several seas, Mediterranean basins and Caspian lakes,—in the emergence of islands which increased in size, or even united with each other to form continents,—in the elevation of mountain-chains, &c. In parallelism with this transformation of the crust of the earth, the organized world presented analogous modifications. The population of the sea, at first entirely pelagic, became combined with a littoral population, then with a terrestrial but exclusively coast population, and lastly with continental populations, varying with flat and mountainous countries. It is this series of phaenomena that we designate under the name of *terripetal evolution*. Either by the successive sequence of organisms, or by the transformation of their

characters, even in cases in which the causes of transformation are unknown to us, this evolution is manifested to us as a perfectly general law of development, which we call the *terripetal law*. As, in general, the inhabitants of coasts are characterized by a higher degree of organization than the inhabitants of the depths of the sea, and the inhabitants of the dry land by a higher degree of organization than those of the waters, this law is intimately connected with a progressive development. . . .

Second fundamental law. 12. Besides this first law, there evidently exists a positive and *independent* law of creation, which manifests itself to us in the simplicity and perfect order of all the simultaneous or successive modifications of the organized world. The external conditions of existence only permitted the investigation of the plan which presided in the creation at each moment and in all the series of time from a perfectly negative point of view. But this second law, thanks to its positive character, furnishes us with the means of following the conducting clue with far more facility and conclusiveness than was permitted by the former, which is so complex. Hence results, in the first place, the strict uniformity in all the creation which existed simultaneously at each moment upon the whole surface of the earth; hence the simultaneous appearance and disappearance of genera and species in all regions and under every zone; hence the constant equilibrium between the plants and animals, the terrestrial and aquatic animals, the Herbivora and Carnivora in each creation; and all this realized far more exactly than could have happened under the influence of the external conditions of existence alone, which may certainly destroy, but can produce nothing. The unfolding of the plan of creation in the series of geological ages has taken place with a perfect consequentiality and in a perfectly independent manner. Systematic and progressive development, and the law which governs it, are facts which can no longer be misconceived. . . .

13. This plan of succession is nowhere more evident than in the vegetable kingdom, in which we see several subkingdoms appear at first and simultaneously, followed by the successive appearance of the superior groups most nearly allied to them in organization, and which only attained their culmination subsequently. The perfectly natural consequence of this, was the comparatively far later appearance of the most highly organized groups of plants—groups which surpass all the others in the number of their genera and

species; and yet, at least as far as we can now judge, the external conditions of existence would have permitted their appearance from the very first. . . .

14. The late appearance of the angiospermous Dicotyledons is undoubtedly, of all causes, that which had the most importance in retarding the appearance of most of the terrestrial animals, such as the insects, birds, and mammalia. . . .

15. Progressive development does not consist only in the fact that new and more perfect types became added to the inferior types which existed before, but also in the circumstance that these latter decreased in importance from their point of culmination, and finally became entirely extinct. . . .

16. All the great phaenomena relating to the order of appearance of the different subdivisions of the organized kingdom result from the laws which we have here developed, and which may be summed up as follows:—a, adaptation to external conditions; b, terripetal movement; c, progressive development, that is to say, the successive appearance of forms with more and more complicated organization. . . .

18. All the phenomena which we deduce from the law of adaptation to the external circumstances of existence, from the law of terripetal evolution, and from that of progressive development, show us a regular progress from the commencement to the close of geological epochs. . . .

The Appearance and Disappearance of Organized Beings

The results at which we have arrived, with regard to the gradual or simultaneous extinction of all the organisms of a single epoch, may be summed up in the following manner:—

1. The creation of new species and the disappearance of older types went on continuously, with the exception of slight oscillations, without being restricted to certain periods of creation, although it is easy to imagine that certain geological events may have, here and there, induced the simultaneous extinction of a larger or smaller number of species.

2. The duration of existence has been very variable, according to the species. Certain specific types have endured 2, 3, 4, or 5 times as long as others, so that some even existed only during a

small fraction of the time necessary for the production of a formation in the geological sense of the word, whilst others survived the deposition of two or three formations, or even more. These phaenomena might take place only at a certain point of the surface of the globe, and not present themselves elsewhere.

3. There are, consequently, no definite formations in the palaeolithic* sense of the word, no definite creations, no successive and well-marked floras or faunas, any more than there exists any formation which simultaneously maintains the same mineralogical characters, the same thickness, and the same lithological and palaeontological characters in all parts of the world.

4. A geological formation, or a geological flora or fauna, is the totality of the sedimentary strata which have been formed upon the whole earth during a certain space of time, or the totality of the animals and plants which have lived during that space of time. It is of little consequence here, whether the lithological character, the thickness, and the limits of demarcation of these strata have been uniform over all the surface of the globe, or have varied in different places, assuming here one aspect, there another; it is of little consequence whether the various species of organisms belonging to this epoch may have lived from its commencement to its termination, only endured for a portion of this time, or passed the limits assigned to this formation.

5. When the deposition of identical strata, according with an identical and constant state of the sea, continued longer in one country than in another, the population of this sea and the organic remains of this population might exist there longer without undergoing modification.

6. When an identical state of the sea reappeared during the deposition of an immediately consecutive formation, or after a longer or shorter interval during which other formations might be deposited, the same marine population might reappear in the same locality and give rise to identical organic *debris*, enclosed in superior strata. Thus are formed what are termed *colonies* in Geology. It is probable, however, that this phaenomenon could only present itself when the same species had continued to live in the interval, perhaps exceedingly reduced in number, in some other locality. We have nevertheless shown how it may happen that remains of

---

* Paleontological.—Editors.

perfectly identical species may pass into rocks of a nature quite different, and deposited by very different seas.

7. There probably exist no formations immediately superposed upon each other, no consecutive faunas and floras, without certain organisms being common to both. The number of common species may vary between 0.01 and 0.10.

8. When, however, in certain localities there have been sudden movements of the soil, heating of the crust of the earth, emissions of sulphurous vapours, carbonic acid, or other injurious gases, long interruptions in the formation of deposits, upheavals of strata, &c., it most frequently happens that the passage of species from one stratum to another is more rare than when the deposits have been formed regularly and without any interruption.

9. The average absolute duration of organisms was sufficiently long to give us no reason for astonishment at the important differences presented by species in this respect, although the history of these species is often told us only by strata of but slight thickness, so that it often happens that we regard as simultaneous, phaenomena which have been separated by long periods of time.

MARSH

George Perkins Marsh (1801–1882), American author and amateur physiographer, long resident in Italy, was one of the first to recognize the significant relationship between deforestation and soil erosion.

INFLUENCE OF THE FOREST ON FLOODS

From *The Earth as Modified by Human Action*, pp. 231–302, New York, 1874.

Inasmuch as it is not yet proved that the forests augment or diminish the precipitation in the regions they principally cover, we cannot positively affirm that their presence or absence increases or lessens the total volume of the water annually delivered by great rivers or by mountain torrents. It is nevertheless certain that they exercise an action on the discharge of the water of rain and snow into the valleys, ravines, and other depressions of the surface, where it is gathered into brooks and finally larger currents, and consequently influence the character of floods, both in rivers and in torrents. . . .

The surface of a forest, in its natural condition, can never pour forth such deluges of water as flow from cultivated soil. Humus, or vegetable mould, is capable of absorbing almost twice its own weight of water. The soil in a forest of deciduous foliage is composed of humus, more or less unmixed, to the depth of several inches, sometimes even of feet, and this stratum is usually able to imbibe all the water possibly resulting from the snow which at any one time covers, or the rain which in any one shower falls upon, it. But the vegetable mould does not cease to absorb water when it becomes saturated, for it then gives off a portion of its moisture to the mineral earth below, and thus is ready to receive a new supply; and, besides, the bed of leaves not yet converted to mould takes up and retains a very considerable proportion of snow-water, as well as of rain.

The stems of trees, too, and of underwood, the trunks and stumps and roots of fallen timber, the mosses and fungi and the numerous inequalities of the ground observed in all forests, oppose a mechanical resistance to the flow of water over the surface, which sensibly

305

retards the rapidity of its descent down declivities, and diverts and divides streams which may have already accumulated from smaller threads of water. . . .

The importance of the mechanical resistance of the wood to the flow of water *over the surface* has, however, been exaggerated by some writers.   Rain-water is generally absorbed by the forest-soil as fast as it falls, and it is only in extreme cases that it gathers itself into a superficial sheet or current overflowing the ground. There is, nevertheless, besides the absorbent power of the soil, a very considerable mechanical resistance to the transmission of water *beneath* the surface through and along the superior strata of the ground.   This resistance is exerted by the roots, which both convey the water along their surface downwards, and oppose a closely wattled barrier to its descent along the slope of the permeable strata which have absorbed it. . . .

The immediate cause of river inundations is the flow of superficial and subterranean waters into the beds of rivers faster than those channels can discharge them.   The insufficiency of the channels is occasioned partly by their narrowness and partly by obstructions to their currents, the most frequent of which is the deposit of sand, gravel, and pebbles in their beds by torrential tributaries during the floods.

In accordance with the usual economy of nature, we should presume that she had everywhere provided the means of discharging, without disturbance of her general arrangements or abnormal destruction of her products, the precipitation which she sheds upon the face of the earth.   Observation confirms this presumption, at least in the countries to which I confine my inquiries; for, so far as we know the primitive conditions of the regions brought under human occupation within the historical period, it appears that the overflow of river-banks was much less frequent and destructive than at the present day, or, at least, that rivers rose and fell less suddenly, before man had removed the natural checks to the too rapid drainage of the basins in which their tributaries originate.   The affluents of rivers draining wooded basins generally transport, and of course let fall, little or no sediment, and hence in such regions the special obstruction to the currents of water-courses to which I have just alluded does not occur.   The banks of the rivers and smaller streams in the North American colonies were formerly little abraded by the currents.   Even now the trees come down almost to the water's edge along the rivers, in the larger forests of the United States, and the surface of the streams seems liable to no great change in level or in rapidity of current.

Inundations in Winter. In the Northern United States, although inundations are not very unfrequently produced by heavy rains in the height of summer, it will be found generally true that the most rapid rise of the waters, and, of course, the most destructive "freshets," as they are called in America, are occasioned by the sudden dissolution of the snow before the open ground is thawed in the spring. It frequently happens that a powerful thaw sets in after a long period of frost, and the snow which had been months in accumulating is dissolved and carried off in a few hours. When the snow is deep, it, to use a popular expression, "takes the frost out of the ground" in the woods, and, if it lies long enough, in the fields also. But the heaviest snows usually fall after midwinter, and are succeeded by warm rains or sunshine, which dissolve the snow on the cleared land before it has had time to act upon the frost-bound soil beneath it. In this case, the snow in the woods is absorbed as fast as it melts, by the soil it has protected from freezing, and does not materially contribute to swell the current of the rivers. If the mild weather, in which great snow-storms usually occur, does not continue and become a regular thaw, it is almost sure to be followed by drifting winds, and the inequality with which they distribute the snow over the cleared ground leaves the ridges of the surface-soil comparatively bare, while the depressions are often filled with drifts to the height of many feet. The knolls become frozen to a great depth; succeeding partial thaws melt the surface-snow, and the water runs down into the furrows of ploughed fields, and other artificial and natural hollows, and then often freezes to solid ice. In this state of things, almost the entire surface of the cleared land is impervious to water, and from the absence of trees and the general smoothness of the ground, it offers little mechanical resistance to superficial currents. If, under these circumstances, warm weather accompanied by rain occurs, the rain and melted snow are swiftly hurried to the bottom of the valleys and gathered to raging torrents.

It ought further to be considered that, though the lighter ploughed soils readily imbibe a great deal of water, yet grass-lands, and all the heavy and tenacious earths, absorb it in much smaller quantities, and less rapidly than the vegetable mould of the forest. Pasture, meadow, and clayey soils, taken together, greatly predominate over sandy ploughed fields, in all large agricultural districts, and hence, even if, in the case we are supposing, the open ground chance to have been thawed before the melting of the snow which covers it, it is already saturated with moisture, or very soon becomes so, and, of course, cannot relieve the pressure by absorbing more water. The consequence is that the face of the country is suddenly flooded with a quantity of melted snow and rain equivalent to a fall of six or eight inches of the latter, or even more. This runs unobstructed to rivers often still-bound with thick ice, and thus inundations of a fearfully devastating character are produced. The ice bursts, from the hydrostatic pressure from below, or is violently torn up by the current, and is swept by the impetuous stream, in large masses and with resistless fury, against banks, bridges, dams, and mills erected near them. The bark of the trees along the rivers is often abraded, at a height of many feet above the ordinary water-level, by cakes of floating ice, which are at last stranded by the receding flood on meadow or ploughland, to delay, by their chilling influence, the advent of the tardy spring.

Another important effect of the removal of the forest shelter in cold climates may be noticed here. We have observed that the ground in the woods either does not freeze at all, or that if frozen it is thawed by the first considerable snow-fall. On the contrary, the open ground is usually frozen when the first spring freshet occurs, but is soon thawed by the warm rain and melting snow. Nothing more effectually disintegrates a cohesive soil than freezing and thawing, and the surface of earth which has just undergone those processes is more subject to erosion by running water than under any other circumstances. Hence more vegetable mould is washed away from cultivated grounds in such climates by the spring floods than by the heaviest rain at other seasons.

In the warm climates of Southern Europe, as I have already said, the functions of the forest, so far as the disposal of the water of precipitation is concerned, are essentially the same at all seasons, and are analogous to those which it performs in the Northern United States in summer. Hence, in the former countries, the winter floods have not the characteristics which mark them in the latter, nor is the conservative influence of the woods in winter relatively so important, though it is equally unquestionable.

If the summer floods in the United States are attended with less pecuniary damage than those of the Loire and other rivers of France, the Po and its tributaries in Italy, the Emme and her sister torrents which devastate the valleys of Switzerland, it is partly because the banks of American rivers are not yet lined with towns, their shores and the bottoms which skirt them not yet covered with improvements whose cost is counted by millions, and, consequently, a smaller amount of property is exposed to injury by inundation. But the comparative exemption of the American people from the terrible calamities which the overflow of rivers has brought on some of the fairest portions of the Old World, is, in a still greater degree, to be ascribed to the fact that, with all our thoughtless improvidence, we have not yet bared all the sources of our streams, not yet overthrown all the barriers which nature has erected to restrain her own destructive energies. Let us be wise in time, and profit by the errors of our older brethren! . . .

The destructive effects of inundations, considered simply as a mechanical power by which life is endangered, crops destroyed, and the artificial constructions of man overthrown, are very terrible. Thus far, however, the flood is a temporary and by no means an irreparable evil, for if its ravages end here, the prolific powers of nature and the industry of man soon restore what had been lost, and the face of the earth no longer shows traces of the deluge that had overwhelmed it. Inundations have even their compensations. The structures they destroy are replaced by better and more secure erections, and if they sweep off a crop of corn, they not unfrequently leave behind them, as they subside, a fertilizing deposit which enriches the exhausted field for a succession of seasons. If, then, the too rapid flow of the surface-waters occasioned no other evil than to produce, once in ten years upon the average, an inundation which should destroy the harvest of the low grounds along the rivers, the damage would be too inconsiderable, and of too transitory a character, to warrant the inconveniences and the expense involved in the measures which the most competent judges in many parts of Europe believe the respective governments ought to take to obviate it.

But the great, the irreparable, the appalling mischiefs which have already resulted, and which threaten to ensue on a still more extensive scale hereafter, from too rapid superficial drainage, are of a properly geographical, we may almost say geological, character, and consist primarily in erosion, displacement, and transportation of the superficial strata, vegetable and mineral—of the integuments, so to speak, with which nature has clothed the skeleton frame-work of the globe. It is difficult to convey by description an idea of the desolation of the regions most exposed to the ravages of torrent and of flood; and the thousands who, in these days of swift travel, are whirled by steam near or even through the theatres of these calamities, have but rare and imperfect opportunities of observing the destructive causes in action. Still more rarely can they compare the past with the actual condition of the provinces in question, and trace the progress of their conversion from forest-crowned hills, luxuriant pasture grounds, and abundant cornfields and vineyards well watered by springs and fertilizing rivulets, to bald mountain ridges, rocky declivities, and steep earth-banks furrowed by deep ravines with beds now dry, now filled by torrents of fluid mud and gravel hurrying down to spread themselves over the plain, and dooming to everlasting barrenness the once productive fields. In surveying such scenes, it is difficult to resist the impression that nature pronounced a primal curse of perpetual sterility and desolation upon these sublime but fearful wastes, difficult to believe that they were once, and but for the folly of man might still be, blessed with all the natural advantages which Providence has bestowed upon the most favored climes. But the historical evidence is conclusive as to the destructive changes occasioned by the agency of man upon the flanks of the Alps, the Apennines, the Pyrenees, and other mountain ranges in Central and Southern Europe, and the progress of physical deterioration has been so rapid that, in some localities, a single generation has witnessed the beginning and the end of the melancholy revolution.

In fine, in well-wooded regions, and in inhabited countries where a due proportion of soil is devoted to the growth of judiciously distributed forests, natural destructive tendencies of all sorts are arrested or compensated, and man, bird, beast, fish, and vegetable alike find a constant uniformity of condition most favorable to the regular and harmonious coexistence of them all.

With the extirpation of the forest, all is changed. At one season, the earth parts with its warmth by radiation to an open sky— receives, at another, an immoderate heat from the unobstructed rays of the sun. Hence the climate becomes excessive, and the soil is alternately parched by the fervors of summer, and seared by the rigors of winter. Bleak winds sweep unresisted over its surface, drift away the snow that sheltered it from the frost, and dry up its scanty moisture. The precipitation becomes as irregular as the temperature; the melting snows and vernal rains, no longer absorbed by a loose and bibulous vegetable mould, rush over the frozen surface, and pour down the valleys seawards, instead of filling a retentive bed of absorbent earth, and storing up a supply of moisture to feed perennial springs. The soil is bared of its covering of leaves, broken and loosened by the plough, deprived of the fibrous rootlets which held it together, dried and pulverized by sun and wind, and at last exhausted by new combinations. The face of the earth is no longer a sponge, but a dust-heap, and the floods which the waters of the sky pour over it hurry swiftly along its slopes, carrying in suspension vast quantities of earthy particles which increase the abrading power and mechanical force of the current, and, augmented by the sand and gravel of falling banks, fill the beds of the streams, divert them into new channels, and obstruct their outlets. The rivulets, wanting their former regularity of supply and deprived of the protecting shade of the woods, are heated, evaporated, and thus reduced in their summer currents, but swollen to raging torrents in autumn and spring. From these causes, there is a constant degradation of the uplands, and a consequent elevation of the beds of water-courses and of lakes by the deposition of the mineral and vegetable matter carried down by the waters. The channels of great rivers become unnavigable, their estuaries are choked up, and harbors which once sheltered large navies are shoaled by dangerous sand-bars. The earth, stripped of its vegetable glebe, grows less and less productive, and, consequently, less able to protect itself by weaving a new network of roots to bind its particles together, a new carpeting of turf to shield it from wind and sun and scouring rain. Gradually it becomes altogether barren. The washing of the soil from the mountains leaves bare ridges of sterile rock, and the rich organic mould which covered them, now swept down into the dank low grounds, promotes a luxuriance of aquatic vegetation that breeds fever, and more insidious forms of mortal disease, by its decay, and thus the earth is rendered no longer fit for the habitation of man.

To the general truth of this sad picture there are many exceptions, even in countries of excessive climates. Some of these are due to favorable conditions of surface, of geological structure, and of the distribution of rain; in many others, the evil consequences of man's improvidence have not yet been experienced, only because a sufficient time has not elapsed, since the felling of the forest, to allow them to develop themselves.   But the vengeance of nature for the violation of her harmonies, though slow, is sure, and the gradual deterioration of soil and climate in such exceptional regions is as certain to result from the destruction of the woods as is any natural effect to follow its cause.

MILLER

Hugh Miller (1802–1856), Scottish stonemason, accountant, and editor. His popular writings on geology have endeared him to geologists and have gained for him a place of honor in the annals of English literature.

A Quarry Laborer as a Geological Observer

From *The Old Red Sandstone*, American ed., pp. 3–13, 56, 65, Boston, 1858.

It was twenty years, last February, since I set out a little before sunrise to make my first acquaintance with a life of labor and restraint, and I have rarely had a heavier heart than on that morning. I was but a slim, loose-jointed boy at the time—fond of the pretty intangibilities of romance, and of dreaming when broad awake; and, woful change! I was now going to work at what Burns has instanced in his "Twa Dogs" as one of the most disagreeable of all employments—to work in a quarry. Bating the passing uneasiness occasioned by a few gloomy anticipations, the portion of my life which had already gone by had been happy beyond the common lot. I had been a wanderer among rocks and woods—a reader of curious books when I could get them—a gleaner of old traditionary stories; and now I was going to exchange all my day-dreams, and all my amusements, for the kind of life in which men toil every day that they may be enabled to eat, and eat every day that they may be enabled to toil!

The quarry in which I wrought lay on the southern shore of a noble inland bay, or frith, rather, with a little clear stream on the one side, and a thick fir wood on the other. It had been opened in the Old Red Sandstone of the district, and was overtopped by a huge bank of diluvial clay, which rose over it in some places to the height of nearly thirty feet, and which at this time was rent and shivered, wherever it presented an open front to the weather, by a recent frost. A heap of loose fragments, which had fallen from above, blocked up the face of the quarry, and my first employment was to clear them away. The friction of the shovel soon blistered my hands; but the pain was by no means very severe, and I wrought hard and willingly, that I might see how the huge strata

313

below, which presented so firm and unbroken a frontage, were to be torn up and removed.   Picks, and wedges, and levers were applied by my brother-workmen; and simple and rude as I had been accustomed to regard these implements, I found I had much to learn in the way of using them.   They all proved inefficient, however; and the workmen had to bore into one of the inferior strata, and employ gunpowder.   The process was a new one to me, and I deemed it a highly amusing one: it had the merit, too, of being attended with some degree of danger as a boating or rock excursion, and had thus an interest independent of its novelty. . . .

The gunpowder had loosened a large mass in one of the inferior strata, and our first employment, on resuming our labors, was to raise it from its bed.   I assisted the other workmen in placing it on edge, and was much struck by the appearance of the platform on which it had rested.   The entire surface was ridged and furrowed like a bank of sand that had been left by the tide an hour before.   I could trace every bend and curvature, every cross hollow and counter ridge of the corresponding phenomena; for the resemblance was no half resemblance—it was the thing itself; and I had observed it a hundred and a hundred times, when sailing my little schooner in the shallows left by the ebb.   But what had become of the waves that had thus fretted the solid rock, or of what element had they been composed?   I felt as completely at fault as Robinson Crusoe did on discovering the print of the man's foot on the sand.   The evening furnished me with still further cause of wonder.   We raised another block in a different part of the quarry, and found that the area of a circular depression in the stratum below was broken and flawed in every direction, as if it had been the bottom of a pool recently dried up, which had shrunk and split in the hardening.   Several large stones came rolling down from the diluvium in the course of the afternoon. They were of different qualities from the Sandstone below, and from one another; and, what was more wonderful still, they were all rounded and water-worn, as if they had been tossed about in the sea, or the bed of a river, for hundreds of years.   They could not, surely, be a more conclusive proof that the bank which had enclosed them so long could not have been created on the rock on which it rested.   No workman ever manufactures a half-worn article, and the stones were all half-worn!   And if not the bank, why then the sandstone underneath?   I was lost in conjecture, and found I had food enough for thought that evening, without once thinking of the unhappiness of a life of labor. . . .

In the course of the first day's employment, I picked up a nodular mass of blue limestone, and laid it open by a stroke of the hammer. Wonderful to relate, it contained inside a beautifully finished piece of sculpture—one of the volutes apparently of an Ionic capital; and not the far-famed walnut of the fairy tale, had I broken the shell and found the little dog lying within, could have surprised me more. Was there another such curiosity in the whole world? I broke open a few other nodules of similar appearance,— for they lay pretty thickly on the shore,—and found that there might. In one of these there were what seemed to be the scales of fishes, and the impressions of a few minute bivalves, prettily striated; in the centre of another there was actually a piece of decayed wood. Of all Nature's riddles these seemed to me to be at once the most interesting, and the most difficult to expound. . . .

My first year of labor came to a close, and I found that the amount of happiness had not been less than in the last of my boyhood. My knowledge, too, had increased in more than the ratio of former seasons; and as I had acquired the skill of at least the common mechanic, I had fitted myself for independence. The additional experience of twenty years has not shown me that there is any necessary connection between a life of toil and a life of wretchedness. . . .

My curiosity, once fully awakened, remained awake, and my opportunities of gratifying it have been tolerably ample. I have been an explorer of caves and ravines—a loiterer along sea-shores— a climber among rocks—a laborer in quarries. My profession was a wandering one. I remember passing direct, on one occasion, from the wild western coast of Ross-shire, where the Old Red Sandstone leans at a high angle against the prevailing Quartz Rock of the district, to where, on the southern skirts of Mid-Lothian, the Mountain Limestone rises amid the coal. I have resided one season on a raised beach of the Moray Frith. I have spent the season immediately following amid the ancient granites and contorted schists of the central Highlands. In the north I have laid open by thousands the shells and lignites of the Oolite; in the south I have disinterred from their matrices of stone or of shale the huge reeds and the tree ferns of the Carboniferous period. I have been taught by experience, too, how necessary an acquaintance with geology of both extremes of the kingdom is to the right understanding of the formations of either. . . .

One important truth I would fain press on the attention of my lowlier readers. There are few professions, however humble, that do not present their peculiar advantages of observation; there are none, I repeat, in which the exercise of the faculties does not lead to enjoyment. I advise the stone-mason, for instance, to acquaint himself with Geology. Much of his time must be spent amid the rocks and quarries of widely separated localities. The bridge or harbor is no sooner completed in one district, than he has to remove to where the gentleman's seat, or farm-steading is to be erected in another; . . . he may pass over the whole geological scale, even when restricted to Scotland. . . .

Shall I venture to say, that the ichthyolites of the Old Red Sandstone have sometimes reminded me of the "fisch of the laithlie flood?" They were hardly less curious. . . . One of their families—that of the *Cephalaspis*—seems almost to constitute a connecting link, says Agassiz, between fishes and crustaceans. They had, also, their families of sauroid, or reptile fishes—and their still more numerous families that unite the cartilaginous fishes to the osseous. And to these last the explorer of the Lower Old Red Sandstone finds himself mainly restricted.

. . . Now, in the fossils of the chalk, with those of the other later formations, down to the New Red Sandstone, we find that the skeleton style of preparation obtains; whereas, in at least three-fourths of the ichthyolites of the Lower Old Red, we find only what we may term the external style. I had marked, besides, another circumstance in the ichthyolites, which seemed, like a nice point of circumstantial evidence, to give testimony in the same line. The tails of all the ichthyolites, whose vertebral columns and internal rays are wanting, are unequally lobed, like those of the dog-fish and sturgeon, (both cartilaginous fishes,) and the body runs on to nearly the termination of the surrounding rays. The one-sided condition of tail exists, says Cuvier, in no recent osseous fish known to naturalists, except in the bony pike—a sauroid fish of the warmer rivers of America. With deference, however, to so high an authority, it is questionable whether the tail of the bony pike should not rather be described as a tail set on somewhat awry, than as a one-sided tail.

OWEN

Sir Richard Owen (1804–1892), English zoologist, professor of comparative anatomy at the Royal College of Surgeons, was strongly opposed to the ideas of Darwin and Huxley. Nevertheless, he was responsible for many advances in knowledge concerning the life of the geologic past. Among his contributions, one of the most interesting is that which demonstrated the presence of birds in Jurassic time.

THE ARCHEOPTERYX FROM SOLENHOFEN

From *Philosophical Transactions of the Royal Society of London*, Vol. CLIII, pp. 33–47, 1864.

THE first evidence of a Bird in strata of the Oxfordian or Corallian stage of the Oolitic series was afforded by the impression of a single feather, in a slab of the lithographic calcareous laminated stone, or slate, of Solenhofen; it is described and figured with characteristic minuteness and care by M. Hermann von Meyer, in the fifth part of the "Jahrbuch für Mineralogie."* He applies to this fossil impression the term *Archeopteryx lithographica;* and although the probability is great that the class of Birds was represented by more than one genus at the period of the deposit of the lithographic slate, and generic identity cannot be predicated from a solitary feather, I shall assume it in the present instance, and retain for the genus, which can now be established on adequate characters, the name originally proposed by the distinguished German palaeontologist.

At the Meeting of the Mathematico-Physical Class of the Royal Academy of Sciences of Munich, on the 9th of November, 1861, Professor Andreas Wagner, communicated the discovery, in the lithographic slate of Solenhofen, of a considerable portion of the skeleton of an animal with impressions of feathers radiating fanwise from each anterior limb, and diverging obliquely in a single series from each side of a long tail.

Upon the report thus furnished to him, Professor Wagner proposed for the remarkable fossil the generic name *Griphosaurus,*

---

* 1861, p. 561.

conceiving it to be a long-tailed Pterodactyle with feathers.   His state of health prevented his visiting Pappenheim for a personal inspection of the fossil; and, unfortunately for palaeontological science, which is indebted to him for many valuable contributions, Professor Wagner shortly after expired.

I thereupon communicated with Dr. Häberlein, and reported on the nature and desirability of the fossils in his possession to the Trustees of the British Museum; they were accordingly inspected by my colleague Mr. Waterhouse, F. Z. S.; and an interesting and instructive selection, including the subject of the present paper, has been purchased for the Museum.

The specimen is divided between the counterpart halves of a split slab of lithographic stone: the moiety containing the greater number of the petrified bones exhibits such proportion of the skeleton from the inferior or ventral aspect.

The lower half of an arched furculum (merry-thought) marks, by its relative position to the wings, the fore part of the trunk.   From this portion of the furculum to the root of the tail measures $4\frac{1}{2}$ inches; the length of the caudal series of vertebrae is 8 inches; but the terminal tail-feathers extend 3 inches further, making the length of the tail 11 inches.   From the end of the tail to the anterior border of the wing-feather impressions is 1 foot $8\frac{1}{2}$ inches.   From the outer border of the impression of the left wing to that of the right wing measures 1 foot 4 inches.   The front margin of the slab of stone has broken away short of the anterior border of the impression of the outspread left wing, and the head or skull of the specimen may have been included in that part of the quarry or stone from which the present slab has been detached.   The preserved parts of the feathered creature indicate its size to have been about that of a Rook or Peregrine Falcon.

Had the manus of *Archeopteryx* been constructed for the support of a membranous wing, the extent to which the skeleton is preserved, and the ordinary condition of the fossil *Pterosauria* in lithographic slate, render it almost certain that some of these most characteristic elongated slender bones of the wing-finger would have been preserved if they had existed in the present specimen. But, besides the negative evidence, the positive proof of the ornithic proportions of the hand or pinion, of the existence of quill-feathers, and the manifest attachment of the principal ones, or "primaries," to the carpal and metacarpal parts of a short terminal segment of the limb, sufficiently evince the true class-affinity of the *Archeopteryx*.

We have here, therefore, plain indications of a large ischio-iliac interspace, answering to that called "great ischiatic foramen or notch," and the smaller ischio-pubic vacuity called "obturator foramen," under conditions of size, formation, and relative position to the acetabulum, known only in the class of birds. The acetabulum itself, moreover, instead of being a bony cup, is a direct circular perforation of the os innominatum, as in birds.

The broad, subquadrate, short, compressed spines of one or two lumbar vertebrae are dimly discernible in front of the sacrum. No trace of the vertebral column in advance of these is visible, nor any part of the sternum; trunk, neck, and head are all wanting. The remains of *Archeopteryx*, as preserved in the present split slab of lithographic stone, recalled to mind the condition in which I have seen the carcase of a Gull or other sea-bird left on estuary sand after having been a prey to some carnivorous assailant. The viscera and chief masses of flesh, with the cavity containing and giving attachment to them, are gone, with the muscular neck and perhaps the head, while the indigestible quill-feathers of the wings and tail, with more or less of the limbs, held together by parts of the skin, and with such an amount of dislocation as the bones of the present specimen exhibit, remain to indicate what once had been a bird.

Were it not for the large proportional size of their cavities, the general configuration of the long bones of the limbs could not have been so well preserved and presented for the requisite comparison. When the bones sank in the soft fine calcareous mud which has hardened into the peculiar stone which the progress of lithographic art has rendered so valuable, the sparry matter in solution, percolating the matrix and entering the cavities of the bones, has slowly crystallized there, and ultimately filled them by a compact body of spar. The degree to which this represents the original bone gives the measure of the pneumatic cavities and cancelli in the skeleton of *Archeopteryx*, and shows that the proportion of the original osseous matter must have been that which we observe in the present day in birds of flight.

The great and striking difference, and that which gives its enig-matical character to this fossil bird's skeleton, is the number, or rather the proportions and distinctness, of the caudal vertebrae; their under surface is exposed, or rather the sparry casts of the cavities of their bodies, the thin crust of the bone adhering to the impressions of the counterpart.

With the exception of the caudal vertebrae, and possibly of the bi-unguiculate and less confluent condition of the manus, the parts of the skeleton preserved in this rare fossil feathered animal accord with the strictly ornithic modifications of the vertebrate skeleton.

Thus we discern, in the main differential character of the by-fos-sil-remains-oldest-known feathered Vertebrate, a retention of a structure embryonal and transitory in the modern representatives of the class, and a closer adhesion to the general vertebrate type. The same evidence is afforded by the minor extent to which the anchylosing process has been carried on in the pinion, and by the apparent retention of two unguiculate digits on the radial side of the metacarpo-phalangeal bones, modified for the attachment of the primary quill-feathers. But when we recall the single unguiculate digit in the wing of *Pteropus*, and the number of such digits, equalling that in *Pterodactylus*, in the fore foot of the Flying Lemur (*Galeopithecus*), the tendency to see only a reptilian char-acter in what may have been the structure of the manus in *Archeop-teryx* receives a due check.

The best-determinable parts of its preserved structure declare it unequivocally to be a Bird, with rare peculiarities indicative of a distinct order in that class. By the law of correlation we infer that the mouth was devoid of lips, and was a beak-like instrument fitted for preening the plumage of *Archeopteryx*. A broad and keeled breast-bone was doubtless associated in the living bird with the great pectoral ridge of the humerus, with the furculum, and with other evidences of feathered instruments of flight.

SHARPE

Daniel Sharpe (1806–1856), English geologist, was in his day an authority on the geology of Portugal. The excerpt from his writings shows his keen interest in the problems of metamorphism.

The Relation between Distortion and Slaty Cleavage

From *Quarterly Journal of the Geological Society of London*, Vol. III, pp. 74–105
1847.

While examining the fossil shells from the Ludlow rocks of Westmoreland, most of which are slightly distorted, it struck me that if we could find out that the changes of form in the shells followed any certain law we might make allowance for them, and thus discover the original form of the shell. Following up this idea, I found that whenever the impressions of several shells occurred on one slab of stone, they were all distorted in the same direction; the change having no reference to the original figure of the shells, but to their position on the stone: it appeared as if every specimen had been contracted in the same direction. This remark enabled me to throw together many shells which I was before disposed to consider distinct species.

Examining other specimens to see how far this observation could be extended, I found that it held generally true, and also that there was a connection between the direction of the cleavage planes and what may be called the axis of distortion. Instances were met with showing the greatest variety in the amount and nature of the changes of form in fossil shells, from a slight alteration in the outline to the most extravagant and complicated distortion in which the proportions of the parts were completely changed; but whenever the rock showed a trace of cleavage, there was evidently some connection between the direction in which the specimens had been altered and the direction of the cleavage. But that connection appeared so complicated that I saw no prospect of understanding it without visiting some localities where the phaenomena might be examined on the spot on a large scale. Besides certain agreements in the direction of the cleavage and the distortion, it appeared to be nearly a general rule that the shells are most distorted in those beds which are most slaty; which confirmed the idea that the two phaenomena were intimately connected. . . . Let us now proceed to the description of the distorted shells, beginning with those found in beds intersected by the cleavage at a high angle. The change is simplest where the shells are flat, and therefore lie entirely in the plane of the bedding; in these cases the shells all appear contracted in a direction perpendicular to the strike of the cleavage across the bed, and this apparently without any lengthening of the shell in the contrary direction, although it is difficult to be sure of this latter point.

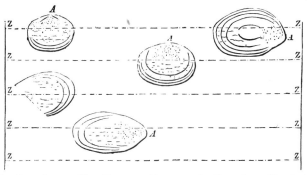

Fig. 15.—Drawing used by Sharpe to illustrate the distortion of fossil shells in slaty beds, 1847.

This change is shown in fig. 3 [15], where several nearly flat brachyopodous shells are lying in various directions on a slab of slaty rock from Aber-y-Wynant, near Dolgelly: the apex of each is marked *A*, the direction of the cleavage across the bed is shown by the lines *Z Z*. Every shell appears shortened in a direction contrary to that of the lines *Z Z*, as if the whole mass had been pressed together in one direction. The shells probably belong to one of the concentrically marked Atrypae, but the longer forms might be mistaken for Lingulae. The surface of each is covered with small wrinkles, parallel to the strike of the cleavage *Z Z*; in this instance the cleavage cuts the bedding at an angle of about 40°: not having the original form of the shell as a guide, we cannot calculate the amount of distortion accurately, but judging from their present forms, they appear shortened at least twenty-five per cent. in a direction perpendicular to the strike of the cleavage across the bed. . . .

From these and similar cases, we learn that the shells have been compressed by a force acting in a direction perpendicular to the planes of cleavage, and that the compression of the mass between the cleavage planes has been counterbalanced by its expansion in a direction corresponding to the dip of the cleavage. . . .

The origin of the oblique pressure on these fossils is easily found: the expansion of the masses of rock in the direction of the dip of the cleavage must cause an oblique pressure on the surface of every bed which is cut obliquely by the cleavage; the fossil shells lying between an expanding mass and a resisting weight of matter have given way before the pressure in the manner described. But as the expansion of the rock in the one direction may have been caused by its compression in the contrary direction, it follows that all the effects yet described may have originated in the compression of the mass of the rock in a direction perpendicular to the cleavage planes. . . .

Having examined all the principal cases of distortion in detail, we may take a broader view of the subject and consider them with reference to the position of the shells in the bedding. The beds of slate vary much in thickness and are frequently separated by thin layers of shells, some of which are impressed on each of their surfaces. When we compare the fossils on the opposite sides of a bed of slate, we find them distorted along lines of which the direction is parallel; but if there has been any oblique pressure, it has acted in opposite directions on the two sides of the bed. . . . It is here obvious that there can be no difference in the distortion of the specimens which come from the upper or the under side of a bed; and it is impossible to distinguish them. The same proof of the expansion of the mass along the planes of cleavage may be found in beds of every thickness. A bed of slate from Tintagel not the third of an inch thick exhibits exactly the same phaenomena.

Similar instances will be found in every bed of fossiliferous slate; and they prove that the mass forming each bed of slate has been expanded in the direction of the dip of the cleavage. A motion of the whole mass upwards in the direction of the cleavage would produce a compression of the fossils in one direction only: but a pressure acting equally but in contrary directions on the opposite sides of the same bed, or in other words, a force pressing from the centre of a bed towards each of its surfaces, proves that an expansion of the mass of each bed has taken place, which, as already stated, may have been caused by pressure in another direction.

From this examination of the distorted fossils under various circumstances, we may conclude that their present forms may be accounted for, by supposing that the rocks in which they are imbedded have undergone compression in a direction perpendicular to the planes of cleavage, and a corresponding expansion in the direction of the dip of the cleavage. . . . Therefore it may be asserted as probable, that all rocks affected by that peculiar fissile character which we usually call slaty cleavage have undergone—

1st.   A compression of their mass in a direction everywhere perpendicular to the planes of cleavage.

2nd.   An expansion of their mass along the planes of cleavage in the direction of a line at right angles to the line of incidence of the planes of bedding and cleavage; or in other words, in the direction of the dip of the cleavage.

No proof has been found that the rock has suffered any change in the direction of the strike of the cleavage planes.   We must therefore presume that the masses of rock have not been altered in that direction.

It has been shown in the first part of this paper, that there was reason to believe *that all slaty rocks had undergone a compression of their mass in a direction perpendicular to the planes of cleavage.* Now that we have discovered the system upon which the cleavage planes are arranged, we may judge of the direction of the pressure which compressed the slates, and endeavour by that means to find out its cause.   I have endeavoured to prove, that though each plane of cleavage runs on for great distances in a uniform direction, the planes are so arranged side by side as to make it probable that they are not true planes, but rather portions of great curves having a common axis and bounded by vertical lines.

It has also been shown, that *the compression of the slaty mass was compensated by its expansion in the direction of the dip of the cleavage, but that no change was observed in the direction of the strike of the cleavage.*   The explanation of these laws is easy, now that the direction of the cleavage and its position relatively to the elevation of the area are ascertained.   The elevation of a mass of rock into a curve by increasing the breadth of the surface of the area, would give room for the expansion of the mass in the direction of the curve, in the proportion of the length of the arc thus formed to its chord; and the curve by the terms of the proposition represents the dip of the cleavage. But as long as the elevation continued uniform over a straight axis, nothing would occur to weaken the resistance to the expansion of the mass in the direction of the axis, which is, as already shown, the direction of the strike of the cleavage. If the elevation should prove sufficient to break up the surface, the fissures would be parallel to the boundaries of the area; and when the mass was once broken in longitudinal fissures, its power of resistance to an expansive force would be farther diminished in the direction of the dip, but would still remain the same along the strike.

Throughout all the preceding pages I have abstained from expressing any opinion on the cause producing the cleavage; and now that I have gone through the subject, I must still leave the immediate agent in the operation undiscovered; although I hope that its discovery may be facilitated by the progress made in ascertaining the circumstances under which it took place.

Pressure appears to have been concerned in the operation; for the cleavage is uniformly at right angles to the direction in which pressure is seen to have taken place; and also the amount of cleavage appears to bear some proportion to the compression suffered by the rock. On the other hand, there are reasons for thinking that pressure could not be the sole agent in the operation, for the cleavage did not take place on the first upheaval of the district, when the crust not having yet given way the pressure might be supposed the greatest, but only after the beds had assumed their present position and the various anticlinal and synclinal axes had been formed.

Heat may have had some share in producing the cleavage: if the elevation was caused by a heated mass below, the conduction of heat must have followed the same direction as the pressure; and each sheet of slate must from its position have received the heat sooner than the sheet above it while the temperature was increasing, and parted with it later while the mass was cooling.

Lastly, Mr. Darwin has suggested an explanation built upon a combination of mechanical and crystalline forces, viz. "that the planes of cleavage and foliation are intimately connected with the planes of different tension to which the area was long subjected, after the main fissures or axes of upheavement had been formed, but before the final cessation of all molecular movement."* And that "this difference in the tension might affect the crystalline and concretionary processes."

These seem to be the agencies among which we have to seek, either separately or in combination, for the immediate cause of slaty cleavage.   I leave others to decide between them, contenting myself with having supplied some of the materials upon which the decision may be built.

---

* *Geological Observations on South America*, p. 168.

BERNHARDI

Reinhard Bernhardi, professor of natural science at the Forst-Akademie zu Dreissigacker in northern Germany, seems to have been the first scientist to propose the idea of a glacial period in the history of the earth.

AN HYPOTHESIS OF EXTENSIVE GLACIATION IN PREHISTORIC TIME

Translated from *Jahrbuch für Mineralogie, Geognosie, und Petrefaktenkunde* (Leonhard), Vol. III, pp. 257–267, 1832.

How did the rock fragments and boulders of northern origin which are to be found in northern Germany and neighboring countries get to their present positions?

. . . The author believes that, more completely than by any hypothesis as yet known to him, these phenomena would be explained by the assumption that the polar ice once reached clear to the southernmost edge of the district which is now covered by those rock remnants; that this, in the course of thousands of years, gradually melted back to its present extent; that, therefore, those northern deposits must be compared to the walls of rock fragments which surround almost every glacier at varying distances, or in other words, that they are nothing other than moraines which that enormous ice sea left behind on its gradual withdrawal.

Should this assumption seem admissible after closer testing, it could also be applied to the mysterious occurrences of similar rock fragments in other regions, for example in the Jura Mountains, etc. Also the eternal névé and glaciers of the Alps must, under the above supposition, have had a far greater spread in times long passed, have reached down much farther into the valleys, and have filled many valleys now free from perennial ice. Thus it was possible that the rock fragments of the High Alps which came on or in the glaciers were finally expelled from them and put down on the edges of the glacier and so arrived at their present locations, which are often separated from their original places by deep valleys and even lakes.

To be sure, one thing which is rightly considered an incontestable fact in the history of the earth's formation seems, at first glance, in

327

direct contradiction to this somewhat generalized hypothesis. It is proven by thousands of examples that a higher temperature formerly prevailed over the earth and especially that the climate in regions of high latitude must have been much warmer than it is at present. But this contradiction is only an apparent one. To the author at least, it seems to disappear as soon as the period of the last major revolution which the crust of the earth underwent is very carefully distinguished from the following—the historical period. That in that prehistoric period the northern region particularly had a warmer climate is entirely incontestable; but it was otherwise in the earlier epochs of the historical period. Here many important data bespeak the contrary. . . .

Esmark has shown very plausibly that great ice fields existed earlier at many places in Norway where now there is no more than seasonal ice. He thought that the numerous rock fragments near the seacoast, which in their occurrence greatly resemble ice dams (moraines), were brought there by those great glaciers, as it does not seem to him unlikely that the Norwegian mountains of olden time were covered with ice clear down to the seacoast and that the sea itself was frozen in that region. If this conjecture, supported by many observations, be correct, it is only a step from there to the previously stated hypothesis of the author. . . .

If, however, it were also proven that Greenland had a somewhat warmer climate several centuries ago, the view advocated here would not be definitely contradicted. This only supposes that the temperature in northern latitudes has decreased considerably from the beginning of our historical period to the present. But it is not necessary for us to postulate a continuous unbroken rise of temperature. More than likely, partial standstills of this temperature-change took place from time to time, and even retrogressions —which perhaps embraced considerable periods of time. Also from local conditions, for example by greater activity of nearby volcanoes, a higher temperature might be induced in earlier times in many countries, for example in Iceland. . . .

Should naturalists who have devoted their attention for a longer time to these striking geological phenomena be pleased to confirm or to contradict the view developed here, either would call for the author's sincerest thanks; for in any case geological research would thereby be advanced!

AGASSIZ

Louis Jean Agassiz (1807–1873), Swiss naturalist, came to America in 1846, was appointed to a professorship at Harvard in 1847, and later founded there the Museum of Comparative Zoology. Although Agassiz's chief fame lies in his glacial theory, his major work was with fossil fish.

Evidence of a Glacial Epoch

Translated from *Études sur les glaciers*, Neuchâtel, 1840.

The simultaneous presence in the lower part of the Alpine valleys of all the phenomena which usually accompany glaciers seems to me to be the most convincing proof that one could demand of the great extension that has been attributed to them. This concurrence of effects due to the different forces included in the general activity of glaciers proves clearly that neither the piles of boulders that have been considered as ancient moraines, the perched blocks, the polished and striated rocks, the *lapiaz*,* nor the troughs of waterfalls could be attributed to any other source than a glacier. For only a glacier produces all of these features at once. . . .

As we have thus succeeded in showing the presence of glaciers as far as the lower part of Alpine valleys, as we have even acquired the certainty that they filled the valleys there to considerable heights above their floors, we have at the same time proven that the entire massif of our Alps has been covered by an immense sea of ice from which great projections descended to the edge of the surrounding low country; that is to say, into the great Swiss plain and the plain of northern Italy, in the same way as the present glaciers of our day send their tongues into the lower valley. But with this difference: instead of being restricted among the isolated peaks and in the highest valleys, the ancient glacier enfolded entire mountain chains and descended into the plain by

---

* Agassiz was using this in the local Swiss sense for sharp furrows in the glaciated region; not in the present geologic sense as solution furrows in limestones.—Editors.

great valleys. This is in general the story told us of the past by moraines. . . .

. . . We are now going to pass to another group of facts which prove that the ice had, at an earlier epoch, a still greater extension.

But before trying to trace the boundaries of the earlier glaciers, let us examine the phenomena by which we can recognize the effect of their presence. We are going to see that here as well, disregarding other less significant facts, the blocks scattered in a certain manner on the surface of the ground and the polished rocks similar to those of the Alpine valleys will serve us as guides everywhere. . . .

There can be no doubt of the Alpine origin of the erratic blocks of the Jura region. Messieurs de Buch, Escher de la Linthe, and Studer have even shown that those of the Vaudoisian and Neuchâtelian Jura came from the Valaisian Alps and the massif of Mount Blanc; those of the Bernaise Jura from the Oberland; and those of Argovie and Zurich from the Petits-Cantons. It is only rarely that one observes mixtures of blocks in these various districts and when one happens to find traces of such a mixture it is always near the boundary of these regions. From this I conclude that the action of the transportation of blocks is repeated sporadically in each of the great corridors which descend from the Alps toward the Jura and toward the plain of northern Italy.

The transportation of these blocks from the Alps to the Jura has intrigued geologists of all time and, as it is evident that the agent which accomplished this must have been endowed with an extraordinary power such as is no longer possessed by the agents of today, it has been necessary to have recourse to pure hypothesis to account for such an extraordinary phenomenon. The hypothesis of great streams was in the greatest favor for a long time, and at first it seemed truly the most natural because it is customary to consider streams as the most energetic agents of transport. Nevertheless we shall see that it is far from accounting for all the phenomena of erratic blocks. Also the partisans of this theory do not agree on the nature of these currents nor on their origin. . . .

Monsieur Lyell, in order to account for the various phenomena which the blocks present, proposed another explanation. He supposes that the transport of the angular blocks is effected on rafts of ice carried along by currents of water in about the same way that the icebergs of the north carry the blocks they deposit on the northern coasts of Europe. ⸱ Monsieur Lyell cites many examples of blocks carried for great distances in this way by masses of ice that are three-quarters submerged by the weight of the blocks. This explanation, although very ingenious, is nevertheless not applicable to the erratic blocks of the Jura for this reason: the erratic blocks of the Jura do not rest directly on the ground.

FIG. 16.—One of the sketches in Agassiz's original portfolio of drawings used in connection with his early studies of glaciation; probably drawn in 1841.

Everywhere that the rounded pebbles, which ordinarily accompany the great blocks, have not been reworked by later action, one notices that they form a bed of several inches and sometimes of many feet on which the angular blocks rest. These pebbles are very much rounded, even polished, and piled up in such a manner that the greatest are at the surface and the smallest, which frequently grade to a fine sand, are at the bottom, immediately upon the polished floor. Thus Monsieur Lyell's mode of transportation would well explain why the blocks are not rounded, granting that they were protected by the ice that covered them;

but it in no way accounts for the presence of the rounded pebbles found below or for the formation of the polished and striated surface on which this bed rests. . . .

Other naturalists, among them Dolomieu and Ebel, suppose that the erratic blocks have been transported from the Alps to the Jura on an inclined slope, but that later revolutions having carried away the ground of this inclined plane and cut the great Swiss Valley, the blocks have remained fixed in the places where one sees them now. This theory is refuted by the fact that the transportation of erratic blocks is the last of the great geologic phenomena which have taken place on the surface of Switzerland. Moreover it is demonstrable that our lakes were already in existence at the time of this transportation.

These considerations without doubt would suffice to convince the most obstinate of the inadequacy of the various theories that we have just reviewed. I hope above all to have demonstrated that the hypothesis of the great streams is no more admissible than the others, the supposition of a transport as violent as this not being in accord with the evident facts. We shall now have to consider whether there are not in the ensemble of phenomena of erratic blocks some details which speak in favor of a slow and peaceful transportation analogous to that which the glaciers of our Alps are effecting today.

We have just seen that the erratic blocks of the Jura generally repose on a bed of cobbles and pebbles lying between them and the surface of the ground which is habitually polished and striated; that these pebbles are well rounded and heaped in such a fashion that the largest are on top while the smallest, which grade to a fine sand, occupy the base and rest directly on the polished surfaces. This arrangement, which is constant, therefore is in opposition to any idea that they were carried into place by currents; for, in that case, the order of the superposition of the rounded pebbles would be reversed. On the other hand, if it is remembered that the present glaciers show at their base an exactly similar bed, which is intermediate between the ice and the floor, and that this bed is the instrument which serves even today to polish and striate the rocks on which the glacier rests, we will be naturally led to assign a similar origin to these pebbles and the fine sand which accompanies the erratic blocks, as soon as other facts prove to us the presence of the ice.

The presence of a fine sand on top of the polished rocks moreover proves that no powerful force has modified, or no important catastrophe affected, the surface of the Jura since the epoch of transportation of the erratic blocks. But as these polished rocks are found along the northern banks of the lakes of Neuchâtel and Bienne we conclude that the Swiss lakes were already in existence at this time. Also the continuity of the moraines on the two banks of Lake Geneva furnish us proof that this basin, as well, is older than the transportation of the blocks, since it preceded the formation of the moraine.

In addition to this bed of rolled pebbles and sand between the erratic blocks and polished rock, one notices on many points of the slope of the Jura stratified deposits of the same debris, which, without doubt, are also connected with the great phenomenon of block transportation but which owe their present arrangement to particular circumstances. . . . I have the conviction that these deposits are formed in the same manner as the stratified moraines, that is to say, in a small pool of water caught at the edge of the ice.

Another phenomenon more important than these stratified deposits is the presence of polished rocks in the Jura. . . . They are continuous surfaces completely independent of the stratification of the beds and of the direction of the Jura Range. . . . They show a polish as uniform as the surface of a mirror, especially where the rock has recently been uncovered. . . . These surfaces are sometimes plane, sometimes undulating, often even traversed by sinuous furrows of varying depth or by longitudinal rounded bosses. But these are never directed parallel to the greatest slope of the mountain. On the contrary, bosses and furrows are oblique and longitudinal—directions which exclude all idea of a current, or the action of atmospheric agents, as cause for this erosion. One very curious fact that likewise could not be explained by water action is that the polish is uniform even when the rock is composed of fragments of different hardness, as for example, the breccias of the Portlandian. . . .

One notices, moreover, on these polished surfaces, where they have been well preserved, the same fine striae that we have mentioned under the present glaciers and on the former surfaces of the Alps. . . .

The simultaneous occurrence, in the Jura, of phenomena which, in the Alps, are evidently associated with the presence of glaciers and which are not met elsewhere in similar circumstances, leads us directly to the following conclusion: that the erratic blocks, the polished surfaces, and the *lapiaz* owe their origin to the action of the ice which at a certain epoch must have covered the flanks of the ranges of our Jura. . . .

The presence of these phenomena among the mountain chains of the Vosges is all the more important because these mountains have never been mentioned as the theater of powerful action by streams. But, even admitting that the Vosges also were higher at a certain epoch than they are now, one cannot avoid associating all of these traces of ice to the single great phenomenon which manifests itself everywhere that one meets erratic blocks, polished and striated surfaces, *lapiaz*, etc. The polished rocks in particular testify to the presence of glaciation in a great number of localities, for they are widely spread not only in the Jura, the Alps, and the Vosges, but throughout the north of Europe. . . .

In England the polished rocks have been observed in various localities. Sir James Hall described them in the environs of Edinburgh. Later Messieurs Sedgwick and Buckland have noticed them in Westmoreland and Cumberland counties. Monsier de Verneuil, who has visited many of these localities, brought me a fragment of magnesian limestone broken from the surface of the ground, which presents exactly the same appearance as the polished rocks of Landeron. I believe there is only one way of accounting for all of these facts and uniting them with the aggregate of known geological phenomena. That is to admit that at the end of the geologic epoch which preceded the upheaval of the Alps, the earth was covered with an immense sheet of ice, in which the mammoths of Siberia were buried and which extended to the south as far as the phenomenon of erratic blocks, surmounting all the inequalities of the surface of Europe as it existed then, filling the Baltic Sea and all the lakes of northern Germany and Switzerland, stretching beyond the borders of the Mediterranean and the Atlantic Ocean, and covering even all of North America and Asiatic Russia; that, at the time of the elevation of the Alps, this ice formation was raised like the other rocks; that the fragments broken from all the crevices during the upheaval fell on its surface and without being rounded, as they underwent no rubbing, moved down the slope of this sheet of ice. In the same way blocks of rock falling on a glacier today are pushed along its edges by continual movements which the ice undergoes in melting and freezing at different hours of the day and in different seasons. . . .

Following the raising of the Alps the world must have regained its warmth. As the ice melted, it formed great funnels in the places where it was thinnest; valleys of erosion were cut in the bottom of crevasses in localities where no stream could exist without being encased in walls of ice; and, when the ice had completely disappeared, great angular blocks were left on a bed of rounded pebbles of which the smallest form the base.

It seems to me that this explanation accounts for and utilizes all the facts that we have studied to this point.

ROGERS

William Barton Rogers (1805–1881), American geologist, was state geologist of Virginia (1835–1841) and professor in the University of Virginia. Rogers' recognition of the need for such an institution led to the founding of Massachusetts Institute of Technology, of which he was the first president.

Fossil Evidence of the Early Paleozoic Age of Certain Rocks in New England

From *Edinburgh New Philosophical Journal*, N.S., Vol. IV, pp. 301–303, 1856.

It is well known that the altered slates and gritty rocks which show themselves uninterruptedly throughout a good part of Eastern Massachusetts, have, with the exception of the coal measures on the confines of this state and Rhode Island, failed hitherto to furnish geologists with any fossil evidence of a Paleozoic age, although, from aspect and position, they have been *conjecturally* classed with the system of rocks belonging to that period. Indeed, the highly metamorphosed condition of these beds generally, traceable, no doubt, to the great masses of igneous material by which they are traversed or inclosed, would naturally forbid the expectation of finding in them any distinguishable fossil forms.

Lately through the kindness of Mr. Peter Wainwright who resides in the neighbourhood, I have been led to examine a quarry in the belt of siliceous and argillaceous slate, which lies on the boundary of Quincy and Braintree, about ten miles south of Boston, and, to my great surprise and delight, I have found it to be a *locality* of *trilobites.*

The fossils are in the form of casts, some of them of great size, and lying at various levels in the strata. So far as I have yet explored the quarry, they belong chiefly, if not altogether, to one species, which, on the authority of Agassiz, as well as my own comparison with Barrande's descriptions and figures, is undoubtedly a *Paradoxidis.*

As the genus *Paradoxidis* is peculiar to the lowest of the Paleozoic rocks in Bohemia, Sweden, and Great Britain, marking the *primordial* division of Barrande and the Lingula flags of the British survey, we will probably be called upon to place the fossiliferous belt of Quincy and Braintree *on or near the horizon of our lowest Paleozoic group,* that is to say, somewhere about the level of the Primal rocks, the Potsdam sandstone, and the Protozoic sandstone of Owen, containing *Dikelocephalus* in Wisconsin and Minnesota. *Thus, for the first time, are we furnished with the data for establishing conclusively the geological age of any portion of this tract of ancient and highly altered sediments,* and what gives further interest to the discovery, *for defining in regard to this region the very base of the Paleozoic column, and that, too, by the same fossil inscriptions which mark it in various parts of the Old World.*

H. D. AND W. B. ROGERS

The brothers, Henry Darwin and William Barton Rogers.

MOUNTAIN BUILDING FORCES EXEMPLIFIED BY THE APPALACHIAN SYSTEM

From *Transactions of the Association of American Geologists and Naturalists*, Vol. 1, pp. 474–531, 1840–42.

Having, in the prosecution of the State Geological Surveys of New Jersey, Pennsylvania, and Virginia, arrived at certain general facts in the structure of the Appalachian chain, involving some new considerations in Geological Dynamics, we propose, in the present memoir, to offer a description and theory of the phenomena. As similar structural features would appear, upon comparison, to prevail in many of the disturbed regions of other countries, and among strata of all geological dates, an exposition of their laws cannot be uninteresting at this time, when every question connected with the elevation of the earth's crust, is receiving so generally the attention of geologists.

Predominance of Southeastern Dips

While the general direction of the Appalachian chain is northeast and southwest, there is a remarkable predominance of southeastern dips throughout its entire length from Canada to Alabama. This is particularly the case along the southeastern or most disturbed side of the belt, where it is strikingly exhibited in the great valley, and in the extensive mountain ridges that bound it on the southeast. But, as we proceed towards the northwest, or from the region of greatest disturbance, the opposite, or northwest dips, which previously were of rare occurrence, and always very steep, become progressively more numerous, and, as a general rule, more gentle.

Upon the correct interpretation of this singular feature depends, we conceive, the clear elucidation of whatever relates to the dynamical actions which the region has experienced, to the stratagraphical arrangement of the rocks, and, as immediately connected with this, to the distribution of their organic remains. The object of the present paper is, to exhibit those general laws of structure, of which the feature in question is but a simple and immediate consequence, and to develope what we have for several years past regarded as the true theory of the flexure and elevation of the Appalachian rocks.

Of the Flexures of the Strata, and the Law of Their Gradation, from Southeast to Northwest

The above-described phenomena of the dips in the Appalachian range may, we think, be readily accounted for by the peculiar character of the flexures of the strata. These flexures, unlike the symmetrical curvature usually assigned to anticlinal and synclinal axes, present, in almost every instance, a steeper or more rapid arching on the northwest than southeast side of every convex bend; and, as a direct consequence, a steeper incurvation on the southeast than the northwest side of every concave turn; so that, when viewed together, a series of these flexures has the form of an *obliquely undulated* line, in which the apex of each upper curve lies in advance of the centre of the arch. On the southeastern side of the chain, where the curvature is most sudden, and the flexures are most closely crowded, they present a succession of alternately convex and concave folds, in each of which the lines of greatest dip on the opposite sides of the axes, approach to parallelism, and have a nearly uniform inclination of from forty-five to sixty degrees towards the southeast. This may be expressed in other words, as a *doubling under or inversion* of the northwestern half of each anticlinal flexure. Crossing the mountain chain from any point toward the northwest, the form of the flexures changes, the close inclined plication of the rocks producing their uniformly southeastern dip gradually lessens, the folds open out, and the northwestern side of each convex flexure, instead of being abruptly doubled under and *inverted*, becomes either vertical or dips steeply to the northwest. Advancing still further in the same direction into the region occupied by the higher formations of the Appalachian series, the arches and troughs grow successively rounder and gentler, and the dips on the opposite sides of each anticlinal axis, gradually diminish and approach more and more to equality, until, in the great coal-field west of the Allegheny mountain, they finally flatten down to an almost absolute horizontality of the strata, at a distance of about one hundred and fifty miles from the chain of the Blue ridge or South mountain.

These general features in the physical structure of the Appalachian region, will be best understood by consulting the *Ideal section*, Plate XVI [Fig. 17], intended to embrace the prevailing character of the different portions of the chain from the Blue ridge to the western coal-field.   Along with this diagram, which embodies the general results of our observations, will be found several *actual sections*, comprising the principal details of structure and topography observed in different parts of the chain, from New

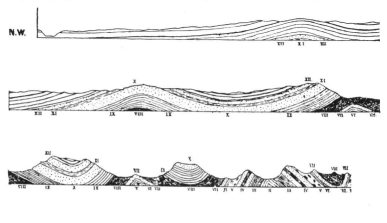

Fig. 17.—"Ideal section across the Appalachian Chain," accompanying the paper by H. D. and W. B. Rogers, 1841.

Jersey to eastern Tennessee.   These cross the belt at nearly equal intervals, and have been selected from a number, all of which equally exhibit the general conditions of structure above described.

To assist in conveying clear conceptions of the diversified and sometimes complicated modes of structure, occasioned by the flexures and foldings of the strata, we deem it important to introduce here two or three new descriptive terms, which seem called for by the necessity of possessing a phraseology adapted to the relationships of the strata about to be detailed.   Using the terms anticlinal and synclinal in their commonly accepted sense, we propose to apply the phrases *anticlinal* or *synclinal* mountain or range, to designate ridges formed respectively by a convex and concave flexure of the strata. Every flexure, of such degree as to fall short of producing an inversion of the rocks on the northwestern side of the anticlinal, and the southeastern side of the synclinal bends, we shall call a *normal* flexure; and the dips corresponding to such flexures, as exhibited in transverse sections, we shall denominate *normal dips*. While the phrases, *anticlinal dip*, and *synclinal dip*, sufficiently express the directions of the beds, due to the concave and convex flexures, we propose the term *monoclinal*, to signify a sameness in the direction of the dip, and shall term a mountain or valley, in which such sameness prevails, a *monoclinal* mountain, or *monoclinal* valley. As briefly expressive of the whole concave and convex flexure, we propose to use the terms *arch* and *trough*.

Conceiving a plane to be extended through the apex or most incurved part of each of the concentric flexures in an anticlinal or synclinal bend, so as to occupy a medial position between the two branches of the curves, we propose to call this plane the *axis-plane*. Where the flexure is perfectly symmetrical on both sides of the plane, and the dip on the one side, therefore, equal to that on the other, it is evident, that the axis-plane will have a vertical position. In the Appalachian region, however, and, as we believe, in nearly all other disturbed chains, where the phenomena of flexure are exhibited on a scale of much extent, these planes are inclined to the perpendicular in a greater or less degree, according to the energy of the inflecting force. In the region before us, the dip of the imaginary plane is almost invariably to the southeast, the amount of the deviation from the vertical altitude diminishing progressively, as we cross the chain towards the northwest. A corresponding law of the axis-planes will, we believe, be found to obtain, in all extensive groups of axes, the general expression of their relation being, that the dip of the axis-planes is always *towards* the region of maximum disturbance. . . .

Flexures Broken, or Passing into Faults

A feature of frequent occurrence in certain portions of the Appalachian belt, is the passage of an inverted or folded flexure into a fault. These dislocations, preserving the general direction of the anticlinal axes, out of which they grow, are usually prolonged to a great distance, having, in some instances,—for example, in southwestern Virginia,—a length of about one hundred miles. These lines of fault occur in all cases, along the northwestern side of the anticlinal, or the southeastern side of the synclinal axis, and never in the opposite situation. This curious and instructive fact is best seen by tracing, longitudinally, some of the principal anticlinal axes of Pennsylvania and Virginia. From a rapidly steepening northwestern dip, the northwestern branch of the arch passes through the vertical position, into an inverted or southeastern dip; and at this stage of the folding, the fault generally commences. It begins with the disappearance of one of the groups of softer strata, lying immediately to the northwest of the more massive beds, which form the northwestern summit of the anticlinal belt. The dislocation increases as we follow it longitudinally, group after group of these overlying rocks disappearing from the surface, until, in many of the more prolonged faults, the lower limestone is brought, for a great distance, with a moderate southeasterly dip, directly upon the carboniferous formations. In these stupendous fractures, of which several instances occur in southwestern Virginia, the carboniferous limestone being brought into close proximity to the great lower Appalachian limestone, a portion of which, even, is occasionally buried, the thickness of the strata ingulfed cannot be less than seven thousand or eight thousand feet.

Of the Distribution of the Axes in Groups

Wherever, in the Appalachian chain, we become minutely familiar with the undulations of the strata, we find it impossible to resist the conclusion, that the axes arrange themselves in natural *groups*, the individual flexures showing a close agreement in their length, mutual distance, straightness, or curvature, and in the extent and style of the arching. In those districts which are crowded with normal axes, such as the Susquehanna and Juniata divisions, many such groups attract our notice. Each of these assemblages of axes being generally distinguished by some special character, we are inclined to regard the comparison and analysis of their several features as of the very highest importance, in those investigations of geological dynamics into which the whole subject of flexures must evidently lead us. . . .

It is a curious and important fact, connected with this group of axes, that in certain cases, chiefly, we believe, in wide and deep

troughs, the included smaller axes or wrinkles, though parallel
to each other, are not parallel to the general synclinal axis of the
basin, in which they occur. This feature is obvious in all the
deep anthracite coal-basins of Pennsylvania, especially near their
terminations. These lesser, subordinate axes, generally have a
strike parallel to that of one of the great flexures bounding the
basin; but, on account of the convergence of the sides of the trough,
they are necessarily more or less oblique to the opposite margin.
They are, therefore, so many long, parallel warpings of the strata,
conforming to one boundary, but abutting acutely against the
other. Sometimes, indeed they cross the basin very gradually,
or pass almost longitudinally, from one side to the opposite,
and die out, as wrinkles on the slopes which bound the basin.
That they have originated in an inequality in the energy of the
linear forces concerned in bending and elevating the rocks along
the principal flexures, and arise, therefore, from an actual warping
of the strata, seems altogether probable. If so, they are secondary
consequences of those more general and extended movements,
which give existence to the grander flexures, in whose folds they
lie. . . .

Of the Origin of the Supposed Subterranean Undulations, and
of the Manner in Which the Strata Became Permanently Bent
and Dislocated

The parallel flexures of the crust, so strikingly exhibited in
the Appalachian chain, and recognizable, we believe, in nearly
all disturbed mountainous districts, we conceive to have originated
in the following manner. We assume, that in every region, where
a system of flexure prevails, the crust previously rested on a
widely extended surface of fluid lava. Let it be supposed, that
subterranean causes competent to produce the result, such, for
example, as the accumulation of a vast body of elastic vapors and
gases, subjected the disturbed portion of the belt to an excessive
upward tension, causing it to give way, at successive times, in a
series of long parallel rents. By the sudden and explosive escape
of the gaseous matter, the prodigious pressure, previously exerted
on the surface of the fluid within, being instantly withdrawn,
this would rise along the whole line of fissure in the manner of an
enormous billow, and suddenly lift with it the overlying flexible
crust. Gravity, now operating on the disturbed lava mass,

would engender a violent undulation of its whole contiguous surface, so that wave would succeed wave in regular and parallel order, flattening and expanding as they advanced, and imparting a corresponding billowy motion to the overlying strata. Simultaneously with each epoch of oscillation, while the whole crust was thus thrown into parallel flexures, we suppose the undulating tract to have been shoved bodily forward, and secured in its new position by the permanent intrusion, into the rent and dislocated region behind, of the liquid matter injected by the same forces that gave origin to the waves. This forward thrust, operating upon the flexures formed by the waves, would steepen the advanced side of each wave, precisely as the wind, acting on the billows of the ocean, forces forward their crests, and imparts a steeper slope to their leeward sides. A repetition of these forces, by augmenting the inclination on the front of every wave, would result, finally, in the folded structure, with inversion, in all the parts of the belt adjacent to the region of principal disturbance. Here, an increased amount of plication would be caused, not only by the superior violence of the forward horizontal force, but by the production in this district of many lesser groups of waves, interposed between the larger ones, and not endowed with sufficient momentum to reach the remoter sides of the belt. To this interpolation we attribute, in part, the crowded condition of the axes on the side of the undulated district, which borders the region where the rents and dykes occur, and to it we trace the far greater variety which there occurs in the size of the flexures.

This theory agrees strikingly with the singularly undisturbed condition of the strata, northwest of our great lines of fault. When describing, under a preceding head, some of these enormous dislocations, especially those of southwestern Virginia, an account was given of the gradual transition of structure, from the normal to the folded or inverted form, and thence, to a successive ingulfing of certain groups of strata, into a line of fault, presenting sometimes, for the distance of seventy miles, an actual inversion of the lower Appalachian limestone or slate, upon either the carboniferous limestone or the next inferior group. The commencement in all cases of these faults, in the steeply folded synclinal part of the flexure, immediately on the northwest of the finally inverted anticlinal curve, would seem to prove conclusively, that the fracture has been due to a profound folding in and inversion of the rocks, carried to the extent of producing an actual snapping asunder of the beds where most incurved, followed by a squeezing downward of the opposite side of the trough, by the horizontal northwestward thrust of the anticlinal portion, causing the lower strata of the latter to lie directly upon geologically higher groups. The enormous mass of rocky material, thus forcibly pressed down and firmly held there, would, we conceive, constitute a vast *subterranean barrier or dam*, capable of arresting, in some degree, the progress of the succeeding waves, and of protecting the region for a moderate distance, towards the northwest, or the leeward side of the fault, from the undulations to which it would otherwise have been exposed. . . .

Of the Date of the Appalachian Axes

It has been stated already, that, excepting in one or two localities, the Appalachian formations constitute an unbroken succession of conforming strata, from the lowest members of the system, which repose immediately on the primary or metamorphic rocks, to the highest of the carboniferous strata. We must therefore conclude, that the elevatory actions, which lifted the entire chain above the level of the ancient sea, and impressed upon it those symmetrical features of structure which we have described, could not have begun, at least with any degree of intensity, until the completion of the carboniferous formation. That the principal movement *immediately* succeeded the termination of this period of gradual operations, or more properly arrested the further progress of the coal-formation, is, we think, clearly proved, by the fact, that nowhere do we meet with any strata, referable to the next succeeding or new red sandstone period, overlying the highest rocks appertaining to the coal; and it can scarcely be supposed, that throughout so vast an area, embracing several enormous basins, in which the upper carboniferous rocks have been preserved, all traces of that newer group, if deposited, should have been so entirely swept away, as not to have left its fragments even in any part of the wide tracts over which the coal-rocks are spread. An additional reason for believing that the elevation and flexure of the strata did not take place as late as the era of the new red sandstone, is to be found in the remarkably undisturbed manner in which a set of rocks of the age, approximately at least, of the European new red group, rest unconformably on the axes which traverse the Appalachian formations. . . .

ROGERS

Henry Darwin Rogers (1809–1866), American geologist, served at various times as state geologist of New Jersey and of Pennsylvania, and as regius professor of geology and natural history at Glasgow.

An Inquiry into the Origin of the Appalachian Coal Strata, Bituminous and Anthracitic

From *Transactions of the Association of American Geologists and Naturalists*, Vol. I, pp. 433–474, 1840–1842.

Of the Limits of the Appalachian Coal Strata

The extensive Appalachian coal formation, embraces all the detached basins, both anthracitic and semi-bituminous, of the mountain chain of Pennsylvania, Maryland, and Virginia, and also the vast bituminous trough, lying to the northwest in Pennsylvania, Ohio, Virginia, Kentucky, Tennessee, and Alabama. I shall endeavor presently to show, that all these coal-fields, extending from the northeastern counties of Pennsylvania, to the northern part of Alabama, and from the great Appalachian valley, westward into the interior of Ohio and Kentucky, include only a portion of the original formation, immense tracts having been destroyed by denudation. A comparison of the coal strata of contiguous basins, has convinced me, that they are only detached parts of a once continuous deposit; and the physical structure of the whole region most satisfactorily confirms this idea, by showing that they all repose conformably on the same rocks; the more or less insulated trough in which they occur, merely being separated by anticlinal tracts of greater or less breadth, from which denuding action has removed the other portions of the formation. This distribution of the coal in a series of parallel and closely connected synclinal depressions, is a direct result of the system of vast flexures, into which the whole of the Appalachian rocks has been bent, by the undulatory movements that accompanied the final elevation of the strata, and terminated the era of the coal.

. . . Considering all of these outlying portions of the formation as subordinate and intimately connected parts of one great bituminous coal-field, the southeastern boundary of which is the escarpment of the Allegheny and Cumberland mountains, the dimensions of the great basin will be nearly as follows: Its length, from northeast to southwest, is rather more than seven hundred and twenty miles, and its greatest breadth about one hundred and eighty miles. Upon a moderate estimate, its superficial area amounts to sixty-three thousand square miles.

There are besides this, however, several smaller basins which lie to the southeast, and are entirely separated from it. These consist of the detached troughs of anthracite, in eastern Pennsylvania, and the solitary outlying basin of semi-bituminous coal in Broad Top mountain, near the Juniata river. . . .

Here then we have a coal formation, which, before its original limits were reduced, measured, at a reasonable calculation, nine hundred miles in length, and in some places more than two hundred miles in breadth. I would ask, is it conceivable, that any lake, bay, or estuary, could have been the receptacle of a deposit so extended, or that any river or rivers could have possessed a delta so vast? . . .

Comparing, in the first place, the rocks of mechanical origin, as they occur in different districts, we almost invariably find them coarsest and most massive towards the southeast, and more and more fine-grained and less arenaceous, as we pursue them across the successive parallel basins northwestward. Thus in the anthracite coal-fields, which are the most southeastern of all, the coal is interstratified with a vast thickness of rough and ponderous grits, and very siliceous conglomerates; but is associated with comparatively very little soft clay slate or shale. In this region, the coal slates themselves, are more than ordinarily arenaceous, and bear a smaller proportion to the sandstones, than in the basins more to the west. At the same time that the coal rocks, viewed in the aggregate, acquire a finer texture, in going westward, the individual strata undergo a corresponding reduction in thickness, while many of them entirely thin away. . . .

. . . Some of the limestone strata of the coal-measures, possess . . . a remarkably wide distribution, ranging without interruption from the vicinity of the Allegheny mountain, to the country west of the Allegheny river. Having ascertained the positions of a number of these fossiliferous beds, I am now engaged in investigating their organic remains. The examinations already made, show that these all belong to *marine* genera, and that the different beds are characterized by their peculiar species. . . . The marine character of their genera,—*Terebratula, Goniatites, Bellerophon, Encrinus*, &c., sufficiently proves that these rocks were originally deposited beneath the waters of an ocean, while at the same time the increasing purity of the limestones, and the multiplication and expansion of the beds westward, clearly show that the ancient ocean augmented regularly in depth in that direction. . . . But the most important result of this mode of tracing the strata, is the evidence we have of the frequent alternation of a tranquil and disturbed condition of the waters. . . .

Of the facts connected with the range of the individual coal-seams, that of their prodigious extent is, itself, one of the most surprising and instructive. As a general rule, this wide expansion characterizes all the beds of both the bituminous and anthracitic basins. It is true, that many seams possess a comparatively local range, but not a few of those which, on first examination, appear of circumscribed extent, cover in reality a very wide area, the error respecting them being caused by fluctuations of thickness, or by their occasionally thinning out and reappearing. Among those which manifest great permanency as to thickness, the vast range of some of the larger ones is truly extraordinary. Let us trace, for example, the great bed, which occurs so finely exposed at Pittsburgh, and along nearly the whole length of the Monongahela river, and which I have called the Pittsburgh seam. . . . The longest diameter of the great elliptical area here delineated, is very nearly two hundred and twenty-five miles, and its maximum breadth about one hundred miles. The superficial extent of the whole coal-seam, as nearly as I can estimate it, is about fourteen thousand square miles.

But the limits here described, though wide, fall very far within those which the bed anciently occupied. To the southeast of the large basin of the Ohio river, there are several other insulated, parallel troughs, which also contain the Pittsburgh seam. . . . If we now take into account the fifty additional miles of breadth which the bed once possessed, its former area must have been at least thirty-four thousand square miles, a superficial extent greater than that of Scotland or Ireland.

Though the above is, perhaps, the greatest extent of surface, which it is in our power positively to assign to this bed of coal, the proofs of a prodigious denudation of the strata, throughout the districts bordering its present outcrop, are so irresistible, that I consider the dimensions here given as bearing actually but a small proportion to the real ancient limits of the stratum. . . .

The general uniformity in the thickness of this superb bed, throughout so vast a region, and at the same time the regular and gentle gradation which it experiences in size, when we trace it from one outcrop to the other, are features not less remarkable than its enormous length and breadth. . . . While we are thus furnished with conclusive evidence, from the fact that its rate of increase is most rapid towards the southeast, that the ancient land with which the stratum was connected must have been situated in that direction, we see that the northeastern part of the coast was the quarter where its materials were supplied in the greatest abundance. To this conclusion I am disposed to appeal, in support of the conjecture already ventured, that this great bed of the main or western coalfield, is but a remnant of a still more expanded stratum, which attained its maximum size, in the enormous seam of which all the anthracite basins present us insulated patches. The singular constancy in the thickness of this Pittsburgh bed, no less than its prodigious range, are circumstances that seem strongly adverse to the theory which ascribes the formation of such deposits to any species of *drifting* action.

. . . The mechanical arrangement of the layers in every coal seam, as seen when viewed edgewise, indicates plainly, that it is a compound stratum, as much as any other sedimentary deposit, each bed being made up of innumerable very thin laminae of glossy coal, alternating with equally minute plates of impure coal, containing a small admixture of finely divided earthy matter. These subdivisions, differing in their lustre and fracture, are frequently of excessive thinness, the less brilliant leaves sometimes not exceeding the thickness of a sheet of paper. . . . If traced out to their edges, all these ultimate divisions of a mass of coal will be found to extend over a surprisingly large surface, when we consider their minute thickness. . . .

. . . Besides the above-mentioned features, all the coal-beds which I have ever examined or seen minutely described, possess another peculiarity in their mechanical constitution, on a less minute scale, which is equally incompatible with the notion of a transportation by currents. I refer here to the subordinate divisions of the coal-beds, some of which are strata of pure coal, some of earthy coal, and some of common shale, all constituting together the compound mass, which we call a coal-seam, but each maintaining its particular position and character as a distinct deposit over an area which is truly astonishing. . . . Only one particular process of accumulation promises to explain the occurrence in such cases, of these thin and uniform sheets of material, of which the thickness is often less than a foot, while their superficial area is many hundred square miles. I cannot conceive any state of the surface, but that in which the margin of the sea was occupied by the vast marine savannahs of some peat-creating plant, growing half immersed on a perfectly horizontal plain, and this fringed and interspersed with forests of trees, shedding their offal of leaves upon the marsh.

Independently of the above argument, based on the breadth and uniform distribution of the layers in the coal, there is another, drawn from the striking deficiency of earthy sedimentary particles. In many of the purest layers, the total proportion by weight of foreign mineral substance, in the coal, is less than two per cent., sometimes barely one per cent., while the ratio by bulk is consequently less than one half of this.

Of the Character of the Strata in Immediate Contact with the Coal-seams

Of the material underlying the coal-beds. The deposit, upon which each seam of coal immediately rests, and which I shall call the floor, is, with a few rare exceptions, wholly distinct in its composition from the roof, or that which reposes directly upon the bed. To Mr. Logan we are indebted for having ascertained the highly important fact, that the floor of every coal-seam in South Wales is composed of a peculiar variety of more or less sandy clay, distinguished by its containing the *Stigmaria ficoides*. . . .

The Stigmaria presented in its structure, according to Lindley and Hutton, a low, dome-shaped, fleshy trunk, or centre, from the edge of which there radiated a number of horizontal branches, supplied with a multitude of slender, cylindrical, and exceedingly long leaves. . . . The plants, according to Dr. Buckland, probably floated on the water.

. . . Both in composition and structure, the roof rock manifests signs of having been deposited by a more or less rapid current. In

place of a single species of fossil plant, it usually includes a prodigious variety, and the delicate ramifications of these, instead of
intersecting the bed in various directions, as the processes of the
Stigmaria do in the fire-clay, lie in a singularly disordered and
fragmentary condition, in planes almost invariably parallel to the
bedding. . . . A further indication of the violence of the currents,
which strewed these coarse materials over the coal, is sometimes to
be detected in the composition of the lowest portion of the overlying
bed of grit or sandstone, in which a large amount of coal, in the
state of powder or sand, is disseminated in the rock, giving it a
dark, speckled appearance. . . . It implies, I conceive, the erosion
of a certain portion of the upper surface of the soft, carbonaceous
mass by the friction of the sandy current.

Theory of the Origin of the Coal Strata

. . . Let us suppose an earthquake, possessing the characteristic
undulatory movement of the crust, in which I believe all earthquakes essentially to consist, suddenly to have disturbed the level
of the wide peat-morasses and adjoining flat tracts of forest on the
one side, and the shallow sea on the other.   The ocean, as usual in
earthquakes, would drain off its waters for a moment from the
great Stigmaria marsh, and from all the swampy forests which
skirted it, and, by its recession, stir up the muddy soil, and drift
away the fronds, twigs, and smaller plants, and spread these, and
the mud, broadly over the surface of the bog.   In this way may
have been formed the laminated slates, so full of fragmentary
leaves and twigs, which generally compose the immediate covering
of the coal-beds.   Presently, however, the sea would roll in with
impetuous force, and, reaching the fast land, prostrate every thing
before it. . . . Upon the dying away of the earthquake undulations, the sea, once more restored to tranquillity, would hold in
suspension at last, only the most finely subdivided sedimentary
matter, and the most buoyant of the uptorn vegetation, that is to
say, the argillaceous particles of the fire-clay, and the naturally
floating hollow stems of the *Stigmariae*. These would at last
precipitate themselves together, by a slow subsidence, and form
a uniform deposit, exhibiting but few traces of any active horizontal currents, such as would arise from a drifting into the sea from
rivers. . . .

. . . Though a vast preponderance of subsidence over elevation is plainly indicated in the prodigious thickness of the coal-measures, each particular coal-seam in which was produced successively at the surface, I cannot conceive that either an alternation of periods of subsidence and repose, or an uninterrupted prolonged depression, will explain the phenomena of the Appalachian coal rocks, as they have been here described. A general subsidence throughout the coal period, of all the great area now occupied by the Appalachian basins, is proved independently of the above evidence derived from the nature of the coal-beds, by the interesting fact, that the lower seams of Ohio and western Pennsylvania, have their eastern limit more than one hundred and fifty miles to the west of the general eastern boundary of the upper ones; and, as we ascend in the formation, the beds extend successively more and more to the east, or in the direction of the ancient land. But considering the many striking instances which I have recorded, of the close approach or actual contact of certain beds of coal and oceanic limestone, we cannot resist the conclusion, that the gradual downward movement was frequently interrupted by a slow upward one. In all the instances that I have cited, where the limestone stratum immediately underlies a coal-seam, it is obvious that an upward movement of the land must have taken place, so gradually as to be unattended by any sensible commotion of the waters.

Of the Gradation in the Proportion of Volatile Matter in the Coal of the Appalachian Basins

There prevails a very interesting law of gradation, in the quantity of volatile matter belonging to the coal, as we cross the Appalachian basins from the southeast towards the northwest. The extraordinary extent of area over which this law obtains, and its intimate connection with corresponding gradations in the structural phenomena of the region, the description and theory of which have been given by my brother and myself in another communication, seem to claim for it a place in the present general account of our coal-measures. The gradation may be thus briefly described. Crossing the Appalachian coal-fields, northwestward from the great valley, to the middle of the main or western trough, by any section between the northeastern termination of the formation in Pennsylvania, and the latitude of Tennessee, we find, as the result of multiplied chemical analyses, a progressive increase in the proportion of the volatile matter, passing from a nearly total deficiency of it, in the dryest anthracites, to an ample abundance in the richest coking coals. . . .

The cause of the different degrees of de-bituminization of the coals, in different parts of their range, I am disposed to attribute to the prodigious quantity of intensely heated steam and gaseous matter, emitted through the crust of the earth, by the almost infinite number of cracks and crevices, which must have been produced during the undulation and permanent bending of the strata. . . . It is easy to conceive, that the coal, throughout all the eastern basins, if thus effectually steamed, and raised in temperature in every part of its mass, would discharge a greater or less proportion of its bitumen and other volatile constituents, as the strata were more or less frequently and violently undulated by earthquake action. It is also obvious, that the more western beds, remoter from the region of active movements, less crushed and fissured, and presenting a greater resistance to permeation by the subterranean vapors, would, in virtue of their mere geographical position in the chain, be much less extensively de-bituminized. The fact that we nowhere, not even in the most dislocated and disturbed districts of the anthracite coal-field, find any traces of true igneous rocks, that, by their contiguity to the coal, could have caused the loss of its bitumen, is a circumstance in their geology, which goes far to confirm the truth of the hypothesis. Precisely in proportion as the flexures of the strata diminish in our progress westward, does the quantity of the bitumen in the coal augment; but it is difficult to conceive how any such law of gradation could have been the result of a temperature transmitted by conduction from the general lava mass beneath the crust, for that would imply a corresponding increasing gradation in the thickness of the crust, advancing westward under the coalfields, whereas such an inference is in direct conflict with the fact of the general diminution westward of the Appalachian rocks, besides being inconsistent with all correct geothermal considerations, which forbid our imagining so unequal a conduction to the surface, of the earth's interior temperature.

DARWIN

Charles Robert Darwin (1809–1882), the great English naturalist, best known for his recognition of the principle of organic evolution, made many important contributions to the science of geology, among which his theory of the origin of coral reefs is outstanding.

The Origin of Coral Reefs and Islands

From *Journal of Researches into the Geology and Natural History of the Various Countries Visited by H. M. S. Beagle*, pp. 554–559, London, 1840.

This island [Keeling or Cocos Island] is, therefore, a lofty submarine mountain, which has a greater inclination than even those of volcanic origin on the land. I will now give a sketch* of the general results at which I have arrived, respecting the origin of the various classes of reefs, which occur scattered over such large spaces of the intertropical seas.

The first consideration to attend to, is, that every observation leads to the conclusion that those lammeliform corals, which are the efficient agents of forming a reef, cannot live at any considerable depth. As far as I have personally seen, I judge of this from carefully examining the impressions on the soundings, which were taken by Captain FitzRoy at Keeling Island, close outside the breakers, and from some others which I obtained at the Mauritius. At a depth under ten fathoms, the arming came up as clean as if it had been dropped on a carpet of thick turf; but as the depth increased, the particles of sand brought up became more and more numerous, until, at last, it was evident the bottom consisted of a smooth layer of calcareous sand, interrupted only at intervals by shelves, composed probably of dead coral rock. To carry on the analogy, the blades of grass grew thinner and thinner, till, at last, the soil was so sterile, that nothing sprung from it.

As long as no facts, beyond those relating to the structure of lagoon islands were known, so as to establish some more comprehensive theory, the belief that corals constructed their habitations,

---

* The sketch was read before the Geological Society, May, 1837.

or, speaking more correctly, their skeletons, on the circular crests of submarine craters, was both ingenious and very plausible. Yet the sinuous margin of some, as in the Radnack Islands of Kotzebue, one of which is fifty-two miles long, by twenty broad, and the narrowness of others, as in Bow Island (of which there is a chart on a large scale, forming part of the admirable labours of Captain Beechey), must have startled every one who considered this subject.

The very general surprise of all those who have beheld lagoon islands, has perhaps been one chief cause why other reefs, of an equally curious structure have been almost overlooked:* I allude to the encircling reefs. We will take, as an instance, Vanikoro, celebrated on account of the shipwreck of La Peyrouse. The reef there runs at the distance of nearly two, and in some parts three miles from the shore, and is separated from it by a channel having a general depth between thirty and forty fathoms, and, in one part, no less than fifty, or three hundred feet. Externally, the reef rises from an ocean profoundly deep. Can any thing be more singular than this structure? It is analogous to that of a lagoon, but with an island standing, like a picture in its frame, in the middle. A fringe of low alluvial land in these cases generally surrounds the base of the mountains; this, covered by the most beautiful productions of a tropical land, backed by the abrupt mountains and fronted by a lake of smooth water, only separated from the dark waves of the ocean by a line of breakers, form the elements of the beautiful scenery of Tahiti—so well called the Queen of Islands. We cannot suppose these encircling reefs are based on an external crater, for the central mass sometimes consists of primary rock, or on any accumulation of sedimentary deposits, for the reefs follow indifferently the island itself, or its submarine prolongation. Of this latter case there is a grand instance in New Caledonia, where the reefs extend no less than 140 miles beyond the island.

The great Barrier which fronts the N.E. coast of Australia, forms a third class of reef. It is described by Flinders as having a length of nearly one thousand miles, and as running parallel to the shore, at a distance of between twenty and thirty miles from it, and, in

---

* Mr. De la Beche, however, seems to have been fully aware of the difficulty. He says, "there are certain situations, where coral reefs run, as it were, in a line with the coast, but separated from it by deep water, which would seem to require a different explanation."—*Geological Manual*, p. 142.

some parts, even of fifty and seventy.   The great arm of the sea thus included, has a usual depth of between ten and twenty fathoms, but this increases towards one end to forty and even sixty. This probably is both the grandest and most extraordinary reef now existing in any part of the world.

It must be observed, that the reef itself in the three classes, namely, lagoon, encircling, and barrier, agrees in structure, even in the most minute details: but these I have not space here even to allude to.   The difference entirely lies in the absence or presence of neighbouring land, and the relative position which the reefs bear to it.   In the two last-mentioned classes, there is one difficulty in undertaking their origin, which must be pointed out.   Since the time of Dampier it has been remarked, that high land and deep seas go together.   Now when we see a number of mountainous islands coming abruptly down to the sea-shore, we must suppose the strata of which they are composed, are continued with nearly the same inclination beneath the water.   But, in such cases, where the reef is distant several miles from the coast, it will be evident upon a little consideration, that a line drawn perpendicularly from its outer edge down to the solid rock on which the reef must be based, very far exceeds that small limit at which the efficient lamelliform corals exist.

In some parts of the sea, as we shall hereafter mention, reefs do occur which fringe rather than encircle islands—the distance from the shore being so small, where the inclination of the land is great, that there is no difficulty in understanding growth of the coral. Even in these "fringing" reefs, as I shall call them in contradistinction to the "encircling", the reef is not attached quite close to the shore.   This appears to be the result of two causes: namely, first, that the water immediately adjoining the beach is rendered turbid by the surf, and therefore injurious to all zoophytes; and, secondly, that the larger and efficient kinds only flourish on the outer edge amidst the breakers of the open sea.   The shallow space between the skirting reef and the shore has, however, a very different character from the deep channel, similarly situated with respect to those of the encircling order.

Having thus specified the several kinds of reefs, which differ in their forms and relative position with regard to the neighbouring land, but which are most closely similar in all other respects (as I could show if I had space), it will, I think, be allowed that no explanation can be satisfactory which does not include the whole series. The theory which I would offer, is simply, that as the land with the attached reefs subsides very gradually from the action of subterranean causes, the coral-building polypi soon raise again their solid masses to the level of the water: but not so with the land; each inch lost is irreclaimably gone;—as the whole gradually sinks, the water gains foot by foot on the shore, till the last and highest peak is finally submerged. . . .

It will at once be evident that a coral reef, closely skirting the shore of a continent, would, in like manner after each subsidence, rise to the surface; the water, however, always encroaching on the land. Would not a barrier reef necessarily be produced, similar to the one extending parallel to the coast of Australia? It is indeed but uncoiling one of those reefs which encircle at a distance so many islands.

Thus the three great classes of reef, lagoon, encircling, and barrier, are connected by one theory. It will perhaps be remarked, if this be true, there ought to exist every intermediate form between a closely-encircled and a lagoon island. Such forms actually occur in various parts of the ocean: we have one, two, or more islands encircled in one reef; and of these some are of small proportional size to the area enclosed by the coral formation; so that a series of charts might be given, showing a gradation of character between the two classes.

THE SETTLING OF CRYSTALS IN MOLTEN LAVA

From *Geological Observations on the Volcanic Islands*, pp. 117–119, London, 1844.

. . . One side of Fresh-water Bay, in James Island, is formed by the wreck of a large crater, mentioned in the last chapter, of which the interior has been filled up by a pool of basalt, about 200 feet in thickness. This basalt is of a gray colour, and contains many crystals of glassy albite, which become much more numerous in the lower, scoriaceous part. This is contrary to what might have been expected, for if the crystals had been originally disseminated in equal numbers, the greater intumescence of this lower scoriaceous part would have made them appear fewer in number. Von Buch[*] has described a stream of obsidian on the peak of Teneriffe, in which the crystals of feldspar become more and more numerous, as

---

[*] "Description des Isles Canaries," pp. 190 and 191.

the depth or thickness increases, so that near the lower surface of the stream the lava even resembles a primary rock. Von Buch further states, that M. Drée, in his experiments in melting lava, found that the crystals of feldspar always tended to precipitate themselves to the bottom of the crucible. In these cases, I presume there can be no doubt that the crystals sink from their weight. The specific gravity of feldspar varies from 2.4 to 2.58, whilst obsidian seems commonly to be from 2.3 to 2.4; and in a fluidified state its specific gravity would probably be less, which would facilitate the sinking of the crystals of feldspar. At James Island, the crystals of albite, although no doubt of less weight than the gray basalt, in the parts where compact, might easily be of greater specific gravity than the scoriaceous mass, formed of melted lava and bubbles of heated gas.

The sinking of crystals through a viscid substance like molten rock, as is unequivocally shown to have been the case in the experiments of M. Drée, is worthy of further consideration, as throwing light on the separation of the trachytic and basaltic series of lavas. Mr. P. Scrope has speculated on this subject; but he does not seem to have been aware of any positive facts, such as those above given; and he has overlooked one very necessary element, as it appears to me, in the phenomenon—namely, the existence of either the lighter or heavier mineral in globules or in crystals. In a substance of imperfect fluidity, like molten rock, it is hardly credible, that the separate, infinitely small atoms, whether of feldspar, augite, or of any other mineral, would have power from their slightly different gravities to overcome the friction caused by their movement; but if the atoms of any one of these minerals became, whilst the others remained fluid, united into crystals or granules, it is easy to perceive that from the lessened friction, their sinking or floating power would be greatly increased. On the other hand, if all the minerals became granulated at the same time, it is scarcely possible, from their mutual resistance, that any separation could take place. A valuable, practical discovery, illustrating the effect of the granulation of one element in a fluid mass, in aiding its separation, has lately been made: when lead containing a small proportion of silver, is constantly stirred whilst cooling, it becomes granulated, and the grains or imperfect crystals of nearly pure lead sink to the bottom, leaving a residue of melted metal much richer in silver; whereas if the mixture be left undisturbed, although kept fluid for a length of time, the two metals show no signs of separating.* The sole use of the stirring seems to be the formation of detached granules.

THE NATURE OF GRANITE CONTACTS

From *Geological Observations on the Volcanic Islands*, pp. 148–150, London, 1844.

After the accounts given by Barrow, Carmichael, Basil Hall, and W. B. Clarke of the geology of this district [Cape of Good Hope], I shall confine myself to a few observations on the junction of the three principal formations. The fundamental rock is granite,† overlaid by clay-slate; the latter is generally hard, and glossy from containing minute scales of mica; it alternates with, and passes into, beds of slightly crystalline, feldspathic, slaty rock. This clay-slate is remarkable from being in some places (as on the Lion's Rump) decomposed, even to the depth of twenty feet, into a pale-coloured, sandstone-like rock, which has been mistaken, I believe, by some observers, for a separate formation. I was guided by Dr. Andrew Smith to a fine junction at Green Point between the granite and clay-slate: the latter at the distance of a quarter of a mile from the spot, where the granite appears on the beach (though, probably, the granite is much nearer underground), becomes slightly more compact and crystalline. At a less distance, some of the beds of clay-slate are of a homogeneous texture, and obscurely striped with different zones of colour, whilst others are obscurely spotted. Within a hundred yards of the first vein of granite, the clay-slate consists of several varieties; some compact with a tinge of purple, others, glistening with numerous minute scales of mica

---

* A full and interesting account of this discovery, by Mr. Pattinson, was read before the British Association in September, 1838. In some alloys, according to Turner ("Chemistry," p. 210), the heaviest metal sinks, and it appears that this takes place whilst both metals are fluid. Where there is a considerable difference in gravity, as between iron and the slag formed during the fusion of the ore, we need not be surprised at the atoms separating, without either substance being granulated.

† In several places I observed in the granite, small dark-coloured balls, composed of minute scales of black mica in a tough basis. In another place, I found crystals of black schorl radiating from a common center. Dr. Andrew Smith found, in the interior parts of the country, some beautiful specimens of granite, with silvery mica radiating or rather branching, like moss, from central points. At the Geological Society, there are specimens of granite with crystallised feldspar branching and radiating in like manner.

and imperfectly crystallised feldspar; some obscurely granular, others porphyritic with small, elongated spots of a soft white mineral, which being easily corroded, gives to this variety a vesicular appearance. Close to the granite, the clay-slate is changed into a dark-coloured, laminated rock, having a granular fracture, which is due to imperfect crystals of feldspar, coated by minute, brilliant, scales of mica.

The actual junction between the granitic and clay-slate districts extends over a width of about 200 yards, and consists of irregular masses of numerous dikes of granite, entangled and surrounded by the clay-slate: most of the dikes range in a NW. and SE. line, parallel to the cleavage of the slate.   As we leave the junction, thin beds, and lastly, mere films of the altered clay-slate are seen, quite isolated, as if floating, in the coarsely-crystallised granite; but although completely detached, they all retain traces of the uniform NW. and SE. cleavage.   This fact has been observed in other similar cases, and has been advanced by some eminent geologists, * as a great difficulty on the ordinary theory of granite having been injected whilst liquefied; but if we reflect on the probable state of the lower surface of a laminated mass, like clay-slate, after having been violently arched by a body of molten granite, we may conclude that it would be full of fissures parallel to the planes of cleavage; and that these would be filled with granite, so that wherever the fissures were close to each other, mere parting layers or wedges of the slate would depend into the granite.   Should, therefore, the whole body of rock afterwards become worn down and denuded, the lower ends of these dependent masses or wedges of slate would be left quite isolated in the granite; yet they would retain their proper lines of cleavage, from having been united, whilst the granite was fluid, with a continuous covering of clay-slate.

Following, in company with Dr. A. Smith, the line of junction between the granite and the slate, as it stretched inland, in a SE. direction, we came to a place, where the slate was converted into a fine-grained, perfectly characterised gneiss, composed of yellowish-brown granular feldspar, of abundant black brilliant mica, and of few and thin laminae of quartz. From the abundance of the mica in this gneiss, compared with the small quantity and

---

* See M. Keilhau's "Theory on Granite," translated in the "Edinburgh New Philosophical Journal," vol. XXIV, p. 402.

excessively minute scales, in which it exists in the glossy clay-slate, we must conclude, that it has been here formed by the metamorphic action—a circumstance doubted, under nearly similar circumstances, by some authors. The laminae of the clay-slate are straight; and it was interesting to observe, that as they assumed the character of gneiss, they became undulatory with some of the smaller flexures angular, like the laminae of many true metamorphic schists.

RECENT UPLIFT OF THE SOUTH AMERICAN COAST

From *Geological Observations in South America*, pp. 14–16, London, 1846.

Not only has the above specified long range of coast (of South America) been elevated within the recent period, but I think it may be safely inferred from the similarity in height of the gravel-capped plains at distant points, that there has been a remarkable degree of equability in the elevatory process. I may premise, that when I measured the plains, it was simply to ascertain the heights at which shells occurred; afterwards, comparing these measurements with some of those made during the Survey, I was struck with their uniformity, and accordingly tabulated all those which represented the summit-edges of plains. The extension of the 330 to 355 feet plain is very striking, being found over a space of 500 geographical miles in a north and south line. A table of the measurements is here given. The angular measurements and all the estimations are by the Officers of the Survey; the barometrical ones by myself:—

	Feet
Gallegos River to Coy Inlet (partly angular meas. and partly estim.)	350
South Side of Santa Cruz (ang. and barom. meas.)	355
North Side of do. (ang. m.)	330
Bird Island, plain opposite to (ang. m.)	350
Port Desire, plain extending far along coast (barom. m.)	330
St. George's Bay, north promontory (ang. m.)	330
Table Land, south of New Bay (ang. m.)	350

A plain, varying from 245 to 255 feet, seems to extend with much uniformity from Port Desire to the north of St. George's Bay, a distance of 170 miles; and some approximate measurements, also given in the following table, indicate the much greater extension of 780 miles:—

	Feet
Coy Inlet, south of (partly ang. m. and partly estim.).....................	200 to 300
Port Desire (barom. m.).............	245 to 255
C. Blanco (ang. m.)..................	250
North Promontory of St. George's Bay (ang. m.).........................	250
South of New Bay (ang. m.)..........	200 to 220
North of S. Josef (estim.).............	200 to 300
Plain of Rio Negro (ang. m.)........	200 to 220
Bahia Blanca (estim.)...............	200 to 300

The extension, moreover, of the 560 to 580, and of the 80 to 100 feet, plains is remarkable, though somewhat less obvious than in the former cases. Bearing in mind that I have not picked these measurements out of a series, but have used all those which represented the edges of plains, I think it scarcely possible that these coincidences in height should be accidental. We must therefore conclude that the action, whatever it may have been, by which these plains have been modelled into their present forms, has been singularly uniform.

These plains or great terraces, of which three and four often rise like steps one behind the other, are formed by the denudation of the old Patagonian tertiary beds, and by the deposition on their surfaces of a mass of well-rounded gravel, varying, near the coast, from ten to thirty-five feet in thickness, but increasing in thickness towards the interior. The gravel is often capped by a thin irregular bed of sandy earth. The plains slope up, though seldom sensibly to the eye, from the summit-edge of one escarpment to the foot of the next highest one. Within a distance of 150 miles, between Santa Cruz to Port Desire, where the plains are particularly well developed, there are at least seven stages or steps, one above the other. On the three lower ones, namely, those of 100 feet, 250 feet, and 350 feet in height, existing littoral shells are abundantly strewed, either on the surface, or partially embedded in the superficial sandy earth. By whatever action these three lower plains have been modelled, so undoubtedly have all the higher ones, up to a height of 950 feet at S. Julian, and of 1,200 feet (by estimation) along St. George's Bay. I think it will not be disputed, considering the presence of the upraised marine shells, that the sea has been the active power during stages of some kind in the elevatory process.

THE IMPERFECTION OF THE GEOLOGIC RECORD OF LIFE DEVELOPMENT

From *The Origin of Species*, pp. 341–345, London, 1859.

I have attempted to show that the geological record is extremely imperfect; that only a small portion of the globe has been geologically explored with care; that only certain classes of organic beings have been largely preserved in a fossil state; that the number both of specimens and of species, preserved in our museums, is absolutely as nothing compared with the incalculable number of generations which must have passed away even during a single formation; that, owing to subsidence being necessary for the accumulation of fossiliferous deposits thick enough to resist future degradation, enormous intervals of time have elapsed between the successive formations; that there has probably been more extinction during the periods of elevation, and during the latter the record will have been least perfectly kept; that each single formation has not been continuously deposited; that the duration of each formation is, perhaps, short compared with the average duration of specific forms; that migration has played an important part in the first appearance of new forms in any one area and formation; that widely ranging species are those which have varied most, and have oftenest given rise to new species; and that varieties have at first often been local. All these causes taken conjointly, must have tended to make the geological record extremely imperfect, and will to a large extent explain why we do not find interminable varieties, connecting together all the extinct and existing forms of life by the finest graduated steps.

He who rejects these views on the nature of the geological record, will rightly reject my whole theory. For he may ask in vain where are the numberless transitional links which must formerly have connected the closely allied or representative species, found in the several stages of the same great formation. He may disbelieve in the enormous intervals of time which have elapsed between our consecutive formations; he may overlook how important a part migration must have played, when the formations of any one great region alone, as that of Europe, are considered; he may urge the apparent, but often falsely apparent, sudden coming in of whole groups of species. He may ask where are the remains of those infinitely numerous organisms which must have existed long before the first bed of the Silurian system was deposited: I can answer this latter question only hypothetically, by saying that as far as we can see, where our oceans now extend they have for an enormous period extended, and where our oscillating continents now stand they have stood ever since the Silurian epoch; but that long before that period, the world may have presented a wholly different aspect; and that the older continents, formed of formations older than any known to us, may now all be in a metamorphosed condition, or may lie buried under the ocean.

Passing from these difficulties, all the other great leading facts in palaeontology seem to me simply to follow on the theory of descent with modification through natural selection. We can thus understand how it is that new species come in slowly and successively; how species of different classes do not necessarily change together, or at the same rate, or in the same degree; yet in the long run that all undergo modification to some extent. . . .

We can understand how the spreading of the dominant forms of life, which are those that oftenest vary, will in the long run tend to people the world with allied, but modified, descendants; and these will generally succeed in taking the places of those groups of species which are their inferiors in the struggle for existence. Hence, after long intervals of time, the productions of the world will appear to have changed simultaneously. . . .

The inhabitants of each successive period in the world's history have beaten their predecessors in the race for life, and are, in so far, higher in the scale of nature; and this may account for that vague yet ill-defined sentiment, felt by many palaeontologists, that organization on the whole has progressed. If it should hereafter be proved that ancient animals resemble to a certain extent the embryos of more recent animals of the same class, the fact will be intelligible. The succession of the same types of structure within the same areas during the later geological periods ceases to be mysterious, and is simply explained by inheritance.

EVEREST

Robert Everest (c.1805–c.1875), an English clergyman resident in India, for whom Mt. Everest was named.

A QUANTITATIVE STUDY OF STREAM TRANSPORTATION

From *Journal of the Asiatic Society of Bengal*, Vol. I, pp. 238–240, 1832.

In the course of last summer, I made some attempts to ascertain the weight of solid matter contained in a given quantity of Ganges water, both in the dry and rainy season, but I found the weight so variable on different days (when little difference might have been expected) that I can hardly consider the observations numerous enough to give a correct average. Such as they are, however, they may not be without interest in the absence of other information on the subject. . . .

1. A quantity of Ganges' water taken 27th May, 1831, gave when evaporated, a solid residuum of 1.084 grains per wine quart.

2. July 21st. There had been little rain for some days, and the river was low for the season: a wine quart contained of soluble matter 2.0 grains; of insoluble 16.2;—total 18.2.

---

5. August 20th. The water had hitherto been taken from the side, but as it was evident that the quantity of matter held in suspension in the middle of the current was much greater than towards the bank, where the water was nearly still, I took two separate portions as before, and obtained, from the middle, 40 grains of insoluble residuum; from the side 20 grains ditto: add for soluble matter, suppose two grains to each, the middle gives 42, the side 22 grs. The river today was at the same height as on the 13th (the maximum).

---

34 grains per wine quart was found to be the average for the rains. Now as a wine quart of water weighs 14544 grains, we have about $\frac{1}{428}$th part of solid matter by weight. But as the specific gravity of this cannot be stated at less than 2, we have $\frac{1}{856}$th part in bulk for the solid matter discharged, or 577 cubic feet per second.

This gives a total of 6,082,041,600 cubic feet for the discharge in the 122 days of the rains:—7.8 grains per wine quart was the weight determined for the five winter months or $\frac{1}{1838}$th part in weight, and $\frac{1}{3676}$th part in bulk, which gives 19 cubic feet per second, or a total of 247,881,600 cubic feet for the whole 151 days of the period:—3.8 grains per wine quart was the weight allowed for the three hot months, which gives a $\frac{1}{3827}$th part by weight, and a $\frac{1}{7654}$th part by bulk, or about 4.8 cubic feet per second for the discharge of solid matter, and a total of 38,154,240 cubic feet for the discharge during the 92 days. The total annual discharge then would be 6,368,077,440 cubic feet.

HUMPHREYS AND ABBOT

Andrew Atkinson Humphreys (1810–1883), American engineer, scientist, and soldier, was chief of the Corps of Engineers after the Civil War.

Henry Larcom Abbot (1831–1927), American army engineer, was consulting engineer to the French and American builders of the Panama Canal.

THE AMOUNT OF SEDIMENT CARRIED BY THE MISSISSIPPI

From Corps of Topographic Engineers, U.S.A., *Professional Paper* 4, Washington, 1861.

Conclusions respecting proportion of sedimentary matter. A comparison of these different results leads to the belief that no material error will result from assuming that the sediment of the Mississippi is to the water, by weight, nearly as 1 to 1500, and, by bulk, nearly as 1 to 2900; provided long periods of time be considered.

Annual amount transported to the gulf. If this be so, and if the mean annual discharge of the Mississippi be correctly assumed at 19 500 000 000 000 cubic feet, it follows that 812 500 000 000 pounds of sedimentary matter, constituting one square mile of deposit 241 feet in depth, are yearly transported in a state of suspension into the gulf.

Observations upon materials rolling along the bottom of the river. Besides the amount held in suspension, the 'Mississippi pushes along into the gulf large quantities of earthy matter.

The well-known fact that rivers in their upper courses transport gravel and sand, and the experiments of Dubuat upon the velocities required to move various materials composing the beds of rivers, and the rate at which fine sand was pushed along the bed of the river Hayne, together with some experiments by Mr. George G. Meade, now Captain Topographical Engineers, on the bar of the Southwest pass in 1838, to ascertain the nature of the earthy matter suspended by the river near the bottom, led to the attempt in 1851 to ascertain by experiment whether any material was pushed along the bottom of the Mississippi in its lower trunk, and what the nature of that material was. . . . A keg similar to that used in collecting water below the surface was sunk to the bottom of the river. The current immediately overturned it, and the valves opening allowed the water to pass freely through. After remaining a few minutes it was drawn suddenly up, and was invariably found to contain material such as gravel, sand, and earthy matter. . . .

No exact measurement of the amount of the annual contributions to the gulf from this source can be made, but from the yearly rate of progress of the bars into the gulf, it appears to be about 750,000,000 cubic feet, which would cover a square mile about 27 feet deep.

Total annual contributions of the river to the gulf. The total yearly contributions from the river to the gulf amount then to a prism 268 feet in height, with a base of one square mile.

To determine the age of the delta from such data, the extent of the area upon which the sedimentary matter is deposited, and the depth below the surface of the former bottom of the gulf, must be known. Neither has been ascertained with sufficient accuracy to make the computation of any value.

SIDELL

William H. Sidell (1810–1873), an American army officer and civil engineer.

On Certain Geological Phenomena of the Mississippi Delta

From Corps of Topographic Engineers, U.S.A., *Professional Paper* 4, Appendix A, Washington, 1861.

The observations taken to ascertain changes of the shores, etc., during the progress of the work [survey of the mouths of the Mississippi], could not be very extensive, but, by inquiry and observation, much information was elicited respecting changes that had occurred in times past and those now in progress. Nevertheless, changes of importance, though not of great extent, occurred during the time of the work. The one most worthy of notice was this: The boat *passed over* a certain place on the Northeast bar, at the commencement of operations, on which there was about 2 feet water. Before their termination a lump at this place projected 2 or 3 feet above water; a change which, by comparison with other known points, was shown to proceed from a *rise* of the *bottom*.

This phenomenon is not uncommon, but, on the contrary, occurs frequently. A channel of entrance may be destroyed by this means, and, until another channel is formed, the bar will be impassable. The pilots and captains of tow-boats give innumerable instances of it. Ballast stones and anchors, which have been thrown overboard or lost, have been brought to the surface. The lumps appear to be forced *through* the ordinary bottom by some power acting from below, but what may be the cause which produces effects so wonderful, future researches must determine.

Another curious circumstance relative to these lumps and salt springs is, that they are only formed in the immediate vicinity of the bars or next to the gulf. The only instance noted in which a spring came up through the marsh, was at a place near the bayou running past the northeast light-house. . . . With this exception, and that of those lying near the mouth of the Balize bayou (which, by-the-by, was once the main pass), the lumps and salt springs are all found near the mouths of the *principal passes*.

It is perhaps proper to mention in this place some experiments that were made to determine if the deposit of sediment were owing solely to the check of velocity of the current on meeting the outside waters. The conclusion was that the effect was not owing solely to this cause. Proper vessels had been provided for the experiments, and in these as many fit substances as were at hand were dissolved in a mixture with the water, each in a separate vessel. These substances were common salt, epsom salt, alum, sea-water, brine from the salt-springs, and sulphuric acid. The river water alone took from ten to fourteen days to settle, while the solutions became perfectly limpid in from fourteen to eighteen hours, or from one-fifteenth to one-eighteenth part of the time. I know not to what cause to attribute the effect, unless it be action of these substances on the vegetable matter contained in the water, which aids in the suspension of the earthy matter. . . .

However, from these experiments we may conclude that the earthy matter is deposited more suddenly than would be the case if it depended on the check of velocity alone; that the bars will be formed just at the debouchés, or where the salt-water is first met; and that the greater the quantity of water brought down, the sooner, on account of the sudden precipitation, will the bars be formed at the debouchés.

OLDHAM

Thomas Oldham (1817–1878), Irish geologist and paleontologist, was head of the Geological Survey of India from 1850 to 1876.

Stream Erosion and Transportation

From *Memoirs of the Geological Survey of India*, Vol. I, pp. 173, 174, 1854.

Another very peculiar feature in the Khasi Hills are the curiously deep and narrow gorges or valleys in which all the rivers, in the Southern portion of the hills, find their course to the plains. The level of the stream under Cherra Poonjee is some 3,000 feet below that of the station. . . .

Now, although believing that marine denudation has exerted its powerful influence in modifying the features of these hills in former times and at different levels, as I have just stated, it is not possible to see how any littoral action, or any such ordinary marine action, could have produced those long, deep, and sinuous gorges here seen. On the contrary, these river gorges appear to me to have been excavated almost entirely by the force of the streams which have flowed and still continue to flow through them. And they appear to me to offer a magnificent instance of the almost incredible power of degradation and removal, which atmospheric force may exert under peculiar and favourable circumstances.

I took an opportunity of visiting one of the streams in these hills after a heavy and sudden fall of rain. The water had then risen only about thirteen feet above the level at which it stood a few days previously; the rush was tremendous—huge blocks of rock, measuring some feet across, were rolled along with an awful crashing, almost as easily as pebbles in ordinary streams. In one night, a block of granite, which I calculated to weigh upwards of 350 tons, was moved for more than 100 yards; while the torrent was actually turbid with pebbles of some inches in size, suspended almost like mud in a rushing stream.

SURELL

Alexandre Surell (1813–1887), French engineer, was one of the first to recognize the seriousness of the problem of soil conservation and to attack it in a practical manner. His study of "torrents" published in 1841 was reprinted in 1870; the excerpt includes a portion of the original text and one of the footnotes added in the second edition.

Torrential Streams and Their Control

Translated from *Étude sur les torrents des Hautes-Alpes*, pp. 22–26, 202–207, 2d ed., Paris, 1870.

The Department of the High Alps displays water courses of a peculiar nature. They are locally known as *torrents*, but they have certain characteristics which are not found in the torrents of other countries.

Their sources are hidden in the recesses of the mountains. They descend toward the valleys and join the brooks or rivers there. When they reach the lower levels, they spread across an extraordinarily large and cone-shaped deposit. This last fact is remarkable; it establishes a marked distinction between torrents and most other water courses.

Measurements of the course of a torrent, ascending from the alluvial cone and continuing into the gorge, . . . verify the general fact . . . that stream beds are concave upward, with the curvature gradually increasing upstream.

At the same time that the stream profile is traced up the gorge, the height of the steep banks may be noted; also, the slope of the plain, in the midst of which the alluvial cone is spread, may be measured. Thus, the elements of another curve, which represents the relief of the terrain crossed by the torrent, are supplied. If these two curves, one representing the stream profile, the other the terrain, are superposed, they provide a diagram from which one may read with perfect clarity the nature of the action that torrents exert on the ground throughout their entire length. Upstream, the curve of the terrain is above that of the stream bed; downstream, it is beneath. Consequently, the two curves must cross each other, and this point of intersection marks the transition from washing away to building up. This point is at the mouth of the gorge and at the top of the alluvial cone.

One sees that the water, compelled to follow at first the relief of the uneven terrain, has little by little destroyed the irregularities of the slopes. It has lowered certain points; it has built up others. Here, it has eroded; there, it has deposited. The obtuse angle formed by the steep slope of the mountain and the level of the plain has been smoothed out in the process, and the water has substituted in this section a curved line for a broken line. The result of all these actions has been to create a new stream which is better for the flowing of the water than the primitive profile of the terrain.

Note particularly: it is not only a question of the removal of certain roughnesses, planed off by the friction of the water. The steep banks of the mouth terminating the catchment basin are gorges, hollowed out sometimes a hundred meters deep. The torrential deposits form the little elevations, whose height above the plain often reaches seventy meters. It is by these figures that we must judge the enormous changes that torrents can produce in their stream profiles.

Considered from this point of view, torrents are a subject for useful comparisons. It is impossible to doubt that the creation of their gradients is entirely their own work. Therefore, they permit us to understand a general phenomenon, which, in other streams, is more difficult to grasp in its entirety, but which with them is immediately evident, as an experiment would be right under our eyes.

Water runs in the bed of a torrent according to the same laws that govern the flow of larger rivers. The stream gradient of a torrent is no different from that presented by the bed of a river or any stream, except that one would have to reduce the horizontal scale while keeping the vertical the same. It is the relation of the abscissa to the ordinate which varies; but the characteristic properties of the curve as well as the laws of its formation remain the same.

Thus torrents do not present phenomena different from those of larger courses of water; they merely show them on an exaggerated scale. The fundamental activity of washing away, of dragging along, then of silting, belongs to all streams; but in rivers, it is spread out over a greater surface, whereas torrents show it concentrated in a very circumscribed region. Deposits made at the mouth of streams when they reach sea level are comparable to deposits made by torrents when they debouch into the plains; deltas are true alluvial cones on which streams wander just as torrents do. On the other hand, streams would not have deltas if they did not arrive charged with debris that they have picked up in the upper part of their courses, and so they also have their catchment basins where they wash away. Thus their two extremities end like those of torrents. But the intermediary trunk, that is to say, "the canal of drainage," which is almost nothing in torrents, constitutes in itself nearly the whole course of rivers.

One place which must be studied especially with regard to the stream gradients of torrents is that where deposition begins. This place is found at the intersection of the two curves [mentioned above]. It is there that the action of the water changes or reverses itself. But in torrents the change occurring there is hardly visible at first sight, even though the consequences are considerable.

In some, the smoothness of the curve is unbroken in this transition and the slopes of the deposits are joined at a tangent with those of the gorge; so well that nothing on the curve, considered alone, would show the point where the phenomenon of erosion begins.

In others, on the contrary, the stream gradient is broken there in a more or less abrupt manner. These give us an example of a bed whose curve has not yet been completely formed. The slopes are imperfect and the process unfinished. One immediately concludes that in such torrents erosion must be proceeding very energetically, whereas that process, in the former, has already been largely completed, being no longer stimulated by the same causes and no longer having the same end to attain. This, in fact, is what observation shows.

One also notices in the first torrents that the slope of the deposits is such that materials brought there would be carried along to the river, if the water did not disperse them in wandering. One can be sure of this fact by comparing that slope with those of other torrents, which, rolling along the same kind of materials, nevertheless deposit no more, simply because they can no longer wander, either as a result of certain artificial works or of the effect of circumstances due to nature. It is believed, moreover, that a similar slope must exist for all kinds of materials; which we will call the *slope limit*.

In the second torrent, the *slope limit* is not yet reached; it is still being developed.

I attack the second problem, that of the control of torrents. . . .

This problem can be solved most effectively and promptly by causing vegetation to grow upon the entire surface of the region. . . . Where it would be impossible to make trees grow at first, we will induce the growth of herbs, shrubs, and bushes. . . . But upstream, where the zones [of defense] spread over all of the catchment basin, the forest must necessarily be restored. We shall choose the most fitting species: we shall have recourse to all known procedures, then to those which still remain to be discovered and which grow out of experience. The object of this work must be to cover the catchment basin with a forest which will grow thicker each day and which, extending farther and farther, will at last spread over every part.

If the vegetation thus developed throughout the area of the zones of defense is guarded against herds, if it is guarded against the depredations of the inhabitants, if it is cared for, kept up, forced by all possible means, it will envelop all the parts of the torrent by a very thick jungle, which will produce two equally salutary effects at the same time: that of arresting the water and that of consolidating the soil. . . .

. . . But the fixation of slopes is of too great importance thus to be abandoned to the caprices of the soil and to the free course of nature. . . .

To induce vegetation on slopes, we should cut them by little canals of irrigation derived from the torrent. These would impregnate the ragged and otherwise arid lands with life-giving moisture; they would also cross the talus slopes and make them more stable. Soon one would see them disappear under the tufts of various plants, grown from seeds in the presence of water.

Finally, while all these plantings will hold back the earth where the torrent has been eroding, we will hinder the washing away by constructing barriers at the foot of the slopes. . . . In most cases, one would find the best materials for their construction among the plantings themselves. Young trees would give posts; the results of pruning and bushes would furnish fagots. These barriers would then be made into fascines, wattled palisades. . . . These barriers would preserve the base of certain slopes until the vegetation covered the entire expanse, and until the torrent itself lost most of its violence. They would serve also to cut off the shallow ravines and to fill the small diffusions; in a word, to smooth the surface of the ground, and to efface completely the innumerable small streams, which spread like hairy fibers of a root and which are truly the root of evil.

In recapitulating the process, we see that it involves five factors:

1. The plan for zones of defense
2. The declaration of public advantage
3. The timbering of the zones
4. The planting of steep slopes
5. The construction of barriers

It remains to speak of the order in which it will be advisable to carry on the work. This order, far from being arbitrary, is one of the principal conditions for success.

I have so often mentioned in the course of this study, the necessity for attacking torrents at their very source that it is unnecessary to refer to it again. Thus, the work should first be undertaken in the highest parts: it should advance from there toward the lower parts. Not only should one commence by planting the catchment basin before occupying oneself with the lower zones; but in the basin itself, one should first go up to the highest ramifications, one should ascend above the last marks of the bed to the slopes, furrowed by ravines that water forms and deforms with each storm. There one should first establish vegetation, which would subsequently be extended toward the base, but would give assurance that the areas behind were well consolidated.

The effect of work undertaken first in higher regions will be to diminish the violence of the torrent downstream from these areas. The slope of the lower regions will then be in less danger, and the construction of barriers will be less difficult. It is plain, moreover, that in reaching these gorges, the consolidation of the zones of defense which control them will be assured only by the protection of the slopes themselves and the defense of their base. It will thus be necessary to begin in these areas first with the construction of barriers and then with the planting of the talus slopes.

Such would be, in general, the method to follow for controlling a torrent. It is for experience to show what modifications may be introduced.*

---

* The general procedure outlined in this chapter has been applied and gradually perfected by the Forestry Administration. The work is in process today in the High Alps, on a great number of torrents. . . .

As for the results, they are no longer debatable. The torrents are reduced where the vegetation produces its result, and that has generally been prompter than was expected.

"Because of vegetation," said M. Gentil, chief engineer in the High Alps, "torrential characteristics have disappeared. The water even during rainfall is less turbulent. No longer are there violent and sudden floods. *Water arriving on the alluvial fan is kept within natural limits.* The people living on the river banks can at least defend themselves from damage.

"The aspect of the mountain has abruptly changed. The soil has acquired such stability that the violent storms of 1868, which have occasioned such disasters in the High Alps, have done little damage in the regenerated areas."

JUKES

Joseph Beete Jukes (1811–1869), English geologist, was a member of the English Geological Survey and during the last nineteen years of his life was director of the Geological Survey of Ireland, serving also as professor of geology in the Royal College of Science.

Distinction Between Contemporaneous and Intrusive Trap

From *Student's Manual of Geology*, pp. 322–324, 2d ed., Edinburgh, 1862.

As it is sometimes not very easy to distinguish between injected sheets of trap and contemporaneous beds of it, it will be useful to examine those circumstances which will enable us to do so.

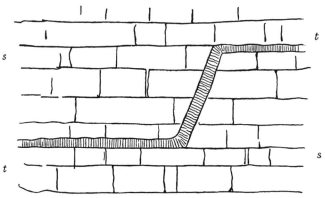

Fig. 18.—Diagram to illustrate Jukes' study of the relation between intrusive trap and the stratified rock which it has invaded, 1862.

If a sheet of trap rock (whether felstone or greenstone), after running for some distance between two certain beds, cut up or down and proceed between other beds, as in fig. 96 [Fig. 18], it is obviously intrusive and not contemporaneous.

If the beds above a sheet of trap be as much altered or "baked" by the igneous rock as those below, or if it send any veins up into the beds above it, it is equally plain that it must be an intrusive sheet.

If, however, the trap runs regularly between two beds, and the bed below the trap be altered, while that above it, composed of

378

equally alterable materials, is quite unaffected, we may conclude that the trap was poured out and flowed over the surface of the lower bed, and that the upper bed was subsequently deposited upon it; in other words, that the trap is contemporaneous and not intrusive as regards the beds in that place.

This conclusion would be confirmed if the upper surface of the trap be rugged and uneven, and if the stratification and lamination of the bed above conformed to these rugosities, as suggested in fig. 97 [Fig. 19].

FIG. 19.—Vertical section drawn by Jukes, 1862, to illustrate the relations between a trap flow and the overlying sedimentary rock. The lamination of the stratified rock, S, conforms to the rugged surface of the trap, T, "in such a way as to shew that it was deposited upon it."

In the "toadstone" of Derbyshire globular masses of its upper surface are often almost completely included in the superincumbent limestone, clearly shewing that the limestone was deposited at the bottom of the sea on the uneven surface of the cooled trap.

If, again, the bed above the trap contained any fragments clearly derived from the erosion of the trap, it would prove the trap to be a contemporaneous one. The Carboniferous Limestone of the county Limerick includes great beds of trap, and the limestone beds immediately above these are often full of little grains and fragments of the trap, shewing that the trap was in existence before those beds of limestone were formed over it.

When beds of trap (whether purely feldspathic or feldspatho-hornblendic) are clearly interstratified with beds of "ash" or "tuff" of the same character, whether that ash were subaerial or submarine ash,* it becomes almost certain that the trap is con-

---

* The student must regard the term "ash", introduced by Sir H. De la Beche as merely an English synonym of the Italian word "tuff", or "tufa", when the latter is applied to igneous materials. The advantage of using the

temporaneous; for that ash is clearly derived from some contemporaneous trap somewhere, and the chances would be greatly
against a sheet of similar trap being subsequently injected into
those ashes, without producing in them great and obvious alteration, or cutting them with dykes and veins so as to clearly shew its
intrusive character.

Even should the ash shew a considerable amount of alteration
from its original state as a mechanical deposit, such, for instance,
as the production of crystals of feldspar through its mass, it would
not be conclusive evidence against its being an "ash," or against
the contemporaneous age of the trap beds associated with it, since
such alteration might be the result of a subsequent general action,
which had produced a greater effect on the "ash" than on the other
rocks, because its nature made it more easily impressible, and more
open and liable to change than the solid igneous or the simple and
more homogeneous aqueous rocks.

Instances also occur of a genuine "ash" looking like a porphyry
from containing crystals of feldspar or hornblende that were not
innate crystals formed in it, but deposited as slightly worn and
rounded crystals along with it.

---

term "ash" is the avoidance of the ambiguity arising from "tufa" being sometimes applied to calcareous or other depositions of a soft friable character.

BUNSEN

Robert Wilhelm Eberhard von Bunsen (1811–1899) the famous German chemist.

GENETIC RELATIONS OF IGNEOUS ROCKS

Translated from *Annales de chimie et de physique*, Third Series, Vol. XXXVIII. pp. 215–289, 1853.

The rocks of Iceland present such a variety that at first it does not seem possible to discover the law which governed their origin. On examining them more closely, however, one soon finds a character common to all these eruptive masses, whatever their age, however varied their mineralogical constitution may be.

There are in Iceland, and probably in almost all the great volcanic systems as well, two principal groups of rocks that are easy to recognize when they are isolated, which unite with each other in all possible proportions. These are the *normal trachytic* rocks on the one side and the *normal pyroxenic* rocks on the other. The minerals peculiar to each of these types of rock, although varying greatly, nevertheless do not alter the general composition. . . .

The ratio existing [in the trachytes] between the silicic acid and the lime, as well as the magnesia, is nearly always constant, whereas that of the aluminium oxide to the iron oxide varies greatly. It is easy to explain this fact by known phenomena. When an alloy of lead and silver cools, the silver found in each of the layers amounts in proportion to the order of solidification. Likewise, alloys of gold and silver are never homogeneous. Facts of the same nature are encountered in the cooling of the silicates, inasmuch as the least fusible of their elements solidified first, then those which were more so, and finally those which were fusible to the highest degree. The form of crystallization does not enter, in this case. Therefore, there is no reason to be surprised that a homogeneous rock in a melted state first permits the solidification of silicates rich in iron oxide and last, of those rich in aluminium oxide. Field observation, moreover, confirms this.

In Iceland, a notable difference is found in the composition of the two extremities of a trachytic column intact in the body of a vast rock. . . .

If these variations in the composition are really due to differential solidification as an effect of temperature and pressure, it is easy to understand that inasmuch as one of the constituent parts of the mixture takes the place of the other in proportion as it diminishes, the sum of the two ought not to vary greatly. . . .

Taking the average of each of these two groups of analyses, numbers but slightly different from those of the individual specimens are obtained. This permits the general composition of the two great sources of normal trachytic and pyroxenic rocks—that is, the most acid and most basic rocks that Iceland has—to be determined; here it is:

	Composition of normal trachytic rocks	Composition of normal pyroxenic rocks
Silicic acid...............	76,67	48,47
Aluminium and iron oxide....	14,23	30,16
Calcium oxide.............	1,44	11,87
Magnesium oxide...........	0,28	6,89
Potassium oxide............	3,20	0,65
Sodium oxide..............	4,18	1,96
	100,00	100,00

According to these analyses, the ratio of oxygen of the acid to that of the bases is, on the average, $::3:0,596$ for the trachytic rocks and $::3:1,998$ for the pyroxenic rocks. All the other, unmodified rocks of Iceland which do not belong to one of these two groups have such a composition that the oxygen of their acid is to that of their base $::3:0,596$ to 1,998. Therefore, it is probable that these rocks are born of the mixture of the two principal species. In view of this fact, one wonders whether all the unmetamorphosed rocks of Iceland were born of this fusion, or rather, in other words, if there were not but two great volcanic centers responsible for the formation of Iceland from most distant times to our day.

. . . To cite but one example, there is, opposite Masfell, in one of the valleys which cuts the chain of Esja in the southeast, a dike of trachyte which cuts the pyroxenic conglomerate. This trachyte, which is white at the center, becomes darker and richer in iron as it approaches the surrounding mass. . . .

Consulting the average analyses given above suffices to show that the interior part of the dike presents the composition of the normal trachytic mass, whereas the rock which encloses it has that of the normal pyroxenic mass, and that the periphery of the dike is formed by the fusion of the two rocks in the ratio of 0,5923 of pyroxenic mass to 1 of trachytic. . . .

For the rest, on examining this dike closely, it is easy to see that it has melted the surrounding rock at all the points where it has touched it. The amount of mixing varies with proximity to the contact, but never reaches the center of the dike. All of these analyses, as well as the field observations, prove that the rocks with composition intermediate between that of the acid and the basic rocks of Iceland cannot be regarded as having been formed all at the same time; instead, they have been formed at any time when there was an injection of one of these rocks through an old mass of the other.

MALLET

Robert Mallet (1810–1881), Irish geologist and civil engineer, president of the Geological Society of Ireland, was largely responsible for placing the study of earthquakes on a truly scientific basis. The paper from which these excerpts were taken is remarkable for its vision of what the science of seismology might accomplish in the future.

The Dynamics of Earthquakes

From "On the Dynamics of Earthquakes; Being an Attempt to Reduce Their Observed Phenomena to the Known Laws of Wave Motion in Solids and Fluids," a paper read February 9, 1846; *Transactions of the Royal Irish Academy*, Vol. XXI, Pt. 1, pp. 51–106, 1846.

The present Paper constitutes, so far as I am aware, the first attempt to bring the phenomena of the earthquake within the range of exact science, by reducing to system the enormous mass of disconnected and often discordant and ill-observed facts which the multiplied narratives of earthquakes present, and educing from these, by an appeal to the established laws of higher mechanics, a theory of earthquake motion.

Now, of all conceivable alternate motions, the only one that will fulfill the requisite conditions observed, namely, that shall move with such an immense velocity as to displace bodies by their inertia, or even shear close off great buttresses from the walls they sustained, or project stones out of their beds by inertia,—that shall have a horizontal, alternate motion, either much quicker in one direction than in the other, or different in its effects, and that shall be accompanied by an upward and downward motion at the same time (a circumstance universally described as attendant on earthquakes);—the only motion that will fulfil these conditions, is *the transit of a wave of elastic compression, or of a succession of these, in parallel or in intersecting lines, through the solid substance and surface of the disturbed country.*

Now, when the original impulse comes from the land, an elastic wave is propagated through the solid crust of the earth and through

384

the air, and transmitted from the former to the ocean water, where the wave is finally spent and lost.

When, on the other hand, the original impulse comes from the bed of the deep ocean, three sorts of waves are formed and propagated simultaneously, namely, one, or several, successively through the land, which constitutes the true earthquake shock or shocks; and coincident with, and answering to every one of these, one or more waves are formed and propagated through the air, which produce the sound like the bellowing of oxen, the rolling of waggons, or of distant thunder, accompanying the shock; and a third wave is formed and propagated upon the surface of the ocean, which rolls in to land, and reaches it long after the shock or wave, through the solid earth has arrived and spent itself.

The plan, or horizontal outline of each of these waves, will be more or less circular or elliptical at first, according as the origin or centre of disturbance is at a single point, or along a line of impulse more or less regular, and the crests of the earthquake waves of every order of which we are about to speak, or, as they may be called, the *earthquake cotidal lines*, will, in their progress of propagation through the earth and sea, alter their curvilinear forms, by changes in their respective velocities, becoming more and more distorted from the original form; but, in every case, these cotidal lines will form closed figures.

It is needless to multiply such facts, which occur abundantly in earthquake narratives, to prove that the earth wave passes parallel to itself in succession through the shaken country. I shall shew hereafter, however, in speaking of *systems of vibrations*, that, under particular conditions as to geological formation, the whole mass of a country may be shaken at one and the same instant.

The earth wave or shock reaches the devoted land at the same instant that the sound of the crash and thunder of the submarine war of elements reaches the ears of its inhabitants, for the wave is itself the bearer of the sounds first transmitted through the solid earth. These are so far exactly the phenomena observed: the shock is felt, and the rolling sounds are heard at the same instant, or as nearly so as can be told. It may, however, occasionally happen that the rolling sounds shall precede considerably the actual shock, because the amount or peculiar character of the disturbance at the centre of impulse may be such as, by several partial disturbances, to set in motion waves of sound through the earth or sea, before any sufficient impulse has been given to propagate a sensible shock; and further, as the velocity of the sound wave through the earth will probably be about 7000 to 10,000 feet per second, or even more, while the velocity of the sound wave through the sea will be about 4700 (both in round numbers), so it will generally occur that the sound will be heard accompanying the shock, as transmitted by the former medium, and still the same sound be heard some time after the shock, thus transmitted more slowly through the latter medium, viz., the sea.

Occasionally waves of sound may be wholly wanting, owing to *no fractures* taking place in the earth's crust; the elastic earth wave, or shock, in such cases, being due merely to compression produced by sudden *flexure*.

So also, if the centre of impulse lie deep in hard, elastic primary rock, extending beneath a country consisting superficially of soft rock, or of diluvial matter of very low elasticity, the sound wave will reach the ear at a very distant point of the surface, by passing horizontally through the elastic rock below, and then vertically through the small distance to the surface occupied by the softer and less elastic materials, thus reaching the ear by a quicker channel than if it passed first vertically up from the deep seat of the disturbance, and then passed horizontally through the superficial deposits; but the latter course is that which the great earth shock or wave *must* take to be felt horizontally at the surface; hence, the sound must be heard, in such circumstances, before the shock; and while the sound, occasionally accompanied by a slight vertical shock, will appear to come up directly from under the feet of the hearers, the principal shock will be felt laterally as coming from a distance—a fact actually recorded by Dolomieu in his Dissertation on the Calabrian Earthquake of 1783, where, on the granite mountains, the sound was heard before the shock was felt; but in the great diluvial plain the shock was felt before the sound was heard.

In similar conditions, the great earth wave itself may, when transmitted from profound depths, be diverted by difference as to elasticity of formation, from its direct course, and, passing horizontally through deep, elastic strata, may reach the surface vertically at distant points, and produce those shocks upwards, from directly below, sometimes spoken of in earthquake narratives.

Again, if the seat of disturbance be at a great depth, the earth wave must reach the surface in its immediate neighbourhood in vertical and diagonal lines of undulation, and produce similar effects as to shock as those last spoken of. This is the case with shaken countries close to volcanoes, where the seat of impulse is often close underneath.

The earth wave which reaches the shore (from an origin beneath the sea) is a real undulation of the surface: it is a wave, whose magnitude depends upon the elasticity of the crust of the earth which has received the original impulse, and upon the nature and force of that impulse; whose vertical height appears to vary from a fraction of an inch to several feet; generally, its vertical height seems to be from two to three feet in great earthquakes, and its length variable, according to the depth and elasticity of the strata which it affects.

———

When the earth wave passes abruptly from a formation of high elasticity to one of low elasticity, or *vice versa*, it will be partly reflected; a wave will be sent back again, producing a shock in the opposite direction; it will be partly refracted, that is to say, its course onwards will be changed, and shocks will be felt upwards and downwards, and to the right and left of the original line of transit of the wave. This is exactly what has been observed to take place. Thus, Dolomieu informs us that, in Calabria, the shocks were felt most formidably, and did most mischief, at the line of junction of the deep diluvial plains with the slates and granite of the mountains, and were felt more in the former than in the hard granite of the latter. Houses were thrown down in all directions along the junction, and fewest of any where these were situated in the mountains.

Here the transit of the wave was from the clay and gravel, which have the lowest possible elasticity, into granite whose elasticity is remarkably high; and hence the shock, after doing great damage by varying its direction, and returning upon itself, at the junction, was at once eased when it got into the elastic material of the mountains. But if the case be converse, if the earth wave pass from highly elastic rock into a mass of clay or sand (suppose lying in a small-sized valley), and pass across this into similar elastic rock at the opposite side, all the former results will follow; but, in addition, the whole mass of clay, or sand and gravel, in the basin or valley will be shaken as a whole by any powerful shock, which will be felt over its whole area at the same instant; in other words, the contents of the basin or valley will be constrained to vibrate as one system, with its walls, namely, the elastic rock of its sides. This gives the solution of the fact frequently recorded of places so circumstanced, and at not very great distances, feeling the shock of the one earthquake at the one moment.

We have thus traced many of the variable and secondary effects of the transit of the great earth wave, and may remark, in concluding this branch of our subject, that earthquakes must be regarded neither as the cause nor as the immediate effects of the elevation of a district of the earth's surface, but merely as the remote effects of elevations occurring at a remote centre, so that the true definition of an earthquake is, *the transit of a wave of elastic compression in any direction, from vertically upwards, to horizontally, in any azimuth, through the surface and crust of the earth, from any centre of impulse, or from more than one, and which may be attended with tidal and sound waves dependent upon the former, and upon circumstances of position as to sea and land.*

Our knowledge is not at present sufficiently advanced as to laws of large waves of elasticity passing through solids, to be able to do more than rudely to predict the many strange alterations of the original wave which will be produced by particular local circumstances, such as by its passing from low to high land, from hard to soft rock, or *vice versa*, round great axes of hard rock, and round great bodies of inland waters, or through masses of softer rock, reposing on much harder, and suddenly reduced as to depth, in which case a single shock will probably be divided into two or three in quick succession, and varying in direction.

When correct data shall have been obtained, when we have found the moduli of elasticity, and of cohesion, and the limits of extensibility, for all our great rock formations, and the changes produced in these by augmentations of temperature,—a work capable of easy performance, but requiring multiplied experiments upon different specimens from every known formation of rock, in order to get the average result for each formation,—and when, in addition, the actual time of transit of the great earth wave or earthquake shock shall have been accurately observed, over a long range, in various formations, and with suitable instruments and precautions, then shall we be in a position not only fully to verify the truth of this theory of earthquake motion, but able also to deliver into the grasp of the computist, a wide domain of physical geology, hitherto an unprofitable waste of uncertainty and conjecture.

We may notice a few of the more direct applications to geology which may be made of our theory, whenever such data are obtained.

However well modern geologists have surveyed and mapped the formations constituting the land which we can see and handle, of the nature of the bottom of the great ocean we know nothing; no human eye ever has or ever can behold it; we cannot even reach its deep abysses with the sounding line; yet the ocean covers nearly three-fourths of our entire globe, and of this vast area the geology is an utter blank. If, however, we are enabled hereafter to determine accurately the time of earthquake shocks, in their passage from land to land, under the ocean bed, we shall be enabled almost with certainty to know the sort of rock formation through which they have passed, and hence to trace out at least approximate geological maps of the floor of the ocean. For, knowing the time of transit of the wave, we can find the modulus of elasticity which corresponds to it, and finding this, discover the particular species of rock formation to which this specific elasticity belongs.

But again, as the height or elevation of the great earth wave is a function of the depth of the solid elastic crust which has been put in motion, future accurate observations of this coefficient will enable us to determine the actual depth from which earthquake dislocation has come, and to which it laterally extends.

While the facility with which one class of our data may be ascertained will be disputed by none, it may, perhaps, occur to some that, as earthquakes are happily rare, and give no notice of their advent, and moreover, are times of such consternation, so but little accuracy is to be hoped for in observations, as to the speed or circumstances of the shock, made during such visitations. This might be partly true, were we dependent upon the nerve or watchfulness of individual observers; but already attention has been given to the contrivance of self-acting instruments (and instruments, though by no means well devised nor self-registering, have been already in use in Scotland, and perhaps elsewhere) for the registration of earthquake shocks; and there can be no doubt that, by *earthquake observatories*, established, with suitable instruments, at distant localities, in South America or Central Asia for instance, where earthquakes, greater or less, are almost daily occurrences, a very complete knowledge of the time of wave transit, and of the amplitude and altitude of the earth waves for given districts, would be soon obtained. No instruments for ascertaining the latter have been yet proposed, but they do not seem by any means difficult to devise. It is almost certain that minute earthquake shocks frequently pass through almost every part of the earth's surface, so slight as to remain unnoticed for want of instruments to detect them. . . . It would, therefore, seem very desirable that suitable instruments for earthquake registration were, at least, added to all the magnetic observatories now so widely extended over the earth, accompanied by proper instructions to the observers,—unless, indeed, separate *geological observatories be established in favourable localities for taking cognizance of all movements of the earth's crust.*

But another, and much more rapid, and perhaps even certain, method, remains to be noticed, for obtaining part of our data as to the *specific period of wave transit,* viz. *by direct experiment,* which in all matters of inductive science may be pronounced, whenever it is possible, better than mere observation.

I have already stated that it is quite immaterial to the truth of my theory of earthquake motion what view be adopted, or what mechanism be assigned to account for the original impulse; so, in the determination of the time of transit of the elastic wave through the earth's crust, if we can only produce a wave, it is wholly immaterial in what way, or by what method, the original impulse be given.

Now, the recent improvements in the art of exploding, at a given instant, large masses of gunpowder, at great depths under water, give us the power of producing, in fact, an artificial earthquake at pleasure; we can command with facility a sufficient impulse to set in motion an earth wave that shall be rendered evident by suitable instruments at the distance, probably, of many miles, and there is no difficulty in arranging such experiments, so that the explosion shall be produced by the observer of the time of transit himself, though at the distance of twenty or thirty miles, or that the moment of explosion shall be fixed, and the wave period registered by chronometers, at *both* extremities of the line of transit.

It is to be remembered, however, that these direct experiments can only give the time of wave transits for the substances forming the uppermost crust of the earth. That earthquake shocks often come from profound depths is in a high degree probable; and while down to a certain depth we may expect to find the density and elasticity of the earth's crust continually increasing, below this again, we must suppose the mineral masses in a more and more softened or even pasty condition, as they approach the lower fluid region, and hence possessed of lower elasticity. While, therefore, we cannot draw direct conclusions as to the time of transit of the wave in the rocks thus circum-stanced at profound depths, from its time of transit in the solid rocks or superficial deposits of the surface, we may reasonably expect to derive information as to some of the physical characters and molecular condition of the deep rocks themselves, by com-paring observations of the actual time of wave transit of natural shocks, coming from great depths, with that of natural or artificial shocks traversing at the surface or near it.

On the other hand, when the modulus of elasticity has been determined for the principal rocks, at various temperatures, augmenting up to their points of fusion, and the same data have been obtained for them in fluid state, we shall be in a position to demand assistance from the mathematician in determining the complex conditions of horizontal and vertical wave motion in a *compound mass*, solid at the surface, and increasing in density and elasticity down to a certain depth; below this gradually becoming a pasty semifluid mass, with probably still increasing density, but diminishing elasticity; and finally becoming a dense elastic liquid susceptible of fluid wave movements, at still profounder depths.

Such a question can scarcely be attempted after the data, already alluded to, have been obtained, without our deriving some additional knowledge as to the constitution of the interior of our planet.

In the progress of this inquiry, and in consulting very many accounts of earthquakes, one thought has been constantly sug-gested to me, which, although not directly belonging to the subject of this paper, may be very briefly noticed. While every part of the earth's surface appears occasionally liable to earthquakes, and while volcanic countries are peculiarly so, though by no means remarkable for being visited with those of greatest violence, the origin, or centre of disturbance, of almost all the greater earthquakes appears to be beneath the sea, and at considerable distances from active volcanoes, as already observed. At the same time, the circumstances of the great sea wave seem to indicate that the centre of disturbance is seldom, if ever, *very* distant from the land. May it not, then, happen that the great general region of local sudden elevation, within which we are to look most commonly for the earthquake's origin, exists as a broad belt surrounding the land; within this belt all the diversified deposits of the detritus of the land are constantly taking place, shifted and modified subsequently by tidal currents, &c.; hence, within this space the isogeothermal planes are in a constant state of fluctuation, now rising, where a thick coat of badly conducting matter is locally deposited, and again rapidly sinking as it is swept away. Such a condition of the sea bottom would seem to be the most likely state of things to give rise to frequent and sudden local elevations, or even submarine eruptions of molten matter, as has been well explained by Herschell and Babbage.

PRATT

John Henry Pratt (c. 1811–1871), English clergyman and mathematician, spent many years in India as archdeacon of Calcutta. His observations and deductions laid the foundation for the later development of the principle of isostasy.

THE ATTRACTION OF THE HIMALAYA MOUNTAINS UPON THE PLUMBLINE IN INDIA

From *Philosophical Transactions of the Royal Society of London*, Vol. CXLV, pp. 53–100, 1855.

It is now well known that the attraction of the Himalaya Mountains, and of the elevated regions lying beyond them, has a sensible influence upon the plumb-line in North India. This circumstance has been brought to light during the progress of the great trigonometrical survey of that country. It has been found by triangulation that the difference of latitude between the two extreme stations of the northern division of the arc, that is, between Kalianpur and Kaliana, is 5°23′42″.294, whereas astronomical observations show a difference of 5°23′37″.058, which is 5″.236* less than the former.

That the geodetic operations are not in fault appears from this; that two bases, about seven miles long, at the extremities of the arc having been measured with the utmost care, and also the length of the northern base having been computed from the measured length of the southern one, through a chain of triangles stretching along the whole arc, about 370 miles in extent, the difference between the measured and the computed lengths of the northern base was only 0.6 of a foot, an error which would produce, even if wholly lying in the meridian, a difference of latitude no greater than 0″.006.

The difference 5″.236 must therefore be attributed to some other cause than error in the geodetic operations. A very prob-

---

* This is the difference as stated by Colonel Everest in his work on the Measurement of the Meridional Arc of India, published in 1847. See p. clxxviii.

able cause is the attraction of the superficial matter which lies in such abundance on the north of the Indian arc. This disturbing cause acts in the right direction; for the tendency of the mountain mass must be to draw the lead of the plumb-line at the northern extremity of the arc more to the north than at the southern extremity, which is further removed from the attracting mass. Hence the effect of the attraction will be to lessen the difference of latitude, which is the effect observed. Whether this cause will account for the error in the difference of latitude in *quantity*, as well as in direction, remains to be considered, and is the question I propose to discuss in the present paper.

To dissect and actually to calculate the attraction of the masses of which the Himalayas, and the regions beyond, are composed, appears, at the very thought of it, to be an herculean undertaking next to impossible. I am fully convinced, however, that no other method will succeed. It is upon this plan that the solution of the problem is conducted in this paper. It will be seen, that by selecting a peculiar law of dissection the calculation is very greatly simplified, and made to depend entirely and solely upon a knowledge of the elevations and depressions, in fact, the general contour of the surface. This information for some part of the mass is already supplied by the maps of the Trigonometrical Survey.

In the following pages I propose, in the first place, to develope my method of calculation, and to deduce a formula by which the attraction can be determined with a precision corresponding to the degree of accuracy to which the contour of the surface is known.

In the second place, I propose to reduce the formula to numbers, and so arrive at such an approximate value of the attraction as the data I have been able to collect will allow.

This approximate value is, as will be seen, larger than 5".236, the error brought to light by the Survey. I make various suppositions with a view, if possible, to reduce my result to this, but without effect. This leads me to look in another direction for an explanation of the cause of discordance, and I arrive at a conclusion which clears up the discrepancy, confirms the calculations of this paper, and illustrates the importance of not disregarding the influence of mountain attraction.

Adding together the results of the last article, we have—

Deflection of plumb-line in meridian at A . . . = 27″.853
Deflection of plumb-line in meridian at B . . . = 11″.968
Deflection of plumb-line in meridian at C . . . = 6″.909

∴Difference of meridian deflections at A and B = 15″.885
Difference of meridian deflections at A and C = 20″.944
Difference of meridian deflections at B and C = 5″.059

The first of these quantities is considerably greater than 5″.236, the quantity brought to light by the Indian Survey. And the values of the deflections at B and C bear a far higher ratio to those at A than has been generally supposed. . . .

The conclusion, then, to which I come is, that there is no way of reconciling the difference between the error in latitude deduced in Colonel Everest's work and the amount I have assigned to the deflection of the plumb-line arising from attraction—and which, after careful re-examination, I am decidedly of opinion is not far from the truth, either in defect or excess—but by supposing, that the ellipticity which Colonel Everest uses in his calculations, although correct as a mean of the whole quadrant, is too large for the Indian arc. This hypothesis appears to account for the difference most satisfactorily. The whole subject, however, deserves careful examination; as no anomaly should, if possible, remain unexplained in a work conducted with such care, labour, and ability, as the measurement of the Indian arc has exhibited.

The Deflection of the Plumb-line in India and the Compensatory Effect of a Deficiency of Matter below the Himalaya Mountains

From *Philosophical Transactions of the Royal Society of London*, Vol. CXLIX, pp. 745–778, 1858.

The Astronomer Royal in a paper published in the Transactions for 1855, suggested that immediately beneath the mountain-mass there was most probably a deficiency of matter, which would produce, as it were, a negative attraction and so counteract the effect of the plumb-line. This hypothesis appears, however, to be untenable for three reasons:—(1) It supposes the thickness of the earth's solid crust to be considerably smaller than that assigned by the only satisfactory physical calculations made on the subject —those by Mr. Hopkins of Cambridge. He considers the thickness to be about 800 or 1000 miles at least. (2) It assumes that this thin crust is lighter than the fluid on which it is supposed to rest. But we should expect that in becoming solid from the fluid state, it would contract by the loss of heat and become heavier. (3) The same reasoning by which Mr. Airy makes it appear that every protuberance outside this thin crust must be accompanied by a protuberance inside, down into the fluid mass, would equally prove that wherever there was a hollow, as in deep seas, in the outward surface, there must be one also in the inner surface of the crust corresponding to it; thus leading to a law of varying thickness which no process of cooling could have produced.

It is nevertheless to this source—I mean a Deficiency of Matter below—that we must look, I feel fully assured, for a compensatory cause, if any is to be found. My present object is to propose another hypothesis regarding the deficiency of matter below the mountain-mass, as first suggested by Mr. Airy; and to reduce my hypothesis to the test of calculation. . . .

I will now state the hypothesis on which my present calculation proceeds. At the time when the earth had just ceased to be wholly fluid, the form must have been a perfect spheroid, with no mountains and valleys nor ocean-hollows. As the crust formed, and grew continually thicker, contractions and expansions may have taken place in any of its parts, so as to depress and elevate the corresponding portions of the surface. If these changes took place chiefly in a vertical direction, then at any epoch a vertical line drawn down to a sufficient depth from any place in the surface will pass through a mass of matter which has remained the same in amount all through the changes. By the process of expansion the mountains have been forced up, and the mass thus raised above the level has produced a corresponding *attenuation* of matter below. This attenuation is most likely very trifling, as it probably exists through a great depth. Whether this cause will produce a sufficient amount of compensation can be determined only by submitting it to calculation, which I proceed to do.

The Tables thus calculated furnish the following results:—

	At Kaliana	At Kalianpur	At Damargida
Deflection in meridian, caused by the mass of the Himmalayas and the mountain region beyond..........................	27″.978	12″.047	6″.790
Ditto, by same mass distributed through a depth of................... 100 miles	26.440	12.111	6.855
Ditto,...................... 300 miles	21.106	11.678	6.866
Ditto,...................... 500 miles	17.066	9.622	6.670
Ditto,...................... 1000 miles	11.199	7.386	5.220

By subtracting each of the last four lines from the first line, we have the following results:—

	At Kaliana	At Kalianpur	At Damargida
Deflection in meridian, caused by the mass of the Himmalayas and of the mountain region beyond......................	27″.978	12″.047	6″.790
Ditto, modified by the supposed attenuation of matter extending down to a depth of 100 miles...............................	1.538	−0.064	−0.065
Ditto,...............................	6.872	0.369	−0.076
Ditto,...............................	10.912	2.425	0.120
Ditto,...............................	16.779	4.661	1.570

It will be seen how much the Deflections are reduced by this hypothesis, especially in the case where the attenuation extends through only 100 miles. In fact, in this case the upheaval of the mountains and the consequent attenuation below produce a slight deviation the other way at the two further stations. The success of the hypothesis may therefore, thus far, be considered to be established, although it remains an hypothesis still; and we must always be in uncertainty, not as to its answering this end, but as to its being true in nature. The existence of the mountain-mass is a fact indisputable. Not so the compensating cause, which is simply conjectural as to its existence, and altogether uncertain as to its extent, if it exist. We have no certain and independent method of determining this; nor of ascertaining even if the hypothesis be valid, how far down the attenuation extends, or what law it follows.

Speculations on the Constitution
of The Earth's Crust

From *Proceedings of the Royal Society of London*, Vol. XIII, pp. 253–276, 1864.

In fact the density of the crust beneath the mountains must be less than that below the plains, and still less than that below the ocean-bed. If solidification from the fluid state commenced at the surface, the amount of contraction in the solid parts beneath the mountain-region has been less than in the parts beneath the sea. In fact, it is this unequal contraction which appears to have caused the hollows in the external surface which have become the basins into which the waters have flowed to form the ocean. As the waters flowed into the hollow thus created, the pressure on the ocean-bed would be increased, and the crust, so long as it was sufficiently thin to be influenced by hydrostatic principles of floatation, would so adjust itself that the pressure on any *couche de niveau* of the fluid should remain the same. At the time that the crust first became sufficiently thick to resist fracture under the strain produced by a change in its density—that is, when it first ceased to depend for the elevation or depression of its several parts upon the principles of floatation, the total amount of matter in any vertical prism, drawn down into the fluid below to a given distance from the earth's centre, had been the same through all the previous changes. After this, any further contraction or any expansion in the solid crust would not alter the amount of matter in the vertical prism, except where there was an ocean; in the case of greater contraction under an ocean than elsewhere, the ocean would become deeper and the amount of matter greater, and in case of a less contraction or of an expansion of the crust under an ocean, the ocean would become shallower, or the amount of matter in the vertical prism less than before. It is not likely that expansion and contraction in the solid crust would effect the arrangement of matter in any other way. That changes of level do take place, by the rising and sinking of the surface, is a well-established fact, which rather favours these theoretical considerations. But they receive, I think, great support from the other fact, that the large effect of the ocean at Punnoe and of the mountains at Kaliana almost entirely disappear from the resultant deflections brought out by the calculations.

This theory, that the wide ocean has been collected on parts of the earth's surface where hollows have been made by the contraction and therefore increased density of the crust below, is well illustrated by the existence of a whole hemisphere of water, of which New Zealand is the pole, in stable equilibrium. Were the crust beneath only of the same density as that beneath the surrounding continents, the water would be drawn off by attraction and not allowed to stand in the undisturbed position it now occupies.

I have, in what goes before, supposed that, in solidifying, the crust contracts and grows denser, as this appears to be most natural, though, after the solid mass is formed, it may either expand or contract, according as an accession or diminution of heat may take place. If, however, in the process of solidifying, the mass becomes lighter, the same conclusion will follow—the mountains being formed by a greater degree of expansion of the crust beneath them, and not by a less contraction, than in the other parts of the crust. It may seem at first difficult to conceive how a crust could be formed at all, if in the act of solidification it becomes heavier than the fluid on which it rests; for the equilibrium of the heavy crust floating on a lighter liquid would be unstable, and the crust would sooner or later be broken through, and would sink down into the fluid, which would overflow it. If, however, this process went on perpetually, the descending crust, which was originally formed by a loss of heat radiated from the surface into space, would reduce the heat of the fluid into which it sank, and after a time a thicker crust would be formed than before, and the difficulty of its being broken through would become greater every time a new one was formed.

The least that can be gathered from the deflections of these coast-stations is, that they present no obstacle to the theory so remarkably suggested by the facts brought to light in India, viz. that mountain-regions and oceans on a large scale have been produced by the contraction of the materials, as the surface of the earth has passed from a fluid state to a condition of solidity— the amount of contraction beneath the mountain-region having been less than that beneath the ordinary surface, and still less than that beneath the ocean-bed, by which process the hollows have been produced into which the ocean has flowed. In fact the testimony of these coast-stations is in some degree directly in favour of the theory, as they seem to indicate, by *excess* of attraction towards the sea, that the contraction of the crust beneath the ocean has gone on increasing in some instances still further since the crust became too thick to be influenced by the principles of floatation, and that an additional flow of water into the increasing hollow has increased the amount of attraction upon stations on its shores.

AIRY

George Bedell Airy (1801–1892), English astronomer, astronomer royal, president of the Royal Society, was the originator of one of the several variations of the theory of crustal balance, later designated as isostasy.

An Hypothesis of Crustal Balance

From *Philosophical Transactions of the Royal Society of London*, Vol. CXLV, pp. 101–104, 1855.

A paper of great ability has lately been communicated to the Royal Society by Archdeacon Pratt, in which the disturbing effects of the mass of high land north of the valley of the Ganges, upon the apparent astronomical latitudes of the principal stations of the Indian Arc of Meridian, are investigated. It is not my intention here to comment upon the mathematical methods used by the author of that paper, or upon the physical measures on which the numerical calculation of his formulae is based, but only to call attention to the principal result; namely, that the attraction of the mountain-ground, thus computed on the theory of gravitation, is considerably greater than is necessary to explain the anomalies observed. This singular conclusion, I confess, at first surprised me very much.

Yet upon considering the theory of the earth's figure as affected by disturbing causes, with the aid of the best physical hypothesis (imperfect as it must be) which I am able to apply to it, it appears to me, not only that there is nothing surprising in Archdeacon Pratt's conclusion, but that it ought to have been anticipated; and that, instead of a positive attraction of a large mountain mass upon a station at a considerable distance from it, we ought to be prepared to expect no effect whatever, or in some cases even a small negative effect. The reasoning upon which this opinion is founded, inasmuch as it must have some application to almost every investigation of geodesy, may perhaps merit the attention of the Royal Society.

Although the surface of the earth consists everywhere of a hard crust, with only enough water lying upon it to give us everywhere a *couche de niveau*, and to enable us to estimate the heights of the mountains in some places, and the depths of the basins in others; yet the smallness of those elevations and depths, the correctness with which the hard part of the earth has assumed the spheroidal form, and the absence of any particular preponderance either of land or of water at the equator as compared with the poles, have induced most physicists to suppose, either that the interior of the earth is now fluid, or that it was fluid when the mountains took their present forms. This fluidity may be very imperfect; it may be mere viscidity; it may even be little more than that degree of

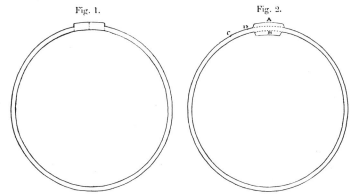

Fig. 1.   Fig. 2.

FIG. 20.—Diagrams accompanying Airy's exposition of his concept of crustal balance, 1855.

yielding which (as is well known to miners) shows itself by changes in the floors of subterraneous chambers at a great depth when their width exceeds 20 or 30 feet; and this yielding may be sufficient for my present explanation. However, in order to present my ideas in the clearest form, I will suppose the interior to be perfectly fluid.

In the accompanying diagram, fig. 1 [Fig. 20], suppose the outer circle, as far as it is complete, to represent the spheroidal surface of the earth, conceived to be free from basins or mountains except in one place; and suppose the prominence in the upper part to represent a table-land, 100 miles broad in its smaller horizontal dimension, and two miles high. And suppose the inner circle to represent the concentric spheroidal inner surface of the earth's crust, that inner spheroid being filled with a fluid of greater density than the crust, which, to avoid circumlocution, I will call *lava*. To fix our ideas, suppose the thickness of the crust to be ten miles through the greater part of the circumference, and therefore twelve miles at the place of the table-land.

Now I say, that this state of things is impossible: the weight of the table-land would break the crust through its whole depth from the top of the table-land to the surface of the lava, and either the whole or only the middle part would sink into the lava.

In order to prove this, conceive the rocks to be separated by vertical fissures at the places represented by the dotted lines; conceive the fissures to be opened as they would be by a sinking of the middle of the mass, the two halves turning upon their lower points of connexion with the rest of the crust, as on hinges; and investigate the measure of the force of cohesion at the fissures, which is necessary to prevent the middle from sinking. Let C be the measure of cohesion; C being the height, in miles, of a column of rock which the cohesion would support. The weight which tends to force either half of the table-land downwards, is the weight of that part of it which is above the general level, or is represented by 50×2. Its momentum is $50 \times 2 \times 25 = 2500$. The momenta of the "couples," produced at the two extremities of one half, by the cohesions of the opening surfaces and the corresponding thrusts of the angular points which remain in contact, are respectively $C \times 12 \times 6$ and $C \times 10 \times 5$; their sum is $C \times 122$. Equating this with the former, $C = 20$ nearly; that is, the cohesion must be such as would support a hanging column of rock twenty miles long. I need not say that there is no such thing in nature.

If, instead of supposing the crust ten miles thick, we had supposed it 100 miles thick, the necessary value for cohesion would have been reduced to $\frac{1}{5}$th of a mile nearly. This small value would have been as fatal to the supposition as the other. Every rock has mechanical clefts through it, or has mineralogical veins less closely connected with it than its particles are among themselves; and these render the cohesion of the firmest rock, when considered in reference to large masses, absolutely insignificant. The miners in Cornwall know well the danger of a "fall" of the firmest granite or killas where it is undercut by working a lode at an inclination of 30° or 40° to the vertical.

We must therefore give up the supposition that the state of things below a table-land of any great magnitude can be represented by such a diagram as fig. 1. And we may now inquire what the state of things really must be.

The impossibility of the existence of the state represented in fig. 1 has arisen from the want of a sufficient support of the table-land from below. Yet the table-land does exist in its elevation, and therefore it *is* supported from below. What can the nature of its support be?

I conceive that there can be no other support than that arising from the downward projection of the earth's light crust into the dense lava; the horizontal extent of that projection corresponding rudely with the horizontal extent of the table-land, and the depth of its projection downwards being such that the increased power of floatation thus gained is roughly equal to the' increase of weight above from the prominence of the table-land. It appears to me that the state of the earth's crust lying upon the lava may be compared with perfect correctness to the state of a raft of timber floating upon water; in which, if we remark one log whose upper surface floats much higher than the upper surfaces of the others, we are certain that its lower surface lies deeper in the water than the lower surfaces of the others.

This state of things then will be represented by fig. 2 [Fig. 20]. Adopting this as the true representation of the arrangement of masses beneath a table-land, let us consider what will be its effect in disturbing the direction of gravity at different points in its proximity. It will be remarked that the disturbance depends on two actions; the positive attraction produced by the elevated table-land; and the diminution of attraction, or negative attraction, produced by the substitution of a certain volume of light crust (in the lower projection) for heavy lava.

The diminution of attractive matter below, produced by the substitution of light crust for heavy lava, will be sensibly equal to the increase of attractive matter above. The difference of the negative attraction of one and the positive attraction of the other, as estimated in the direction of a line perpendicular to that joining the centres of attraction of the two masses (or as estimated in a horizontal line), will be proportional to the difference of the inverse cubes of the distances of the attracted point from the two masses. . . .

The general conclusion then is this. In all cases, the real disturbance will be less than that found by computing the effect of the mountains, on the law of gravitation. Near to the elevated country, the part which is to be subtracted from the computed effect is a small proportion of the whole. At a distance from the elevated country, the part which is to be subtracted is so nearly equal to the whole, that the remainder may be neglected as insignificant, even in cases where the attraction of the elevated country itself would be considerable. But in our ignorance of the depth at which the downward immersion of the projecting crust into the lava takes place, we cannot give greater precision to the statement.

In all the latter inferences, it is supposed that the crust is floating in a state of equilibrium. But in our entire ignorance of the *modus operandi* of the forces which have raised submarine strata to the tops of high mountains, we cannot insist on this as absolutely true. We know (from the reasoning above) that it will be so to the limits of *breakage* of the table-lands; but within those limits there may be some range of the conditions either way. It is quite as possible that the immersion of the lower projection in the lava may be too great, as that the elevation may be too great; and in the former of these cases, the attraction on the distant stations would be negative.

Again reverting to the condition of *breakage* of the table-lands, as dominating through the whole of this reasoning, it will be seen that it does not apply in regard to such computations as that of the attraction of Schehallien and the like. It applies only to the computation of the attractions of high tracts of very great horizontal extent, such as those to the north of India.

HALL

James Hall (1811–1898), American geologist and paleontologist, was in charge of the Survey of the Fourth Geological District of New York from 1837 to 1843 and for three years, beginning in 1855, was state geologist of Iowa. Both before and after that assignment he was engaged as paleontologist and geologist in Albany, and from 1866 until his death he was state geologist of New York. To him goes the credit for placing the Paleozoic stratigraphy of eastern America on a sound scientific basis.

Correlation of Paleozoic Rocks of New York with Those of Europe

From *Geology of New York*, Pt. IV, comprising the "Survey of the Fourth Geological District," chap. XXIV, 1843.

. . . From the very fully illustrated work of Mr. Murchison, we are made acquainted with many fossils holding the same relative position in the rocks of England that similar species do in this country. The general lithological characters of many of the successive strata correspond with those of New-York, and it is very natural that we should endeavor to find sufficient resemblance to identify them as of the same geological periods. For the great systems this has already been done, and there remains no doubt but the sedimentary rocks of New-York correspond with those of the Silurean and Old Red systems, as described in the *Silurian Researches*. If the Devonian is to be regarded as a distinct system, we shall find its representative in the Chemung and Portage groups, with, perhaps, a part of the Hamilton Group. In New-York, however, as already stated, no subdivisions can be made which are entitled to the name of systems. . . .

The tabular arrangement [on page 407] corresponds, very nearly, with the relative position of the rocks of the two systems in Great Britain and New-York.

The Relation of Mountains to Regions of Thick Sedimentary Accumulation*

From *Natural History of New York, Paleontology*, Vol. III, pp. 67–73, 1859.

An approximate measurement of all the strata along the Appalachian chain gives an aggregate thickness of forty thousand feet,

---

* This early recognition of the development of folded mountains from geosynclines seems to have been first made known in Hall's presidential

while the same formations in the Mississippi valley measure scarcely four thousand feet; in this, also, are included the Carboniferous limestones, which do not exist in any eastern section.

Subdivisions of the Rocks of the New-York System	Subdivisions of the Silurian and Old Red Systems in Great Britain
Old Red sandstone	Old Red sandstone
1. Chemung group	
2. Portage group	Upper and Lower Ludlow rocks, including the Devonian System of Phillips
3. Genesee slate	
4. Tully limestone	
5. Hamilton group	
6. Marcellus shale	
7. Corniferous limestone	
8. Onondaga limestone	
9. Schoharie grit	
10. Cauda-galli grit	
11. Oriskany sandstone	
12. Upper Pentamerus limestone	Wenlock rocks
13. Encrinal limestone	
14. Delthyris shaly limestone	
15. Pentamerus limestone	
16. Water-lime group	
17. Onondaga salt group	
18. Niagara group	
19. Clinton group	
20. Medina sandstone	
21. Oneida conglomerate	Caradoc sandstone
22. Grey sandstone	
23. Hudson-river group	
24. Utica slate	Llandeilo flags
25. Trenton limestone	
26. Birdseye and Black-river limestones	These formations are not as fully recognized in Great Britain as in New-York.
27. Chazy limestone	
28. Calciferous sandrock	
29. Potsdam sandstone	

In the Mississippi valley we have numerous points where the Lower Silurian strata are exposed, and at some points there is a thickness of five hundred feet of the Potsdam sandstone. From this base we follow the series upwards to the top of the mounds,

address at the Montreal meeting of the American Association for the Advancement of Science in 1857, but that address was not published until 1882.

Neither in the Montreal address nor in this memoir is the word "geosyncline" used, but the idea is clearly presented.—EDITORS.

capped by the Niagara limestone; and we there attain an elevation above the Mississippi waters of one thousand feet, which is the whole thickness of the formations from the Potsdam sandstone to the Niagara limestone.   The actual measurement of the same set of strata in the Appalachian region would give us more than sixteen thousand feet; and even making large allowances for excess in the measurements, we certainly have, in the Appalachians, more than ten times the thickness of the entire series in the west.   Still we have no mountains of this altitude; that is to say, we have no mountains whose altitude equals the actual vertical thickness of the strata composing them.

In the west there has been little or no disturbance, and our highest elevations of land mark essentially the aggregate thickness of the strata which produce the elevation.   In the east, though we prove step by step that certain members of the series, with a known thickness, are included in these mountains, the altitude never reaches the aggregate amount of the formations.   Reasoning from the facts adduced, and without prejudice or theory, the result certainly does not agree with our anticipations; for on the one hand, we find in a country not mountainous, elevations corresponding essentially to the thickness of the strata; while in a mountainous country, where the strata are immensely thicker, the mountain heights bear no comparative proportion to the thickness of the strata.

We have seen that one simple and intelligible sequence of strata, from the Potsdam sandstone to the end of the Coal measures, covers, with small exceptions, the entire country from the Atlantic slopes to the base of the Rocky mountains; that the same geological formations occupy the mountain chain and the plateau.   But while the horizontal strata give their whole elevation to the highest parts of the plain, we find the same beds folded and contorted in the mountain region, and giving to the mountain elevation not one-sixth of their actual measurement.

We are accustomed to believe that mountains are produced by upheaval, folding and plication of the strata; and that from some unexplained cause, these lines of elevation extend along certain directions, gradually dying out on either side, and subsiding at one or each extremity.   In these pages, I believe I have shown conclusively that the line of accumulation of sediments has been along the direction of the Appalachian chain; and, with slight variations at different epochs, the course of the current has been essentially the same throughout. The line of our mountain chain, and of the ancient oceanic current which deposited these sediments, is therefore coincident and parallel; or, the line of the greatest accumulation is the line of the mountain chain. In other words, the great Appalachian barrier is due to original deposition of materials, and not to any subsequent action or influence breaking up and dislocating the strata of which it is composed.

To be satisfied of this, it seems only necessary to compare the eastern and western exposures of the formations; for here the valleys, cutting through the rocks of the several groups down to the lower limestones, or to the Potsdam sandstone, present mountain ranges of several thousand feet on either side; while in the valley of the Mississippi, where the strata have thinned, the same denuding action has produced low cliffs or sloping banks of one or two hundred feet in height. Therefore had the country been evenly elevated without metamorphism or folding of the strata, making the lowest palaeozoic rocks the base line, in the States bordering the Atlantic we should have had higher mountains and deeper valleys, wherever the series was complete. At the same time, the great plateau on either side of the Mississippi river would have presented the features it now does, of valleys extending to the Lower Palaeozoic beds, with cliffs of the height represented by the actual thickness of the beds which there constitute the entire series.

The gradual declination of the country westward is due primarily to the thinning out of all the formations which have accumulated with such great force in the Appalachian region. It is also susceptible of proof, that no beds of older date have contributed to elevate the later ones, or to form a part of the mountain chain.

We have in the east one example where the conditions of elevation correspond with those in the Mississippi valley. The Catskill mountains are composed almost entirely of strata in a horizontal or very slightly inclined position; the Hudson-river group, which constitutes a few feet of their elevation at the base, is disturbed, and the succeeding beds lie upon this unconformably. These mountains, therefore, rising to a height of 3800 feet above tidewater, mark in their altitude simply the vertical thickness of the strata.

At this point of our inquiry, several questions of importance present themselves: First, what has been the cause of this folding and plication of the strata; secondly, having been thus folded and plicated, what influence has this action exerted upon the elevation of the parts, or of the whole; and thirdly, what effects are due to the metamorphism which accompanies this mountain chain?

It has been long since shown that the removal of large quantities of sediment from one part of the earth's crust, and its transportation and deposition in another, may not only produce oscillations, but that chemical and dynamical action are the necessary consequences of large accumulations of sedimentary matter over certain areas. When these are spread along a belt of sea bottom, as originally in the line of the Appalachian chain, the first effect of this great augmentation of matter would be to produce a yielding of the earth's crust beneath, and a gradual subsidence will be the consequence. We have evidence of this subsidence in the great amount of material accumulated; for we cannot suppose that the sea has been originally as deep as the thickness of these accumulations. On the contrary, the evidences from ripple-marks, marine plants, and other conditions, prove that the sea in which these deposits have been successively made was at all times shallow, or of moderate depth. The accumulation, therefore, could only have been made by a gradual or periodical subsidence of the ocean bed; and we may then inquire, what would be the result of such subsidence upon the accumulated stratified sediments spread over the sea bottom?

The line of greatest depression would be along the one of greatest accumulation; and in the direction of the thinning margins of the deposit, the depression would be less. By this process of subsidence, as the lower side becomes gradually curved, there must follow, as a consequence, rents and fractures upon that side; or the diminished width of surface above, caused by this curving below, will produce wrinkles and foldings of the strata. That there may be rents or fractures of the strata beneath is very probable, and into these may rush the fluid or semifluid matter from below, producing trap-dykes; but the folding of strata seems to me a very natural and inevitable consequence of the process of subsidence.

The sinking down of the mass produces a great synclinal axis; and within this axis, whether on a large or small scale, will be produced numerous smaller synclinal and anticlinal axes. And the same is true of every synclinal axis, where the condition of the beds is such as to admit of a careful examination.* I hold, therefore, that it is impossible to have any subsidence along a certain line of the earth's crust, from the accumulation of sediments, without producing the phenomena which are observed in the Appalachian and other mountain ranges.

That this subsidence was periodical, we have the best possible evidence in the unconformability of the Lower Helderberg group upon the Hudson-river group; showing that previous to the deposition of these limestones, there were already foldings and plications, the consequence of a subsidence along the line of accumulation. Subsequently to the deposition of the latter formations, or at intervals during their accumulation, there have been other periods of subsidence, and consequently of folding and plication; so that these are not synchronous, nor are they conformable with each other.

This successive accumulation, and the consequent depression of the crust along this line, serves only to make more conspicuous the feature which appears to be the great characteristic, that the range of mountains is the great synclinal axis, and the anticlinals within it are due to the same cause which produced the synclinal; and as a consequence, these smaller anticlinals, and their correspondent synclinals, gradually decline towards the margin of the great synclinal axis, or towards the margin of the zone of depression which corresponds to the zone of greatest accumulation.†

This affords a partial explanation of the fact already observed, that the mountain elevations in the disturbed regions bear in their altitude a much smaller proportion to the actual thickness of the formations, than do the hills in undisturbed regions. Furthermore it so happens that so soon as disturbance takes place and anticlinals are formed, the beds are weakened at the arching, and

---

* I am indebted to Sir William Logan for this latter suggestion, as a result of his very accurate and extensive observations on the relations of anticlinal and synclinal axes.

† This mode of depression, which is the result of accumulation, and the production of numerous synclinal and anticlinal axes offers a satisfactory explanation, as it appears to me, of the difference of slope on the two sides of the anticlinals which have been so often pointed out as occurring in the Appalachian range, where the dips on one side are uniformly steeper than on the other.

become more liable to denuding action. Thus the anticlinals are often worn down to such an extent as to form low grounds or deep valleys; while the synclinal, protected in the downward curving of the beds, remains to form the prominent mountain crest. This is very generally true in many parts of the Appalachian range; and it is only where some heavier or stronger bedded rock occurs, protecting the anticlinals, that they form the higher mountain elevations. Similar features will be observed in other mountain ranges.*

It nowhere appears that this folding or plication has contributed to the altitude of the mountains: on the other hand, as I think can be shown, the more extreme this plication, the more it will conduce to the general degradation of the mass, whenever subjected to denuding agencies. The number and abruptness of the foldings will depend upon the width of the zone which is depressed, and the depth of the depression, which is itself dependent on the amount of accumulation.

We have, therefore, this other element of depression to consider, when we compare mountain elevations with the thickness of the original deposition.

It is possible that the suggestion may be made, that if the folding and plication be the result of a sinking or depression of the mass, then these wrinkles would be removed on the subsequent elevation; and the beds might assume, in a degree at least, their original position. But this is not the mode of elevation. The elevation has been one of continental, and not of local origin; and there is no more evidence of local elevation along the Appalachian chain, than there is along the plateau in the west. As it is, a large mass of the matter constituting the sediments of this mountain range still remain below the sea level, as a necessary consequence of the great accumulation; while in the plateau of the west, we have a much greater proportion above the level of the sea.

---

* The sections of the Geological Survey of Great Britain exhibit numerous examples of this kind. On the geological map of Great Britain, a section across the country presents us with Snowden summit as a synclinal, the height of which is much less than the thickness of the strata from the Longmynd to the Caradoc; while, had the bedded trap of Moel Wyn and Aran Mowddwy, and its superincumbent strata, been sufficiently strong to have resisted denudation, the anticlinal axis would have presented a mountain far higher than Snowden.

So far, therefore, as our observation extends, we are able to deduce some general principles in regard to the production of this mountain range; To explain its existence, we are to look to the original accumulation of matter along a certain line or zone, the direction of which will be the direction of the elevation. The line of the existing mountain chain will be the course of the original transporting current. The minor axes or foldings must be essentially parallel to the great synclinal axis and the line of accumulation. The present mountain barriers are but the visible evidences of the deposits upon an ancient ocean bed; while the determining causes of their elevation existed long anterior to the production of the mountains themselves. At no point, nor along any line between the Appalachian and Rocky mountains, could the same forces have produced a mountain chain, because the materials of accumulation were insufficient; and though we may trace what appears to be the gradually subsiding influence of these forces, it is simply in these instances due to the paucity of the material upon which to exhibit its effects. The parallel lines of elevation, on the west of the Appalachians, are evidenced in gentle undulations, with the exception of the Cincinnati axis, which is more important, extending from Lake Ontario to Alabama, and is the last or most western of those parallel to the Appalachian chain.

PHYSICAL CONDITIONS IN PALEOZOIC SEAS

From Presidential Address, Montreal, 1857, *Proceedings of the American Association for the Advancement of Science*, Vol. XXXI, pp. 29–63, 1882.

It might be presumptuous to hope to advance any new views upon this subject; still there is one which seems to me not unlikely to receive sufficient attention in the increasing interest and attractions of palaeontology. This science instead of being studied in connection with geology, and as an exponent of geological phenomena, has come to be regarded as a great object of independant investigation; and rocks are looked upon merely as the conservatory or museum of the natural history of the past geological periods. Certainly I ought not to oppose a view which exalts palaeontological science; and still, gentlemen, I would not in this study lose sight of the attendant physical conditions, which are equally important to be regarded. The progressive steps in the development of animal life have not been made without great physical changes; and we can never arrive at a true history of our planet till both are studied in connection.

The relation of the physical conditions to the vital force and to its development is a problem of vast importance. It should always be borne in mind, that the expression of animal life is coincident with physical conditions, and that the vital force is never developed independent of physical conditions. This is true without giving any undue importance to these conditions, or any countenance to materialistic views.

The subdivisions into formations, groups and systems, have always been marked by physical changes. Where a group of strata has its materials like those of the preceding group, there will the forms most resemble those of the preceding period; and where the lithological change is greatest, there will be found the greatest change in the nature of the organisms. When an intervening formation cuts off all the species of a preceding group, and the next, or third, is similar to the first in lithological character, so far will the fossils resemble each other not only in generic forms, but in the characteristics of species. We have then the converse of this proposition, that uniform physical conditions give uniform fauna. And I may here say, that while this holds in a large majority of instances, it does not hold true in all; because there is a law still beyond that of physical condition which affects the geographical distribution of species.

———

The Hudson River group presents us on the one hand with a series of soft shales becoming coarser and alternating with sandstones above, and on the other with irregular masses of limestone and finally immense masses of coarse sandstones or conglomerates with great bands of shale. This group attains in Canada a thickness of 2000 feet, and in Pennsylvania it has thickness, according to Prof. Rogers, of 6000 feet. Its northeastern extremity stretches out to Gaspé, and on the south side of the St. Lawrence, and, throughout Canada and the Green mountains of Vermont, forms in its metamorphic and folded condition, a great feature of the country. In its continuation southward it forms a marked feature in the Appalachian chain gradually dying out as the range declines to the southward. Along this line, which I shall term its line of trend, it presents no striking variations—nothing but what might be expected in any sedimentary group of rocks. It is only in pursuing it to the westward that we find interesting and important changes. Tracing it through Canada, west by the Islands of Lake Huron and by Green Bay to the Mississippi River and more southerly still by Cincinnati and the mouth of the Ohio, we learn that the coarser sediments gradually disappear, the finer mud alone having been transported, and finally that calcareous matter becomes an important ingredient in the formation and sandstone has become an exception. Or to give the best expression to the change, we here in New York and Canada speak of the shales and sandstones of the Hudson River group terminated by the Oneida or Shawangunk conglomerate. In Ohio the same group is known as the "Blue limestone." In Wisconsin and Illinois, this group, which in New York and the east is thousands of feet in thickness, has thinned so that it measures only from sixty to one hundred feet, while in Iowa I have not been able to obtain a thickness of fifty feet. Here we have an extensive group of strata changing from fine and coarse sedimentary materials and passing through all the intermediate phases until it becomes a shaly limestone.

Now in looking at the present continent as at that time an open sea, I conceive that the most rational explanation of these phenomena is that a powerful current has brought these sedimentary materials of sand, clay and pebbles from the northeast and has distributed them along the line now marked by the Appalachian chain; or, if you please to assume that there has been a continent to the east of our own which has been subsequently submerged, then the materials have been distributed along its coast. On whatever ground we view the deposit, we must admit, I conceive, that these materials have been distributed by a current, and that the direction of this current has been in the line of greatest accumulation and coarser materials; and that towards the westward, where we find less accumulation and finer materials, but still a wide ocean, the current gradually diminished until it essentially ceased and the fine materials were slowly spread over the broad area which they occupy in their diminished thickness. Such I conceive to have been the physical conditions of this ancient ocean-bed, an ocean disturbed by currents as extensive and as powerful as any we can point to in modern oceans.

DANA

James Dwight Dana (1813–1895), American geologist, was professor at Yale, first of natural history and later of geology and mineralogy, from 1850 to 1890. His *Manual of Geology* (1863), *Textbook of Geology* (1864), and *System of Mineralogy* (first edition, 1837, fifth edition, 1868) were the most important treatises in this field of science written in America during his lifetime.

ORIGIN OF THE MINERAL CONSTITUTION OF IGNEOUS ROCKS

From *American Journal of Science and Arts*, Second Series, Vol. II, pp. 335–355, 1846.

It has been a difficult problem for solution, why volcanic regions should have a centre of solid feldspathic rocks, unstratified and compact, while the exterior consisted mainly of basaltic lavas. Scrope, Von Buch, and other writers on volcanoes, have mentioned instances of this structure; and it seems to characterize generally the large volcanic mountains. It is well exhibited when the elevations are cut through by gorges; and when not, the clinkstone appears often at the summit of the cone or dome. The explanations we here venture, proceed on two principles:

1. *The motion which belongs to a boiling fluid.*
2. *The less fusibility of feldspar than the other ingredients.*

In the great boiling pools, there will necessarily be a rising of the fluid, in the hotter part, and a flow away towards either side, producing a kind of circulation. This is no hypothesis, as the fact may be witnessed in any boiling cauldron; and the lavas of Kilauea are a visible example of it. The ebullition in lavas on the earth, proceeds principally from the vapors of water and sulphur, which are constantly rising through them, inflating them more and more as they ascend, and finally escaping in bubbles at the surface. Now the feldspar being the less fusible part of the lavas, would thicken somewhat, wherever the temperature became too low for complete fusion: the more liquid portion would then ascend most easily, being carried along by the inflating vapors, and much of the feldspar would thus be left behind, and it might be in a nearly pure state. The centre of the volcano under this action, becomes necessarily feldspathic. The summit might therefore eject either basaltic or feldspathic rocks from the material of the vent; though when the action was violent and deep, it would eject feldspathic rocks alone.

At the same time the *basaltic* lavas, descending laterally in this system of circulation along the sides of the great central conduit, may pass out as flank eruptions through fissures. Besides, there will also be basaltic ejections from sources of lavas at a distance from the central conduit, where they have not been subjected to the separating process described; and this may be the more common source.

Mountains with a feldspathic centre, and basaltic layers forming the circumference, are therefore quite intelligible without supposing the feldspar to have been first thrown up, or appealing to a different system of fissures for their origin, and the examples which the moon presents, are more extensive than is necessary to explain the widest facts on the earth.

In these remarks we have spoken of the lavas as consisting mainly of feldspar and augite, their more common constitution; but we use the terms in a general sense, understanding by feldspar one or another of the feldspar family of minerals, and by augite the remaining fusible material, whether ordinary augite (silicate of lime, magnesia and iron) or silicates of one or more of these bases or alumina in other combinations.

There is some difficulty in applying this hypothesis to particular cases, on account of our ignorance of the actual fusibilities of the materials of the lava, in the condition in which they are placed; for we know that an infusible mineral may be held in fusion, far below the temperature at which it fuses: or, previous to the commencement of cooling it may be in some other combination.

We should infer that the process which separates the feldspar would also separate any excess of the more infusible mineral quartz. This may not follow: still it is a remarkable fact that the quantity of quartz contained in trachyte is often in great excess, as analyses have shown. But why is not the infusible mineral chrysolite also detained? The fact appears to be, that it is of subsequent formation. The small proportion of silica it contains implies a deficiency of this substance, while, as we have stated, in the feldspathic rocks there is often an excess. It may, therefore, under certain circumstances proceed from the basaltic material, for its elements are the same in different proportions. Subsequent investigations may give us more light on this point.

The general principle which we have above brought forward, is well illustrated in the fact that the scoria or surface glass of any vent, where it occurs, is the most fusible part of the lava, consisting in general of ferruginous or alkaline silicates, and containing no magnesia. On account of the diminished heat, this material alone remains sufficiently fluid to be inflated and borne up to the surface by the rising vapors; and this takes place in spite of superior gravity.

We hence comprehend the *rapid cooling* which characterizes ejected lavas, *for only a part of the material is in complete fusion.*

The actual nature of the *cooled igneous rock* may be more correctly understood, if we consider that the minerals present will depend, not only on heat and pressure and the causes above alluded to, but also on rate of cooling. The effect of slow cooling is exemplified in the feldspathic centre of a volcanic mountain. Being wholly enclosed by rocks, the heat of fusion passes off slowly, and owing to the pressure of its own superincumbent portions, the rock is compact. Whatever augite may be present, instead of appearing as augite, will take the form of hornblende, a mineral which requires slow cooling, and differs from augite in the crystalline form which it thus receives. In corroboration of this statement, hornblende is common in trachytes and such feldspathic rocks. The same remarks apply to mica: and other minerals also may form according to the elements present. . . .

Farther we observe that with a still more gradual rate of cooling, the whole feldspathic rock becomes crystalline in texture like a granite or syenite, and it is well known that granite-like or syenitic rocks or peaks occur in some volcanic regions, whose interior has been laid open by denudation. Many minerals too might crystallize under these circumstances, which with more rapid cooling would not be distinguishable.

The opinion that the nature of the resulting rock is directly connected with the nature of the rock which entered into fusion, cannot be maintained if the above views be true. On the contrary, it appears that while the result may thus be varied, the mode of distributing minerals in a volcanic focus by the boiling process, may produce from the same material, rocks of a predominant feldspathic character in one place, and rocks of a hornblendic or augitic character in other places.   Simple feldspathic granites may be fused and ejected as feldspathic rocks, like those of porphyry dikes.   But it is an interesting fact, that the rock of most dikes is of the augitic (or hornblendic) kind, like the dikes of volcanoes that rise from sources in which the *separating* process could not have been operating.

We also arrive at the important conclusion, that rocks perfectly compact in texture may be of subaërial origin, as we have pressure from the fluid lavas themselves in the volcanic focus.

Another deduction proceeds from the facts stated;—that the same igneous rocks may occur of all ages, provided the atmosphere or waters of the earth were not too warm for the more rapid rate of cooling required for uncrystalline rocks.   Scorias, basalt, trap, porphyry, syenite, granite, have no relations to one epoch rather than another, beyond what may depend on the circumstance just mentioned.   Whenever therefore in the history of the world, the variations in heat, pressure, and rate of cooling, now possible, may have taken place, similar rocks to those of the present day may have been in progress:—and as far as the variations of former times, in these respects, may now take place, former rocks and minerals may still be in progress.   In this statement it is implied that the necessary elements are present in the fused material.

ON THE ORIGIN OF CONTINENTS AND MOUNTAIN RANGES

From *American Journal of Science and Arts*, Second Series, Vol. III, pp. 94–100, 1847.

In a paper on the Volcanoes of the Moon, read before the Association of Geologists and Naturalists, in September last, some suggestions were thrown out with regard to the Origin of Continents, drawn from the condition of a cooling globe.   It was observed that the portions of the earth now constituting the great areas of land, were free, or nearly so, from volcanic action, even in the Silurian period: while the oceans appear to have been regions of eruption.   Hence it was inferred that contraction must have taken place to the greatest extent over the parts now oceanic, just as any cooling sphere becomes depressed on the side which cools last.   This was shown to correspond with the actual history of our globe, inasmuch as an increasing depth in the ocean cavity would necessarily leave more and more land above water in successive epochs as accords with observations.   It was observed that the hypothesis was *farther* borne out by facts: for while it appears that the land has, on the whole, been increasing in extent, even through the tertiary era and subsequent to it, the ocean's bottom has actually subsided several thousand feet within a late period, as shown by the coral islands scattered over the wide Pacific. . . .

In order to understand the bearing of the facts, we should bring to mind the effects of contraction. The more prominent are as follows:—

1. Depressions, provided the contraction be unequal in different parts.

2. Apparent elevations, as a consequence of the depressions; that is, elevations as compared with the lowest level, or with a body of water occupying the depressions.

3. Fissures.

4. Ejection of igneous matter, at times, through fissures.

5. Upheaval along a line of fissure, the surface adjoining being more or less raised.

6. Upliftings and foldings from lateral pressure.—An arc of the exterior surface being greater than any corresponding arc below the surface, a depression of the hardened exterior, produced by the cooling beneath, would in some instances cause lateral displacements.

7. An *unequal rate* of subsidence over given areas in different periods.—Contraction tends to occasion a strain upon the cooled and unyielding exterior, accompanied generally by a consequent diminished rate of subsidence, or a cessation of it. This strain increases until it results in fractures; and following this crisis, subsidence would for a while be more rapid in rate. The strain, or state of tension, might also occasion elevations in some places, within or without the area; and at the time of fissuring, there might be other upheavals. It follows, hence, that—

a. There would be prolonged intermissions in the subsidence of given areas; and this must have been the fact throughout the history of the globe.

b. There must have been oscillations in the land as compared with a water level, the water at times rising gradually over land that, during a previous period, had emerged; and the reverse.

c. There might be in the same epoch, under such circumstances, an unequal retreat of the ocean from the coasts of different continents, or a rise in one place and a retreat in others: for the changes by contraction are supposed to have been every where in progress at the same time, and throughout different in character and extent.

d. Changes of level may in some cases have been *gradual,* and in other cases *paroxysmal;* for the opening of large fissures would often be of the latter character.

8. In an elliptical area of contraction, there will be two systems of fissures at right angles with one another, as follows from the calculations of Wm. Hopkins, Esq. But if the area is bounded on one side by a region participating but little in the contraction, the effects would be most decided on the borders of such a region; and they would consist in extensive fissures ranging along the area, and an attending swelling of the surface, or else a rising of the strata into folds by lateral pressure.

The effects of lateral pressure might in many parts be local or of very limited extent. A contracting area might be made up of several separate areas of contraction not acting together upon any particular line. Even supposing a whole quarter of our globe to exert laterally all the force possible, by a uniform contraction continued until the surface was depressed eight miles in depth, the whole effect would be equivalent to a lateral dislocation of only twelve miles. And in this calculation, we make no allowance for upliftings over the contracting area, which would diminish the action; nor for a diminution of breadth in the surface of the area, which diminution must be going on if the surface is losing heat. In the remarks which follow relating to this point, America, therefore, is not instanced as an example of what *must every where* have happened, but of what *has here* happened.

From these explanations, we proceed to the application of them.

If the reader will place before him a good map of North America, he will perceive at once the effects which have been alluded to exhibited on a grand scale, on both sides of the continent. On the *Atlantic* side, the Appalachians, from Maine to Georgia, consist of rock strata, which have been variously folded up into ridges, as has been made out with great beauty and fullness by Professors W. B. and H. D. Rogers. These folds are in several series, but are nearly uniform or parallel in position. As should be expected from the nature of the cause, the plications are more frequent and abrupt on the side of the chain nearest the ocean, and gradually die out westward just beyond the limits of the Appalachians. As another result of proximity to the contracting area, the rocks on the eastern side have been most altered by fire. To so great a degree has the heat operated, (which escaped by the opened cavities and fissures, and was distributed laterally by the aid of the contained and incumbent waters), that it is difficult in New England to distinguish the true igneous rocks from those that are metamorphic.

On the *Pacific* side of the continent, we observe the Rocky Mountain range rising with a gentle swell from the coast. From the mouth of the Kansas to the top, and on the opposite or western side, the average slope is hardly twelve feet to the mile. The summit is about eight thousand feet high. But there are ridges which add five or six thousand feet to the chain: these form a crest to parts of the range but are not properly the range itself, though often so recognized. The Rocky Mountains appear, then, to be another effect of contraction, viz. a gradual swelling of the surface, accompanied by fissures and dislocations over its area. These dislocations are very marked in the sandstone, just east of the summit. Thus each great oceanic depression, the Atlantic and Pacific, has its border range of heights thrown up by the very contraction which occasioned the depression; and between lies a vast plain, scarcely affected at all by these changes, the great central area of the continent. This view is farther sustained by finding that the effects of fire are most apparent on the ocean side of the mountains, precisely as about the Appalachians, yet to a more remarkable extent. Indeed, there are no remains of volcanoes, or their ejections, to the east of the summit; while to the west, the country of Oregon is in many parts buried beneath basaltic or other volcanic rocks, and several existing volcanic cones have been described. Still farther, we observe a second, a third, and even a fourth parallel range of heights from the summit of the mountains to the coast; and the third (the Cascade range) rivals the Rocky Mountains in the height of some of its snowy peaks. Vast fissures were opened to the fires below, as these ranges indicate, and some of the vents have not yet ceased action. Here, then, are the natural effects of proximity to a region of contraction—the Pacific—in which the remains of igneous action everywhere abound.

It has been well established that the Appalachian folds or plications were made since the coal period, for the coal beds are enclosed in the folds; and the rising of the Rocky chain was also subsequent to that era. The effect of contraction in producing these elevations, was therefore comparatively little felt in the very earliest ages, when the surface of the depressed (or igneous) portion was itself somewhat yielding, but subsequently, when it had become stiffened to a considerable depth by cooling. There appears hence to be a perfect harmony between the results and the causes adduced.

If these conclusions are correct, we must give up the popular idea (at least as a general theory) of the elevation of mountains by a force below causing at the time an irruption of igneous matter; for the irruption is in general an effect of a very different action, as has been urged by Prévost. This may be as true of the Urals, as of the Rocky Mountains and Andes.

Even the trap dykes of New England and New Jersey, whose general course corresponds with that of the Appalachians, may be a result of the contraction in progress subsequent to the coal era. The dip of the new red sandstone accompanying them is probably another effect. The Ozark mountains, forming a line parallel with the Appalachians, beyond the Mississippi, may be referred to the same system of changes.

On the Nature of Volcanic Eruptions

From *United States Exploring Expedition, during the years* 1838, 1839, 1840, 1841, 1842, *under the command of Charles Wilkes, U. S. N.,* Vol. X, pp. 221–222, 366–369, Philadelphia, 1849.

Volcanoes no Safety-Valves.—From these considerations we may doubt whether volcanoes are ever "*safety-valves*," as they have been often called, and are almost universally considered by writers on these subjects. We may strongly doubt whether action so deep-seated as that of the earthquake must be, can often find relief in the narrow channels of a volcano, miles in length. The conduit of Mokua-weo-weo is almost three miles long, down to the level of the sea. Assuredly if while Kilauea is open on the flanks of Mount Loa,—a vast gulf three and a quarter miles in diameter,—lavas still rise and are poured out at an elevation of more than 10,000 feet above it, Kilauea is no safety-valve, even to the area covered by the single mountain alone. . . . Volcanoes are in fact indexes of danger, and the absence of them is the best security. . . .

Phases of Volcanic Action.—Moreover there can be no truth, at least as regards Mount Loa, in the principle reasoned out at length, in an able article on volcanoes, by Bischof,* that the phases of volcanic action depend on water gaining access to the central fires of the globe; for the evidence is certainly conclusive that the main action of waters is comparatively near the surface.

The phases of volcanic action at Kilauea are simply as follows:—

1. The centres of action, when most quiet, are reduced to a single one, which occasionally overflows. This overflowing raises the bottom of the crater: the lavas continue to boil over and go on accumulating, and elevating the area of action; the pressure is consequently gradually increasing; the action becomes after a while more intense, from the increasing pressure, and increasing height to which vapours ascend before escaping; new centres of ebullition add to the effect; finally, after the bottom is raised 400 feet above its lower level, these centres are numerous, the ebullition is violent, the overflowing almost incessant;—at last the increased pressure, in addition to the force of rising vapours proceeding from the increased action, cause a rupture through the mountain's sides and the liquid rock flows out.

Conclusions.—The foregoing discussions have led to the following conclusions:

I. That Mount Loa and similar summits (among which we would include most of the islands of the Pacific examined by us, besides all others of the same general character) were formed from successive eruptions of molten rock, alternating sometimes with cinder or fragmentary ejections.

II. That the eruptions are in general the result of a rising or ascent of the lavas, owing to the inflation by heat of such vaporizable substances as sulphur and water, the overflow or lateral outbreak taking place in consequence of the increased pressure from gravity, and from the elasticity of the confined vapours, and the contraction of the earth's surface is no more necessary for

---

* *Natural History of Volcanoes* by G. Bischof, Jameson's Edinburgh Journal xxvi., 1839; American Journal of Science, xxxvi., 249, 250.

an eruption, than the contraction of the sides of a boiling pot of water to make it boil over. . . .

III. That eruptions will usually take place from the central vent in case the sides of the mountain will sustain the pressure of the column of lava; and when incapable of sustaining this pressure, lateral outbreaks occur. Fissures are also frequently opened by local action through the pressure of vapours.

IX. That the ordinary eruptions and usual action of a volcano proceed principally from water gaining access to the branch or branchlets belonging to a particular vent, and not to a common channel below: the fresh waters of the island are the principal source of the vapours of Kilauea: . . .

X. That pulsations in the central igneous fluids of the globe have but little influence, if any at all, on the action of volcanoes; for vents in the same vicinity are not contemporaneously affected, and the phases of volcanic action are fully accounted for by a more superficial action.

The Origin of Valleys in the Pacific Islands

From *United States Exploring Expedition, during the years* 1838, 1839, 1840, 1841, 1842, *under the command of Charles Wilkes*, U. S. N., Vol. X, pp. 380–392, Philadelphia, 1849.

The causes operating in the Pacific, which have contributed to valley-making, are the following:

1. Convulsions from internal forces, or volcanic action.

2. Degradation from the action of the sea.

3. Gradual wear from running water derived from the rains.

4. Gradual decomposition through the agency of the elements and growing vegetation.

The *action of volcanic forces* in the formation of valleys, is finely illustrated in the great rupture in the summit of Hale-a-Kala (Kauai). The valleys formed by the eruption are as extensive as any in the Hawaiian Group, being two thousand feet deep at their highest point, and one or two miles wide. They extend from the interior outwards towards the sea. Above they open into a common amphitheatre, the remains of the former crater, the walls of which are two thousand feet high.

Other examples of volcanic action are seen in the pit craters of Mount Loa, among which Kilauea stands pre-eminent. . . .

The many fissures which are opened by the action of Kilauea, might be looked upon as valleys on a smaller scale, and the germs of more extensive ones. But with few exceptions, these fissures as soon as made are closed by the ejected lava, and the mountain is here no weaker than before. Those which remain open, may be the means of determining the direction of valleys afterward formed.

Action of the Sea.—The action of the sea in valley-making is often supposed to have been exerted during the rising of land; and as such changes of level have taken place in the Pacific, this cause, it would seem, must have had as extensive operation in this ocean as any where over the world.

But in order to apprehend the full effect of this mode of degradation, we should refer to its action on existing shores. At the outset we are surprised at finding little evidence of any such action now in progress along lines of coast. The waves tend rather to fill up the bays and remove by degradation the prominent capes, thus rendering the coast more even, and at the same time, accumulating beaches that protect it from wear. If this is the case on shores where there are deep bays, what should it be on submarine slopes successively becoming the shores, in which the surface is quite even compared with the present outline of the islands? Instead of making bays and channels, it could only give greater regularity to the line of coast.

. . . The effects of the sea in making valleys have been much exaggerated, as is obvious from this appeal to existing operations, the appropriate test of truth in geology.

The action of a rush of waters in a few great waves over the land, such as might attend a convulsive elevation, though generally having a levelling effect, might it is true produce some excavations; yet, it is obvious on a moment's consideration, that such waves could not make the deep valleys, miles in length, that intersect the rocks and mountains of our globe. . . .

Running water of the land, and gradual decomposition.—Of the causes of valleys mentioned in the outset we are forced to rely for explanations principally upon running streams: and they are not only gouges of all dimensions, but of great power, and in constant action. There are several classes of facts which support us in this conclusion.

We observe that Mount Loa, whose sides are still flooded with lavas at intervals, has but one or two streamlets over all its slopes, and the surface has none of the deep valleys common to the other summits. Here volcanic action has had a smoothing effect, and by its continuation to this time, the waters have had scarcely a chance to make a beginning in denudation.

Mount Kea, which has been extinct for a long period, has a succession of valleys on its windward or rainy side, which are several hundred feet deep at the coast and gradually diminish upward, extending in general about half or two-thirds of the way to the summit. But to westward it has dry declivities, which are comparatively even at base, with little running water. A direct connexion is thus evinced between a windward exposure, and the existence of valleys; and we observe also that the time since volcanic action ceased is approximately or relatively indicated, for it has been long enough for the valley to have advanced only part way to the summit. Degradation from running water would of course commence at the foot of the mountain, where the waters are necessarily more abundant and more powerful in denuding action, in consequence of their gradual accumulation on their descent.

. . . The valleys of Mount Kea alone, extending some thousands of feet up its sides, sustain us in saying that time only is required for explaining the existence of any similar valleys in the Pacific. As in Tahiti, these valleys in general radiate from the centre, that is, take the direction of former slopes; they often commence under the central summits, and *terminate at the sea level, instead of continuing beneath it.* . . .

A brief review of the action and results of flowing waters will render the origin of these features intelligible.

a. Suppose a mountain, sloping like one of the volcanic domes of the Pacific. The excavating power at work proceeds from the rains or condensed vapour, and depends upon the amount of water and rapidity of slope.

b. The transporting force of flowing water increases as the sixth power of the velocity,—double the velocity giving sixty-four times the transporting power. The rate will be much greater than this on a descending slope, where the waters add their own gravity to the direct action of a progressive movement.

c. Hence, if the slopes are steep, the water gathering into rills, excavates so rapidly that every growing streamlet ploughs out a gorge or furrow; and consequently the number of separate gorges is very large and their size comparatively small, though of great depth.

d. But if the slopes are gradual, the rills flow into one another from a broad area, and enlarge a central trunk which with incessant additions from either side, descends towards the sea. The excavation above, for a while, is small: the greater abundance of water below, during the rainy seasons, causes the denudation to be greatest there, and in this part the gorge or valley most rapidly forms. In its progress, it enlarges from below upward, though also increasing above, while at the same time the many tributaries are making lateral branches.

e. Towards the foot of the mountain, the excavating power gradually ceases when the stream has no longer in this part a rapid descent,—that is, whenever the slope is not above a few feet to the mile. The stream *then consists of two parts, the torrents of the mountains and the slower waters below,* and the latter is gradually lengthening at the expense of the former.

f. After the lower waters have nearly ceased excavation, a new process commences in this part, that of widening the valley. The stream which here effects little change at low water, is flooded in certain seasons, and the abundant waters act *laterally* against the inclosing rocks. Gradually, through this undermining and denuding operation, the *narrow bed becomes a flat strip of land between lofty precipices,* through which in the rainy seasons, the streamlet flows in a winding course. . . .

g. The torrent part of the stream, as it goes on excavating, is gradually becoming more and more steep. The rock-material operated upon, consists of layers of unequal hardness, varying but little from horizontality and dipping towards the sea, and this occasions the formation of cascades. Whenever a soft layer wears more rapidly than one above, it causes an abrupt fall in the stream. . . .

h. As the gorge increases in steepness, the excavations above deepen rapidly,—the more rapid descent more than compensating, it may be, for any difference in the amount of water. Moreover, as the rains are generally most frequent at the very summits, the rills in this part are kept in almost constant action through the year, while a few miles nearer the sea they are often dried up or absorbed among the cavernous rocks. The denudation is consequently at all times great about the higher parts of the valley, (especially after the slopes have become steep by previous degradation) and finally an abrupt precipice forms its head.

i. The waters descending the ridges either side of the valley or gorge, are also removing these barriers between adjacent valleys, and are producing as a *first* effect, a thinning of the ridge at summit to a mere edge; and as a *second* its partial or entire removal, so that the two valleys may become separated by a low wall, or terminate in a common head,—a wide amphitheatre enclosed by lofty mountain walls.

With literal truth may we speak of the valleys of the Pacific islands, as the furrowings of time, and read in them marks of age. Our former conclusions with regard to the different periods which have passed over the several Hawaiian Islands since the fires ceased and wear begun, is fully substantiated. We also learn how completely the features of an island may be obliterated by this simple process, and even a cluster of peaks like Orohena, Pitohiti and Aorai of Tahiti, be derived from a simple volcanic dome or cone.

RAMSAY

Sir Andrew Crosbie Ramsay (1814–1891), Scottish geologist, was director general of the Geological Survey of Great Britain and professor of geology in University College and the Royal School of Mines.

MARINE DENUDATION IN SOUTHERN WALES

From *Memoirs of the Geological Survey of Great Britain*, Vol. I, pp. 297–299, 326–328, London, 1846.

It may be considered as an axiom in geology, that the materials of all sedimentary strata, mechanically deposited, have been derived from the wreck of pre-existing rocks. This waste and re-formation of rocks, is partly the result of atmospheric influences and the agency of running water, and also, to a great extent, is attributable to the ordinary destructive action of the sea on certain coasts.

The subject of this paper is intimately connected with the latter part of this proposition, in so far that many of the results sought to be explained and established, are referred to the influence of this powerful geological agent. . . .

. . . To whatever causes the existing features of any country may be assigned, it is evident that without data by means of which we may form true conceptions of that form, it is impossible to reason correctly either of the manner of action of these bygone operations, or of the magnitude of their effects. Such data are to a certain extent supplied by the construction of geological sections on a true scale, vertically and horizontally, having for their base line the level of the sea. Without these, it is vain to attempt precise argument on a variety of phenomena, either as regards questions of physics or natural history. If a false scale be adopted, the form and elevations of the country, the thickness, inclinations, and general arrangement of its component strata, are all distorted, producing results equally exaggerated and irreconcilable. And, as a further consequence, it is impossible correctly to estimate the original extent of the broken and disjointed

strata, and thus to attempt to form any approximate conception
of the amount of matter destroyed and rearranged, during the
processes that reduced them to their present fragmentary condition. During the progress of the Geological Survey of Great
Britain, in South-Western England and in South Wales, a number
of sections have been constructed on a true scale of six inches to a
mile; and on evidence afforded by them the following reasonings
are adduced.

What first strikes the eye on examining certain of these sections,
is the remarkable curvature and distortion to which the strata
composing all the formations, from the top of the coal measures
downwards, have been subjected. Following these breaks and
curves, the same series of rocks are seen repeatedly to rise to the
surface, sometimes in rapid, sometimes in wide-spreading undulations. When, in accordance with the curves indicated by surface
dips, vast masses of rock are carried in these sections deep down
into the earth, far below our actual cognizance, it is yet impossible
to doubt their underground continuity, when we find again and
again, the same set of beds diving downwards in one district, and
(perhaps somewhat modified) again outcropping to the surface in
remote parts of the country. The abuse of this fact, now familiar
to every geologist, in earlier times led to the hypothesis of the
original universal continuity of all strata over the entire circumference of the globe. But if the inference now drawn be
legitimate, a little reflection will show, that in the case of curved
and conformable strata, the same arguments that apply to the
continuity of rocks below the surface, may often safely be employed
to prove the original connexion of contorted strata, the upturned
edges of which may frequently be far apart. Attention being
given to the physical relations of all the rocks in any country,
such restoration of masses of rock to the form they once possessed,
is within the limits of safe inference. And if, in the cases above
noticed, this original continuity of distant masses, and their
spreading over tracts where they have left not a trace, be once
granted, then the vast amount of matter we shall be able to show
has been removed from such tracts, may well make us cautious in
disbelieving the probable or possible destruction of other masses,
once resting above the rocks that compose the present surface, but
of the former existence of which above that surface, we have at
first sight no direct evidence. Outliers, cape-like projections, and

anticlinals of various strata, so common on our maps of geological England, sufficiently illustrate the first proposition; and the frequent occurrence of vast thicknesses of strata, disposed vertically or at high angles, afford perfect evidence that such strata were not originally discontinued at their present outcrop, since such supposition would involve the necessity of asserting, that the rocks in question were deposited in successive layers, forming together, at their extreme edge, a wall or highly inclined plane, often many thousand feet in height.

When the high ground constituting the restored portions of the sections was, in the progress of geological events, removed, the land again uprose to attain its present elevation; and if, during the progress of this general upheaval, occasional oscillations of level occurred (and this seems to have been the case), then the destroying agency for the waves would act, in favourable conditions, with increased power. And thus, in the long lapse of geological time, as the land slowly reached its existing height, the restless action of the sea gave to our hills and valleys the normal outlines of their present forms, since but slightly modified by atmospheric agencies, the loose drifty deposit that covers the hills and valleys so formed, being, as it were, but the dregs of the matter removed from the rising land during its last elevation. . . .

The line of greatest waste on any coast, is the average level of the breakers.   The effect of such waste is obviously to wear back the coast, the line of denudation being a level corresponding to the average height of the sea.   Taking *unlimited* time into account, we can conceive that any extent of land might be so destroyed, for though shingle beaches and other coast formations will apparently for almost any ordinary length of time protect the country from the further encroachments of the sea, yet the protections to such beaches being at last themselves worn away, the beaches are in the course of time destroyed, and so, unless checked by elevation, the waste being carried on for ever, a whole country might gradually disappear.

If to this be added an *exceedingly slow depression* of the land and sea bottom, the wasting process would be materially assisted by this depression, bringing the land more uniformly within the reach of the sea, and enabling the latter more rapidly to overcome obstacles to further encroachments, created by itself in the shape of beaches. By further gradually increasing the depth of the surrounding water, ample space would also be afforded for the outspreading of the denuded matter. To such combined forces, namely, the *shaving away* of coasts by the sea, and the spreading abroad of the material thus obtained, the great *plain* of shallow soundings which generally surrounds our islands is in all probability attributable. . . .

The waves acting equally on cliffs of unequal hardness soon produce great irregularities of outline, the harder rocks standing out in bold promontories, while the softer materials yielding more rapidly to the shock of the breakers, slowly originate bays, creeks and arms of the sea. If to this we add the influence of exposure to prevalent winds, the indented character of a coast often becomes very remarkable,* the general depth of the plain, so to speak, formed by denudation continuing the same, unless varied by oscillations of level. Besides those peculiarities of outline dependent on unequal hardness of rocks, the form of the escarpment bounding the surface beneath the sea, would of course be much modified by the dip and geological position of its component rocks, so that it sometimes happens that a soft material on account of the seaward slope of its beds, resists the power of the breakers longer than a harder substance placed under less favorable conditions. But as the chances are equal to these varieties being thrown into favourable or unfavourable positions, it will be found that in a great majority of cases, the promontories on a coast are formed of the less yielding material.

. . . By modifications of these causes islands are formed, which afterwards by further upheaval become the tops of hills. The flat sea bottoms surrounding such newly upheaved islands, are again subjected to denudations, and again unequal hardness and exposure produce further inequalities.

Thus by endless oscillations of level, the contour of a country assumes its varied outline. It is remarkable how frequently the hilltops and the higher lands of Wales are more or less tipped with rock of superior hardness, such land having been saved or upheld, as it were by reason of that hardness in the midst of the surrounding denudations. From the same operations most of the greater

---

* See the west coast of Scotland and Ireland. Doubtless the Western Isles are but the fragments of a larger land.

hollows have been scooped out by the destruction of softer materials. . . .

. . . Reasoning back from these considerations to the axiom with which we started, it will be remembered that all strata mechanically deposited were formed by the waste of pre-existing rocks.

In South Wales the Silurian rocks attain a thickness of at least 12,000 feet. The greatest thickness of the old red sandstone is between 7,000 and 8,000 feet; and the coal measures attain in their greatest development, a thickness of not less than 12,000 feet. Simply estimating their cubic contents in the area they now occupy, and adding to this the amount removed by denudation, and that existing beneath the level of the sea, it is evident that the quantity of matter employed to form these strata was many times greater than the entire amount of solid land they now present above the waves. Now, though in South Wales a small proportion of this material may have been used twice over in the formation of its older strata, yet, from the almost perfect conformity of these deposits, it is evident that this is the exception, and not the rule. To form, therefore, so great a thickness, a mass of matter of nearly equal cubic contents must have been won by the waves and the outpourings of rivers from neighbouring lands, of which, perhaps, no original trace now remains. . . . When we consider the solid cubic contents of the aggregate strata as they existed when entire, or even as they now are (for a far greater proportion of their mass lies beneath than above the level of the sea), we cannot but conclude, when compared with the little surface they now present above water, that the time occupied in their formation must have been infinitely longer than the time required to destroy all that now remains exposed to the elements, or all that existed even prior to the tertiary denudations. If the same agencies be still at work, that which has been achieved once may be performed again. Why, then should we wonder at the destruction of the old land depicted in the restored sections. Yet the matter torn from above the present surface was *far greater* than all which still remains above the level of the sea. . . . What, then can we conclude, but that the time requisite to remove these mountains was at least equal or greater in amount than that which may yet pass ere the existing land of the same district be utterly worn away.

As we estimate time, it is vain to attempt to measure the duration of even small portions of geological epochs. Within the historical period no great authentic change has been effected on the coasts of Wales. On many an available headland, the cliffs are still crowned with ancient fortified retreats, whose origin is lost in the mists of antiquity. If then, we cannot contemplate the far distant period when the present land shall be utterly destroyed, so also of the time occupied in that last great denudation in days we may almost call but little antecedent to our own, if it were possible to express so vast a period in figures, they could convey no impression to the mind save one almost approaching to infinity.

On the Probable Existence of Glaciers and Icebergs in the Permian Epoch

From *Quarterly Journal of the Geological Society of London*, Vol. XI, pp. 185–205, 1855.

The sedimentary strata which contain the fragments of striated and polished rocks to which I am about to call attention belong to the inferior portion of that which has been defined as the "Permian Group" by Sir Roderick Murchison, the true geological horizon of which in England was first explained by Professor Sedgwick, in his celebrated memoir, "On the geological relations and internal structure of the Magnesian Limestone and the lower portions of the New Red Sandstone series," &c. It is of the last-named division of this series that they probably form a part.

In the summer of 1852 I traced the boundaries of the Permian breccias that run between the Bromsgrove Lickey and the Clent Hills, having previously visited similar rocks on the flanks of the Abberley and Malvern range. Though much struck with the size and angularity of the fragments, and with the marly paste in which they are imbedded, I did not then venture to propose to myself the solution of these and other peculiarities, at which I have since arrived, viz. that they are chiefly formed of the moraine matter of glaciers, drifted and scattered in the Permian sea by the agency of icebergs. But when, in connexion with my duties on the Geological Survey, I began in 1854 to inspect these rocks near Enville, and afterwards revisited the equivalent strata in South Staffordshire, and on the Abberley and Malvern Hills, their true nature gradually dawned on me, and on the 18th of July I wrote to our late deeply lamented President, announcing what (if true) I considered a discovery of considerable value. Though I was unaware of the circumstance at the time, it appears that two authors had previously hinted at the possible agency of ice in two epochs,—paleozoic and secondary. In the "History of the Isle of Man" (1848), p. 89, in describing the conglomerate of the Old Red Sandstone, Mr. Cumming compares it to "a consolidated ancient boulder-clay formation," and continues, "Was it so, that those strange trilobitic-looking fishes of that aera (the *Coccosteus*, *Pterichthys*, and *Cephalaspis*) had to endure the buffeting of icy waves and to struggle amidst the wreck of ice-floes and the crush of bergs? These are questions which we may perhaps venture to ask, but which we dare not hope to have solved till we know something more than at present we know of the history of the boulder-clay formation itself." It may be remarked as a curious coincidence, that, when in Worcestershire I arrived at the conclusion that the Permian breccias are also boulder-clays, my thoughts at once reverted to the more ancient Old Red conglomerates of Scotland, and I stated at the time to my colleague Mr. Howell that they might afterwards turn out to have had a similar origin.

In a paper published in the Quarterly Journal of the Geological Society, February 1850, pp. 96 and 97, Mr. R. Godwin-Austen observes that "the great blocks of porphyry of the middle beds of the New Red series in the West of England, included in sands and marls indicating no great moving power, seem to require some such agent as that of floating ice to account for their position." In the following observations I hope to carry this subject considerably further, and to show, not only that there were icebergs of Permian date, but also partly to indicate the district whence the glaciers descended that gave these icebergs birth.

The Breccias on one horizon; and extent of the area which they occupy.—I have now described these Breccias as occurring in ten localities, exclusive of small outliers, or mere minor separations of the same mass by local faults. Though occurring at intervals, there can be little doubt that they all belong to one Permian horizon.

Character of the stones in the Breccia; and whence derived.—The lithological nature of the imbedded fragments has already been described. Everywhere, in spite of exceptional fragments in the Malvern district, they seem to be derived from one set of rocks; they are all enclosed in the same red marly paste, and they are mostly angular or subangular. A well-rounded waterworn pebble is, in places, of rare occurrence. The surfaces of a great majority of the pebbles are much flattened, numbers are highly polished, and, when searched for, many of them are observed to be distinctly grooved and finely striated. The striae in some are clear and sharp, and run parallel to or cross each other at various angles; while in others, though you see their remains, age and surface-decomposition have impaired their sharpness and roughened the original polish of the stone.

I have stated that (if lithological character be any guide) the fragments (with rare exceptional pieces) seem to have been derived from the conglomerates and green, grey, and purple Cambrian grits of the Longmynd and from the Silurian quartz-rocks, slates, felstones, felspathic ashes, greenstones, and Upper Caradoc rocks of the country between the Longmynd and Chirbury. The south end of the Malvern Hills is from forty to fifty miles, the Abberleys from twenty-five to thirty-five miles, Enville from twenty to thirty miles, and South Staffordshire from thirty-five to forty-five miles distant from that country. The question then arises, by what process were so many angular and subangular fragments transported so far; many of them being a foot, and some two, three, or even four feet in diameter; the whole in places forming a deposit of several hundred feet in thickness? Why also are they angular, and not well-rounded, like the pebbles of the great conglomerate-beds of the Bunter Sandstone; and why have they flattened sides, and often polished and striated surfaces?

Glacial origin of the Breccia.—They were therefore deposited in water with considerable regularity, and, as we have seen, over a large area. It is altogether unlikely that the stones were poured into the sea by rivers in the manner in which some conglomerates are formed on steep coasts, where mountain-ridges nearly approach the shore, 1st, because the fragments, being derived almost exclusively from the Longmynd country, if the sea then washed its old shores, no river-currents passing out to sea could carry such large fragments from thirty to fifty miles beyond their mouths and scatter them promiscuously along an ordinary sea bottom; and, 2ndly, if the rivers merely passed from the Longmynd across a lower land to the sea, transporting stones and blocks of various size, these would have been waterworn on their passage seaward after the manner of all far-transported river-gravels, whereas many of the stones are somewhat flat, like slabs, and most of them have their edges but little rounded. Neither could ordinary marine currents move and widely distribute fragments so large that some of them truly deserve the name of boulders; and, except in the case of earthquake-waves, which here and there produce an occasional debacle on a shore, I have no faith in violent currents of sea-water (such as have been sometimes assumed to result from imagined sudden great upheavals of land), washing across hundreds or thousands of square miles, and bearing along and scattering vast accumulations of debris far from the parent rocks. This is an assumption without proof. It is also unlikely, and I think impossible, that large debris of this kind could be distributed over so wide an area by the sifting process which Mr. Darwin has shown probably to take place on the east coast of South America, in consequence of movements communicated into deep water during long-continued heavy gales. Neither have they been moved along sea-shores, or subjected to breaker action, like the stones of the Chesil Bank, or of the conglomerate of the Upper New Red Sandstone, all the pebbles of which are true pebbles, spherical or oval, and smoothed by long attrition.

If, then, they were not distributed by any of these agents, there remains but one other means of transport and distribution—the agency of ice.

1st. There is in proof, the great size of many of the fragments, —the largest observed weighing (by a rough estimate) from a half to three-quarters of a ton.

2nd. Their forms. Rounded pebbles are exceedingly rare. They are angular or subangular, and have those flattened sides so peculiarly characteristic of many glacier-fragments in existing moraines, and also of many of the stones of the Pleistocene drifts, and the moraine matter of the Welsh, Highland, Irish, and Vosges glaciers.

3rd. Many of them are highly polished, and others are grooved and finely striated, like the stones of existing Alpine glaciers, and like those of the ancient glaciers of the Vosges, Wales, Ireland, and the Highlands of Scotland; or like many stones in the Pleistocene drift.

It has been said that in any breccia or conglomerate the stones may be scratched. In other ancient breccias I have never observed it; and I think that in the Permian fragments the experienced eye will have no difficulty in recognizing the peculiar characteristics of glacial scratching.

I conceive, therefore, that the peculiar forms, polish, and markings of many of the stones indicate that these characteristics have been produced by the agency of ice of the nature of glaciers, for mere coast-ice would have picked up and drifted away numerous rounded pebbles from the beach, and not a great majority of angular flattened stones, such as form the breccias wherever they occur.

If this conclusion be correct, and if the parent rocks whence the stones were derived be properly identified, then it follows that the ancient territory of the Longmynd and the adjacent Lower Silurian rocks, having undergone many mutations, at length gave birth to the glaciers, which, flowing down some old system of valleys, reached the level of the sea, and, breaking off into bergs, floated away to the east and south-east, and deposited their freights of mud, stones, and boulders in the neighbouring Permian seas.

DAUBRÉE

Gabriel August Daubrée (1814–1896), French geologist and mining engineer, is known for his work on underground water and as the great leader in experimental geology.

EXPERIMENTS ON THE ACTION OF SUPERHEATED WATER IN THE FORMATION OF SILICATES

From the translation by T. Eggleston, *Annual Report of the Smithsonian Institution*, 1861, pp. 228–304, 1862.

The principal difficulty consists in finding walls and fastenings that will resist for a sufficient length of time the enormous tension which steam acquires when the temperature is raised towards a dull red heat. Water and the matters that are to react are placed in a glass tube, which is then sealed. This glass tube is next introduced into a very thick tube of iron, which is closed in a forge at one of its ends. The other end is often closed by means of a screw, having a square head, which can be turned with a wrench. Between the head of the screw, which should be made with great precision, and the end of the tube a washer of very pure copper is placed; it should be thin enough to be crushed by the pressure when the tube is closed, and penetrate into the grooves made for that purpose. For closing the second extremity, however, I have now adopted another plan in preference; I introduce hot a very strong bolt, which, if the welding is skilfully done, becomes part of the tube. A workman must be very skilful to succeed in this operation; for it is essential that the greater part of the tube remain cold, in order that the water in the interior may not, by evaporating, hinder the operation. To counterbalance in the interior of the glass tube the tension of the steam, which might burst it, water is poured around it, between its sides and those of the iron tube which surrounds it. In this way the principal strain is put upon the latter tube, which offers the most resistance. This apparatus, like those which de Senamont used, is laid on the dome, or in the conduits of the retort-furnace of a gas-works in contact with the masonry, which is at a dull red heat, and is buried beneath a thick bed of sand. At a temperature which is a little below nascent red heat water reacts very energetically on certain silicates.

In this manner ordinary glass, at the expiration of a few days, gives two, and often three, distinct products: 1. A white and altogether opaque mass, which results from the complete transformation of the glass; it is porous, sticks to the tongue, and would look like kaolin if it were not for its very decidedly fibrous structure. The substance has lost a great part of its weight, nearly half its silica, and a third of its alkali; a new silicate has been formed which has fixed the water, and belongs, by its composition, to the family of zeolites. 2. An alkaline silicate which has been dissolved, carrying with it the alumina. 3. Often there are developed, in addition, numerous perfectly limpid and colorless crystals, which have the ordinary pyramidal form of quartz, and which, in reality, is nothing more than crystallized quartz. . . . What renders this transformation of glass still more remarkable in a geological as well as a chemical point of view is, that it is obtained with a very small quantity of water, which in weight is not equal to a third of that of the transformed glass.

Volcanic glass, known under the name of obsidian, acts in a manner similar to the artificial. Pieces of obsidian, heated under the same conditions, are changed into a gray product of a crystalline nature, having the aspect of a fine-grained trachyte. Its powder, examined under the microscope, presents exactly the characters of crystalline feldspar. . . .

With the fragments of obsidian on which I operated there were pieces of vitreous feldspar detached from a trachyte of Drachenfels, and, also, a piece of oligoclase from Sweden. These last two minerals underwent no appreciable change. . . . We see here a kind of confirmation of the preceding experiment on the stability of the silicates, which have, perhaps, originally crystallized in conditions very similar to those in which they were again placed. . . .

To examine, as far at least as the presence of glass will permit, how the solutions of natural silicates which we usually find in water act when superheated, I used water from the hot springs of Plombières, which is comparatively rich in silicates of potash and soda. . . .

. . . After an experiment that was stopped at the end of only two days the sides of the tube were already covered with a silicious coating in the form of crystallized quartz, and also of chalcedony. As the glass was as yet only altered on its surface, this deposit must have come, almost all of it at least, from the decomposition of the alkaline silicate contained in the water of Plombières. . . .

A new proof of the facility with which minerals of the feldspar group can be produced in the presence of water is furnished by the following experiment. . . . Kaolin, perfectly purified from all feldspathic debris, by washing, having been treated in a tube with water from Plombières, this earthy mass was transformed into a solid substance, confusedly crystallized in little prisms which scratched glass. . . . It is a double silicate of alumina and an alkali, having all the characters of a feldspar; it is mixed with a little crystallized quartz. . . . On the surface and in the interior of the whitish mass resulting from the transformation of the tube I found a great quantity of very small crystals, but of a perfectly distinct form, having great brilliancy and perfectly transparent; they had different tints of green, and many of them that olive-green tint which is peculiar to peridot. . . . In fine, they have the composition of a *pyroxene*, having a base of lime and iron; and from their transparency, they belong to the variety *diopside*. . . . All of them, by their aspect, immediately recall the best known crystals of diopside.

The fossil vegetables having undergone modifications under the influence of the same agents as stony matters, it is proper to see what becomes of wood in superheated water. Fragments of spruce were transformed into a black mass, having a bright lustre, perfectly compact—in a word, presenting the aspect of a pure anthracite. . . . It differs from the carbons formed at high temperatures in the fact that, like the diamond, it does not conduct electricity. The veins of silver at Kongsberg, in Norway, which are encased in gneiss, contain anthracite which very much resembles the artificial anthracite we have just mentioned. . . . At lower temperatures, but in conditions otherwise analogous, wood is transformed into a kind of lignite or coal. In these experiments I obtained, as I have before said, liquid and volatile products resembling natural bitumens, and possessing even their characteristic odor.

To recapitulate: superheated water has a very energetic influence on the silicates; it dissolves a great many of them, destroys some combinations of multiple bases, and forms new ones, either hydrated or anhydrous; in fine, it causes those new silicates to crystallize far below their point of fusion. In these changes the silicic acid, set at liberty, isolates itself under the form of crystallized quartz. Transformations so complete are, moreover, obtained with a very small quantity of water. In general we can distinguish this law, that near the point of nascent red heat the affinities of the wet way acquire, so far as concerns the production of silicates, the same character with those of the dry way.

————————

The results which we have just given enable us to account for what takes place in the crystallization of silicious rocks in general, as well the eruptive as the metamorphic. Let us first examine the former of these, commencing with the lavas.

Whatever may be the molecular state of the water in lavas, it intervenes to cause them to crystallize much in the same way as in the experiments of the laboratory for transforming obsidian into crystallized feldspar, and for producing pyroxene in perfect crystals. Thus, in one as in the other case, the water appears to favor the *elimination* of substances which would remain mixed, and to permit the crystallization of silicates at a temperature very much lower than their point of fusion. It is again through the influence of this kind of mother water that the same silicates crystallize in a succession which is often opposed to their relative order of fusibility.

FRACTURES RESULTING FROM TORSIONAL STRAINS

Translated from *Comptes rendus des séances de l'Académie des sciences,*
Vol. LXXXVI, pp. 77–83, 283–289, 428–432, 1878.

A plate of the substance to be examined, in the form of a very elongate rectangle, is caught by one of its shorter sides between two jaws of wood screwed together to act as a vise. The other extremity is fitted into a tap wrench, where it is wedged evenly by an insertion of pasteboard. In making the wrench turn about a horizontal axis, one brings about a torsion which is not long in causing a breakage.

A first series of attempts, made on plates of gypsum having a thickness of twelve millimeters, gave a small number of breaks.

Nevertheless, in certain cases, these breaks had a marked tendency to be parallel with each other, while others were nearly perpendicular to them.

With the plates of glass, the tests have been more significant. These plates had a length of 80 to 90 centimeters, with a width varying from 30 to 120 millimeters, and a thickness of 7 millimeters. For each experiment, the plate was enveloped in gummed paper, which kept the fragments formed from separating. Without this precaution it would have been difficult to ascertain the arrangement of the fractures.

In each of these rectangular plates of glass, fissures were produced in great number at the time of breaking.

In spite of their curves and variations, these present, in their general effect, a pattern in which an entirely geometric regularity is quickly recognized. This regularity stands out particularly when one is at some distance from the plate or examines a series of them so as to gain a general impression of the results.

(1) The fissures in question consist of irregular surfaces, of rather variable form, whose traces on the main surfaces of the plates, which we shall designate here as *affleurements* [outcrops], deviate slightly from a straight line.

(2) In addition, the fissures group themselves along two directions or systems, which are equally inclined on the axis. These two systems, which can be styled *conjugate*, thus form a net, the meshes of which are more or less crowded, depending on the plates. Many of the lines intersect to form lozenges.

In general, the two conjugate systems cross at very wide angles, whose values seem to depend on the relative dimensions of the two sides of the plate. This angle, which is sometimes approximately a right angle, is reduced in other cases to seventy degrees or below.

In the deformation by torsion, the results of which have just been seen, each longitudinal fiber takes on a spiral form and each of the great plane faces of the plate becomes uneven. At the moment of rupture, the angle of torsion does not exceed twenty degrees.

The characters of faults have been studied in a much more complete manner in the metalliferous veins which arise from their filling, thanks to the numerous galleries that follow them in all directions. . . . The outcrops represented on these maps resemble, in an incontestable manner, in their general effect and even in many of their details, the artificial fissures which have just been considered. . . .

The joints which traverse a single massif of rocks are often oriented parallel to three or even four different directions. These multiple systems of joints have long been observed by workers in slate quarries, who have a great interest in recognizing them and who have used a special name for each direction. . . .

In general, it has not been admitted . . . that all the systems of joints forming these networks can be contemporaneous. Meanwhile, returning to the afore-cited experiments, one sees many systems of fractures oriented differently and yet produced simultaneously by a single strain. This can have taken place often in nature.

In the conjectures that have been made on the origin of faults, it has been supposed . . . that they were evolved by actions of very long duration. . . . At the same time, one general fact arises from the experiments just reported; that is, that slow deformations and gradual strains, when they have sufficiently surpassed the limit of elasticity of the rocks on which they function, can end in systems of fracture, suddenly produced, and presenting, together with the character of evident parallelism, other similarities with natural breaks common in the earth's crust. A sudden movement readily explains certain phenomena such as slickensides which are not to be conceived as a result of slow action, the importance of which, in other respects, is manifest everywhere in the physical history of the globe.

EXPERIMENTS ON THE FOLDING AND FAULTING OF STRATA

Translated from *Comptes rendus des séances de l'Académie des sciences*, Vol. LXXXVI, pp. 733–736, 1878.

The apparatus consists of an iron frame of rectangular form, which is destined to receive the beds to be compressed. These beds are placed parallel to one of the wider sides of the frame, which has torsion screws to produce a perpendicular pressure on the beds. A second side, contiguous to the first, has torsion screws that serve to exercise a pressure on the beds parallel to their direction. The first can be called screws of *vertical pressure*, and the second, screws of *horizontal pressure*. The pressures are exerted either on the face or on the edges of the beds, through the agency of *pressure plates* of wood or iron.

This arrangement, simple as it is, allows one to produce a great variety of effects.

On partially closing the frame by two ends which transform it into a rectangular parallelepipedon, or on giving it a circular section, still more common conditions are reproduced; for, aside from the vertical pressures, horizontal pressure in two directions perpendicular to each other can be exerted in the very plane of the bed.

In order to establish a certain resemblance to natural conditions, it is important to choose suitable subjects on which to exert the pressures.   Instead of the layers of clay or cloth of Hall's experiments, I used beds of various kinds, some of metal, galvanized iron and especially laminated lead of various thicknesses; others of wax mixed with various substances such as plaster, resin, turpentine. Taking proper proportions, mixtures of very different consistency can be obtained, from the plastic state of modeling wax to the brittle state of casting wax and harder.   These substances were used either in the form of thick tablets or in the form of sheets of different thicknesses superimposed in a way reminiscent of a group of sedimentary beds.

Results of the experiments; major foldings and convolutions taken as examples.   I am going to indicate succinctly the results of the experiments and mention some of the natural types whose forms they reproduce.

(1) Homogeneous beds of equal thickness have been submitted to vertical pressures which were uniform over the entire extent of the beds.   The horizontal pressures then gave rise to rather uniform folds of which the number and shape varied with the pressures exercised.   After a simple arch which formed first, continued pressing gave rise to sinuosities which succeeded each other, assuming more and more numerous inflections as the pressures increased.   These bendings, with concavities alternately directed up and down, bring about a series of *synclinal* and *anticlinal* lines, in the terms adopted by geologists.

Configurations of this sort are extremely common in nature.

(2) This regularity in the folding ceases when the vertical pressures are not uniformly spread over the whole extent of the beds.   If these can yield more easily at one side than at the other,

instead of having a regular sinuosity, one can reach an arrangement where, at the side of the least pressure, numerous sharp folds appear, while on the opposite side the beds are barely bent.   Under this sole condition of inequality in the vertical pressures, there is an asymmetry in the folding; moreover this asymmetry can be provoked indifferently, either on the side of the mobile pressure plate or on the side of the fixed resistance.

(3) It is not only the difference in the vertical pressures exercised on the different points that influences the intensity of the folding: inequalities in the thickness of the beds have a well-marked influence as well. . . .

(4) In the folding experiments, especially in the cases of asymmetry from vertical pressures or from irregularity in the thicknesses, when the pressure continues to act, the sinusoidal or serpentine forms without overturn are gradually deformed and pass into folds with a *reversal* of beds.   The direction of these reversals varies; the convexity sometimes adjoins the side of pressure, sometimes the side of resistance.

FAVRE

Alphonse Favre (1815–1890), Swiss geologist, professor of geology at Geneva, was a great student of Alpine structure and a pioneer in the attempt to reproduce mountain structures in the laboratory.

EXPERIMENTAL REPRODUCTION OF MOUNTAIN STRUCTURES BY LATERAL COMPRESSION

Translated from *Comptes rendus du Congrès internationale géologique,* Paris, 1878, pp. 35–38, 1880.

Mountains of crystalline rocks have most certainly a different origin than mountains formed of sedimentary terrains. The experiments, the results of which I have the honor of showing you, are applicable solely to the latter.

When it was believed that the water of the sea had been raised above the level of the mountains which contained marine fossils, no hypotheses on the origin of mountains were made, but since this idea has been abandoned, new ideas on this subject have arisen. They can be divided into three theories, according to the direction of the forces which have operated: (1) from bottom to top; (2) from top to bottom; and (3) horizontally.

The first, that of upheaval, goes back to great antiquity; it is now abandoned.

The second, that of sinking, has been sustained in Switzerland by J.-A. Deluc. He thinks the mountains are the solid parts of the surface of the earth, which have withstood a general lowering. This theory still has some adherents.

The third, that of compression, was enunciated for the first time in 1795, by H.-B. de Saussure. In developing it, Élie de Beaumont designated it by the name of lateral crushing.

To see whether experiment would be an aid to the theoretical ideas, I have attempted to compress bands of plastic clay in the following manner. A band of thick rubber, twelve centimeters wide and forty centimeters long, was stretched to sixty centimeters; a layer of soft clay, two to six centimeters thick, was spread on it, care being used to get as much adhesion as possible.

The clay was sometimes composed of superimposed beds, but most often what seems to be beds in the plates presented herewith is only an appearance of stratification produced on the exterior by lines drawn horizontally before the contracting. To each end

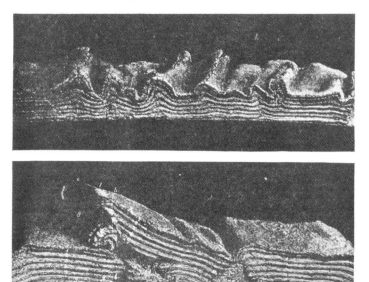

Fig. 21.—Two of the illustrations used by Favre in describing his experiments with deformation by lateral compression, 1880.

of the rubber is fixed a strong board, against which the clay rests. Letting the rubber return slowly and gradually to its original dimension (forty centimeters), the clay is compressed and one sees reproduced, at its surface and on its sides, a configuration and a structure entirely similar to those observed in the Alps, the Jura, the Appalachians, etc.

LESLEY

J. Peter Lesley (1819–1903), American geologist, was state geologist of Pennsylvania from 1874 to 1893 and during much of that time served also as professor at the University of Pennsylvania. His writings contain the first comprehensive description and regional classification of the physiography of the eastern United States.

Appalachian Structure and Topography

From *Manual of Coal and Its Topography*, pp. 121–187, Philadelphia, 1856.

The Science of Topography, like every other science, proceeds to deduce from a few elementary laws an infinity of forms, by which these few laws may or actually do express themselves upon the surface of the earth. The possible is always an infinite series, the actual a limited and fortuitous selection from it—not always, therefore, the most striking, perfect, and complete. Here and there, now and then, one such occurs and we call it a typical phenomenon, because, like the Apollo or the Venus in human Fine Art, it fills out the expression of those laws or creative ideas, the elucidation of which alone is science and the standard of value for a fact.

In the coal regions of America from which most of the illustrations in this book have been obtained, one group of topographical expressions are in great perfection, but others are entirely wanting. For alpine forms, that is, for the aiguille or needle peak—for the jagged crest line—for the cirque of névé wherein glaciers form—for roches moutonnées or polished bossy slopes where glaciers have passed—for moraines or piles of fragments left by glaciers in their retreat—for viae malae or profound gorges with fitting walls and dark recesses cleft by earthquake action, and too high above the later sediments to be filled up—for perpendicular horizon outlines, thousands of feet long against the sky—for thread-like waterfalls pendant from the edges of pastures, hundreds of feet above their base—for rosaries of lakes filling with mud and sand, protecting others below them, as clear as crystal, and lined at the bottom with the thinnest strata of impalpable powder, and of filled up lakes, now fertile meadows, supporting secluded hamlets—for such phenomena as these, we must go to the Alps, to the Pyrenees, to the Andes, to the Himalayas.

In the coal measures we have, on the contrary, long and regular mountain crests vanishing horizontally in parallel lines into the horizon, broken at intervals with gaps and curving in and out in zigzags including inner sets of zigzags, system within system, down through the series of the rocks. We have all the phenomena of drift and denudation, deluvial scratches, eddy hills and terraces, in great perfection.

In that Appalachian region, as it is now called, which, as a belt from fifty to one hundred miles in width, between the edge of the Bituminous Coal region and the South Mountain or Blue Ridge, stretches from the Delaware to the head waters of the Tennessee, we recognize a country where, from the comparative deficiency of fossils and of coal, and from the repeated and elongated outcrops of a few valuable mineral deposits, the science of geology transforms itself into the science of topography. Nowhere else on the known earth is its counterpart for the richness and definiteness of geographical detail. It is the very home of the picturesque in science as in scenery. Its landscapes on the Susquehanna, on the Juniata and Potomac are unrivalled of their kind in the world. Equally beautiful to the eye of the artist is a faithful representation of their symmetrical, compound, and complicated curves upon a map.

All topography resolves itself into a discussion of the *Mountain*. The surface of the earth may be considered as a congeries of mountains (and hills are but smaller mountains) touching each other at their bases. Valleys have only a negative topographical existence and represent the absence or removal of mountain land. Mountains are solid portions of the earth's crust, while valleys are but vacua in it, taken possession of by the water and the air. The propriety of this view, opposed though it be to the economical history of the human race, will become more evident when the denudation of valleys is discussed. In geology it is important to true views to consider all changes to be mountainous, and valleys to be an incidental consequence.

A Mountain has *three Elements*, top, side, and end, and the primary discussion of a mountain is that of its slopes, its crest-line, and its termini.   Gaps are irregularities in its crest-line, as terraces are in its slopes.   The cross-section of a mountain shows why its slopes, and usually also why its crest-line and its termini are not only what they are, but could not be different.   It is a deep set feeling among men that if there be accidental forms upon earth they are to be found in mountains.   There could not be a greater mistake; for if there be natural forms unalterably predestined by the direction and intensity of natural forces they are those of mountains.   Not a wrinkle in the side, not a notch in the crest, not a flexure in the trend of a mountain or a hill but is an evidence of laws which have operated upon it with the nicest precision. Not a ravine, not a rod of cliff, not a waterfall, but exists in the immediate vicinity of its own explanations.   The place where a stream breaks through, however apparently accidental, was determined by positive relationships between the rocks of the locality; nor can any investigation be more exciting than that which rewards itself with perpetual discoveries of cause and effect in a wilderness of apparent lawlessness and unexplained confusion.

The Mountain Slope.   To illustrate the influence which its interior structure has upon the form of a mountain, the accompanying series of cross sections are introduced.   The first set show how a stratum of sand-rock or other hard material inclosed in softer stuff, arranges the height and slopes of its mountain to suit its own dip.   When it is vertical the mountain is low, sharp, and symmetrical; at 60°, it is higher with a front side long; at 30° higher still, with a long, gently sloping back, and short steep front covered with angular fragments from a range of cliffs above; when horizontal, the mountain is at its maximum height, forming a table land with precipices and steep slopes in front.   It is needless to suggest the infinite variations of this simplest law of mountain form.

When two such sand-rocks lie in neighborhood they of course form a double mountain, subject to the same vicissitudes of external aspect in view of similar changes of internal structure.   It is not so easy to show the effect of these changes under another influence, that of subsidence beneath a universal plane of denudation.   The attempt, however, is made in the following sets of cross sections.   It must be understood that a mountain whose rocks stand vertical or horizontal at one point may show them much inclined at others (its form will change to suit), and also, that as mountains are but fragments of the upper layers of the earth's crust preserved from the general denudation and translation by lying lower than the rest, in hollows or synclinals, as they are technically called, these synclinals as they rise above or sink below the average level or line of denudation will give up to destruction more or less of their contents. In other words, when a geologist traverses one of these geological basins lengthwise he finds the highest rocks in its deepest parts, and the lowest rocks in its shallowest or highest parts. The cross sections given below are arranged to show the coming in of higher and higher rocks as the basin sinks, and the consequent changes of form which the mountain or mountains undergo.

In every *Shallow synclinal* there is a high, narrow, flat-topped mountain, with precipices on both sides looking down upon the lowlands. Such is the structure of the Catskill, Towanda, Blossburg, etc., and other mountains in the north, and the Cumberland Mountains of Tennessee and Alabama. As the basins deepen, other higher sand-rocks come in above, and double the precipices and slopes; finally the whole is cleft in two, and the drainage, after traversing the parted mountain often for many miles, breaks out sideways into the plain. It is evident that it was due to the direction of the original currents setting along the centre of the geological basin before the cutting developed the present mountain.

The *Sharp synclinal* shows this more clearly, with this difference, however, that its longitudinal central cutting is never a ravine, but always a valley. The hard rocks pass off in diverging crests terraced and gashed, inclosing one another, and giving place to mountain within mountain, in a series that has no limit but the number of hard beds in the system of formations. Their precise inclination with the horizon makes no essential difference in the action so long as the synclinal remains a simple one; but the moment this becomes compound then all manner of complications inaugurate themselves upon the surface and present a thousand puzzles to the skill of the geologist, as in the following set (fig. 30) [Fig. 22], which more or less nearly represent the wrinkled compound synclinals of the anthracite coal region.

The *Anticlinal* structure is the reverse of the synclinal, and has its own definite system of forms equally subordinated to the general laws of denudation, and in many respects curious inverted parodies of those above. In fact, it may be seen in the next set (fig.

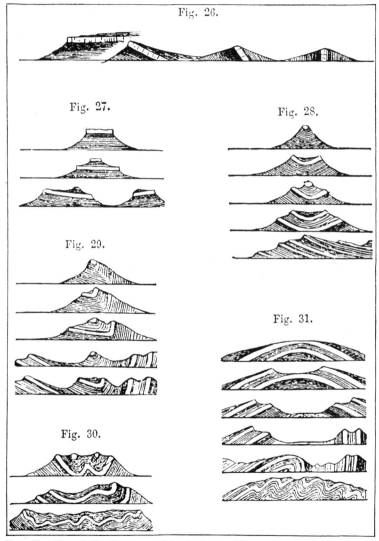

Fig. 26.

Fig. 27.

Fig. 28.

Fig. 29.

Fig. 31.

Fig. 30.

Fig. 22.—Six of the figures used by Lesley to illustrate his paper on Appalachian structure and topography, 1869.

31) [Fig. 22], the moment the huge back of an anticlinal mountain splits in two, we lose the anticlinal as a character of the mountain

while it remains in the valley, until from the centre rises another mountain, to be again split lengthwise in its turn. In this case we represent the anticlinal as rising slowly, just as before we represented the synclinal as slowly sinking. The ridges on each side of a split anticlinal are called *monoclinal,* and are in fact the same as the ridges into which a parted synclinal mountain divides itself. Hence the forms are common to both.

The Crest Line of a mountain is normally either a point or a horizontal line, according to whether it is a mountain of ejection under air, or a mountain of elevation and denudation under water. The type of the volcano is a cone; the type of any other mountain is a double sloping wall. Sedimentary mountains are the turned up edges of some hard sediment. The central hardest or massivest layer of that sediment will of course form its crest. If there be subordinate layers, they will form terraces; and, at the gaps, ribs. If several central layers near together be equally able to resist denudation the crest will move from one to the other along the line according to the cross-cutting. So long as the dip maintains itself steadily the mountain crest will be a straight and level line; when the dip changes, the crest line will fall off accordingly and suffer change in height and evenness. When the dip falls completely over or reverses itself, the crest line will double back upon itself, and do so as many times as the dip changes. At such a recurve coves are formed; here valleys head up; here gaps usually break through the side walls; here also the mountain attains its maximum in height. These doublings, if anticlinal, are in one direction; if synclinal, in the other. A series of them can only occur where a system of parallel anticlinal waves and synclinal basins pass under an outcrop, as in the following fine instance [Fig. 23].

It represents the Buffalo Mountains, *L* of Union County, Pennsylvania; Jack's Mountain, *K*; the anticlinal Kishicoquillas Valley (Logan the Indian lived at *G*); the upper end of the synclinal Standing Stone Valley *H*, and the Seven Mountains back of it; the Bald Eagle Mountain sweeping from Bellefonte *J*, to Muncy *D*, on the Susquehanna West Branch; the Nittany Valley *A*, with its anticlinal limestones appearing again in the Nippenose or Oval Valley *B* and the smaller Oval valley *C*; the Nittany synclinal mountains south of it with their small

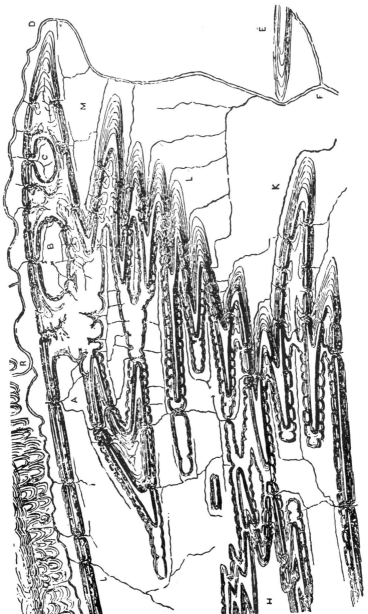

Fig. 23.—Sketch map of a portion of the Appalachian Mountains, reproduced as "Fig. 36" in Lesley's paper on Appalachian structure and topography, 1869.

included anticlinal vales; and finally the simple anticlinal Montour's Ridge *E* apart from all the rest.

This magnificent group of *Zigzags* (studied out by Mr. Alex. McKinley) has probably no rival yet known upon the surface of the earth.   Until its structure was explained it was a confused wilderness of mountains covered with forest and interpenetrated with a network of ravines; some of its crests long horizontal lines of singular evenness, and others broken by a series of sharp notches or deep gaps.   Its most singular feature is the terrace structure of its valleys running into a *reversed ship's keel structure* in the mountains which inclose them; a structure which needs no description when the reader turns from a perusal of the map to the third cross-section in Figs. 28 and 29 [Fig. 22].   These mountains are all composed of the Second Great Sand-rock described in the last chapter, the lower Silurian, Shawangunk Grit, No. IV.   The terraces are made by the lower of its two hard members, the keels and crests by the upper.   The included valleys are all of No. II Trenton Limestone, with No. III or Hudson River slates around their edges.   The outside country from Lock Haven *R*, round to Sunbury *F*, consists of Upper Silurian through which the Lower Silurian projects in the broad-backed anticlinal Montour's Mountain, *E*.

This map, rough as it is, will reward an attentive examination. The notching of the *southern* leg of each anticlinal zigzag by the rush of denudation across it, from within outward, or *vice versa*, is particularly interesting.   The peculiar drainage of the head of the anticlinal, not inwards towards the central valleys, but sideways through into the outside synclinal coves is also very instructive, and speaks volumes for the elucidation of the denuding action.

PHYSIOGRAPHIC PROVINCES OF EASTERN UNITED STATES

From *Transactions of the American Philosophical Society*, N.S., Vol. XIII, pp. 305–312, 1869.

Very little of the topography of America has been published. No large areas have been mapped in such a manner as to show the features of the surface. . . . Accordingly, I compiled a map of the United States, fifteen feet square, giving my own interpretations of the published topography of every State which exhibited marked features of relief,—basing these interpretations, first, upon the authentic topography of the State of Pennsylvania; secondly, upon my own topographical notes in other States; thirdly, on numerous local and county maps; fourthly, on the better marked portions of the State maps; and lastly, on the colored geological maps of the several States, or parts of States, already published.

The result was very striking. Familiar as I have been for years with the topography of every part of the United States, and of Canada, east of the Mississippi River, I was surprised at the beauty of the whole representation now for the first time made to the eye. The correlation of parts was very fine, in a geological sense. . . . But the charm of the map lay in its unmistakable utterances respecting different topographical types of earth-surface or strongly contrasted systems of erosion, lying in masses, side by side, or running for long distances in parallel belts.

Thus, for instance, the eye takes in at a sharp glance the whole Blue Ridge, Highland, and Green Mountain belt, full of short sharp ridges, in parallel order, but in echelon arrangement, with irregular summit lines, rising into knobs and peaks from 3000 to near 7000 feet above the sea.

Behind it runs the belt of the Appalachians, composed of interminably long and narrow barrow-mountains, with level summits, seldom 1000 feet in height, looped and gophered in an intricate and artificial style, with lens-shaped coves in the northern part; and on the other hand, in the Southern States, terminating in pairs of perfectly straight ridges, cut off short by faults.

Behind these, again, lies the Great Cumberland-Alleghany-Catskill Plateau, with its horizontal geology and its quaquaversal, arborescent drainage-system, boldly contrasting with the Appalachian topography in front of it, and settling the questions of mode and agency in favor of slow aërial denudation.

Still further west, the low finger-shaped bounding ridges and central plains of the Blue Grass country of Kentucky and Ohio shows another but allied type.

And in the east, the wide belt of low sand-hills, southeast of the Blue Ridge, and the immense cretaceous and tertiary flats of the tidewater country, crenulated with bays and covered with dismal swamps, presents a fifth, differing from all the rest. . . .

It is a pity we have none of the minute work yet published, which has been done here and there along borders of the great coal area, for it would help greatly to explain the true nature of the erosive agency which has relieved an expanse of continent, amounting to one or two hundred thousand square miles (its original limits are unknown), from a superincumbent weight of Coal Measures at least two thousand and perhaps, if capped with Permian measures, five thousand feet in height. It is evident from the map, that the whole of this wastage (with the exception of a small section in the north, drained by the Susquehanna and Delaware Rivers southeastwardly, and by the Genesee and Schoharie Rivers northwardly), has passed off into the Valley of the Mississippi; and thus we can account for a certain due proportion of the Cretaceous and Tertiary deposits of the Southwest. But as these were also products in some measure of an ancient system of short rivers which ravined the eastern slope of the Rocky Mountains along a face of, say, two thousand miles,—and probably also of another, similar, but opposite and shorter system of more important rivers flowing from the Laurentian mountains of Canada westwardly into the same Cretaceous sea,—we must not assign rashly too high a value to the subsidy of deposit paid in by the coal area of the Eastern United States.

We see also that the Cretaceous and Tertiary deposits of the Atlantic coast must have had another origin. And yet this origin could not be wholly different from the first. For, as the map will show at a glance, the whole Devonio-Silurian (Appalachian) belt, in front of the edge of the arborescently and westwardly drained area of the present Coal Measures, has suffered a still more considerable loss of superincumbent material, the whole of which loss must be represented by the semi-continental area of flat Cretacio-tertiary tidewater country, through which the rivers, which at first produced it, are still flowing, and extending themselves, as they push forward their deltas, along the present seaboard. Now from this Appalachian belt, not from two to five thousand feet of rocks, as in the case of the Coal Measures, but from twenty to thirty thousand feet of superincumbent Upper Silurian, Devonian and Carboniferous measures have been swept away. . . .

. . . Erosion by wind and rain, sunshine and frost, slow chemical solution, and spring and fall freshets, has done the whole work. I have long taught that it could not have been accomplished under water level by oceanic currents, because the ocean is a maker and not a destroyer. But I must now abandon wholly the idea to which I have clung, with slowly relaxing grasp, so many years, that a complete erosion theory demanded some such forces as would have been supplied by the extra efficiency of an ocean translated across the upheaved surface through the air. At the same time, the above considerations make me all the less willing to admit the ice-cake theory of erosion as even approximately true. For if aerial erosion has been going on uninterruptedly ever since the uprise of the Coal Measure continent, how little of the whole effect can have remained over to be still produced at the time when the Glacial epoch set in and the supposed ice-cake began to take the work in hand.

The erosion of the wonderful Green River ravines, in the Colorado country, to a depth of five thousand feet, is not more evidently the product of ordinary meteorological causes than is the erosion of any given segment of the Appalachian or Alleghany districts exhibited upon the map. . . .

I believe that this chemical element of erosive energy has been slighted even in discussions the most recent. I ascribe to it nine-tenths of the wastage of the Blue Grass area. I believe that the erosion went on, chiefly underground, along those narrow belts where the lime-rock formations approached the surface at whatever level the surface happened at the time to stand. The caverns grew; their roofs fell in; streams washed the debris continually away. As the surface thus kept falling piecemeal into the cavernous traps everywhere laid for it below, the general level of the area was slowly and insensibly let down to its present grade above the sea. While falling thus, the opposing outcrops of the upper rocks retired from one another, eastward and westward, and let the Blue Grass area between them widen slowly to its present size. And still the work goes bravely on. . . .

DAWSON

John William Dawson (1820–1899), Canadian geologist and naturalist, principal of McGill University from 1855 to 1893, where he also occupied the chair of natural history during the earlier years of his administration. He was the first president of the Royal Society of Canada, and in 1884 he received the honor of knighthood. His *Acadian Geology*, published in Edinburgh in 1855, was the most important work of his earlier years.

Marine Alluvial Soils

From *Acadian Geology*, pp. 23–24, Edinburgh, 1855.

The western part of Nova Scotia presents some fine examples of *marine alluvial soils*. The tide-wave that sweeps to the northeast, along the Atlantic coast of the United States, entering the funnel-like mouth of the Bay of Fundy, becomes compressed and elevated, as the sides of the bay gradually approach each other, until in the narrower parts the water runs at the rate of six or seven miles per hour, and the vertical rise of the tide amounts to sixty feet or more. In Cobequid and Chiegnecto Bays, these tides, to an unaccustomed spectator, have rather the aspect of some rare convulsion of nature than of an ordinary daily phenomenon. At low tide, wide flats of brown mud are seen to extend for miles, as if the sea had altogether retired from its bed; and the distant channel appears as a mere stripe of muddy water. . . .

The rising tide sweeps away the fine material from every exposed bank and cliff, and becomes loaded with mud and extremely fine sand, which, as it stagnates at high water, it deposits in a thin layer on the surface of the flats. This layer, which may vary in thickness from a quarter of an inch to a quarter of a line, is coarser and thicker at the outer edge of the flats than nearer the shore; and hence these flats, as well as the marshes, are usually higher near the channels than at their inner edge. From the same cause, the more rapid deposition of the coarser sediment, the lower side of the layer is arenaceous, and sometimes dotted over with films of mica, while the upper side is fine and slimy, and when dry has a shining and polished surface. The falling tide has little

461

effect on these deposits, and hence the gradual growth of the flats, until they reach such a height that they can be overflowed only by the high spring tides. They then become natural or salt marsh, covered with the coarse grasses and *carices* which grow in such places. So far the process is carried on by the hand of nature; and before the colonization of Nova Scotia, there were large tracts of this grassy alluvium to excite the wonder and delight of the first settlers on the shores of the Bay of Fundy. Man, however, carries the land-making process farther; and by diking and draining, excludes the sea water, and produces a soil capable of yielding for an indefinite period, without manure, the most valuable grains and grasses.

Carboniferous Fossils of the Joggins Section

From *Acadian Geology*, pp. 160–163, Edinburgh, 1855.

Group XV. is one of the most interesting in the section, in consequence of the discovery in it, in 1852, by Sir Charles Lyell and the writer, of the bones of a reptile,* *Dendrerpeton Acadianum* [Fig. 24a], those of another small reptile, and the shell of a land snail (Pupa?) [Fig. 24b]. These remains are of great interest, as they are the first reptilian animals found fossil in the carboniferous rocks of America, and the only land snail whose remains have ever been found in rocks of that age; in fact, the only evidence yet obtained of the existence of animals of that tribe at so early a period. These interesting remains were all found in the interior of an erect tree, mingled with the sand, decayed wood, and fragments of plants which had fallen into it after it became hollow. The bed of argillaceous sandstone, nine feet in thickness, which enclosed this tree, contains a number of erect plants [Fig. 24c]. Three erect trees in the form of sandstone casts and erect calamites were observed in it, with many Stigmaria roots. There was also a tree not in the form of a cast, but of a mass of coaly fragments surrounded by a broken and partly crushed cylinder of bark; the whole being evidently the remains of a trunk which has been reduced to little more than a pile of decayed pieces of wood before the sand was deposited; consequently it must have been either an older or more perishable plant than those which stand as pillars of sandstone. The wood of this tree shows, in the cross section,

---

* Now identified as an amphibian.—Editors.

a cellular tissue, precisely similar to that of the *Coniferae;* the longitudinal section shows only elongated cells, but is very badly preserved. A tree of this description is not likely to have been more perishable than the *Sigillariae,* which, in the same situation,

a

b

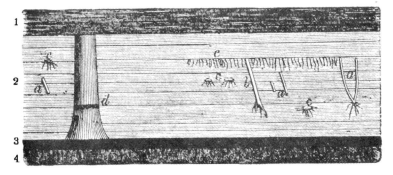

c

Fig. 24.—Three of the illustrations accompanying Dawson's description of the Carboniferous fossils found at Joggins, 1855.

remained until nine feet of sandy mud had accumulated. I suspect, therefore, that this stump may be the remains of a coniferous forest, which preceded the *Sigillariae* in this locality, and of which only decaying stumps remained at the time when the latter were buried by sediment. This is the more likely, as the appearances indicate that this tree was in a complete state of decay at the very commencement of the sandy deposit. . . .

On the whole, we can scarcely err in affirming that the habitat of the *Dendrerpeton Acadianum* and its associates was a peaty and muddy swamp, occasionally or periodically inundated, and in which growing trees and Calamite brakes were being gradually buried in sediment, while others were taking roots at higher levels, just as now happens in the alluvial flats of large rivers.

NEWBERRY

John Strong Newberry (1822–1892), American geologist and paleontologist, was at various times engaged in the early exploratory surveys of the western United States and later was state geologist of Ohio and professor at Columbia University.

CIRCLES OF DEPOSITION IN AMERICAN SEDIMENTARY ROCKS

From *Proceedings of the American Association for the Advancement of Science*, Vol. XXII (1873), pp. 185–196, 1874.

In the United States the geological column is composed of the following elements: at the base we have the Laurentian and Huronian groups, forming the Eozoic system, and composed of crystalline rocks, once limestones, sandstones, shales, etc., but now much metamorphosed and disturbed, and their fossils obliterated. . . . Upon the Eozoic rocks we find, between the Atlantic and the Mississippi, the various strata which compose the palaeozoic systems, the Lower Silurian, Upper Silurian, Devonian and Carboniferous. Of these the Lower Silurian consists, beginning at the base, of, 1st, the *Potsdam sandstone*, generally a coarse, mechanical shore deposit; 2d, the *Calciferous sand-rock*, a mixed mechanical and organic sediment, more sandy towards the east, more calcareous and magnesian towards the west, which we must class as an off-shore deposit; 3rd, the Trenton limestone group, consisting of the Chazy, Bird's-eye, Black River and Trenton limestones. . . . This is plainly an open sea deposit. . . . 4th, the *Hudson group*, consisting of shales and impure limestones, mixed mechanical and organic sediments, the deposits of a shallowing and retreating sea. This member completes the circle of the deposits of the Lower Silurian and ends the history of the first submergence of the Eozoic continent.

The Upper Silurian system is composed at base of the *Medina sandstone*, locally a conglomerate to which the term Oneida has been applied, a shore deposit corresponding to the Potsdam; above this, the *Clinton group*, which is composed of limestones and shales, and the peculiar Clinton iron ore, evidently an offshore deposit; still higher, the *Niagara group;* below, shaly, and showing a shallowing of the Clinton sea; above, a great and widespread mass corresponding in position to the Trenton group of the Lower Silurian circle. . . . The Niagara limestone is overlaid by the *Salina* and *Helderberg* groups. Of these the Salina is evidently the deposit from a shallow and circumscribed basin like the Caspian, Dead sea or Salt Lake, where the salts held in solution . . . were precipitated by evaporation, with a considerable portion of introduced earthy matter. The *Water-lime group*, which overlies the Salina and forms the base of the Helderberg series, is an earthy magnesian limestone. . . . Notwithstanding some local irregularities of deposition, the Helderberg group corresponds in character and position with the Hudson of the Lower Silurian and completes the Upper Silurian series by a return to land conditions.

The two circles of deposition which have been described are grouped together under the term Silurian, but as each is complete in itself, and is a record of a totally distinct round of changes, and as the fauna of the two systems have almost nothing in common, it will, I think, be generally conceded that it was an error to combine them under one name. . . . *

. . . It will probably clarify and simplify the theory now advanced, to claim as the *essential* elements of each circle of deposition resulting from an invasion of the sea, but three distinct sheets of sediments, viz.: the mechanical, organic and mixed, the products respectively of the advancing, abiding and retreating sea. The lines of separation between these are more or less sharply defined according to the rapidity of the submergence, and the nature of the materials acted upon by the shore waves.

---

* The interests of science and the cause of justice would both be served if we could agree to call the Lower Silurian by Prof. Sedgwick's name, *Cambrian*, leaving Murchison adequate honor in retaining his names, *Silurian* and *Devonian*, for the overlying systems.

LE CONTE

Joseph Le Conte (1823–1901), American geologist, while still a young man abandoned the life of a country physician and after studying under Agassiz for two years was a successful teacher of natural science in Georgia and South Carolina until the end of the Civil War. In 1869 he became a member of the faculty of the new University of California where, at first, he taught botany, zoology, and geology, without assistance or laboratory appliances. Although he is probably better known to the general public through his philosophic essays, chiefly on evolution, than because of his geologic work, his many years of residence in the Cordilleran region inevitably increased his geologic interest, and he made a profound impress upon earth science. Especially during the later decades of his life, he was an outstanding figure among American workers in this field.

On the Origin of Normal Faults and of the Structure of the Basin Region

From *American Journal of Science*, Third Series, Vol. XXXVIII, pp. 256–263, 1889.

I have already, in a previous paper (Am. Geol., vol. iv, p. 38), given reasons for thinking that the general structure of the earth is that of *a solid nucleus* constituting nearly its whole mass, a solid crust of inconsiderable comparative thickness, and a subcrust liquid layer, either universal or over large areas, separating the one from the other. In this paper I assume such a general constitution. I assume also that the crust rests upon the subcrust liquid as a *floating body*. We may well assume this because, broken as we know the crust to be, if it were not so it would long ago have sunk into the subcrust liquid. I have also, in the previous article already alluded to, shown that this condition of flotation would be the necessary result of the increasing density of the earth as we go down. I now wish to apply these two assumptions to the explanation of Normal Faults and of the origin of the Structure of the Basin region. . . .

Theory of Faults.—The explanation of the *reverse* faults seems obvious enough. They occur, as we have already said, mostly in strongly folded regions. Such folds can only be produced by lateral pressure. The pressure when extreme often produces overfolds.   If such overfolds break, the dip of the fissure will be toward the direction from which the pressure came and the hanging wall be *pushed* forward and upward over the footwall by the sheer force of the lateral thrust. . . .

But the explanation of normal faults which are by far the most common is not so obvious.   I will give very briefly what seems to me the simplest explanation—an explanation which I have used in my class lectures for many years.

Suppose then the earth-crust in any place to be *not crowded together* by lateral pressure, as in the formation of mountains of the Appalachian type, but *uplifted into an arch* by intumescence of the subcrust liquid.   Such local intumescence of the subcrust liquid may be the result (a) of elastic force of steam incorporated in the magma in more than usual quantity by the access of water from above, or (b) of hydrostatic pressure transferred from a subsiding area in some other perhaps distant place.   Such an arch being put upon a stretch would be broken by long fissures more or less parallel to one another and to the axis of uplift into oblong prismatic crust-blocks many miles in extent.   After the outpouring of liquid lava or the escape of elastic vapors had relieved the tension, these crust-blocks would again be re-adjusted by gravity.   If the blocks are *rectangular prisms*, some may float bodily higher and some sink bodily lower, giving rise to level tables separated by fault cliffs as in the Plateau region already explained. But if the fissures are more or less *inclined*, as is more commonly the case, then it is evident that the crust-blocks will be either rhomboidal (*a, b, f, g,*) or wedge shaped (*c, d, e,* Fig. 25 A).   These in the arching of the crust would be separated from one another, Fig. 25 B.   But after the relief of tension by outpouring of lava or by the escape of steam, they would of course readjust themselves by gravity in new positions.   Now by the laws of flotation how would such blocks adjust themselves?   It is quite evident that every rhomboidal block would tip over on the *overhanging side* and heave up on the obtuse angle side producing in every case *normal* faults, and every wedged-shaped block would sink bodily lower or float bodily higher according as the base of the wedge were upward or downward, producing again in every case normal faults (Fig. 25 C).   A thick board sawn in the manner represented in Fig. 25 A and the separated blocks placed together and floated on water would take exactly the positions represented in Fig. 25 C.

The explanation is complete. Of course erosion will modify the fault-scarps thus formed, by sculpturing their faces and by reducing their heights and slopes or even in some cases effacing them altogether. But if the fracturing and faulting have been geologically recent and on a large enough scale they may still remain and give rise to very conspicuous orographic features.

It is in this way that the orographic features of the Basin region have been formed; and the scale has been so grand and the features are so conspicuous that the resulting structure has been appro-

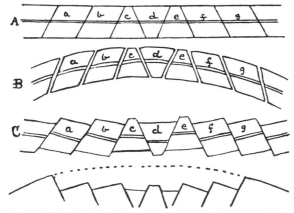

F<small>IG</small>. 25.—Four of LeConte's drawings to illustrate his concept of the origin of "the Basin system," 1889.

priately called *Basin structure*. The Basin region, according to the researches of King, Gilbert and Russell, is traversed by numerous north and south ridges several thousand feet high, with intervening valleys which are now or have been occupied by lakes. Although greatly modified by contemporaneous igneous ejections and subsequent erosion, these mountains consist essentially of gentle monoclinal slopes terminated by fault-scarps. They are in fact a succession of tilted and displaced crust blocks. The simplest and most beautiful illustration of this structure is found in the northern part of the Basin region in S.E. Oregon.

How the Basin system was formed.—Now regarding the Sierra and Wahsatch as belonging to the Basin system, we may imagine how the whole system was formed. At the end of the Tertiary the whole region from the Wahsatch to the Sierra, inclusive, was lifted by intumescent lava into a great arch, the abutments of which were the Sierra on the one side and the Wahsatch on the other as shown in the dotted line (Fig. 25). The arch broke down and the broken parts readjusted themselves by gravity into the ridges and valleys of the Basin region, leaving the raw faces of the abutments overlooking the Basin and toward one another. It must not be supposed, however, that this took place at once, but *gradually;* the lifting, the breaking down and the readjustment going on together *pari passu;* each readjustment probably giving rise to an earthquake.

Process still going on.—There are many evidences that the process of adjustment of these crust-blocks is still going on. In the Sierra we find evidence in the still deepening channels of the rivers and especially in the occasional readjustment of the walls of the eastern fault of the Sierra block. . . . Finally, during a camp of two or three weeks in 1887 in Warner Mountains, where the structure described by Russell is finely displayed, I found abundant evidences of local subsidence still in progress. Many small lakes in that region, probably of the type produced by block-tilting have apparently been formed during the present century. In Blue Lake, for example, I found stumps of pines standing in water fifty or more feet deep, rotted off level with the surface, *but perfectly sound below.*

Two kinds of mountains.—I must not be understood as maintaining that the Sierra, the Wahsatch and the Basin ranges were entirely formed at the end of the Tertiary. Some of the Basin ranges, for example those of S.E. Oregon, were indeed wholly formed at that time and in the manner already explained. But the Sierra, the Wahsatch and many of the Basin ranges existed before that time. The Sierra was born from the sea by the folding and upswelling of thick sediments at *the end of the Jurassic.* Many ranges in the Basin region were formed at the same time and in the same way. The Wahsatch was similarly formed, probably about the end of the Cretaceous. But at the end of the Tertiary the greatly eroded land surface previously formed in this region was arched and broken and readjusted, forming these ranges *in their present condition*, as already explained.

In an article published in 1872,* entitled "Theory of the

---

formation of the greater features of the earth's surface," I showed that mountain ranges were formed by lateral pressure acting upon thick sediments folding and swelling up the mass along the line of yielding. In another article published in 1878,* I further developed the same views and tried to show that even the Basin ranges—claimed by Gilbert as belonging to a different type and formed in a different way—were no exception; that *they also* were formed by lateral pressure; only that in this case the crust of the earth being rigid would not yield by mashing, but only by arching —the blocks of the broken arch readjusting themselves to form the orographic features already described; and therefore that mountain ranges ate all of one type and formed in one way, viz: by lateral pressure. I now feel compelled to modify this statement. It is evident from the *character of the faults* i.e. normal instead of reverse faults, that the arch was not formed by lateral *pressure* but by *tension of lifting*. Therefore, I now believe that mountain ranges are of two types: (1.) Those formed by lateral crushing and folding, and (2.) those formed by adjustment of crust-blocks. The one produces reverse faults, the other normal faults. The best types of the one are the Appalachian, the Alps, and the Coast ranges of California; the best types of the other are the Basin ranges. Very often the two types are mixed, or one is superposed on the other—the one or the other predominating. This is the case with the Sierra, the Wahsatch, and to some extent with many of the Basin ranges.

---

* This Journal, vol. xvi, p. 95.

THOMSON

Sir William Thomson, Lord Kelvin (1824–1907), British physicist and
electrical engineer, was professor of natural philosophy at Glasgow University.
In addition to his brilliant work in pure science, he was responsible for the
research which made possible the construction of the first transatlantic cables.

On the Secular Cooling of the Earth

From *Transactions of the Royal Society of Edinburgh*, Vol. XXIII, pp. 157–169,
1864.

8. It must indeed be admitted that many geological writers
of the "Uniformitarian" school, who in other respects have taken
a profoundly philosophical view of their subject, have argued in a
most fallacious manner against hypotheses of violent action in
past ages. If they had contented themselves with showing that
many existing appearances, although suggestive of extreme
violence and sudden change, may have been brought about by
long-continued action, or by paroxysms not more intense than
some of which we have experience within the periods of human
history, their position might have been unassailable; and certainly
could not have been assailed except by a detailed discussion of their
facts. It would be a very wonderful, but not an absolutely
incredible result, that volcanic action has never been more violent
on the whole than during the last two or three centuries; but it is
as certain that there is now less volcanic energy in the whole earth
than there was a thousand years ago, as it is that there is less
gunpowder in a "Monitor" after she has been seen to discharge
shot and shell, whether at a nearly equable rate or not, for five
hours without receiving fresh supplies, than there was at the
beginning of the action. Yet this truth has been ignored or
denied by many of the leading geologists of the present day,
because they believe that the facts within their province do not
demonstrate greater violence in ancient changes of the earth's
surface, or do demonstrate a nearly equable action in all periods.

9. The chemical hypothesis to account for underground heat might
be regarded as not improbable, if it was only in isolated localities
that the temperature was found to increase with the depth; and,

indeed, it can scarcely be doubted that chemical action exercises an appreciable influence (possibly negative, however) on the action of volcanoes; but that there is slow uniform "combustion," "eremacausis," or chemical combination of any kind going on, at some great unknown depth under the surface everywhere, and creeping inwards gradually as the chemical affinities in layer after layer are successively saturated, seems extremely improbable, although it cannot be pronounced to be absolutely impossible, or contrary to all analogies in nature. The less hypothetical view, however, that the earth is merely a warm chemically inert body cooling, is clearly to be preferred in the present state of science.

10. Poisson's celebrated hypothesis, that the present under-ground heat is due to a passage, at some former period, of the solar system through hotter stellar regions, cannot provide the circum-stances required for a palaeontology continuous through that epoch of external heat. For from a mean of values of the conductivity, in terms of the thermal capacity of unit volume, of the earth's crust, in three different localities near Edinburgh, which I have deduced from the observations on underground temperature instituted by Principal Forbes there, I find that if the supposed transit through a hotter region of space took place between 1250 and 5000 years ago, the temperature of that supposed region must have been from 25° to 50° Fahr. above the present mean temperature of the earth's surface, to account for the present general rate of underground increase of temperature, taken as 1° Fahr. in 50 feet downwards. Human history negatives this supposition. Again, geologists and astronomers will, I presume, admit that the earth cannot, 20,000 years ago, have been in a region of space 100° Fahr. warmer than its present surface. But if the transition from a hot region to a cool region supposed by Poisson took place more than 20,000 years ago, the excess of temperature must have been more than 100° Fahr., and must therefore have destroyed animal and vege-table life. Hence, the farther back and the hotter we can suppose Poisson's hot region, the better for the geologists who require the longest periods; but the best for their view is Leibnitz's theory, which simply supposes the earth to have been at one time an incandescent liquid, without explaining how it got into that state. If we suppose the temperature of melting rock to be about 10,000° Fahr. (an extremely high estimate), the consolidation may have taken place 200,000,000 years ago. Or, if we suppose the tempera-

ture of melting rock to be 7000° Fahr. (which is more nearly what it is generally assumed to be), we may suppose the consolidation to have taken place 98,000,000 years ago.

11. These estimates are founded on the Fourier solution demonstrated below. The greatest variation we have to make on them, to take into account the differences in the ratios of conductivities to specific heats of the three Edinburgh rocks, is to reduce them to nearly half, or to increase them by rather more than half. A reduction of the Greenwich underground observations recently communicated to me by Professor EVERETT of Windsor, Nova Scotia, gives for the Greenwich rocks a quality intermediate between those of the Edinburgh rocks. But we are very ignorant as to the effects of high temperatures in altering the conductivities and specific heats of rocks, and as to their latent heat of fusions. We must, therefore, allow very wide limits in such an estimate as I have attempted to make; but I think we may with much probability say that the consolidation cannot have taken place less than 20,000,000 years ago, or we should have more underground heat than we actually have, nor more than 400,000,000 years ago, or we should not have so much as the least observed underground increment of temperature. That is to say, I conclude that LEIBNITZ's epoch of "emergence" of the "consistentior status" was probably between those dates.

---

24. How the temperature of solidification, for any pressure, may be related to the corresponding temperature of fluid convective equilibrium, it is impossible to say, without knowledge, which we do not yet possess, regarding the expansion with heat, and the specific heat of the fluid, and the change of volume, and the latent heat developed in the transition from fluid to solid.

25. For instance, supposing, as is most probably true, both that the liquid contracts in cooling towards its freezing-point, and that it contracts in freezing, we cannot tell, without definite numerical data regarding those elements, whether the elevation of the temperature of solidification, or of the actual temperature of a portion of the fluid given just above its freezing-point, produced by a given application of pressure, is the greater. If the former is greater than the latter, solidification would commence at the bottom, or at the center, if there is no solid nucleus to begin with, and would proceed outwards, and there could be no complete

permanent incrustation all round the surface till the whole globe is solid, with, possibly, the exception of irregular, comparatively small spaces of liquid.

26. If, on the contrary, the elevation of temperature, produced by an application of pressure to a given portion of the fluid, is greater than the elevation of the freezing temperature produced by the same amount of pressure, the superficial layer of the fluid would be the first to reach its freezing-point, and the first actually to freeze.

27. But if . . . the liquid expanded in cooling near its freezing-point, the solid would probably likewise be of less specific gravity than the liquid at its freezing-point. Hence the surface would crust over permanently with a crust of solid, constantly increasing inwards by the freezing of the interior fluid in consequence of heat conducted out through the crust. The condition most commonly assumed by geologists would thus be produced.

---

30. It is, however, scarcely possible, that any such continuous crust can ever have formed all over the melted surface at one time, and afterwards have fallen in. The mode of solidification conjectured in §25, seems on the whole the most consistent with what we know of the physical properties of the matter concerned. So far as regards the result, it agrees, I believe, with the view adopted as the most probable by Mr. HOPKINS.* But whether from the condition being rather that described in §26, which seems also possible, for the whole or for some parts of the heterogeneous substance of the earth, or from the viscidity as of mortar, which necessarily supervenes in a melted fluid, composed of ingredients becoming, as the whole cools, separated by crystallising at different temperatures before the solidification is perfect, and which we actually see in lava from modern volcanoes; it is probable that when the whole globe, or some very thick superficial layer of it, still liquid or viscid, has cooled down to near its temperature of perfect solidification, incrustation at the surface must commence.

31. It is probable that crust may thus form over wide extents of surface, and may be temporarily buoyed up by the vesicular character it may have retained from the ebullition of the liquid in some

---

* See his Report on "Earthquakes and Volcanic Action." British Association Report for 1847.

places, or, at all events, it may be held up by the viscidity of the
liquid; until it has acquired some considerable thickness sufficient
to allow gravity to manifest its claim, and sink the heavier solid
below the lighter liquid.   This process must go on until the sunk
portions of crust build up from the bottom a sufficiently close
ribbed solid skeleton or frame, to allow fresh incrustations to
remain bridging across the now small areas of lava pools or lakes.

32. In the honey-combed solid and liquid mass thus formed,
there must be a continual tendency for the liquid, in consequence
of its less specific gravity, to work its way up; whether by masses
of solid falling from the roof of vesicles or tunnels, and causing
earthquake shocks, or by the roof breaking quite through when
very thin, so as to cause two such hollows to unite, or the liquid
of any of them to flow out freely over the outer surface of the
earth; or by gradual subsidence of the solid, owing to the thermo-
dynamic melting, which portions of it, under intense stress, must
experience, according to views recently published by my brother,
Professor JAMES THOMSON.*   The results which must follow from
this tendency seem sufficiently great and various to account for
all that we see at present, and all that we learn from geological
investigation, of earthquakes, of upheavals and subsidences of
solid, and of eruptions of melted rock.

33. These conclusions, drawn solely from a consideration of
the necessary order of cooling and consolidation, according to
BISCHOF'S result as to the relative specific gravities of solid and of
melted rock, are in perfect accordance with what I have recently
demonstrated† regarding the present condition of the earth's
interior—that it is not, as commonly supposed, all liquid within a
thin solid crust of from 30 to 100 miles thick, but that it is on the
whole more rigid certainly than a continuous solid globe of glass
of the same diameter, and probably than one of steel.

---

* Proceedings of the Royal Society of London 1861, "On Crystallization
and Liquefaction as influenced by Stresses tending to Change of Form in
Crystals."

† In a paper "On the Rigidity of the Earth," communicated to the Royal
Society a few days ago.   April, 1862.

STERNBERG

Hermann Sternberg (1825–1885), German engineer, was professor of engineering science and head of the school of engineering in the Institute of Technology at Karlsruhe throughout the greater part of his life.

STREAM PROFILES

Translated by A. O. Woodford from "Untersuchungen über Längen- und Querprofil geschiebefuhrender Flüsse," *Zeitschrift für Bauwesen*, Vol. XXV, pp. 438–406, 1875.

Stream valleys and stream beds have their forms determined almost entirely by the work of running water. . . . Flowing water has the capacity, under favorable circumstances, to carry solid bodies (debris particles) with it, if the frictional resistance, which opposes their movement, is overcome by the impact of the water. With debris particles of similar surface forms, the resistance increases as the cube, the impact of the water only as the square, of their diameters. It follows that for a given water velocity all the debris of various sizes on the stream bed must fall into two groups, of which one moves with the current, the other is left behind. The pebbles or other particles left behind raise the stream bed until through increase of water or of gradient velocities are attained which will move the particles along. It is therefore clear that at any point on a stream bed the size of the debris particles corresponds to a certain velocity, as a rule the maximum velocity, of the water flowing over it. But as the debris moves along, wear and comminution take place according to definite laws dependent on the physical properties of the rock. In general, therefore, the maximum velocity of the water necessary for the progress of the bed load becomes less the farther the load has traveled from its place of production, *i.e.*, from rocky mountains. At a given point on a stream course the quantity of water discharged remains essentially unchanged, and extraordinarily long periods of time must have elapsed after the initiation of valley formation, so that there have developed the appropriate maximum water velocities, the sorting of the load according to particle size, the progress of the load along the bed, etc., in regularly continued or periodically recurring fashion, and thus all the characteristics of the stream are in equilibrium and mutually determined. That such equilibria prevail in actual streams according to general laws is shown in a striking fashion by their longitudinal profiles: as long as their beds lie in alluvium, the profiles are lines concave upward, which become ever steeper toward the source regions. The lines are broken at places where a stream bed is for any reason less mobile than the load which is being carried over it, whether the obstruction be bedrock or hardened and cemented debris masses. Below such irregularities—breaks in slope and stream rapids—the proper curve of the long profile is again continued.

Knowledge of the laws prevailing [in debris transport] is of general value. It will be especially useful to the hydraulic engineer. . . to be able to predict the changes of long profile which must occur as the result of stream control works which alter the properties of the stream and especially the width of the water surface. The lowering of the water surface and the deepening of the stream bed, almost invariably observed in the course of a few years after the completion of control works, undoubtedly belong here.

HUXLEY

Thomas Henry Huxley (1825–1895), English naturalist and expositor of the Darwinian theory of evolution, was extraordinarily effective in developing an understanding of geological and biological science among people who were not professional workers in those fields. The quotation is from "a course of six lectures to working men, delivered in the theatre of the Royal School of Mines."

The Geological History of the Horse

From *Nature*, Vol. XIV, pp. 33–34, 1876.

In the highest group of Vertebrates, the Mammalia, the perfection of animal structure is attained. It will hardly be necessary, indeed it will be impossible in the time at our disposal, to give the general characters of the group, but our purpose will be answered as well by devoting a short time to considering the peculiarities of a single well-known animal, the evidence as to the origin of which approaches precision.

The horse is one of the most specialised and peculiar of animals, its whole structure being so modified as to make it the most perfect living locomotive engine which it is possible to imagine. The chief points in which its structure is modified to bring about this specialisation, and in which, therefore, it differs most markedly from other mammals, we must now consider.

In the skull the orbit is completely closed behind by bone, a character found only in the most modified mammals. The teeth have a very peculiar character. There are, first of all, in the front part of each jaw, six long curved incisors or cutting teeth, which present a singular dark mark on their biting surfaces, caused by the filling in of a deep groove on the crown of each tooth, by the substances on which the animal feeds. After the incisors, comes on both sides of each jaw a considerable toothless interval, or *diastema*, and then six large grinding teeth, or molars and premolars. In the young horse a small extra premolar is found to exist at the hinder end of the diastema, so that there are, in reality, seven grinders on each side above and below; furthermore, the male horse has a tusk-like tooth, or canine, in the front part of the diastema immediately following the last incisor. Thus the horse has, on each side of each jaw, three incisors, one canine, and seven grinders, making a total of forty-four teeth.

It is, however, in the limbs that the most striking deviation from the typical mammalian structure is seen, the most singular modifications having taken place to produce a set of long, jointed levers, combining great strength with the utmost possible spring and lightness.

The humerus is a comparatively short bone inclined backwards: the radius is stout and strong, but the ulna seems to be reduced to its upper end—the olecranon or elbow; as a matter of fact, however, its distal end is left, fused to the radius, but the middle part has entirely disappeared: the carpus or wrist—the so-called "knee" of the horse—is followed by a long "cannon-bone," attached to the sides of which are two small "splint-bones"; the three together evidently represent the metacarpus, and it can be readily shown that the great cannon-bone is the metacarpal of the third finger, the splint-bones those of the second and fourth. The splint-bones taper away at their lower ends and have no phalanges attached to them, but the cannon-bone is followed by the usual three phalanges, the last of which, the "coffin-bone," is ensheathed by the great nail or hoof.

The femur, like the humerus, is a short bone, but is directed forwards; the tibia turns backwards, and has the upper end of the rudimentary fibula attached to its outer angle. The latter bone, like the ulna, has disappeared altogether as to its middle portion, and its distal end is firmly united to the tibia. The foot has the same structure as the corresponding part in the fore-limb—a great cannon-bone, the third metatarsal; two splints, the second and fourth; and the three phalanges of the third digit, the last of which bears a hoof.

Thus, in both fore and hind limb one toe is selected, becomes greatly modified and enlarged at the expense of the others, and forms a great lever, which, in combination with the levers constituted by the upper and middle divisions of the limb, forms a sort of double C-spring arrangement, and thus gives to the horse its wonderful galloping power.

In the river-beds of the Quaternary age—a time when England formed part of the Continent of Europe—abundant remains of horses are found, which horses resembled altogether our own species, or perhaps are still more nearly allied to the wild ass.

The same is the case in America, where the species was very abundant in the Quaternary epoch—a curious fact, as, when first discovered by Europeans, there was not a horse from one end of the vast continent to the other.

In the Pliocene and older Miocene, both of Europe and America, are found a number of horse-like animals, resembling the existing horse in the pattern and number of the teeth, but differing in other particulars, especially the structure of the limbs. They belong to the genera *Protohippus*, *Hipparion*, &c., and are the immediate predecessors of the Quaternary horses.

In these animals the bones of the fore-arm are essentially like those of the horse, but the ulna is stouter and larger, can be traced from one end to the other, and, although firmly united to the radius, was not ankylosed with it. The same is true, though to a less marked extent, of the fibula.

But the most curious change is to be found in the toes. The third toe though still by far the largest, is proportionally smaller than in the horse, and each of the splint bones bears its own proper number of phalanges; a pair of "dew-claws," like those of the reindeer, being thus formed, one on either side of the great central toe. These accessory toes, however, by no means reached the ground, and could have been of no possible use, except in progression through marshes.

The teeth are quite like those of the existing horse, as to pattern, number, presence of cement, &c.; the orbit also is complete, but there is a curious depression on the face-bones, just beneath the orbit, a rudiment of which is, however, found in some of the older horses.

On passing to the older Miocene, we find an animal, known as *Anchitherium*, which bears, in many respects, a close resemblance to Hipparion, but is shorter-legged, stouter-bodied, and altogether more awkward in appearance. Its skull exhibits the depression mentioned as existing in Hipparion, but the orbit is incomplete behind, thus deviating from the specialised structure found in the horse, and approaching nearer to an ordinary typical mammal. The same is the case with the teeth, which are short and formed roots at an early period; their pattern also is simplified, although all the essential features are still retained. The valleys between the various ridges are not filled up with cement, and the little anterior premolar of the horse has become as large as the other grinders, so that the whole forty-four teeth of the typical mammalian dentition are well developed. The diastema is still present between the canines and the anterior grinding teeth—a curious fact in relation to the theory that the corresponding space in the horse was specially constructed for the insertion of the bit; for, if the Miocene men were in the habit of riding the Anchitherium, they were probably able to hold on so well with their hind legs as to be in no need of a bit.

The fibula is a complete bone, though still ankylosed below to the tibia; the ulna also is far stouter and more distinct than in Hipparion. In both fore and hind foot the middle toe is smaller, in relation to the size of the animal, than in either the horse or the Hipparion, and the second and fourth toes, though still smaller than the third, are so large that they must have reached the ground in walking. Thus, it is only necessary for the second and fourth toes, and the ulna and fibula to get smaller and smaller for the limb of Anchitherium to be converted into that of Hipparion, and this again into that of the horse.

Up to the year 1870 this was all the evidence we had about the matter, except for the fact that a species of Palaeotherium from the older Eocene was, in many respects, so horse-like, having, however, well-developed ulna and fibula, and the second and fourth toes larger even than in Anchitherium, that it had every appearance of being the original stock of the horse. But within the last six years some remarkable discoveries in central and western North America, have brought to light forms which are, probably, nearer the direct line of descent than any we have hitherto known.

In the Eocene rocks of these localities, a horse-like animal has been found, with three toes, like those of Anchitherium, but having, in addition, a little style of bone on the outer side of the fore foot, evidently representing the fifth digit. This is the little Orohippus, the lowest member of the Equine series.

It may well be asked why such clear evidence should be obtainable as to the origin of mammals, while in the case of many other groups—fish, for instance—all the evidence seems to point the other way. This question cannot be satisfactorily answered at present, but the fact is probably connected with the great uniformity of conditions to which the lower animals are exposed, for it is invariably the case that the higher the position of any given animal in the scale of being, the more complex are the conditions acting on it.

. . . The accurate information obtained in this department of science has put the *fact* of evolution beyond a doubt; formerly, the great reproach to the theory was, that no support was lent to it by the geological history of living things; now, whatever happens, the fact remains that the hypothesis is founded on the firm basis of palaeontological evidence.

SORBY

Henry Clifton Sorby (1826–1908), English geologist and petrographer, was president of Firth College in his native Sheffield. Building on the foundations laid earlier by William Nicol, he "opened out a new and vast field for petrographical investigation . . . by means of the microscope." (Geikie)

On the Origin of Slaty Cleavage

From *New Philosophical Journal*, Edinburgh, Vol. LV, pp. 137–148, 1853.

For several years I have devoted myself entirely to investigating the physical structure of rocks, both on a large scale, as seen in the field, and by preparing sections of extreme thinness, capable of being examined with the highest powers of the microscope. This latter subject has hitherto attracted little or no attention, though the inspection of two or three thin sections will sometimes solve most important geological problems. Amongst other branches of the study, I have applied this method of research to ascertain the origin of slaty cleavage, which, being obviously due to some peculiarity of structure, I thought might, in all probability, be solved by that means. The examination of thin sections of slate rocks with high powers, and a comparison with those of similar mineral composition not possessing cleavage, have led me to form a theory to account for their difference of structure, materially different from any yet propounded, and which, in my opinion, not only does so most satisfactorily, but also explains perfectly every fact that I am acquainted with, connected with the subject. To enter fully into the whole would require a long treatise, and I shall therefore, on the present occasion, merely give a short outline of my general conclusions.

Professor Phillips and Mr. Daniel Sharpe have shewn that the organic remains found in slate rocks indicate a change of their dimensions; and it was their observations which first led me to test the mechanical theory, as applied to explain the microscopical structure. I am fully prepared to substantiate their observations, and have also ascertained a number of other facts, proving, in an equally conclusive manner, that slate rocks have undergone a great change in their mechanical dimensions, which change is invariably related to the direction and intensity of the cleavage, and is such that the cleavage lies in the line of greatest elongation, and in a plane perpendicular to that of greatest compression.

A most careful examination of very numerous contortions of the beds in slate rocks, in North Wales and Devonshire, has led me to conclude that they indicate a very considerable amount of lateral pressure, the thickness of the contorted beds being very different in one part to what it is in another. . . .

In slaty rocks of very mixed structure,—as for instance some in the north of Devonshire,—the greatly contorted beds are those which have only an indistinct or imperfect cleavage, and are of such a nature as not to have so readily undergone a change of dimensions as beds above and below them. I have frequently seen cases where such beds are contorted, so as to indicate a very great amount of lateral pressure and change of dimensions, whilst the finer beds just above and below them are most distinctly seen not to have been contorted at all. . . .

If a thin section of a rock not having cleavage be examined, which has a similar mineral composition to those which, when having it, form good slates, it will be seen that the arrangement of the particles is very different. For instance, the well-known Water of Ayr stone has no cleavage, but shews more or less of bedding. It consists of mica and a very few grains of quartz sand, imbedded in a large proportion of decomposed feldspar; the peroxide of iron being collected to certain centres, and having the characters of peroxidised pyrites. The flakes of mica do not lie in the plane of bedding, but are inclined tolerably evenly at all angles, so that there is no definite line of structural weakness, independent of that due to bedding; which results chiefly from alternations of layers of somewhat different composition, and not from the arrangement of the ultimate particles. This is however totally different in a rock of similar composition having cleavage. If a section be examined, cut perpendicular to cleavage, in the line of its dip, it will be seen that though some of the minute flakes of mica lie perpendicular to the cleavage or at high angles to it, by far the larger part are inclined at low, so that the majority lie within 20° on each side of it. In fact they are most numerous nearly in the plane of cleavage, and gradually but rapidly diminish in quantity in passing to higher angles, so that there are twenty times as many nearly in the plane of cleavage as to 45° to it, and

very few at 90°. Where a section is examined, cut perpendicular
to cleavage, in the line of the strike, it is seen that the arrangement
is similar, but there is not near so rapid a diminution of the members in passing from the line of cleavage, so that there are comparatively several times as many more inclined at about 45° to it,
than when the section is in the line of dip, and those at still higher
angles are also much more numerous. In a section in the plane
of cleavage, but few flakes are cut through so as to have a greatly
unequiaxed form; but they are similarly arranged with respect to
the line of dip, though not in so marked a manner. It is not
merely the larger flakes of mica that are thus arranged, but the
whole of those unequiaxed particles which existed in the rock
before the cleavage was developed.

When a cleavage crack in the thin sections is examined, it is
clearly seen that the cleavage is due to the above described
arrangement of the particles, which it follows most perfectly;
not passing straight forwards, but turning about according to the
manner in which the ultimate particles lie in every part. It
therefore appears that the fissile character of slate is due to a line
of structural weakness, brought about by the manner or arrangement of the ultimate, unequiaxed particles. The natural cleavage
cracks, of course, bear the same relation to this arrangement
as those so often seen in many crystalline bodies do to that of their
ultimate atoms. They appear, in general, to have been mainly
due to meteoric agencies; their position having been determined
by the structural weakness. In accounting, then, for so-called
slaty cleavage, it is only requisite to shew how such particles could
have had their position so changed that their arrangement should
be altered from that found in rocks not having cleavage to that in
those having it; which explanation must of course be such as would
agree with every other fact connected with the subject.

. . . Having mixed some scales of oxide of iron with soft
pipeclay, in such a manner that they would be inclined evenly
in all directions, like the flakes of mica in Water of Ayr stone, I
changed its dimensions artificially to a similar extent to what has
occurred in slate rocks. Having then dried and baked it, I rubbed
it to a perfect flat surface, in a direction perpendicular to pressure
and in the line of elongation, which would correspond to that of
dip of cleavage, and also, as it were, in its strike, and in the plane
of cleavage. The particles were then seen to have become

arranged in precisely the same manner as theory indicates that they would, and as is the case in natural slate, so much so, that, so far as their arrangement is concerned, a drawing of one could not be distinguished from that of the other. Moreover, it then admitted of easy fracture into thin flat pieces in the plane corresponding to the cleavage of slate, whereas it could not in that perpendicular to it. Even in clay which has but few very unequiaxed particles, a most distinct lamination is produced by changing its dimensions, as described above, but it would not cleave perfectly, no more than will natural slate of similar mineral composition, and moreover one cannot obtain their firm, uniform structure.

It is a fact well worthy of remark, that, on each side of the larger rounded grains of mica, in the line of cleavage, in well-cleaved slates, the particles are arranged evenly at all angles, over small triangular spaces, having their bases towards the grain. This is just the part which would be protected from change of dimensions by its presence; and this fact is therefore very good evidence of the slate having had originally such a structure as would be changed into its present, if its dimensions had been altered in the manner and to the extent indicated by the breaking up of other rounded grains of mica seen in the same thin section.

What I therefore contend is, that there is abundance of proof that slate rocks have suffered such a change of dimensions, as would necessarily alter the arrangement of their ultimate particles from what is found in rocks not having cleavage to that in those which have, and hence develop a line of structural weakness in the direction in which it really does occur.

Some slates have a very poor cleavage, although their mineral composition is similar to that of such as often have a most perfect. In these the green spots indicate a comparatively small change of dimensions; and in others having no cleavage, the contortions and spots shew that little or none has occurred. Whence it should appear that the perfection of cleavage depends both upon the ultimate mineral composition, and the amount of change of dimensions of the rock. . . .

Perhaps it may be said, How is it possible that hard rocks could have had their dimensions changed to the extent described? To this I would reply, If the rocks be examined, it will be seen that it really has occurred, and I would suggest that solidity is but a comparative property, and that the intensity of the forces in action during the elevation of a range of mountains, could gradually change the dimensions of rocks; for it is well known that many hard and brittle substances will admit of such movements, as for instance the ice of glaciers, and hard and brittle pitch.

. . . Mr. Sharpe's theory, of course, only differs from mine in his assuming that the particles have been really *compressed;* whereas I am persuaded, that in general they have only suffered a *change of position.*

On the Microscopical Structure of Crystals, Indicating the Origin of Minerals and Rocks

From *Quarterly Journal of the Geological Society of London,* Vol. XIV, pp. 453–500, 1858.

In this paper I shall attempt to prove that artificial and natural crystalline substances possess sufficiently characteristic structures to point out whether they were deposited from solution in water or crystallized from a mass in the state of igneous fusion; and also that in some cases an approximation may be made to the rate at, and the temperature and pressure under which they were formed.

The various facts described above [in discussion of the structure of artificial crystals] will, I think, warrant the following general conclusions:—

1. Crystals possessing only cavities containing water more or less saturated with various salts were formed by being deposited from solution in water.

2. The relative size of the vacuities in normal fluid-cavities depends on the temperature and pressure at which the crystals were formed, and may in some cases be employed to determine the actual or relative temperature and pressure.

3. Crystals containing only glass- or stone-cavities were formed by being deposited from a substance in the state of igneous fusion.

4. Crystals containing only gas- or vapour-cavities were formed by sublimation or by the solidification of a fused homogeneous substance, unless they are fluid-cavities that have lost all their fluid.

5. Other circumstances being the same, crystals containing few cavities were formed more slowly than those containing more.

6. Crystals possessing fluid-cavities containing a variable amount of [minute] crystals, and gradually passing into gas-cavities, were formed under the alternate presence of the liquid and a gas.

7. Crystals in which are found both cavities containing water and cavities containing glass or stone were formed, under great pressure, by the combined action of igneous fusion and water.

8. Crystals having the characters of 6 and 7 combined were formed, under great pressure, by the united action of igneous fusion and water alternating with vapour or a gas, so as to include all the conditions of igneous fusion, aqueous solution, and gaseous sublimation.

Such then are the general principles I purpose to apply in investigating the origin of minerals and rocks. It will be perceived at once that, in one way or other, they may be brought to bear on almost every branch of physical and chemical geology. In this communication I shall illustrate the subject by applying them to some of the leading branches of inquiry, without attempting to treat each in a complete manner.

By many experiments, I have proved most conclusively that the fluid in the cavities in the quartz of granites and elvans is *water*, holding in solution the chlorides of potassium and sodium, the sulphates of potash, soda, and lime, sometimes one, and sometimes the other salt predominating. Since the solution has often a most decided acid reaction before, or even after, having been evaporated to dryness, there must be an excess of the acids present. This occurrence of *free* hydrochloric and sulphuric acid is, I think, a very interesting fact, when we bear in mind how very characteristic they are of modern volcanic activity. . . .

On the whole, then, the microscopical structure of the constituent minerals of granite is in every respect analogous to that of those formed at great depths and ejected from modern volcanos, or that of the quartz in the trachyte of Ponza, as though granite had been formed under similar physical conditions, combining at once both igneous fusion, aqueous solution, and gaseous sublimation. The proof of the operation of water is quite as strong as of that of heat; and, in fact, I must admit, that in the case of coarse-grained, highly quartzose granites there is so very little evidence of igneous fusion, and such overwhelming proof of the action of water, that it is impossible to draw a line between them and those veins where, in all probability, mica, felspar, and quartz have been deposited from solution in water, without there being any definite genuine igneous fusion like that in the case of furnace slags or erupted lavas.   There is, therefore, in the microscopical structure a most complete and gradual passage from granite to simple quartz-veins; and my own observations in the field cause me to entirely agree with M. Élie de Beaumont (Note sur les émanations vol-caniques et métallifères, Bulletin de la Société Géologique de France, 2 série, t. iv. p. 1249) in concluding that there is also the same gradual passage on a large scale.

In my opinion, the water associated with thoroughly melted igneous rocks at great depths does not dissolve the rock, but the rock dissolves the water, either chemically as a hydrate, or physically as a gas.   In the case of those obsidians and pitchstones which, when heated to redness, give off water having a strong acid reaction, it may probably be in the form of a hydrate, retaining its water when heated under pressure.   It is also sufficiently probable that, as suggested by M. Angelot (Bulletin de la Société Géologique de France, 1 sér. t. xiii. p. 178), fused rock, under great pressure, may dissolve a considerable amount of the vapour of water, in the same manner as liquids dissolve gases.   In either case, if the fused rock passed by gradual cooling into anhydrous crystalline compounds, the water would necessarily be set free; and, if the pressure was so great that it could not escape as vapour, an intimate mixture of partially melted rock and liquid water would be the result.   It is difficult to form any very definite opinion as to the actual amount of this water, and to decide whether or no it exercised an important influence over the crystalline processes that took place during the consolidation of such rocks as granite.   The comparatively large quantity of alkaline chlorides and sulphates, dissolved in these portions caught up in the growing crystals, indicates that the amount cannot have been *unlimited;* but, bearing in mind the facts I alluded to when describing the fluid-cavities in the blocks ejected from modern volcanos, and knowing, as we do, that the action of highly heated water is so very energetic, I cannot think that its influence was unimportant. On the contrary, seeing that the fluid-cavities in the quartz of quartz-veins contain the selfsame salts and acids as those in the granite, as though it had been deposited from portions of the liquid which had passed from the granite up fissures, I think the amount, though limited, must nevertheless have been *considerable*, and that its presence will serve to account for the connexion between granite and quartz-veins, and the very intimate relation of both to the metamorphic rocks, and explain many peculiarities in the arrangement of the minerals in the cavities in granite or in the solid rock, even if it was not the effective cause of their elimination and crystallization.

With respect, then, to minerals and rocks formed at a high temperature, my chief conclusions are as follows. At one end of the chain are erupted lavas, indicating as perfect and complete fusion as the slags of furnaces, and at the other end are simple quartz veins, having a structure precisely analogous to that of crystals deposited from water. Between these there is every connecting link, and the central link is granite. When the water intimately associated with the melted rock at great depths was given off as vapour whilst the rock remained fused, the structure is analogous to that of furnace slags. If, however, the pressure was so great that the water could not escape as vapour, it passed as a highly heated liquid holding different materials in solution up the fissures in the superincumbent rocks, and deposited various crystalline substances to form mineral veins. It also penetrated into the stratified rocks, heated, sometimes for a great thickness, to a high temperature, and assisted in changing their physical and chemical characters, whilst that remaining amongst the partially-melted igneous rock served to modify the crystalline processes which took place during its consolidation. These results are all derived from the study of the microscopical structure of the crystals; but my own observations in the field lead me to conclude that they agree equally well with the general structure of the mountains themselves, and serve to account for facts that could not have been satisfactorily explained without the aid of the microscope.

THIN SECTIONS AND THEIR VALUE TO THE GEOLOGIST

Translated from *Bulletin de la Société géologique de France*, Second Series,
Vol. XVII, pp. 571–573, 1859–60.

The work about which I have the honor of speaking to the Geological Society of France is not yet completely finished, but at the moment I wish to call attention to certain important facts to which the microscopical examination of thin slices of rocks leads.

It is known that the microscopic structure of shells, teeth. and fossil woods has been studied by many eminent observers. For this, they have prepared plates thin enough to become transparent and have examined them with microscopes of great magnifying power. It is the same method which I have applied to the study of the physical structure of all rocks, and it has led me to many new and interesting results.

To prepare the thin plates, one side of small fragments of rocks is ground smooth; first emery and a zinc tray are used, then a very smooth stone until the surface of the rock is perfectly flat and polished. Then this rock is fixed on a glass by this surface, by means of Canada balsam, and the opposite side is worn down with emery on various stones until its thickness is reduced to $\frac{1}{100}$ or $\frac{1}{1000}$ of an inch. After these operations it is well polished on both sides and sufficiently transparent to be examined with a microscope magnifying several hundred times. Another glass plate is then placed on top with Canada balsam, so that it is protected and more transparent.

The study of thin plates of rocks constitutes an entirely distinct branch of geology and demands additional knowledge. In the ten years since I began this, I have prepared many hundreds of these plates, and nevertheless much remains to be done. If I wished to describe all the results to which their examination led, I should have to go into each branch of geology; I shall content myself at present, therefore, with some remarks on the conclusions to which I have been drawn.

The method of investigation with a microscope is very useful for the study of limestones; for the organic structure of shells and debris of mollusks which compose them may be easily recognized; often it is perfectly preserved. Thus one can determine not only the nature of the particles which form the limestone, but even their relative proportion. It proves also that limestones which, to the naked eye, appear the same nevertheless differ completely. Some are composed of fragments of shells and corals, and present mechanical mixtures analogous to sands. Others, on the contrary, are formed of microscopic particles resulting from the complete decomposition of the shells or corals, and they are analogous to clays.

We can also render a very good account of the constitution of shales and find out not only the substances which compose them but also their mode of arrangement. The slates are seen to present two types of very different cleavage: the one resulting from the compression of a rock that has acted like plastic matter, the other arising from a system of fractures and compressed crevices which prove that the rock has yielded to pressure as a somewhat rigid material. By the study of the structure of the rock, one can even state its physical condition at the time when it was submitted to the force which dislocated the crust of the earth.

These researches throw much light on the structure of the metamorphic rocks and allow us to appreciate the changes that they have undergone.

The ancient and modern igneous rocks alike present remarkable peculiarities, and we can realize that sometimes there are many more minerals there than could possibly be found with the naked eye; we even distinguish those which have been formed at the moment the melted rock solidified from those resulting from a later action of water.

The granitic rocks particularly present many curious facts; an immense number of cavities are distinguished there, which enclose water as well as saline solutions; and these substances must have been in a liquid state in these rocks at the time they were formed. The cavities containing these fluids are like those known in quartz; only they are too small to be distinguished by unaided sight. With a very great magnification it is possible to see them, however, with greatest perfection; in the quartz of granite and some metamorphic rocks they are sometimes so numerous that there are more than one thousand million in a cubic inch.

FOUQUÉ and MICHEL-LÉVY

Ferdinand André Fouqué (1828–1904), French geologist and mineralogist, professor in the Collège de France.

Auguste Michel-Lévy (1844–1908), French mining geologist and mineralogist, collaborator with Fouqué in many important studies in petrography and volcanology.

THE ARTIFICIAL PRODUCTION OF FELDSPARS

Translated from *Comptes rendus des séances de l'Académie des sciences*, Vol. LXXXVII, pp. 700–702, 1878.

We have the honor of presenting to the Academy the result of our first work on the artificial reproduction of feldspars, with the aid of a process practically identical with that which gave birth to the crystallization of the same minerals in the eruptive rocks poured out at high temperature without notable intervention of modifying volatile elements.

In a platinum crucible, in the Schloesing furnace, at a temperature near that of the fusion of platinum, we melt either natural porphyritic feldspars or artificial mixtures of the chemicals that constitute them: silica and alumina in the state of dried chemical precipitates, melted carbonates of sodium and potassium, calcined calcium carbonate. In the two methods of operation, the results are identical.

Of all the feldspars, oligoclase is the most fusible, then come labradorite, albite, orthoclase, and microcline, and finally anorthite, which is most refractory. This order of fusibility, which is the same for the artificial mixtures, led us to work first with oligoclase, labradorite, and albite; our experiments on orthoclase and anorthite are not yet finished.

Once the mixture is melted and transformed into a homogeneous material which, by sudden cooling, gives an isotropic glass, we take it quickly to a Bunsen burner equipped with a blast and leave it for forty-eight hours at a temperature as little below that of fusion as possible. After this delay, we let the crucible cool without other precaution.

In the fused state, the material occupies a small space, for example, a fourth of the capacity of a ten-gram crucible; during the heating in the blast, the material generally inflates and forms a bulbous and voluminous mushroom of porcelain appearance. Under a lens, its crystalline nature is barely suspected; but the examination, with a polarizing microscope by parallel light, of the thin plates cut through different sections with a bow saw, allows recognition that there has been crystallization by and large, throughout. . . .

Summarizing, we have obtained the above feldspars in simple conditions and rather close to purely igneous fusion, with the chemical composition, the optical properties, and even the structure that they generally affect in a great number of eruptive rocks.

Thus the feldspars crystallize with so much facility that they should have been obtained and recognized frequently in laboratory experiments. The failure to use the microscope and the uncertainty which has long prevailed over the optical properties of feldspars in parallel polarized light explain how they have been able to pass unperceived. However, the easy crystallization of the feldspars was to have been foreseen, considering their extreme abundance in eruptive rocks and in such varied conditions as they are observed.

THE ARTIFICIAL PRODUCTION OF A CRYSTALLINE IGNEOUS ROCK

Translated from *Comptes rendus des séances de l'Academie des sciences,* Vol. LXXXVII, pp. 779–781, 1878.

Our last experiments were on a mixture of natural, porphyritic labradorite and augite ($\frac{3}{4}$ labradorite, $\frac{1}{4}$ augite). This mixture, melted first into a black, entirely amorphous glass, has been submitted to a reheating prolonged for seventy-two hours at a temperature lower than that of the fusion of the experimental material, which is rather low. We have obtained a crystalline rock identical with one of the most widespread types of natural volcanic rock. The product in question is, for example, so much like the common, olivine-free lavas of Etna that, even under the microscope, it would be impossible to distinguish the slightest difference between the artificial product and the natural volcanic materials. . . .

Finally, the crystallized magma resulting from our experiment shows still another mineral in crystallized state: ferric oxide, which appears in the form of cubes and regular octahedrons, and which, as in the natural rocks, has crystallized before the pyroxene and the feldspar.

Although the experimental material has crystallized fairly generally throughout, there remain some interstices between the crystals filled with amorphous matter.

Nothing is lacking then in the similitude of our product with the augitic labradorites [basalts] of modern volcanoes.

BREWER

William Henry Brewer (1828–1910), American geologist and botanist, was professor of agriculture at Yale University, and was largely responsible for the establishment of the first agricultural experiment station in the United States. Naturally enough, he was a pioneer in the study of soils and fine-textured sediments.

The Deposition of Clay in Salt Water

From *American Journal of Science*, Third Series, Vol. XXIX, pp. 1–5, 1885.

I have carried on a long series of experiments on the sedimentation of clays, and the finer portions of soils and the pulverized rocks, mostly in tall precipitating flasks in which the materials were first agitated with the respective liquids and were then allowed to stand at rest under various conditions as to light, temperature, etc.

There is considerable difference as to details in the behavior of various clays in water. With some of them, if agitated and thoroughly diffused through the liquid and then allowed to stand at rest, the turbidity fades gradually and regularly in density from the bottom to the top, and the liquid gradually grows clearer until it becomes as clear as natural waters ever do. Usually however, and with the great majority of clays (if the water be pure enough), the deposition is in quite a different manner. After some time, it may be in a few hours or it may be only after some days, the suspended material disposes itself in layers or strata which are more or less obvious because of the different degrees of turbidity of the liquid. . . .

In the presence of mineral acids and various salts (and numerous substances not classed as salts or acids), the behavior of suspended clays is very different. If a small quantity of mineral acid, or some saline substance be added to muddy water, the strata described are either not formed at all, or are fewer and settle more quickly. If the quantity of dissolved material is sufficient, the clay curdles or flocculates and immediately falls to the bottom. If now the clear saline (or acid) liquid be decanted from the sediment, an equal quantity of distilled water be added and the mud again diffused by agitation, the clay again allowed to settle, the liquid again decanted when clear, and this process repeated over and over, the saltness (or acidity as the case may be) of the liquid growing less with each dilution, we may study the behavior of the same portion of clay in solutions of different (and known) degrees of strength, and by repetition with the same clay in the same vessel, over a sufficiently long period of time, we may imitate the conditions which take place in the erosion and transportation of muds by rivers and their ultimate deposit in the sea.

From such experiments we may say in a general way, that the more saline the suspending water, the more rapid the sedimentation, but the rapidity of precipitation is not directly proportionate to the quantity of salts dissolved. Reducing the saltness one-half does not double the time of precipitation, and the precipitation is comparatively rapid until the solution is very weak indeed. With some clays the precipitation is as complete in thirty minutes in sea water as in thirty months in distilled water. This completeness of precipitation refers to the actual clearing of the liquid, and not the rate of deposition of the first and heavier portions.

The behavior of these finer suspensions is analogous to that of a colloid. The diffusion through water is like that of a colloid, and when the finer portions are evaporated slowly and at low temperatures, they are at first very bulky and colloidal in appearance, shrinking enormously on drying into a mass curiously like some organic colloids. If the clay is not strictly colloidal, it is indeed very like it, and its behavior toward water very similar.

OCHSENIUS

Carl Ochsenius (1830–1906), German mining engineer and geologist, was long engaged in professional work in Chile. His theory of the origin of extensive deposits of rock salt is today one of the widely accepted principles of geology.

THE DEPOSITION OF ROCK SALT

From *Proceedings of the Academy of Natural Sciences*, Philadelphia, Vol. XL, pp. 181–187, 1888.

As is well known the ocean-water, from which all primitive rock-salt masses have been formed, contains on the average $3\frac{1}{2}\%$ fixed *i.e.* saline constituents, of which $2\frac{1}{2}\%$ is sodium chloride, the remainder consisting of magnesium compounds, calcium sulphate, potassium chloride, sodium bromide and small quantities of boron, iodine and lithium salts, as well as traces of every other element, of which indeed there exists one or the other compound, soluble in water and much more so in sea-water.

The open sea precipitates no salt, but in bays partially cut off from it, a deposition can take place under certain circumstances, in such a way that gypsum forms the base, and anhydrite the uppermost layer of the salt deposit; this is plainly seen in every large rock-salt bed. In considering the mode of formation of such deposits we are met on all sides by three questions, which hitherto have remained somewhat inexplicable:—1st the absence of fossils in the salt, whilst neighbouring rocks often contain them well preserved and in abundance, 2nd the small quantities of easily soluble magnesium and potassium salts, though they were contained in the sea-water and 3rd the replacement of these latter by one of the most insoluble constituents, viz. sulphate of lime in the form of a cap of anhydrite, the so-called *Anhydrithut*. These facts can, however, be explained, if we take a hydrographical element, viz. the bar, into account in the process of formation. When a nearly horizontally running bar cuts off a bay from the sea, so that only as much sea-water runs in over it as is compensated by evaporation from the surface of the lagoon, and the so partially separated portion receives no large additions of fresh—, *i.e.* rain or running water, a deposition of salt takes place in the way to be described.

In such a bay the following phenomenon may be observed:— The sea-water running in evaporates, and by the amount of salt it adds, the solid constituents of water, warmed by the sun, sink as they get specifically heavier from the larger amount of salt, and in the course of time, a vertical circulation setting in, the whole aqueous contents become enriched in saline matter and rise in temperature. The greater part of the deliquescent magnesium salts however remains in the upper layers while chloride of sodium is found preponderating below. As the saltness increases, organisms possessing free locomotive power, are compelled to seek a new habitat and make into the open sea against the currents and waves sweeping over the bar; those without free movements die off and generally leave only indistinct remains in the strata, which are next deposited. The formation of the latter commences with the precipitation of oxide of iron and carbonate of lime, as soon as the concentration has proceeded so far as to double the amount of saline matter in the lagoon and then ceases until the solution contains five times as much salts, when a second layer of carbonate of lime settles, this being brought about by a double decomposition between the soda and gypsum held in solution in producing calcium carbonate and sodium sulphate. At the same time gypsum begins to deposit and constitutes the basis proper. As soon as the saline solution has increased its weight of salts eleven times, its specific gravity reaches 1.22 and the precipitation of chloride of sodium begins in the form of the well known foliated crystalline masses, accompanied by some calcium sulphate etc., added from the sea-water running in.

Though generally speaking the sediments follow in reverse order of their solubilities, as Usiglio* has shown in his exhaustive experiments, it often happens that small quantities of easily soluble salts are mechanically included in the others; thus magnesium sulphate is frequently found contaminating rock-salt, and especially there, where clayey mud washed in, and was deposited at the same time. Then again some substance, only scantily represented in sea-water remains longer in solution than we should be led to expect from laboratory experiments. This is

---

* *Ann. Chim. et Phys.* 27, 172.

especially the case with borates, magnesium borate in particular, as well as with silica and titanic acid. As the depositing process continues, the greater part of the deliquescent salts remains dissolved in the upper layers and constitutes the mother-liquors (*Mutterlaugen*) which contain, along with sodium chloride, the potassium and magnesium compounds etc. We have then in the mother-liquors above the rock-salt, approximately arranged in order of solubility; sulphate of magnesium, chloride of potassium, chloride of magnesium, borates, bromides, lithium salts, an iodine compound probably magnesium iodide, and calcium chloride. In the course of the continued growth of the rock-salt beds and likewise of the mother-liquors, the latter attain the level of the bar and commence flowing out seawards directly over it, as soon as their specific gravity can overcome the current of the inflowing sea-water. After this stage is reached, ordinary sea-water can only have access through the upper portion of the bar-mouth, the lower part being occupied by the outgoing mother-liquors.

At this point the last stage of the process begins viz., the deposition of the uppermost bed of sulphate of calcium in the form of the *Anhydrithut*. Portions of the concentrated mother-liquors get mixed with surface-water washed in, and this, from the increased amount of the hygroscopic chlorides of magnesium and calcium, lessens the superficial evaporation of the bay, and hence the influx of sea-water diminishes gradually. The sulphate of lime in the sea-water that has flown in, is now precipitated, the other salts mixing with the mother-liquor and flowing out with them over the bar. As the gypsum falls through the concentrated mother-liquors, its water of combination gets abstracted, and a seam of anhydrite is by degrees deposited. Sometimes a compound is formed of gypsum with sulphates of magnesium and potassium (the latter by double decomposition of sulphate of magnesium and chloride of potassium) viz., polyhalite, a mineral occurring in the upper strata of many salt deposits. The bay meanwhile assumes the character of a bittern-lake and influences the surrounding shores, the organisms inhabiting the littoral waters dying off, and the neighbouring rock disintegrating to dust, which is blown into the lake, forming the material for the salt clay; this offers a good explanation for the increased thickness of the salt-clay seams often observed in the upper layers of salt deposits.

A regular succession of these briefly described phenomena will rarely be found in nature. Every alteration in the height of the bar, resulting from storms and other disturbances, naturally affects the precipitations about to take place, by accelerating or retarding them, or even redissolving some of the layers already *in situ*. In some cases where the *Anhydrithut* was never formed, the bar not having retained its original height long enough, the salt-clay plays the part of protecting covering; however, even under these circumstances the resulting series of deposits are so characterized as to point clearly to their mode of origin.

Salt beds deposited from aqueous solutions under the above-named conditions, are found in all geological epochs as far back as the Archaean rocks; this is shown by the superposition of Silurian strata to the salt in the Salt Range in India.* The existence of primitive salt beds points conclusively to the presence of shores, *i.e. terra firma* at the time of formation. At the present day the first of the above stated agents is found in operation in several localities on the East coast of the Caspian, especially in the great bay of Adschi Darja, whose narrow mouth, Kara-boghaz ("black abyss") is partially cut off from the Caspian by a bar. The bay is one of the saltiest of this inland sea, and receives no supplies of water at all from the land, only its evaporation being balanced by a corresponding influx of sea-water. . . .

To go back to the time when the first signs of the anhydrite cap make their appearance, we find that an increase in the altitude of the bar, sufficient to cut off the influx of sea-water, causes the mother-liquors to stagnate and under favorable conditions of temperature to solidify. Such a process has taken place in the Egeln-Stassfurt basin, and in some other localities of the old North-German Permian salt-sea.

---

* This is now considered an overthrust.—EDITORS.

SUESS

Eduard Suess (1831–1914), Viennese professor of geology, is justly famous for his monumental treatise, *Das Antlitz der Erde.*

On Mountain Structures and the Forces Which Form Them

Translated from *Die Entstehung der Alpen*, Vienna, 1875.

A majestic mountain range, the Alps, adorns the center of our continent. Every year groups of investigators compete in the study of its structure. But if one of them is asked how the Alps came to exist, he is forced to admit that although a great number of fragmentary structural plans have been developed with great conscientiousness in the last decade, opinions are still very divergent concerning the nature of the erective force.

It is true that with the progress of the observations, one has learned to measure on an ever greater scale the disturbance present in the original beds. But it is just as true that the Alps were not produced by a force different from that which produced the other mountain ranges of the earth's surface, and that consequently the investigation of the origin of their formation must be synonymous with the effort to ascertain the origin of the relief of the earth's surface in general, that is, insofar as the outer forms are not changed through the destructive or constructive influences of the atmosphere, water, ice, or other secondary phenomena.

The Alps divide toward the east into several mountain ranges, particularly the Carpathians and the central mountains of Hungary. The high hills of Croatia and the Dinaric Alps join these as similar chains. The northern end of the Apennines also merges indivisibly with the Lake Alps. To the northwest the Jura Mountains are extended in front of them like a fore wall. All these mountains from the Juras to the Apennines and to the Carpathians are likewise distinguished by the steady preponderance of certain lines of strike. I include all these branches and offshoots under the name, Alpine System.

The older mountains of the Balearic Islands, the eastern edge of the Central Plateau of France, the south end of the Vosges and the Black Forest, and the southern border of the Bohemian massif mark the western and northern edge of a wide area within which the folded ranges of the Alpine System are developed with wonderful regularity. They stretch their bends from one of these older mountains to another. As soon as the south point of the Bohemian massif is passed, the whole mountain swings to the northeast, following, in a gentle curve, the slope of the older Moravian mountain area, and then spreads out in the bow of the Carpathians. . . .

Thus from France to Poland, now more, now less, clearly, the outline and even the steepness of slope of the opposing older mountains are reflected in the structure and course of the northern border of the Jura, the East Alps, and the Carpathians. Thus is betrayed the resistance of these older masses against a force, acting here from the mountain chains and in a direction which does not vary essentially from the horizontal. . . .

From these first considerations, it appears ever more clearly that uniform motion of great masses in a horizontal direction has had a much more important influence on the present structure of the Alpine System than the formerly much overrated, vertical motion of certain parts, i.e. the distinct upheaval by a radial force acting from the interior of the planet to its surface.

If, in accordance with the preceding, not the outpushing of great central eruptive masses, but a more or less horizontal and regular universal movement, influenced in its action by the resistance of the older mountains, should be the origin of the uprising of our mountain chains, the further question at once arises, as to whether the source of this movement is to be sought within the individual mountain ranges, or whether some single, common origin is the source of this great phenomenon for all parts of the Alpine System, from the Apennines clear to the Carpathians. Surely the course of the chains favors the second hypothesis. All the enumerated chains show, in their trends, the efforts to form curves bent toward the northwest, north, or northeast. So marked a difference appears in their structure between the northern and southern flanks that hardly a doubt remains as to the homogeneity and identity of the motive force throughout. What might be called an outer and an inner side may be recognized for every one of the carefully studied mountain ranges. . . .

The twisted course of the various branches of the Alpine System, which gives way to a certain extent to obstacles, forbids our granting that the chains have been produced solely by the sinking of a widened Mediterranean basin, and by the outpush of the sinking edge. . . .

The resistance which the masses immediately to the northwest have opposed to the development of the Alpine System, the obstruction to the Jura Mountains of the Serre and the Black Forest, and the bowing of the East Alps at the south end of Bohemia have been mentioned at the close of the first chapter. It was noticed particularly that the south end of the Bohemian massif not only governs the altered trend of the mountains, but also deflects the course of the fault lines in the interior of the Calcareous Alps. In contrast to this, farther to the northeast, the folded outer chains of Bieskiden seem to be pushed over the flat coal basin of Ostrau, something, it might be said, like waves which break on a shallow strand. . . .

Perforce we look for certain great natural laws in the geographical arrangement of the mountain chains; perhaps, in final form, toward a geometric network and a definite age sequence. We soon see that the strike of the chains is in no way parallel to great circles, but is even deflected by obstacles. We recognize the frequent, if not exclusive, formation in geosynclines, the unity of great fold systems like those of the Alps, the subordinate role of volcanoes, and the persistence of mountain-building forces with their activity continuing through many geological epochs in the same great chains. After we have renounced a geometric system and admitted the one-sidedness of the movement, we find that a shifting toward the north predominates uniformly in numerous mountain ranges, whether considered young or old, from the Cordilleras clear to the Caucasus. We should like to evolve a law of movement of the upper earth mass toward the pole. Even this is wrong. Farther east there are some twists in direction, then the moving force turns toward the south in the mighty towering ranges of Inner Asia. Thus we have a picture of the face of our earth which does not correspond at all with our conception of orderly beauty, but, for that reason, does correspond all the more closely with the truth. . . .

The simplest form of mountain structure consists of a fracture running perpendicular to the direction of contraction, the advancement of broken-off pieces in the direction of contraction, and also, perhaps, the appearance of volcanic rocks in the fracture. Thus parts which mutually correspond can remain on both sides of the rift. . . .

The second, and most common, mountain form begins with the outline of a major fold striking transversely to the contraction and inclining in the direction of the contraction. Only then does the fracture follow in the fold along the line of greatest tension. Hereupon, by continuation of the same force as in the first case, the forward-lying part of the major fold moves farther in the direction of contraction, and piles up the sediments in front of it in wide irregular folds; while the part lying behind sinks down and volcanoes appear between its fragments. The Apennine and Carpathian branches of the Alpine System are examples.

It may happen that such a major fold, stopped in its forward movement by another mass, may be deflected and a succeeding great fold pressed against its inner side. Besides many other complications, this case, which occurs in the East Alps, seems to have as a concomitant that the arch of the first major fold remains more completely preserved than otherwise as a so-called anticlinal axis. Further, the various major groups, where their development is less restricted, may diverge and then fan out in one-sided branches. This is the relationship of the Hungarian mountain chains to the Alps; the partially anticlinal structure of the major fold loses itself, in this instance, in the Carpathians.

In a third form of mountain structure, instead of the development of a single or a few major folds, a great number of parallel folds are formed. These take in a great width, but normally end with a steep fault on the inner side of the innermost fold. The Jura Mountains, the mountains between the Taunus and the Belgian coal basin, and the Appalachians belong in this group. In these cases, volcanic eruptions are lacking at the faulted edge. It might be postulated that they are due to a motion of the shallower zones of the earth. However, in this as in the preceding cases, the long duration or the repeated occurrence of the same force is recognizable in various ways, and especially by folding of sediments which were once transgressively deposited over earlier folds.

STRUCTURAL FEATURES OF THE EARTH

Translation by Nevil N. Evans of a lecture delivered before the Natural History Society of Görlitz, *Canadian Record of Science*, Vol. VII, pp. 272–290, 1897.

We observe two phenomena through which the contraction is evident: either horizontal movement—that is, folding; or vertical contraction—that is, subsidence.

According to the predominance of one or other of these two movements, we see the surface of the earth laid out in long folds, as in the Alps and the Ural; or we have flat table-lands, as in the Sahara and Central Russia, or lines of subsidence, as in the Dead Sea, or whole regions depressed, as on the western side of the Apennines. . . .

From this it is, however, seen that the relief of the earth's surface does not by any means always correspond with the deeper structure. Therefore, in order to obtain a correct understanding of the facts, the structure, that is, the lines of folding or the lines of fault, must be kept in view. These lines are the determinants, not the relief. . . .

All these curves, beginning with the bend at Gibraltar, that is with the portion lying in Spain, I say all these curves from Gibraltar to Kamtschatka and the Aleutian Islands are distinguished by being folded towards the south. They produce with one another a curiously formed but very sharp boundary against the table-lands to the north of which they lie. To these table-lands belong all Africa south of the Atlas; Arabia with Palestine and Syria, and the peninsula of East India. . . .

Let us glance for a moment at other parts of the earth's surface. Both North and South America exhibit the remarkable phenomenon of being in the main folded towards the west, that is towards the Pacific Ocean. . . .

In the north of Venezuela series of folds occur which run from east to west and which reach their clearest expression in the contour of the Island of Trinidad. It seems that these folds find their continuation in Tobago and the Lesser Antilles. With approximate certainty one follows through the Lesser Antilles the trace of a mountain system which comes over from the Virgin Islands to Porto Rico, and in San Domingo splits into two parts.

One part finds its continuation in Jamaica and the other in Southern Cuba. . . .

If we now consider the main features of this picture [i.e. European structure], the whole middle and north of Europe is seen to consist of a series of folds or scales of the crust of the earth thrust one over the other towards the north in such a way that the northern folds are the oldest, that they were broken down, that this process of degradation was succeeded by new foldings from the south and that each time the new folds abutted against or were hemmed in by the horsts, that is, the projecting remains of the preceding folds.

READE

Thomas Mellard Reade (1832–1909), English geologist and civil engineer, is most widely known for his treatise on the *Origin of Mountain Ranges*, but his persistent attention to quantitative data and his emphasis upon the role played by solution in the process of erosion characterize his most significant contributions to the progress of geological science.

THE IMPORTANCE OF SOLUTION AS A FACTOR IN EROSION

From *American Journal of Science*, Third Series, Vol. XXIX, pp. 290–300, 1885.

My former calculations dealt almost exclusively with the amount of matter annually removed in river water from the surface of England and Wales, and from some of the river basins of Europe. I now propose laying before you calculations of a similar nature relating to some of the larger rivers of the two Americas. This done we shall be able to take a wider survey of the subject, and to ascertain how far the provisional generalizations to which previous investigations led are confirmed or otherwise by the greater experience since gained.

First then we will see what the Father of Waters, the Mississippi, tells us. I may observe that for a long while I found great difficulty in obtaining answers to my various questionings. Years elapsed and letters innumerable were written before I could alight upon any analyses of the waters of the Mississippi, reliable or otherwise. At last through the kindness of Prof. J. W. Spencer, of the State University of Missouri, I was supplied with the following analysis:

Analysis of Mississippi water near Carrolton, a few miles above New Orleans [is shown in the table on page 510].

According to this analysis the proportion of total solids in solution is by weight $\frac{1}{3615}$. If we take the mean annual discharge of the Mississippi at 541,666,666,666 tons* in round figures, there are 150 million tons of solids in solution per annum poured into the

---

* Report of Humphreys and Abbott, (1876), p. 146, 19,500,000,000,000 cubic feet at 36 feet to the ton.

Gulf of Mexico by the Mississippi, a truly remarkable quantity, which if reduced to rock at 15 feet to the ton is represented in round numbers by 80 square miles, 1 foot thick. . . .

If we take the drainage area of the Mississippi proper at 1,244,-000 square miles, the calculated amount of solids in solution, according to the analysis, will be 120 tons, removed from each square mile of surface per annum. From the surface of England and Wales I have shown that 143.5 tons per annum are removed

IN A GALLON (56,000 GRAINS)

	Grains
Potash sulphate, " chloride, Calcium chloride,	3.154
Silica,	2.455
Alumina,	1.753
Calcium carbonate, Magnesium "	7.307
Organic matter,	0.818
Total solid residue,	15.487

in solution and from the Danube basin 90 tons, so that this is a mean and probably correct.

It has been estimated that the basin of the Mississippi is lowered at the rate of one foot in 6,000 years, but this rate has been calculated from the removal of sediment alone; if we add to the matter removed mechanically that in solution it will raise the rate to one foot in 4,500 years. What stronger evidence can we have of the importance of chemical action in geological investigation; an importance that has hitherto been strangely overlooked.

Not less surprising considering the apparent insolubility of silica by ordinary agencies is the fact that in round numbers from 23,000,000 to 24,000,000 tons of silica are poured into the sea annually by this river, while there are 70,000,000 tons of carbonate of lime and magnesia. There is also an exceptional quantity of alumina and a low percentage of sulphates in this water.

RICHTHOFEN

Baron Ferdinand von Richthofen (1833–1905), German geologist and petrographer, was an indefatigable worker in the field, extending knowledge through his exploratory surveys, especially in China and Central Asia.

A CLASSIFICATION OF VOLCANIC ROCKS

From California Academy of Science, *Memoirs*, Vol. I, Pt. 2, pp. 5–94, 1868.

The following classification, in which existing names are retained, as nearly as could be done with convenience, is chiefly founded on observations made in the Carpathians and in the States of California and Nevada. Until recently, all volcanic rocks, at least those of more frequent occurrence, were comprehended in the terms: trachyte, phonolite, trachydolerite, dolerite, basalt; while, besides, separate names were used to distinguish modifications of texture, such as pumice-stone, obsidian, pearlite; or varieties somewhat more distinct in mineral composition, such as leucitophyre. . . . In order to establish a more natural system, we have, not to *make* the groups but to *find* them. Dropping all of those *a priori* principles which may be conceived having an artificial basis, we must endeavor to discover whether any great divisions are established by nature herself, and if so, of what character they are. We may then apply, as second in the order of their importance, those results which are obtained in the laboratory or geological cabinet, for defining and subdividing those groups. . . .

The following is the classification, the approach of which to a natural system of volcanic rocks, I will endeavor to set forth in the course of this paper:

ORDER FIRST: *Rhyolite*
    Family 1.   *Nevadite*, or granitic rhyolite
    "   2.   *Liparite*, or porphyritic rhyolite
    "   3.   *Rhyolite proper*, or lithoidic and hyaline rhyolite
ORDER SECOND: *Trachyte*
    Family 1.   *Sanidin-trachyte*
    "   2.   *Oligoclase-trachyte*

ORDER THIRD: *Propylite*
  Family 1.   *Quartzose propylite*
  "   2.   *Hornblendic propylite*
  "   3.   *Augitic propylite*
ORDER FOURTH: *Andesite*
  Family 1.   *Hornblendic andesite*
  "   2.   *Augitic andesite*
ORDER FIFTH: *Basalt*
  Family 1.   *Dolerite*
  "   2.   *Basalt*
  "   3.   *Leucitophyre*

Relation of Volcanic Rocks to Ancient Eruptive Rocks

All rocks which, bearing evidence of an intrusive or eruptive origin, preceded in age the Tertiary period, may, by principles similar to those which we applied in tracing the natural system of volcanic rocks, be divided in two great classes, for which we may use the terms "granitic rocks" and "porphyritic rocks," derived from the mode of texture predominating in either class. . . . The annexed table will show the mutual relation of these two classes and their subdivisions, and of either of them to volcanic rocks.

It appears that this general classification is based upon as natural principles as are within reach of our still limited knowledge of eruptive rocks, and therefore may at least approach the natural system.

Marine Abrasion and Transgression

Translated from *China*, Vol. II, pp. 766 et seq., Berlin, 1882.

Among the formative factors which play an important role in the geological history of North China, the periodic appearance of transgressive deposition is marked in a special manner. In the majority of cases, the beds of the overlapping formation are not, as would be expected in the extension of sea over land, laid down on a mountainous floor formed by high ranges and erosion valleys; but they rest, widespread and uniform, on a surface especially prepared as it were for the deposit, being flattened out for the most part, occasionally wavy, and sometimes in terraces. There also project from it, more or less, a number of high, resistant, mountain ridges which have been partially or completely covered by the horizontal

FIRST CLASS: GRANITIC ROCKS.	SECOND CLASS: PORPHYRITIC ROCKS.	THIRD CLASS: VOLCANIC ROCKS.	ESSENTIAL INGREDIENTS.
Order First—Granite. Fam. 1st. Granite. Fam. 2d. Granitite. Fam. 3d. Syenitic Granite.	*Order First—Felsitic Porphyry.* Fam. 1st. Quartzose Porphyry. Fam. 2d. Varieties without Quartz.	*Order First—Rhyolite.* (Quartziferous Varieties.) Fam. 1st. Nevadite. Fam. 2d. Liparite. Fam. 3d. Rhyolite proper. (Varieties without Quartz.)	Quartz, Orthoclase, Oligoclase, Biotite, (Hornblende.)
Order Second—Syenite. Only Family: Syenite.	*Order Second—Porphyrite.* Only Family: Porphyrite.	*Order Second—Trachyte.* Fam. 1st. Sanidin-Trachyte. Fam. 2d. Oligoclase-Trachyte.	Oligoclase, Orthoclase, Hornblende, (Biotite), (Quartz.)
Order Third—Diorite. Fam. 1st. Diorite. Fam. 2d. Rocks intermediate between Diorite and Diabase.	*Order Third—Melaphyr.* Fam. 1st. Melaphyr. Fam. 2d. Rocks intermediate between Melaphyr and Augitic Porphyry.	*Order Third—Propylite.* Fam. 1st. Quartzose-Propylite. Fam. 2d. Hornblendic Propylite. Fam. 3d. Augitic Propylite.	Oligoclase, Hornblende, (Titaniferous Magnetic Iron-ore.) Oligoclase, Hornblende, Titaniferous Magnetic Iron-ore.
Order Fourth—Diabase. Fam. 1st. Gabbro and Hypersthenite. Fam. 2d. Diabase	*Order Fourth—Augitic Porphyry.* Only Family: Augitic Porphyry.	*Order Fourth—Andesite.* Fam. 1st. Hornblendic Andesite. Fam. 2d. Augitic Andesite.	Oligoclase, Labrador, Augite, Hornblende, Titaniferous Magnetic Iron-ore.
		Order Fifth—Basalt. Fam. 1st. Dolerite. Fam. 2d. Basalt.	Labrador, Augite, Titaniferous Magnetic Iron-ore.

FIG. 26.—Richthofen's classification of igneous rocks, 1868.

beds. The surface of deposition cuts across the formations antedating the transgressing system, regardless of their position. As their beds are highly tilted and folded, these folds are planed down along the surface, while all parts which once projected above the latter have disappeared. At many places, the amount of denudation is extraordinarily great. A few troughs still preserved are often the only remnants of sedimentary formations which had many thousand feet of thickness and were folded together in a series of arches. Not only have the arches formed by these been cut away, but such parts of still older rocks as projected among them were removed in the same way. The facts mentioned in the description of the base of the Sinisian beds show this clearly.

In such cases it is clear that the encountered sedimentary formations were pressed into mountains after their deposition and that these formed the mainland—for denudation could not take place on the bottom of the sea.

Therefore, an enormous cutting away of the then-existing land preceded every period of transgression, forming a surface which has the tendency to approach, as much as possible, a rolling plain— although this is only accomplished rather imperfectly. If we seek the agents which might bring about such an extraordinarily great amount of erosion and produce denudation surfaces of the described form, those offered by the atmospheric activities of weathering and by the mechanical force of flowing water are entirely inadequate. Where the first alone are active, they produce, as we have seen, a decomposition of the soil, the depth of which depends first on climatic factors, then on the shape of the surface, and third on the nature of the rock. The unevenness of the contact between the solid and decomposed rock would be in no way lessened by this, but considerably increased. Then, if agents which are in a position to take away this cover become active, the form of this generally very uneven surface will appear. Just as little would flowing water have the power to denude great mountainous areas into flat plains. The base form of its erosion is the groove. In the beginning it strives to deepen the channels and to increase the unevenness in a lateral direction. Only when this has been accomplished to a certain extent, will it tend to widen the groove, to smooth the gradient, to destroy the side walls, and, in further sequence, to connect the bottoms of neighboring grooves by removing the separating ridges. But this stage is only reached locally and to a negligibly small extent. The change of level between land and sea works constantly against it.

The cutting down of a widespread mountainous land to a surface approaching a plain would thus never occur through the separate or united activities of land agents, and would remain an unattainable goal of their activity. Neither atmospheric force, flowing water, nor ice would ever be able to achieve that goal on a great scale.

Among all the mechanically destructive agents, there is only one which might accomplish regional abrasion on a wide scale. We have indicated it earlier in the description of individual cases. It is the work of waves of the surf working towards the interior of a continent. In order to estimate its possible potentialities, we start with an elementary consideration because this agency, which takes first place by far among all the forces working on the outside of the earth's crust and may wear down continents, is hardly sufficiently valued in its mighty functions of transformation.

The work of the surf on the coast has frequently been the object of thorough study, locally, and on a small scale where it was immediately evident. On flat coasts, which were built of alluvium, it is slight and limited essentially to a rearrangement of the silty and sandy portions, with wind entering into reciprocal action with the sea waves. As the sea draws back, dune-building moves forward toward the sea; as it presses landward again, it devours the old dunes and takes part of their material to form, with the aid of the wind, new dunes farther inland.

On a steep coast, the three cases of a constant sea level, a raising, and a sinking of the land, are likewise to be considered separately. The possibility of periodic changes must also be considered.

(1) Action of the surf with constant sea level. On rock coasts where no alteration of sea level takes place during an extensive period, the surf works destructively along a horizontal zone which begins at the level of the ebb and extends above that of the flood tide. The action is strongest between the line of half tide and the upper limit of the surf waves. The weathering and loosening of rock by sea salts, carbonic acid, the formation of ozone, and the

gripping of plants and animals—to which must be added the action of frost in higher latitudes—aids the mechanical action of the striking billows. It will carve a concave surface along this zone in the wall of rock, and the softer the latter is, the faster and wider it will cut. . . .

The terrace formed in this way is not horizontal but slopes seaward from the innermost level which lies between half-tide and full-tide level. . . .

The extensive terraces which are formed in this way on a rocky coast would never be able to exceed a certain width if the sea level remained constant, as the waves lose force from friction while rolling up the inclined surface.

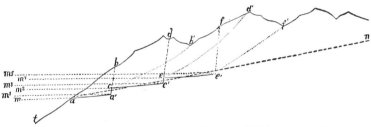

Fig. 27.—Diagram used by Richthofen to illustrate his ideas concerning marine abrasion, 1882.

(2) Action of surf with negative displacement of level. Surf terraces cannot be formed where a rocky coast of homogeneous stone rises regularly and permanently, and the climatic conditions undergo no change during the displacement. . . . This result can be modified if the movement is not uniform, the rock not homogeneous but with differing resistance at different heights, or if any of the other factors mentioned alter periodically.

(3) Action of the surf with positive displacement. The result is entirely different if the rocky coast attacked by the surf is sinking slowly and steadily. . . .

We can assume that the motion is intermittent at definite periods, although this seldom occurs in nature. If now after the first period, the land sinks by the amount $a'c$, with the position of the sea between m^2 and m^3 in the second, the terraces cc' would be formed and the section $bcc'd$ cut away. If the time interval is considered as infinitely small, i.e. the sinking continual, the cut surface *an* would develop and the whole hill *afn* would be taken away. . . .

Regional abrasion can therefore be accomplished only by the advancing surf line. Where the abraded material is not constantly conveyed to great distances by other agents, transgressive beds must be formed as a consequence. Therefore, as a rule, where transgression is very regular over wide stretches, the surface of deposition is formed by regional abrasion.

POWELL

John Wesley Powell (1834–1902), American geologist, was director, in sequence, of the Second Division of the United States Geological and Geographical Survey of the Territories (1871–1877), the United States Geographical and Geological Survey of the Rocky Mountain Region (1877–1879), and the United States Geological Survey (1881–1894). His "boat trip down the Colorado" through the Grand Canyon in the summer of 1869 was a journey "unequalled in the annals of geographical exploration for the courage and daring displayed in its execution." (Emmons)

A Classification of Streams and Valleys

From *Exploration of the Colorado River of the West*, Washington, 1875.

The Wasatch is a great trunk, with a branch called the Uinta. Near the junction, the two ranges have about the same altitude, and the gulches of their summits are filled with perpetual snow; but toward the east, the Uinta peaks are lower, gradually diminishing in altitude, until they are lost in low ridges and hills.

Through this range Green River runs, and a series of cañons forms its channel.

To a person studying the physical geography of this country, without a knowledge of its geology, it would seem very strange that the river should cut through the mountains, when, apparently, it might have passed around them to the east, through valleys, for there are such along the north side of the Uintas, extending to the east, where the mountains are degraded to hills, and, passing around these, there are other valleys, extending to the Green, on the south side of the range. Then, why did the river run through the mountains?

The first explanation suggested is that it followed a previously formed fissure through the range; but very little examination will show that this explanation is unsatisfactory. The proof is abundant that the river cut its own channel; that the cañons are gorges of corrasion. Again, the question returns to us, why did not the stream turn around this great obstruction, rather than pass through it? The answer is that the river had the right of way; in other words, it was running ere the mountains were formed; not before

518

the rocks of which the mountains are composed, were deposited, but before the formations were folded, so as to make a mountain range.

The contracting or shriveling of the earth causes the rocks near the surface to wrinkle or fold, and such a fold was started athwart the course of the river. Had it been suddenly formed, it would have been an obstruction sufficient to turn the water in a new course to thé east, beyond the extension of the wrinkle; but the emergence of the fold above the general surface of the country was little or no faster than the progress of the corrasion of the channel. We may say, then, that the river did not cut its way *down* through the mountains, from a height of many thousand feet above its present site, but, having an elevation differing but little, perhaps, from what it now has, as the fold was lifted, it cleared away the obstruction by cutting a cañon, and the walls were thus elevated on either side. The river preserved its level, but mountains were lifted up; as the saw revolves on a fixed pivot, while the log through which it cuts is moved along. The river was the saw which cut the mountains in two.

Recurring to the time before this wrinkle was formed, there were beds of sandstone, shale, and limestone, more than twenty-four thousand feet in thickness, spread horizontally over a broad stretch of this country. Then the summit of the fold slowly emerged, until the lower beds of sandstone were lifted to the altitude at first occupied by the upper beds, and if these upper beds had not been carried away, they would now be found more than twenty-four thousand feet above the river, and we should have a billow of sandstone, with its axis lying in an easterly and westerly direction, more than a hundred miles in length, fifty miles in breadth, and over twenty-four thousand feet higher than the present altitude of the river, gently rounded from its central line above to the foot of the slope on either side. But as the rocks were lifted, rains fell upon them and gathered into streams, and the wash of the rains and the corrasion of the rivers cut the billow down almost as fast as it rose, so that the present altitude of these mountains marks only the difference between the elevation and the denudation.

It has been said that the elevation of the wrinkle was twenty-four thousand feet, but it is probable that this is not the entire amount, for the present altitude of the river, above the sea, is nearly six thousand feet, and when this folding began we have reason to believe that the general surface of this country was but slightly above that general standard of comparison.

———————

I have endeavored above to explain the relation of the valleys of the Uinta Mountains to the stratigraphy, or structural geology, of the region, and, further, to state the conclusion reached, that the drainage was established antecedent to the corrugation or displacement of the beds by faulting and folding. I propose to call such valleys, including the orders and varieties before mentioned, *antecedent valleys*.

In other parts of the mountain region of the west, valleys are found having directions dependent on corrugation. I propose to call these *consequent valleys*. Such valleys have been observed only in limited areas, and have not been thoroughly studied, and I omit further discussion of them.

A part of the district in which my observations were made has since been much more thoroughly studied by Mr. Archibald R. Marvine, one of the geologists of the First Division of the "Geological and Geographical Survey of the Territories." In his report of June 19, 1874, he says:

"It is true that the structure of the lower rocks has begun to affect the courses of the streams, and in places to a considerable extent. Meeting a softer bed a cañon will often have its course directed by it, and follow it for some distance, leaving the adjacent harder beds plainly indicated by the ridges, and sometimes the sinuosities of structure are very curiously followed by a stream in all its windings, but it soon breaks away and runs independently of the bedding. Many of the smaller ravines have had their positions determined by the structure; but in a broad sense the drainage is from the main mountain crest eastward, independent of structure. Thus, while in places geological features may find expression in surface form, yet, as often, there may be no conceivable relation between topography and geology. The subaqueous erosion, in smoothing all to a common level, destroys all former surface expression of geological character, and the present erosion has not yet been in progress sufficiently long to recreate the lost features."

I fully concur with Mr. Marvine in the above explanation of the valleys in the main Rocky Mountains of Colorado, as my own observations in that country had led me to the same conclusion.

There can be no doubt that the present courses of the streams were determined by conditions not found in the rocks through which the channels are now carved, but that the beds in which the streams

Fig. 28.—Panoramic views by W. H. Holmes, in the Atlas of Colorado, United States Geological and Geographical Surveys of the Territories (Hayden Survey), 1881. *Top:* The Elk Mountains; Snowmass and Capital in the center. *Middle:* The Twin Lakes; Lake Fork of the Arkansas, showing the great moraines. *Bottom:* Southwestern border of the Mesa Verde, showing the Sierra El Late in the distance.

had their origin when the district last appeared above the level of the sea, have been swept away. I propose to call such *super-imposed valleys.* Thus the valleys under consideration, if classi-fied on the basis of their relation to the rocks in which they

originated, would be called *consequent valleys*, but if classified on
the basis of their relation to the rocks in which they are now found,
would be called *superimposed valleys*.

In this and the foregoing chapter I have attempted to describe
the agencies and conditions which have produced the more important topographic features in the Valley of the Colorado. . . .
The primary agency in the production of these features is upheaval,
i.e., upheaval in relation to the level of the sea, though it may possibly be down-throw in relation to the center of the earth. . . .

The second great agency is erosion. . . .

. . . We may consider the level of the sea to be a grand base
level, below which the dry lands cannot be eroded; but we may
also have, for local and temporary purposes, other base levels of
erosion, which are the levels of the beds of the principal streams
which carry away the products of erosion. (I take some liberty
in using the term level in this connection, as the action of a
running stream in wearing its channel ceases, for all practical purposes, before its bed has quite reached the level of the lower end
of the stream. What I have called the base level would, in fact,
be an imaginary surface, inclining slightly in all its parts toward
the lower end of the principal stream draining the area through
which the level is supposed to extend, or having the inclination of
its parts varied in direction as determined by tributary streams.)
Where such a stream crosses a series of rocks in its course, some
of which are hard, and others soft, the harder beds form a series
of temporary dams, above which the corrasion of the channel
through the softer beds is checked, and thus we may have a series
of base levels of erosion, below which the rocks on either side of
the river, though exceedingly friable, cannot be degraded. In
these districts of country, the first work of rains and rivers is to
cut channels, and divide the country into hills, and, perhaps,
mountains, by many meandering grooves or water-courses, and
when these have reached their local base levels, under the existing
conditions, the hills are washed down, but not carried entirely
away.

With this explanation I may combine the statements concerning
elevation and inclination into this single expression, that the more
elevated any district of the country is, above its base level of
denudation, the more rapidly it is degraded by rains and rivers.

GEIKIE

Sir Archibald Geikie (1835–1924), British geologist, was professor at the University of Edinburgh, director-general of the Geological Survey of Great Britain and Ireland, and president of the Royal Society. His *Founders of Geology* (2nd ed., 1905) is an excellent account of "the lives and work of some of the masters to whom we mainly owe the foundation and development of geological science."

On Denudation Now in Progress

From *Geological Magazine*, Vol. V, pp. 249–254, 1868; abstract of part of a paper read before the Geological Society of Glasgow on March 26.

The extent to which a country suffers denudation at the present time is to be measured by the amount of mineral matter removed from its surface and carried into the sea. An attentive examination of this subject is calculated to throw some light on the vexed question of the origin of valleys and also on the value of geological time. Of the mineral substances received by the sea from the land, one portion, and by far the larger, is brought down by streams, the other is washed off by the waves of the sea itself.

I. The material removed by streams is two-fold; one part being chemically dissolved in the water, the other mechanically suspended or pushed along by the onward motion of the streams. The former, though in large measure derived from underground sources, is likewise partly obtained from the surface. In some rivers the substances held in solution amount to a considerable proportion. The Thames, for example, carries to the sea every year about 450,000 tons of salts invisibly dissolved in its waters. But the material in mechanical suspension is of chief value in the present enquiry. The amount of such material annually transported to the sea by some of the larger rivers of the globe has been the subject of careful measurement and calculation. Much has been written of the vastness of the yearly tribute of silt borne to the ocean by such streams as the Ganges and Mississippi. But, as was first pointed out by Mr. Tylor, "the mere consideration of the number of cubic feet of detritus annually removed from any tract of land by its rivers does not produce so striking an impression upon the mind as the statement of how much the mean surface level of the district in question would be reduced by such a removal."[*] When the annual discharge of sediment and the area of the river-basin are both known, the one sum divided by the other gives the fraction by which the area drained by the river has its general level reduced in one year.   For it is clear that if a river carries so many millions of cubic feet of sediment every year into the sea, the area of country drained by it must have lost that quantity of solid material, and if we could restore the sediment so as to spread it over the basin, the layer so laid down would represent the fraction of a foot by which the basin had been lowered during a year. . . .

But besides the materials held in suspension there must also be taken into account the quantity of sand and gravel pushed along the bottom.   In the case of the Mississippi this was estimated by the United States Survey at 750,000,000 cubic feet.   In our own rivers it is probably on the whole proportionally greater.   Indeed the amount of coarse detritus carried down even by small streams is almost incredible. . . .

Comparing the measurements which have been made of the proportion of sediment in different streams we shall probably not assume too high an average if we take that of the carefully elaborated Survey of the Mississippi.   This gives an annual loss over the area of drainage equal to $\frac{1}{6000}$ of a foot.   If then a country is lowered by $\frac{1}{6000}$ of a foot in one year, should the existing causes continue to operate undisturbed as now, it will be lowered 1 foot in 6000 years, 10 feet in 60,000 years, 100 feet in 600,000 years, and 1000 feet in 6,000,000 years.   The mean height of the Continents, according to Humboldt's calculation,[†] is in Europe 671, North America 748, South America 1151, and Asia 1132 English

---

[*] Tylor, Phil. Mag., 4th series, v. 260 (1853).   My attention was first called to this very obvious and instructive method of representing the results of denudation by some remarks of Mr. Croll in the Phil. Mag. for February, 1867. Mr. Tylor's earlier publication was afterwards pointed out to me by Professor Ramsay.   Mr. Croll, following up the line of argument suggested in his former paper, has gone into further detail upon this subject in a memoir published in the Phil. Mag. for this month (May), which will be of essential service to geology.

[†] Asie Central, tome i. 168.

feet. Under such a rate of denudation therefore Europe must disappear in little more than four million of years, North America in about four millions and a half, South America and Asia in less then seven millions. These results do not pretend to be more than approximative, but they are of value inasmuch as they tend to shew that geological phenomena, even those of denudation, which are often appealed to as attesting the enormous duration of geological periods, may have been accomplished in much shorter intervals than have been claimed for them. . . .

The material carried to sea by rivers has been spoken of in the previous part of this paper as having been removed from the general area of drainage, which in consequence is thereby reduced in level. It is of importance to look at the subject from this point of view in order to obtain some adequate idea of the extent of the loss which the land is constantly undergoing before our eyes. But it is obvious that the material so removed does not come equally from the whole area of drainage. Very little may be obtained from the plains and watersheds; a great deal from the declivities and valleys. It may not be easy to apportion its share of the loss to each part of a district, but the sum total of denudation is not affected thereby. If we allow too little for the loss of the table-lands, we increase the proportion of the loss sustained by the slopes and valleys, and *vice versa*. There can be no doubt that the erosion of the slopes and water-courses is very much greater than that of the more level grounds. Let it be assumed that the waste is nine times greater in the one case than the other (in all likelihood it is more); in other words, that while the plains and table-lands have been having one foot worn off their surface, the declivities and river-courses have lost nine feet. Let it be further assumed that one-tenth part of the surface of a country is occupied by its water-courses and glens, while the remaining nine-tenths are covered by the plains, wide valleys, or flat grounds. Now, according to the foregoing data, the mean annual quantity of detritus carried to the sea is equal to the yearly loss of $\frac{1}{6000}$ of a foot from the general surface of the country. The valleys, therefore, are lowered by $\frac{1}{1200}$ of a foot, and the more open and flat land by $\frac{1}{10800}$ of a foot. At this rate it will take 10,800 years before the level ground has had a foot pared off its surface, while in 1200 years the valleys will have sunk a foot deeper into the framework of the land.   By the continuance of this state of things a valley 1000 feet deep may be excavated in 1,200,000 years. We may take other proportions, but the facts remain, that the country loses a certain ascertainable fraction of a foot from its general surface per annum, and that the loss from the valleys and water-courses is much larger than that fraction, while the loss from the level grounds is much less.

It seems an inevitable conclusion that those geologists who point to deep valleys, gorges, lakes, and ravines, as parts of the primeval architecture of a country, referable to the upheavals of early geological time, ignore the influence of one whole department of natural forces.   For it is evident that if denudation in past time has gone on with anything like the rapidity with which it marches now, the original irregularities of surface produced by such ancient subterranean movements must long ago have been utterly effaced. That the influence of these underground disturbances has often controlled the direction in which the denuding forces have worked, or are now working, is obvious enough; but it is equally clear that under the regime of rain, frost, ice, and rivers, there must have been valley-systems wherever a mass of land rose out of the sea, irrespective altogether of faults and earthquakes.   No one who has ever studied rocks in the field is likely to overlook the existence of faults and other traces of underground movement.   But he meets everywhere with proofs of the removal of vast masses of rock from the surface, which no amount of such movements will explain.   At their present rate of excavation the "gentle rain from heaven," and its concomitant powers of waste, will carve out deep and wide valleys in periods, which by most geologists, will be counted short indeed.   And if an agency now in operation can do this, it seems as unnecessary as it is unphilosophical to resort to conjectural cataclysms and dislocations for which there is no evidence, save the very phenomena which they are invented to explain. . . .

In accordance with the views now expressed, existing lakes as a rule must be of comparatively modern origin.   We see that the streams which enter them push yearly increasing deltas into the water.   The rate at which the deltas grow shows that they cannot, in a geological sense, be very old.   If the delta of the Rhone has crept a mile and a half into the lake of Geneva during eight centuries, a thousand years must represent no insignificant fraction of the interval since the river began to push its detritus into the lake. Had the lake basin, therefore, been of ancient origin, it must necessarily have been long ago filled up with sediment; and once, in that condition, no power of running water could re-excavate it so as to turn it into a lake again. If the immense mass of lakes scattered over the temperate and northern parts of our hemisphere be due in any large measure to underground forces, there must have been in recent geological times an amount of dislocation, upheaval, and depression, of which there is no other evidence, and which indeed is directly contradicted by the actual facts. We are driven, therefore, to seek some explanation which will account for these rock basins on the admission that they are of recent date, and cannot be due to underground agency. The theory of Professor Ramsay—that they were scooped out by the ice of the Glacial period—harmonizes these postulates, and furnishes a most important element in the elucidation of the history of the earth's surface.

II. The detritus wasted from the land is carried away not only by streams, but in part also by the waves and currents of the sea. Yet if we consider the abrasion due directly to marine action, we shall be led to perceive that its extent is comparatively small. In what is called *marine denudation*, the part played by the sea is rather that of removing what has been loosened and decomposed by atmospheric agents than that of eroding the land by its own proper action. Indeed, when a broad view of the whole subject is taken, the amount of denudation which can be traced to the direct effects of the sea alone is seen to be altogether insignificant. Yet even if we grant to the action of the waves and tides all that is usually included under marine denudation, the sum total of waste along the sea-margin of the land is still trifling compared with that effected by the meteoric agents upon the interior. . . .

. . . Let us suppose that the sea eats away a continent at the rate of ten feet in a century—an estimate which probably attributes to the waves a very much higher rate of erosion than can as the average be claimed for them,—then a slice of about a mile in breadth will require about 52,800 years for its demolition, ten miles will be eaten away in 528,000 years, one hundred miles in 5,280,000 years. But we have already seen that on a moderate computation such a continent as Europe will, at the present rate of subaërial waste, be worn away in about 4,000,000 years.   Hence, before the sea could pare off more than a mere marginal strip of land between 70 and 80 miles in breadth, the whole land would be washed into the ocean by atmospheric denudation.*

---

* The action of meteoric agents and of the sea is independent of subterranean movements, and must go on whether a land is upheaved or depressed.   These movements will in some cases favour subaërial denudation, in others marine denudation, as shewn in the paper of which the above is an abstract.   But in taking a generalized view of the subject their influence may be disregarded.

DE LA NOË AND DE MARGERIE

Gaston de la Noë (1836–1902) was lieutenant-colonel of engineers in the geographic service of the French army and later became brigadier general and director of that service.

Emmanuel de Margerie (1862–   ), French geologist and geographer, was from 1919 to 1933 director of the geological survey of Alsace and Lorraine. He is widely known for his translation into French of Eduard Suess's *Das Antlitz der Erde*.

Concerning the Fashioning of Slopes

Translated from *Les formes du terrain*, chap. 3, Service géographique de l'armée, Paris, 1888.

We will here investigate the manner in which water transports movable materials when it runs on slopes, reserving for another chapter the examination of its action when it is concentrated in brooks, rivers, etc.

It is a fact of daily observation that rain water running over the surface of the ground, carries along the loose materials which it collects. There is no storm without muddy water; slopes and artificial benches are rapidly lowered; on certain steep slopes, for example Vignoble near Salins, the disaggregated materials of the soil are constantly carried along, so that the vinegrower is obliged each year to carry back in his basket the soil washed down to the lower part of his vineyard. This action is furthered by the decrease in the carrying power of the water with respect to the materials it bears.

The effect of rain on the slopes increasingly lessens their gradient. In a particular plot of ground this decrease will be all the more rapid as disintegration of the surface proceeds more swiftly, as the decomposed matter becomes finer particles, as the slope is greater and as the volume of water is greater. Moreover, as the least slope on which the materials can be dragged along is determined by their size, if there is no limit to the division of these particles, the declivity of the slope will have only the limit necessary for movement of the water filled with the finer particles. Obviously, whatever may have been the size of the materials when first detached from the

529

bedrock, they can at length, under the repeated action of the agents
of decomposition and also by the mutual shocks which they undergo

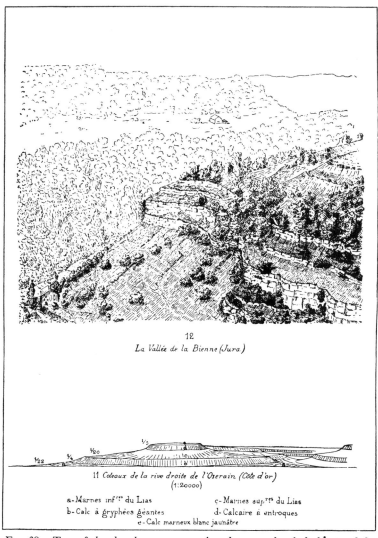

12
La Vallée de la Bienne (Jura)

11 Coteaux de la rive droite de l'Oserain (Côte d'or)
(1:20000)

a-Marnes inf^{res} du Lias   ç-Marnes sup^{res} du Lias
b-Calc à gryphées géantes   d-Calcaire à entroques
e-Calc marneux blanc jaunâtre

FIG. 29.—Two of the sketches accompanying the paper by de la Noë and de
Margerie concerning the fashioning of slopes, 1888.

in even a short transport, be reduced in their turn to such fine
particles that one must admit impartially that every land, whatever

its mineralogical composition, is capable of reaching a very feeble slope, on the sole condition that the action of rain is sufficiently prolonged. In a word, the action of the rain finally flattens all the slopes.

If we consider the slopes of valleys cut across rocks of different kinds and subjected during the same period to the same action of disintegration and removal, it is indeed evident that the declivity of slopes of each of these valleys depends solely on the rate at which the rock composing it is disintegrated: the faster this rate, the gentler the slope. If we consider now the case of a slope on which rocks of different kinds outcrop, each part of the profile will present a different slope, which will be in inverse relation to the rate of disintegration of the rock to which the part corresponds.

Figure 11 [Fig. 29], which reproduces the right profile of the valley of the Oserain seen from Mt. Auxois, shows these differences.

Another very striking example is shown in fig. 12 [Fig. 29], where one sees a slope formed of successive grades fairly near each other. The rocks which outcrop on the surface of the ground are stratified beds, comparatively thin and alternately resistant and easy to disintegrate; the removal of the latter has successively destroyed the base which supported the overlying beds and, by the progressive retreat of the latter, has forced the slope profile to assume the form of a giant staircase. The shelflike arrangement of certain amphitheaters, such as the cirque of Gavarnie, is the result of a similar operation on a much greater scale.

We must remark here that the degradation of easily decomposed rock, instead of producing a gentle slope directly uniting the top and bottom of the two adjacent resistant beds, appears, as in fig. 13 [Fig. 30], as a rather deep excavation of the profile which has the effect of leaving the higher rock overhanging. Where the stratification is horizontal, this excavation of even one bed produces in the middle of the general escarpment a horizontal groove of a length sometimes notable, that certain observers have attributed wrongly to the action of a watercourse which must at some time have run at a corresponding height on the flank of the valley. This point of view is immediately contradicted by the observed fact that where the beds are crooked, as for example in the transverse valleys of the Jura, the groove follows the contour exactly: it has consequently

developed along the outcrop of a single bed and arises from the
feeble resistance of the latter to atmospheric agents.

In reality, the law which governs the profile of slopes is not quite
so simple as we have said, because of the greater volume of water
which the lowest beds receive; as a result of this excess, if a single
rock forms two distinct beds, one at the summit and the other
toward the base of the slope, the first would present a slope steeper
than the second.

However, the bed which forms the base of the slope provides an
exception which must be noted (fig. 14) [Fig. 30]; as we have said, in
the case of any particular bed, its profile can only be lowered by

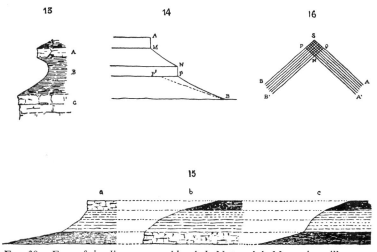

FIG. 30.—Four of the diagrams used by de la Noë and de Margerie to illustrate
their ideas concerning the fashioning of slopes, 1888.

turning around foot *B* of the slope, while the profile of the rocks
which rise above it lie parallel to it, since point *P*, which forms the
foot of it, is displaced. In fact, we have seen that the profile was
established by the foot: the profile *BP'* will then be established
independently of the one which exists above: if the degradation of
NP has not proceeded so swiftly, rock *NP* will find itself overhanging on a part of variable *P'P* in accordance with the relative rate
of disintegration, and it will give way, producing a corresponding
mound on the lower slope, whence it will subsequently be washed
away. This action of *sapping* will occur wherever a resistant rock
rises above a rock easy to disintegrate; and this circumstance will

accelerate the modeling of slopes by hastening the removal of rocks the cohesion of which would have offered long resistance to meteoric agents, if the latter had acted alone to effect their disintegration and to prepare them for removal.

It is easy, with what we have just said, to explain the varied forms which characterize the profile of slopes composed of several superimposed rocks. If they succeeded each other in such order that the rate of disintegration of each regularly decreases from the base to the summit, the corresponding profile will be concave upward (fig. 15a) [Fig. 30]; it will be convex in the opposite order (fig. 15b); it will be concave-convex (fig. 15c), when there is superposition of beds in an order which is at once apparent; and one can imagine all sorts of intermediate combinations giving to slopes the most varied profiles. The fact must not be overlooked, however, that all things being equal, the most gentle slopes correspond to the lowest rocks, for the reason we have just given above; this state of affairs contributes to the enlargement of the valley floors when the rock lying at the base of the slope presents a less notable resistance than all the others.

In the above examples we have given an angular appearance to profiles, but more often in nature the profile is on the contrary a continuously curved line, for the reentrant angles have been filled with debris and the ridges have been blunted by erosion. In a general way, moreover, we can see that sharp ridges can only occur as exceptions: let us consider (fig. 16) [Fig. 30] two slopes AS, BS, meeting each other as S in an acute angle; as we have said above, if all the circumstances are the same and above all if the rock is uniform, the disaggregation will be effected along two hands limited by lines such as $A'P$, $B'Q$, parallel to the primitive surface AS, BS. One sees then that the neighboring region of the ridge, that is to say the one represented in section by the rectangle $PNQS$, will undergo a double attack, by surface SQ and surface SP simultaneously, and consequently it will be more rapidly disintegrated and therefore susceptible to more rapid removal than the rest of the surface of the slopes.

Classification of Rocks According to the Inclination of Their Talus

It would not be possible to establish a classification of rocks based on the relative value of the inclination of their talus that would be true in all imaginable cases. For this inclination depends on the *resistance of the rock, the nature of the agents which produce its disag-*

gregation, and the duration of their action.   But in certain regions one agent predominates while another may be completely absent; if we consider in particular the effects produced by ice, it is evident that a particular rock, often frozen in a high latitude, would occupy a very different rank in the series than that which it would occupy near the equator where it rarely freezes.   We can therefore make only a very broad classification, considering primarily what happens in a region having a medium climate like France.

The rocks which finally establish themselves under a very gentle inclination are those susceptible of forming with water a fluent mud, capable (because of its fluidity) of flowing over very feeble slopes. Of this type are the shales, if they are exposed in such a way as to imbibe water which is continually seeping into them because of the alternating states of dryness and wetness which fissure the clay and permit water to be infiltrated.   The marls, which are a mixture of clay and lime, display quite gentle slopes, but nevertheless steeper than the preceding rocks; they hold a sort of middle place between the clay rocks (shales) and the limestones.   The compact limestones, in fact, can maintain themselves sometimes in vertical escarpments; the break-up of their elements by water is very slow, in general, for in our climate it is not accelerated either by the action of ice or by the action of sapping, which results from the carrying away of the lower, less resistant rocks; and in many instances, the outer side of the escarpment recedes while remaining vertical.   It is generally the limestones which in nature have the most remarkable escarpments and the boldest and most accentuated forms.   Nevertheless, certain of the sandstones present analogous aspects, but in general these have a more massive structure; more commonly the sandstones are present beneath the talus at a mean inclination which is only a little greater than that of the marls and marly limestones.   Finally, from the point of view of the slope, the granites can be ranked nearly in the same group as the sandstones of which we have just spoken.   Nevertheless the peaks of high granitic mountains present needlelike and toothed ridges which seem to be in contradiction to what we have just said; this is because the decomposed parts are elevated as fast as they are disintegrated, with such rapidity that the surface is, in a manner of speaking, constantly reestablished.   It is thus that shales are able to maintain themselves in a very steep slope if they are sapped at the foot and carried away before the contiguous parts have had time to be saturated with dampness: such are the cliffs of a part of the coast of Calvados, made up of Oxford clay. . . .

We had reason for saying at the beginning of the present section that it is impossible to make a general classification of rocks according to the inclination of their talus, since this varies according to altitude, latitude, and method of destruction which the rocks undergo. One may say that in general the slopes are steeper in dry climates and gentler in wet climates, a fact which is easily understood if one looks back to the considerations developed above concerning the role of rain water in the reduction of the profile of slopes; thus the Quaternary alluvium of the Sahara, outcropping in the *goûr* or isolated buttes and on the sides of the dry *ouadis*, appears along slopes rarely attained by corresponding rocks in our climate.

The Role of Vegetation

We know that following the deforestation of a slope, rain water immediately carries away the movable materials of its surface and causes disasters of a type previously unknown. Evidently, therefore, the vegetation intervenes to maintain the surface of the soil at a certain inclination; without it the disaggregation, and especially the carrying away of disaggregated materials by the savage waters, would continue until the leveling of the relief of the continents would be complete. But to permit vegetation to establish itself, the rapidity of its development must be greater than the facility for carrying away the movable materials. This particular state of equilibrium is reached more or less quickly according to the nature of the subsoil and according to the mass of exterior circumstances more or less favorable to vegetation. Thus, in warm, damp climates, vegetation will compete in a certain sense with the abundance of rain, striving to maintain a notable slope of the talus in spite of the tendency of the latter. In a general way, the vegetation thus gives to the rocks a slope of equilibrium and assures the stability or fixation of forms of the earth which without it would continue to wear away and to be flattened indefinitely. The notion of equilibrium of slopes is thus inseparable from the idea of the vegetation which determines it.

The proofs of the conservation role of the plant carpet are shown everywhere. When the engineer wants to maintain the talus of artificial trenches and embankments, he plants it with substances capable of rapid growth and thus anchors the soil, thanks to their protecting mantle. In spite of this precaution, if the talus has a very steep slope and if atmospheric conditions have been unfavorable, degradation will occur. A striking proof of this action of vegetation is furnished us by the well-known example of torrents in the Alps. In regions where there has been deforestation the voluntary degradation of the soil has recommenced and we know what ravages have been the consequence. Near Prony, clearings made in the basin of the Po since the end of the Middle Ages have produced an analogous effect, which appears in a notable enlargement of the delta of this river.

POŠEPNÝ

Ferencz Pošepný (1836–1895), Bohemian geologist and mining engineer, was a professor in the School of Mines at Przibam. His views exerted a profound influence upon all students of the genesis of ore deposits during the latter part of the nineteenth century. The excerpt included here was translated by the secretary of the International Engineering Congress held in Chicago in 1893.

GROUND WATER AND THE DEPOSITION OF ORE BODIES

From *Transactions of the American Institute of Mining Engineers*, Vol. XXIII, pp. 197–369, 1893.

Surface phenomena exhibit clearly a constant circulation of liquids, and corresponding phenomena, so far as they are observable underground, indicate the persistence of this condition, so that we must infer a subterranean circulation connected with that of the surface. . . .

The Vadose Underground Circulation. When in a given terrain, by wells or other openings, the *ground-water* (that is, the water-level, *Grundwasserspiegel, nappe d'eau*) has been reached at several points, it is found that these points are in a gently inclined plane, dipping towards the deepest point of the surface of the region, or towards a point where an impermeable rock outcrops. The ground-water is not stagnant, but moves, though with relative slowness, according to the difference in height and the size of the interstitial spaces, down the plane mentioned, and finds its way, in the first instance, directly into the nearest surface-stream, or, in the second instance, forms a spring, which takes indirectly a similar course. Thus stated, free from all complications, the phenomenon exhibits clearly the law of circulation. The atmospheric moisture evidently descends; and even the movement of the upper layer of the ground-water is only apparently lateral, but really downwards, and is determined (for equal sectional areas of the rock-interstices) by the difference in height between the water-level and the surface-outlet.

For that part of the subterranean circulation, bounded by the water-level, and called the vadose or shallow underground circulation, the law of a descending movement holds good in all cases, even in those complicated ones which show ascending currents in parts. The total difference in altitude between the water-level and the surface-outlet is always the controlling factor.

When these two controlling levels are artificially changed, as often happens in mining, the law still operates. In sinking a shaft through permeable ground, it is of course necessary to lift continuously the ground-water. The water-level thus acquires an inclination towards the shaft, which may thus receive not only the flow of the immediate vicinity but even also that of the neighboring valley-systems. A shaft imparts to the previously plane water-level a depression, giving it the form of an inverted conoid with parabolic generatrix. An adit produces a prismatic depression in the water-level; and so on for other excavations. On the other hand, a bore-hole, from which the water is not removed, does not effect the water-level.

Atmospheric waters falling upon impermeable rocks at the surface cannot penetrate them, but must join the existing surface-circulation. The rocks are usually covered with more or less detritus, in the interstices of which the ground-water can move; and the water-level is in most cases at the boundary between the permeable surface-formation and the impermeable rock below.

These relations are complicated by the occurrence of fissures (which the ground-water of course fills), and by the communication of such fissures in depth with permeable formations, which come to the surface somewhere at a lower level, though at great distance. In such cases, as is well known, a siphon-action is set up, and the ground-water of one region may find an outlet far away, even beyond a mountain range.

Peculiar conditions are created by the occurrence of relatively soluble rocks, such as rock-salt, gypsum, limestone and dolomite, in which, by the penetration of meteoric waters and the circulation of the ground-water, connected cavities are formed, constituting complete channels for the vadose circulation.

Artesian wells present an analogous case, also explained hitherto by the principle of hydrostatic pressure. The outcrop of the permeable layer has been assumed to be necessarily higher than the mouth of the well, in order to account for the rising of the water above the latter level. The cause has been conceived as the operation of communicating pipes, the drill hole being one leg, and the permeable layer the other, and it has been overlooked that the latter is no open pipe, but a congeries of rock-interstices, in which the water has to overcome a great resistance, and that, perhaps, in level regions no hydrostatic head at all can be demonstrated. Certainly the powerful factor of the higher temperature, and in some cases the gaseous contents, of the ascending water, were omitted from the calculation.

The Deep Underground Circulation. Thus far, we have considered only such processes as take place in the region above water-level, and are still, in some cases, open to our observation. As we descend to a deeper region, there is less hope of encountering formative processes still active. When we penetrate by mining into the depths, we artificially depress the water-level, and create conditions unlike those which attended the formation of the deposits.

But, if we compare the deposits formed below water-level, under proportionally greater pressure and at higher temperature, with those of the upper region, it appears beyond doubt that the former also must have been produced by deposition from fluid solutions.

When we compare the low solubility of certain ingredients of the deposits with the spaces in which they occur, often in large quantity, it is impossible to assume that they could have been precipitated from solutions existing in these spaces only. We must concede that immense volumes of solutions must have flowed through the space—in other words, that the deposits were precipitated from liquids circulating in these channels.

The formation of these cavities has been already discussed, and referred to mechanical and chemical causes. It remains to consider the manner of their filling. We have seen that the uppermost layer of the ground-water has an apparently lateral, but really descending movement; and it is very natural to imagine that this top layer slides, as it were, upon a lower mass, which is apparently stagnant. According to this conception, the deep region would be comparable to a vessel filled with various permeable, impermeable, and soluble materials, over which water is continually passed, so that, from the moment when all the interstices have been once filled, only the uppermost water-layer has any movement.

But with the increase of depth, the pressure of the water-column increases, as does the temperature. The warm water certainly tends to rise, if not prevented by interstitial friction, as is, no doubt, generally the case. But where the warmed water finds a half-opened channel communicating with the upper region, it will experience much less friction on the walls, and must evidently ascend. It might thus be conceived that the ground-water descends by capillarity through the rock-interstices over large areas, in order to mount again through open channels at a few points.

Daubrée's experiment confirms our view that the portion of the ground-water lying below water-level is not stagnant, but descends by capillarity, and since it cannot be simply consumed in depth, receives there through a higher temperature a tendency to return towards the surface, which tendency is most easily satisfied through open channels. Stated summarily: The ground-water descends in the deep regions also through the capillaries of the rocks; at a certain depth it probably moves laterally towards open conduits, and, reaching these, it ascends through them to the surface.

The solvent power of the water increases with temperature and pressure, and also with the duration of its underground journeying. Hence, while it is descending, it can dissolve or precipitate only the more soluble substances, which, as the conditions of their solubility (temperature and pressure) gradually disappear in the ascent, must be deposited in the channels themselves.

Origin of Ore-deposits in the Deep Region. We have seen that the mineral springs which ascend to the surface are dilute metallic solutions, and that at their outflow (the only point where we can directly observe their activity) they form deposits containing metals, among other things, and exhibiting a structure which occurs in ore-deposits likewise. We have followed to a not inconsiderable depth one ore-deposit which occurs upon an ascending spring, and have found that, apart from changes conditioned by the vicinity of the surface, it continues in character. Finally, we have encountered mineral springs in many places where mining has followed ore-deposits in depth. Joining these several links of observation, we cannot avoid the conclusion that the ore-deposits found in the deep region are the products of mineral springs, the more so since many of them have a structure and form which can only be explained as the result of precipitation from liquids circulating in channels. The deposits from these liquids contain substances which are foreign to the surface and to the shallow region, and hence could not have been brought into circulation by the descending ground-water, but must have come from a deep-region, where higher temperature and pressure (the two factors increasing the solubility of all substances) exist.

PUMPELLY

Raphael Pumpelly (1837–1923), American geologist, was engaged by the Chinese Government in 1863 to examine the coal fields west of Peking.   Journeying in northern China and Mongolia, he subsequently crossed Siberia to St. Petersburg in 1865.   Later, he surveyed the copper regions of Michigan, and in 1871 he became state geologist of Missouri.   For many years thereafter he was associated with the United States Geological Survey.

The Relation of Secular Rock-disintegration to Loess, Glacial Drift, and Rock Basins

From *American Journal of Science*, Third Series, Vol. XVII, pp. 133–144, 1879.

Loess is a calcareous loam.   It is easily crushed in the hand to an almost impalpable powder, and yet its consistency is such that it will support itself for many years in vertical cliffs 200 feet high.   A close examination shows that it is filled with tubular pores branching downwards like rootlets, and that these tubes are lined with carbonate of lime.   It is to these that it owes its consistency and its vertical internal structure.   It is wholly unstratified, and often where erosion has cut into it, whether one foot or one hundred yards, the walls are absolutely vertical.   Its vertical internal structure causes it to break off in any vertical plane but in no other.   Hence when a cliff is undermined the loess breaks off in immense vertical plates leaving again a perpendicular wall.

It is divided into beds varying in thickness from one foot to two or three hundred which thin out to nothing at the borders and are separated by parting planes.   These planes are marked by angular debris near the mountains, and by elongated upright calcareous concretions elsewhere.

The loess, first well known in the valley of the Rhine and in France, was recognized later in other parts of Europe and in the Mississippi Valley.   It was always looked upon as a subaqueous formation.   Fourteen years ago I observed and afterward described some of the great loess-basins along the boundaries between China and Central Asia.   I was led to the conclusion, chiefly from topographical reasons, that the loess of these valleys had been deposited in a series of great lakes; and subsequent observers took the same view. But Richthofen, extending his fruitful journeyings over a wide area, found the loess occupying the loftiest passes of northeastern China, over 8,000 feet above the sea, and he proves, in the most conclusive manner, that an aqueous origin is impossible. For the former theories of loess formation which required inconceivable conditions and are full of contradictions, Richthofen substitutes an exceedingly simple and thoroughly consistent explanation. I hope I may be excused for giving it in brief terms:

Whenever, from any cause, the winds blowing toward an interior portion of a continent are drained of their moisture on the way, as by the elevation in their path of lofty condensing mountains, the region thus deprived of its rain-bringing clouds soon has its evaporation in excess of its rainfall. Its streams dry up; its soluble and insoluble products of disintegration are no longer carried to the ocean; the region becomes what Richthofen calls a central area, in contrast to peripheral regions which are drained directly into the ocean. The destruction of the vegetation lays bare the surface, and the products of disintegration are blown and sorted by the wind and washed by the occasional rains from the hills down into the valleys. This material is very nutritive and supports the grass of the steppes; the dust left by the winds and the hill-wash are arrested by the grass which they gradually bury while forming the soil for new growths. In this way, portions of the country become buried in their own and their neighbor's debris. Great thicknesses thus gradually accumulate undergoing a transformation into loess by the rootlets and stems of the vegetation. . . .

The one weak point of Richthofen's theory is in the evident inadequacy of the current disintegration as a source of material. When we consider the immense area covered by loess to depths varying from 50 to 2,000 feet, and the fact that this is only the very finest portion of rock-destruction, and again that the accumulation represents only a very short period of time, geologically speaking, surely we must seek a more fertile source of supply than is furnished by the current decomposition of rock surface.

It seems to me that there are two important sources: I. The silt brought by rivers, many of them fed by the products of glacial attrition flowing from the mountains into the central region. Where the streams sink away, or where the lakes which receive them have dried up, the finer products of the erosion of a large territory are left to be removed in dust storms.

II. The second, and I believe, the more important source is in the residuary products of a secular disintegration which we will now consider.

In all regions where the soil is protected by a luxuriant vegetation the greater part of the insoluble products of disintegration remains *in situ*. Considerable portions of the continents have remained above water during long geological periods. Where this has been the case, and where the region thus exposed enjoyed a peripheral climate with a protecting vegetation and abundant generation of carbonic acid, the feldspathic rocks have been profoundly affected; granite and gneisses being decomposed often to the depth of several hundred feet. . . .

Other things being equal, the granular rocks of a region will be nearly or quite reduced to a loose mass when the compact rocks like porphyries and basalts have still a large proportion of their mass represented by the rounded cores. . . .

Over a large part of Europe and America this accumulation has been removed by glacial action. We may assume that during the gradual approach of the glacial epoch the ground in the northern half of the northern temperate zone, as the mean annual temperature fell below the freezing point, became perpetually frozen and when covered by the glacial ice, the thickness of the glacier instead of being measured from the upper surface down to the former soil should be measured down to the bottom of the residuary mass of disintegrated rock. It seems to me that only on this supposition can we explain the enormous amount of the ground moraine that covers our northern country, and the predominance in it of the debris of local rocks. The whole of this loose material must have participated in the movement of the ice, though far more slowly than the overlying and more nearly pure ice. Thus at any given point, the material on and near the surface would represent localities considerably farther northward than that at the bottom. We can only thus account for the disappearance of the vast amount of residuary disintegrated material that must have existed over the surface of the Archæan feldspathic rocks of eastern British America. But in Northern Asia, north of the 40th degree of latitude there are no traces of a general glacial action such as existed in Northeastern America and Northern Europe. The evidence indeed is all the

other way.   And yet, while the rocks of Southern Asia show extensive residua of disintegration, the results of a secular decomposition
protected from erosion by an abundant vegetation, the feldspathic
rocks of Northern China and of Central Asia are as free from this
as are those of northeastern America.

The only answer to the question, what has become of them? is,
that they have been blown and sifted and assorted by the winds,
the heavier fragments remaining to be reduced by weathering and
to form the stony steppes, the sand drifting in billowy waves over
the country, and forming sand-deserts, while the fine dust floating
in the air, an impalpable powder is deposited far and near, and,
under the influence and protection of the steppe grasses, is transformed into the loess.

There are few problems in dynamical geology that have been
considered more difficult to solve than the origin of rock-basins.
Wherever my route between the great wall of China and the
Siberian frontier lay through a region of crystalline rocks, I found
that one of the characteristic features of the surface was the prevalence of basin-shaped depressions of all sizes hollowed out of the
rock.   In other words, if filled with water they would have been
lakes without outlets and with unbroken sides of rock.   In some
exceptional instances it was clear that systems of intersecting dykes
had been less acted upon by the basin-making process than the
intervening rocks, and the basins were formed in these last.

The rock basins of Scotland have been graphically described by
Geikie, and while a great many of the countless lakes and lakelets in
the region of the crystalline rocks of North America and Northern
Europe are valleys of erosion with dams of glacial moraine material,
vast numbers of them undoubtedly fill rock-basins.   An ingenious
explanation of the formation of these rock-bound depressions given
by Ramsay and accepted by Geikie as the only reasonable hypothesis, is that the rock-basins of the loch and fiord kinds are due to
unequal sculpturing by glaciers.

While this may be admitted, with some reservations, as a satisfactory explanation for many fiords and lochs cut in the declivities
of a mountain range or of a steep coast, it is useless in regard to the
lake-basins of flat countries, like Finland and British America.
Geikie endeavors to use this hypothesis to explain also the rock-
basins observed by me in Central Asia; but it is still more useless

here as an explanation, because there are absolutely no traces of glaciation in Central Asia, outside of the high mountain chains. These Asiatic depressions are rough and ragged, and the debris contained in them consists of ragged angular fragments of the local rocks, while the glaciated basins of America and Europe are smoothed and polished, and the debris they contain consists of the rounded and scored material of the drift.

The basins of Asia were emptied by wind, and those of Northern Europe and America were emptied by ice, but the wind and ice were only immediate agents employed in rapidly emptying basins which had been long forming by a process common to both—the secular decay of the rock.

PROCTOR

Richard Anthony Proctor (1837–1888), English astronomer, was never associated with an institution or government bureau as teacher or research worker, but he was an honorary member of the Royal Astronomical Society and the author of several popular books on astronomy.

Meteoric Systems and the Origin of the Earth

From *Other Worlds Than Ours*, pp. 214–216, 2nd ed., London, 1870.

We know that the materials composing meteors, and we conclude, therefore, that those composing comets, do not differ from those which constitute the earth and sun, and presumably the planets also. Therefore, under the continual rain of meteoric matter, it may be said that the earth, sun, and planets are *growing*. Now the idea obviously suggests itself, that the whole growth of the solar system, from its primal condition to its present state, may have been due to processes resembling those which we now see taking place within its bounds. It is of course obvious, that if this be so, the number of meteoric and cometic systems must have been enormously greater originally, than it is at present. Countless millions of meteoric systems, travelling in orbits of every degree of eccentricity and inclination, travelling also in all conceivable directions around the centre of gravity of the whole, would go to the making up of each individual planet. A marked tendency to aggregate around one definite plane, and to move in directions which, referred to that plane, corresponded to the present direction of planetary motion, would suffice to account for the present state of things. The effect of multiplied collisions would necessarily be to eliminate orbits of exaggerated eccentricity, and to form systems travelling nearly on the mean plane of the aggregate motions, and with a direct motion. Further, where collisions were most numerous, there would be found not only the most circular resulting orbits, not only the greatest approach to exact coincidence of such orbits with the mean plane of the whole system, but the bodies formed out of the resulting systems would there exhibit rotations coinciding most nearly with the mean plane of the entire system.

It seems to me, that not only has this general view of the mode in which our system has reached its present state a greater support from what is now actually going on than the nebular hypothesis of Laplace, but that it serves to account in a far more satisfactory manner for the principal peculiarities of the solar system.

GEIKIE

James Geikie (1839–1915), Scottish geologist, brother of Sir Archibald, whom he succeeded as professor at Edinburgh, is best known for his contributions to glacial geology. This excerpt from his voluminous writings presents one of the earliest recognitions of the fact of multiple glaciation during the Great Ice Age.

On Changes of Climate during the Glacial Epoch

From *Geological Magazine*, Vol. VIII, pp. 545–553, 1871.

The intercalated beds consist, as I have just said, of silt, clay, sand, and gravel—sometimes the coarser, and at other times the finer grained ingredients predominating. In certain localities they have yielded Arctic shells of species identical with those which occur so abundantly in the later Glacial deposits of our maritime regions. It is quite certain, therefore, that some inter-Glacial beds are of marine origin. . . . But some inter-Glacial beds are no less certainly of freshwater origin. A few years ago I described in this MAGAZINE* a very interesting section at Crofthead, which showed most clearly a set of lacustrine beds resting upon and covered by the Till. From these beds the skull of Bos primigenius and other mammalian remains were obtained.

From the fact thus briefly indicated, we are entitled to conclude that that section of the Glacial epoch which is represented by the Scottish Boulder-clay was not one unbroken age of ice. It was certainly interrupted by several intervening periods of less Arctic conditions, during the prevalence of which the ice-sheet must gradually have melted away from the low grounds, and given place to streams, and lakes and rivers. At such periods a vegetation like that of cold temperate regions clothed the valleys with grasses and heaths, and the hillside with pine and birch. Reindeer wandered across the country, while herds of the great ox and the mammoth frequented the grassy vales. . . . What evidence we have, points

* Vol. V. p. 393.

to the existence of local glaciers in our higher valleys—to moderate summers and severe winters. . . .

And yet we might be committing a gross error were we to assume that Scotland during inter-Glacial times, never enjoyed milder conditions than now obtain in the forest regions and barrens of North America. We must ever bear in mind that the inter-Glacial deposits are the veriest fragments. . . . Moreover, we must not forget that if really warm climates ever did supervene during inter-Glacial times, every such warm period must have been followed by temperate, cold-temperate, and Arctic conditions. And these last would consequently be the most fully represented of the series.

SHALER

Nathaniel Southgate Shaler (1841–1906), American geologist, professor of geology and paleontology in Harvard University, was the first director of the Kentucky Geological Survey and prosecuted much of his field studies under the auspices of the United States Geological Survey.

The Change of Sea Level

From *Bulletin of the Geological Society of America*, Vol. VI, pp. 141–166, 1895.

The great importance of changes in the distribution of land and sea has in a general way long been recognized. Of late years interest in the question has been increased by the studies of geologic climate which have been undertaken, as well as by the extension of our knowledge concerning the movements of faunas and floras over the submerged and emerged portions of the earth's crust. From the time of Strabo down to the present day all those who have looked intelligently on shoreline phenomena have recognized the inconstancy in the relations of sea and land. Almost all the students of such phenomena have, in their thinking, made the assumption that the changes in the positions of the shoreline were due to the simple process of up or down movement of the crust where the changes in elevation have occurred. Strabo saw, and briefly indicated in his writings, that we must take into account not only the swayings of the land itself, but those movements of the deep-sea bottoms which, by displacing the ocean waters, might cause them to flow over or recede from the land masses.

Yet, further, we note that the accumulation during glacial epochs of great ice-masses about either pole would in two ways serve to change the ocean plane—by the withdrawal of water from the ocean and by the irregular attractions of a gravitative nature which the masses would exert on the seas. If, as J. Adehemar has pointed out, the ice-cap were accumulated about one pole to the thickness which is sometimes estimated as having occurred in the last glacial period, the displacement of the earth's center of gravity might amount to half a mile or more. Although this estimate has little value in a quantitative way, it is evident that ice-cap displacement must be reckoned as among the considerable influences which from time to time affect the equation of causes which determine the level of the sea.

It seems to me to be evident that the position of a shoreline at any time and place is determined by an exceedingly complicated equation, in which there enter as factors not only the positive up-and-down movement of the area of the lithosphere on which the coast lies and the axial rotation or movement about a fulcrum line in relation to the shore, but also the form of the bottom of the deeper seas, the water-displacing value of other lands in their several movements of up-and-down going, the attraction of ice-caps, which rapidly vary in importance, that of mountains and other high-lying lands, which varies in a less rapid way, as well as other influences which have not been taken into account. It may be noted in passing that no reckoning has been made as to the progressive effect arising from the importation of sediments into the sea. The phenomena which are associated with this action are too complicated for discussion in this paper, where limitations as to length must be considered. It seems well, however, to advert to them.

So far we have not succeeded in obtaining data which will enable us to determine, even in an approximative way, the amount of detritus annually contributed to the bottom of the sea and there built into clastic rocks; yet we perceive that, in addition to the contribution from the lands borne in by the atmospheric agents and the waves and tides, there is probably a yet vaster amount contributed to the floor of the deep by volcanic ejections. As there is probably not more than about 10 per cent of the detrital accumulations which is occupied by the fluid, the effect of the deposition, except so far as it is compensated by the movements of the sea-bottom and the land masses, is to lift the plane of the oceanic waters in what may be termed a geologically rapid manner. This will be seen by a consideration of the following facts: The average downwearing of the land—that is, the rate of exportation of its materials to the sea-floor—is probably at least as rapid, taking into account both atmospheric and seashore actions, as 1 foot in 3,000 years. The inquiries of various naturalists concerning the Javanese volcanoes appear to indicate that 100 cubic miles or more of detrital matter has been poured forth from these vents during the last hundred years, practically all of which has found its way to the sea-floor. The total amount of this contribution of sediments to the oceans from this small but very active group of volcanoes during the time mentioned probably exceeds the supply afforded by all the rivers of North and South America.

It thus appears to me evident that the water-displacing value of the marine sediments accumulated in the course of a million of years might, if other sources of change were excluded, suffice to affect the general sealevel to the amount of some score or perhaps hundreds of feet. Owing, however, to the other sources of instability of the land, this influence is perhaps of small value in the equation which determines the position of a shoreline.

Although I do not regard the facts above noted, which appear to show the prevailing low position of the coastlines of the world, as of decisive value, they seem to me to raise the presumption that along a large part of continental coasts the average movement in relatively modern times has been of a nature to bring the coasts inwardly toward the centers of the continents. This may possibly be due to movements of the land-masses themselves, but it appears to me more likely that we should recur to the hypothesis of Strabo, briefly set forth in the opening paragraph of this paper, and account for them, at least in part, by alterations in the depth of the ocean floors.

Origin of Fiords and Estuaries

From *Bulletin of the Geological Society of America*, Vol. VI, pp. 152–153, 1895.

It seems to me unquestionable that where a coast exhibits a system of valleys affording drainage to large rivers, the basins of which have been the seat of abundant and recent degradation, and where the rivers in place of having normal deltas enter the sea by estuaries, the shapes of which can only be explained by the supposition that the margins of the marine reëntrant are the steeps of the old river valley, we are justified in assuming the subsidence of the land.

It is true that there is some difficulty encountered when we come to apply this topographic index to glaciated districts. Thus in the fiord zone, where the plane of the sea cuts the glacially worn, hard rocks, we may always assume that the ice-streams could have excavated the channels for some distance outward beyond the point where the sea, if freed from the ice, would have come in contact with the land. It is very likely, indeed, that much of the fiord topography was due to ice erosion, accomplished below the ocean level. Where, however, the valleys are broad, in the manner of that of the Saint Lawrence below Montreal and in other similar instances, the most reasonable supposition generally is that the indentation is due to the flooding of a river trough. This supposition can often be verified by the fact that rivers occupying preglacial channels converge in a normal digitating manner toward the axis of the valley. This evidence is beautifully shown in the case of such flooded valleys as that of the Chesapeake.

DUTTON

Clarence Edward Dutton (1841–1912), American army officer and geologist, was detailed for duty with the Powell Survey in 1875 and spent the next ten years in geological research in the plateau region of the western United States. In 1886, as a member of the United States Geological Survey, he made a special study of the Charleston earthquake. He returned to military duty in 1890 and while in Central America and Texas, he made frequent excursions to the volcanoes of Mexico. After his retirement from active service in 1901, "his active mind was much employed in the further study of volcanoes and earthquakes." (J. S. Diller)

THE GREAT DENUDATION

From United States Geological Survey, *Second Annual Report*, 1880–1881, pp. 95–103.

The Rate of Erosion

Erosion, viewed in one way, is the supplement of the process by which strata are accumulated. The materials which constitute the stratified rocks were derived from the degradation of the land. This proposition is fundamental in geology—nay, it is the broadest and most comprehensive proposition with which that science deals. It is to geology what the law of gravitation is to astronomy. We can conceive no other origin for the materials of the strata, and no other is needed, for this one is sufficient and its verity a thousand times proven. Erosion and "sedimentation" are the two half phases of one cycle of causation—the debit and credit sides of one system of transactions. The quantity of material which the agents of erosion deal with is in the long run exactly the same as the quantity dealt with by the agencies of deposition; or, rather, the materials thus spoken of are one and the same. If, then, we would know how great have been the quantities of material removed in any given geological age from the land by erosion, we have only to estimate the mass of the strata deposited in that age. Constrained by this reasoning, the mind has no escape from the conclusion that the effects of erosion have indeed been vast. If then, these operations have achieved such results, our wonder is transferred to the immensity of the periods of time required to accomplish them; for the processes are so slow that the span of a lifetime seems too small to render those results directly visible. As we stand before the terrace cliffs and try to conceive of them receding scores of miles by secular waste, we find the endeavor quite useless. There is, however, one error against which we must guard ourselves. We must not conceive of erosion as merely sapping the face of a straight serried wall a hundred miles long; the locus of the wall receding parallel to its former position at the rate of a foot or a few feet in a thousand years; the terrace back of its crest line remaining solid and uncut; the beds thus dissolving edgewise until after the lapse of millions of centuries their terminal cliffs stand a hundred miles or more back of their initial positions. The true story is told by the Triassic terrace ending in the Vermilion Cliffs. This terrace is literally sawed to pieces with cañons. There are dozens of these chasms opening at intervals of two or three miles along the front of the escarpment and setting far back into its mass. Every one of them ramifies again and again until they become an intricate net-work, like the fibers of a leaf. Every cañon wall, throughout its trunk, branches, and twigs, and every alcove and niche, becomes a dissolving face. Thus the lines and area of attack are enormously multiplied. The front wall of the terrace is cut into promontories and bays. The interlacing of branch cañons back of the wall cuts off the promontories into detached buttes, and the buttes, attacked on all sides, molder away. The rate of recession therefore is correspondingly accelerated in its total effect.

The largeness of the area presents really no difficulty. The forces which break up the rocks are of meteoric origin. The agency which carries off the debris is the water running in the drainage channels. Surely the meteoric forces which ravage the rocks of a township may ravage equally the rocks of the county or state, provided only the conditions are uniform over the larger and smaller areas. And what is the limit to the length of a stream, the number of its branches and rills, and to the quantity of water it may carry? It is not the area, then, which oppresses us by its magnitude, but the vertical factor—the thickness of the mass removed. But upon closer inspection the aspect of this factor also will cease to be forbidding.

For if the rate of recession of a wall fifty feet high is one foot in a given number of years, what will be (*ceteris paribus*) the rate of recession in a wall a thousand feet high? Very plainly the rate will be the same.* If we suppose two walls of equal length, composed of the same kind of rocks, and situated under the same climate, but one of them much higher than the other, it is obvious that the areas of wall-face will be proportional to their altitudes. In order that the rates of recession may be equal, the amount of material removed from the higher one must be double that removed from the other, and since the forces operating on the higher one have twice the area of attack, they ought to remove from it a double quantity, thus making the rates of recession equal. In the same way it may be shown that the rate of recession is substantially independent of the magnitude of the cliff, whatever its altitude. Here a momentary digression is necessary.

We have hitherto spoken of the recession of cliffs as if it comprised the whole process of erosion, and have hardly alluded to the possible degradation of the flat surfaces of plateaus, terraces, and plains. Is it meant that there is no degradation of the horizontal surfaces, and that the waste of the land is wholly wrought by the decay of cliffs? Approximately that is the meaning, but some greater precision may be given to the statement.

Erosion is the result of two complex groups of processes. The first group comprises those which accomplish the disintegration of the rocks, reducing them to fragments, pebbles, sand, and clay. The second comprises those processes which remove the debris and carry it away to another part of the world. The first is called disintegration; the second, transportation. We need not attempt to study these processes in all their scope and relations, but we may advert only to those considerations which are of immediate concern. When the debris produced by the disintegration of rocks is left to accumulate upon a flat surface it forms a protecting mantle to the rocks beneath, and the disintegration is greatly retarded, or even wholly stopped. In order that disintegration may go on rapidly the debris must be carried away as rapidly as it forms. But the efficiency of transportation depends upon the

---

* The geologist will no doubt recognize that this is a simple and unqualified statement of a result which is in reality very complex, and sometimes requiring qualification. But a candid review of it in the light of established laws governing erosion will, I am confident, justify it for all purposes here contemplated. Though some qualifying conditions will appear when the subject is analyzed thoroughly, they are of no application to this particular stage of the argument. The statement is amply true for the proposition in hand, and it would be hardly practicable, and certainly very prolix, to give here the full analysis of it.

declivity.   The greater the slope the greater the power of water to transport.   When the slope is greater than 30° to 33° ("the angle of repose") loose matter cannot lie upon the rocks, and shoots down until it finds a resting place.   Hence the greater the slope the more fully are the rocks exposed to the disintegrating forces, and the more rapidly do they decay.   This relation is universal, applying to all countries, and explains how it comes about that the attack of erosion is highly effective against the cliffs and steep slopes, and has but a trifling effect upon flat surfaces.

Reverting to the main argument, it now appears that erosion goes on by the decay and removal of material from cliffs and slopes; that the recession of high cliffs is as rapid as the recession of low ones, and that the quantity of material removed in a given time increases with the altitudes of the cliffs and slopes.   In other words, the thickness of the strata removed in a given period of erosion should be proportional to the amount of *relief* in the profiles of the country.   But in the Plateau country, and especially in the Grand Cañon district, these reliefs are very great.   It is a region of giant cliffs and profound cañons, and, as will ultimately appear, it has been so during a very long stretch of geological time.   The thickness of the strata removed from it is only proportional to the values of those conditions which favor rapid erosion.   In the foregoing discussion it may appear that the area of denudation in the Grand Cañon district, though large, and the thickness of the strata denuded, though very great, are not so excessive as to impose such a heavy burden upon the credulity as the first announcement of the figures portended.

Thus the stratification, the outliers, the faults and flexures, and the drainage all yield their quotas of testimony to the great fact of denudation, and indicate that at some initial epoch the whole Mesozoic system and the lower Eocene once extended over the entire platform of the Grand Cañon district, with a thickness varying somewhat, no doubt, but on the whole differing but little, from that which we now find in the terraces of the High Plateaus.

Base Levels of Erosion

In his popular narrative of Explorations of the Colorado River, Powell has employed the above term to give precision to an idea which is of much importance in physical geology. The idea in some form or other has, no doubt, occurred to many geologists, but, so far as known to me, it has not before received such definite treatment nor been so fully and justly emphasized. It may be explained as follows.

Whenever a smooth country lies at an altitude but little above the level of the sea, erosion proceeds at a rate so slow as to be merely nominal. The rivers cannot corrade their channels. Their declivities are very small, the velocities of their waters very feeble, and their transporting power is so much reduced that they can do no more than urge along the detritus brought into their troughs from highlands around their margins. Their transporting power is just equal to the load they have to carry, and there is no surplus left to wear away their bottoms. All that erosion can now do is to slowly carry off the soil formed on the slopes of mounds, banks, and hillocks, which faintly diversify the broad surrounding expanse. The erosion is at its base-level or very nearly so. An extreme case is the State of Florida. All regions are tending to base-levels of erosion, and if the time be long enough each region will, in its turn, approach nearer and nearer, and at last sensibly reach it. The approach, however, consists in an infinite series of approximations like the approach of a hyperbola to tangency with its asymptote. Thus far, however, there is the implied assumption that the region undergoes no change of altitude with reference to sea-level; that it is neither elevated nor depressed by subterranean forces. Many regions do remain without such vertical movements through a long succession of geological periods. But the greater portion of the existing land of the globe, so far as known, has been subject to repeated throes of elevation or depression. Such a change, if of notable amount, at length destroys the pre-existing relation of a region to its base-level of erosion. If it is depressed it becomes immediately an area of deposition. If it is elevated new energy is imparted to the agents and machinery of erosion. The declivities of the streams are increased, giving an excess of transporting power which sweeps the channels clear of debris; corrasion begins; new topographical features are literally carved out of the land in high relief; long rapid slopes or cliffs are generated and vigorously attacked by the destroying agents; and the degradation of the country proceeds with energy.

It is not necessary that a base-level of erosion should lie at extremely low altitudes. Thus a large interior basin drained by a trunk river, across the lower portions of which a barrier is slowly rising, is a case in point. For a time the river is tasked to cut down its barrier as rapidly as it rises. This occasions slackwater in the courses above the barrier and stops corrasion, producing ultimately a local base-level. Another case is the Great Basin of Nevada. It has no outlet, because its streams sink in the sand or evaporate from salinas. Its valley bottoms are rather below base-level than above it. The general result of causes tending to bring a region to an approximate base-level of erosion is the obliteration of its inequalities.

During the progress of the great denudation of the Grand Cañon District the indications are abundant that its interior spaces have occupied for a time the relation of an approximate base-level of erosion. Throughout almost the entire stretch of Tertiary and Quaternary time the region has been rising, and in the aggregate the elevation has become immense, varying from 11,000 to 18,000 feet in different portions. But it seems that the movement has not been at a uniform rate. It appears to have proceeded through alternations of activity and repose. Whether we can point to more than one period of quiescence may be somewhat doubtful, but we can point decisively to one. It occurred probably in late Miocene or early Pliocene time, and while it prevailed the great Carboniferous platform was denuded of most of its inequalities, and was planed down to a very flat expanse. Since that period the relation has been destroyed by a general upheaval of the entire region several thousands of feet.

ISOSEISMAL LINES AND EARTHQUAKE INTENSITY

From "The Charleston Earthquake of August 31, 1886," United States Geological Survey, *Ninth Annual Report*, 1889, pp. 209–528.

In selecting a scale of intensity based upon the sensible effects of the earthquake the Rossi-Forel scale was finally chosen. The first series of circulars distributed immediately after the earthquake embraced a scale established by Prof. C. G. Rockwood, containing only five degrees of intensity. It was found, however, that in most cases it was possible to distinguish, with a fair degree of confidence, intermediate degrees of intensity, and as the Rossi-Forel scale ranged through ten degrees, it was thought best to employ it in all subsequent circulars, and also to re-estimate the intensities given in the Rockwood scale in conformity with Rossi-Forel. This re-estimate was found to be quite practicable in most

Fig. 31.—Dutton's map showing isoseismal lines for the Charleston earthquake, 1886.

cases. It was not, however, free from difficulty, and one of a curious kind is worth recording. After estimating some hundreds of reports, I found to my surprise that my own subjective standard

(I can think of no other term to characterize it) had undergone a gradual change. Effects which were adjudged to indicate an intensity of four or five in the earlier stages of the process were later on adjudged to indicate a lower and lower intensity. This necessitated beginning the process over again. In spite of precautions the same defect repeated itself. It was not until the fourth or fifth trial that this unsteadiness of the mental machine could be prevented.

[Fig. 31] exhibits the isoseismals. Probably no two persons taking the same data independently would draw exactly similar lines. Nevertheless it is believed that all of the principal features of the drawing here given would appear in one form or another if a similar work were undertaken by any other person.

It is to the irregularities of these lines that we must look for instruction or suggestion. But they suggest little. Several areas of earthquake shadow are indicated, of which perhaps the most notable is in West Virginia. A closer grouping of the isoseismals along the coast than in the interior also appears, the explanation of which has already been proposed. It will be perceived that the intervals between isoseismals is not very unequal from the epicentrum to the remotest points of sensibility. If this be a true representation of the variation of intensity, and if the intervals between any two correspond to a difference of intensity equal for all intervals, then it would follow that the intensity was, on the average, inversely proportional to the distance in a simple reciprocal ratio. But according to universal laws governing all radiant energy or energy moving away from a center, it should diminish as the *square* of the distance increases, provided no energy is dissipated in the transmission. But if energy be dissipated, the rate of diminution must be still greater. Here is suggested at once the need of a much more definite understanding as to the relative degrees of intensity the several isoseismals ought to express. As the case now stands there is no assignable relation of one isoseismal to another. To ascertain what relations they might express we must have recourse to theory.

We are to consider how the intensities at different distances would compare with each other on the assumption that no energy is lost in the transmission.

The following table will show the ratio of the intensity at distant points to that at the epicentrum, on the assumption that the depth of the focus is 12 miles:

Distance from epicentrum, *miles*	Ratio of intensity	Ratio of amplitude	Distance from epicentrum, *miles*	Ratio of intensity	Ratio of amplitude
12	.5	.7	300	.0016	.04
24	.2	.447	400	.0009	.03
36	.1	.316	500	.00058	.024
48	.0588	.242	600	.00040	.020
60	.0389	.197	700	.00029	.017
72	.027	.164	800	.000225	.015
84	.020	.141	900	.000178	.0133
96	.0156	.125	1,000	.000144	.0120
100	.0142	.119	1,100	.000119	.0109
200	.0036	.06	1,200	.0001	.01

From this it will appear that the differences of intensity between the outer isoseismals are very small, while those between the inner isoseismals are many times greater. If our estimate of the depth of the focus is correct, then the intensity at the most distant points at which the earthquake was felt must have been less than the six-thousandth part of the intensity at the epicentrum. Even this enormous difference makes no allowance for any loss of energy in transit, which must correspondingly magnify the disproportion. It is evident that without accurate measurements of intensity and a large number of them the outer isoseismals can convey but little meaning. But an important question may be raised here. Is it not possible that the intensity of every earthquake may be considered as approximately the same at the extreme distances at which it is felt? If this were true we should then have one ordinate, we should be able to find the whole curve, and thus compare all earthquakes with each other. The answer to this question is so qualified as to be of little value at present. It will appear in the following chapter that when different waves are compared the effects they produce are dependent not wholly upon their intensity (which will be defined to be the amount of energy per unit area of wave front); but with equal energies the effects will differ with the varying periods, wave lengths, and amplitudes, in such a way that no exact comparison is possible until we know all these factors which go collectively to make up the intensity. Furthermore, the intensity of a single wave may be great and yet produce a smaller perceptible effect than the cumulative result of a considerable number of waves each of much smaller intensity. Thus the duration of the shocks may, and in most cases probably does, constitute an important factor. And, finally, we do not know what proportion of the energy is lost in the propagation of the waves. This loss of energy must modify the fundamental law of variation of intensity with the distance which we are here assuming.

Notwithstanding these difficulties, all of which must qualify our conclusions, it is believed that they may possibly be insufficient to wholly vitiate them and to render them absolutely worthless. It certainly seems as if the smallest intensities which seismoscopes of uniform pattern would record upon the outermost isoseismals may represent energies in different earthquakes which are not very unequal. If this be so, then the mean distance of such an isoseismal from the origin would become a measure of the total energy of the earthquake, not however in absolute, but in relative, terms. The total energy would in that case be proportional to the square of the greatest distance at which such an intensity was manifested. If we can determine any two ordinates of the intensity curve, or one ordinate and the depth of the focus, we can determine the entire curve.

These considerations suggest a method by which we may seek to establish some system of isoseismals which will be more rational than the arbitrary one now in general use. By the present method the isoseismals express the wall-cracking power, the chimney-breaking power, the man-frightening power, etc. They really express very little of what we wish to know. Although they indicate that one area was more forcibly shaken than another, they do not convey the remotest idea *how much* more forcibly. What we really want is a system which will express graphically the ratio which the energy displayed in one region bears to that displayed in another in terms of some unit of energy. I believe such a system may in future become possible though I concede the present difficulty of procuring the data for it. The mathematical theory upon which it may be founded is soon stated. Its practical exemplification must be a matter of future research.

A system of isoseismals as here contemplated would express the relation between two variables, viz: (1) the distance from the epicentrum and (2) the intensity, or some function of the intensity, measured in units of energy. The independent variable must obviously be the distance from the epicentrum; the function must be the intensity. For every observation a map and a scale gives us the distance. It remains to measure the intensity. The only possible source of information at present available for this purpose is the seismograph. It measures amplitudes and wave periods. If we also know either the wave speed or the wave length the record would be complete. Here is a serious difficulty at once. Although the Charleston earthquake has probably settled the first near approximation to the true speed of a deeply seated wave, it does not affect the question of the speed of those transformed surface vibrations which the seismograph measures. We may hope however, as the result of such experiments as those begun by Mallet and prosecuted by Ewing and Milne in Japan and by Fouqué and Michel-Lévy in France, to ascertain something more definite about wave speeds in the surface soils and in different kinds of rocky strata near the earth's surface.

There is another difficulty to be met, viz: the fact that amplitude and wave period both vary in closely adjoining localities. These differences are due presumably to surface conditions, *i.e.*, to the kind of rock or soil which constitute the superficial masses. Time and experience, however, with multiplied observations, may enable us to judge what allowance must be made for such perturbations. In the long run and with many observations such discordances may be treated as accidental errors, which may be in greatest part eliminated.

The question as to the dissipation of energy in transmission is one which may be disposed of with little comment. The proportion of energy so lost is probably small in the depths of the earth. Taking an analogy from light, the case, so far as regards reflection, is probably analogous to that of rays passing through a transparent medium and falling at its boundary upon a rough black surface.

The foregoing considerations may be of utility in determining future methods of observing earthquake phenomena. They indicate the importance of establishing a large number of seismoscopes, or instruments for merely noting the fact of a tremor of an intensity not less than some definite degree and the time of its occurrence; secondly, a moderate number of seismographs located in carefully selected places and in groups, which may give indications of amplitude and wave period. The first class of instruments are cheap and simple; the second, costly and elaborate. Both may be necessary to the complete scientific investigation of the earthquake.

Isostasy

From *Bulletin of the Philosophical Society of Washington*, Vol. XI, pp. 51–64, 1889.

If the earth were composed of homogeneous matter its normal figure of equilibrium without strain would be a true spheroid of revolution; but if heterogeneous, if some parts were denser or lighter than others, its normal figure would no longer be spheroidal. Where the lighter matter was accumulated there would be a tendency to bulge, and where the denser matter existed there would be a tendency to flatten or depress the surface. For this condition of equilibrium of figure to which gravitation tends to reduce a planetary body, irrespective of whether it be homogeneous or not, I propose the name *isostasy*. . . . An isostatic earth, composed of homogeneous matter and without rotation, would be truly spherical. If slowly rotating it would be a spheroid of two axes. If rotating rapidly within a certain limit, it might be a spheroid of three axes.

But if the earth be not homogeneous—if some portions near the surface be lighter than others—then the isostatic figure is no longer a sphere or spheroid of revolution, but a deformed figure bulged where the matter is light and depressed where it is heavy. The question which I propose is: How nearly does the earth's figure approach to isostasy.

Mathematical statistics alone will not enable us to answer this question with a sufficient degree of approximation. It does, indeed, enable us to fix certain limits to the departure from isostasy which cannot be exceeded. This very problem has been treated with great skill by Prof. George Darwin.

But this problem may be approached from another direction with more satisfactory results. Geology furnishes us with certain facts which enable us to draw a much narrower conclusion. There are several categories of fact to which we may turn. One of the most remarkable is the general fact that where great bodies of strata are deposited they progressively settle down or sink seemingly by reason of their gross mechanical weight, just as a railway embankment across a bog sinks into it. The attention of the earlier Appalachian geologists was called, as soon as they had acquired a fair knowledge of their field, to the surprising fact that the paleozoic strata in that wonderful belt, though tens of thousands of feet in thickness, were all deposited in comparatively shallow water. . . . No conclusion is left us but that sinking went on *pari passu* with the accumulation of the strata. When the geology of the Pacific coast was sufficiently disclosed, the same fact confronted us there. As investigation went on the same fact presented itself over the western mountain region of the United States. One of the most striking cases is the Plateau Country. This great region, near 100 000 square miles in area, lying in the adjacent parts of Colorado, Utah, New Mexico, and Arizona, discloses from 8 000 to 12 000 feet of mesozoic and cenozoic strata. Here the proof is abundant that the surface of the strata was throughout that vast stretch of time never more than a few feet from sea level. Again and again it emerged from the water a little way, only to be submerged. . . . In short it may be laid down as a general rule that where great bodies of sediment have been deposited over extensive areas their deposition has been accompanied by a subsidence of the whole mass.

The second class of facts is even more instructive, and stands in a reciprocal relation to those just mentioned. Wherever broad mountain platforms occur and have been subjected to great erosion the loss of altitude by degradation is made good by a rise of the platform. . . .

It seems little doubtful that these subsidences of accumulated deposits and these progressive upward movements of eroded mountain platforms are, in the main, results of gravitation restoring the isostasy which has been disturbed by denudation on the one hand and by sedimentation on the other. The magnitude of the masses which thus show the isostatic tendency are in some cases no greater than a single mountain platform, less than 100 miles in length, from 20 to 40 miles wide and from 2500 to 3500 feet mean altitude above the surrounding lowlands. . . . It is extremely probable that small or narrow ridges are not isostatic with respect to the country round about them. Some volcanic mountains may be expected to be non-isostatic, especially isolated volcanic piles.

It remains to inquire what is the resulting direction of motion. The general answer is, towards the direction of least resistance. The specific answer, which must express the direction of least resistance, will, of course, turn upon the configuration of the deposition on the one hand, and of denudation on the other, and also upon the manner in which the rigidity or viscosity varies from place to place. Taking, thus, the case of a land area undergoing denudation, its detritus carried to the sea and deposited in a heavy littoral belt, we may regard the weight of each elementary part of the deposited mass as a statical form acting upon a viscous support below. Assuming that we could find a differential expression applicable to each and every element of the mass and a corresponding one for the resistance offered by the viscosity, the integration for the entire mass might give us a series of equipotential surfaces within the mass. The resultant force at any point of any equipotential surface would be normal to that surface. A similar construction may be applied to the adjoining denuded area, in which the defect of isostasy may be treated as so much mass with a negative algebraic sign. The resultants normal to the equipotential surfaces would, in this case, also have the negative sign. The effective force tending to produce movement would be the arithmetical sum of the normals or of a single resultant compounded of the two normals. From this construction we may derive a force which tends to push the loaded sea bottoms inward upon the unloaded land horizontally.

This gives us a force of the precise kind that is wanted to explain the origin of systematic plications.

LOSSEN

Karl August Lossen (1841–1893), German geologist, member of the Geological Survey of Prussia and professor at the University of Berlin, was noteworthy for his work on dynamic metamorphism.

THE DISTINCTION BETWEEN CONTACT AND
REGIONAL METAMORPHISM

Translated from *Jahrbuch der königlicher preussischen geologischen Landesan-
stalt und Bergakademie*, Berlin, 1883, pp. 619–642, 1884.

I have frequently emphasized the importance of those metamorphic areas in which eruptive rocks, enclosed between the beds
and merely passive in the folding and mountain-building process,
have plainly undergone transformation in substance and structure
to approximately the same degree as the adjoining beds; such may
be the case either in the contact zone of eugranitic eruptive masses
caught in the folding or as a result of the dislocation process.*

Such metamorphosed eruptive rock is of the greatest value for
the study of metamorphism if one of the family of "hard rocks of
definitely determined mineralogical composition, chemical content,
and structure" can be established definitely as its mother rock. In
the primary minerals and primary structures of the solid rock, we
possess a well-known constant on which we can base our decision;
as a definite measure of change, the type and character of those
secondary minerals and structures which determine approximately
the nature of metamorphic rocks can be ascertained.

Frequently it is merely the certain, incontestable fact of pseudo-
morph formation that serves as proof of the transformation in such
rocks, much more commonly than the fossils, which survive
in metamorphosed sediments only under especially favorable
circumstances.

For example, I would cite the feldspar pseudomorphs filled with
secondary orthoclase mosaic and in part also with tourmaline,
which I described in the biotite-rich hornfels zone (gabbro contact

---

* Dislocation process—Lossen's term for regional metamorphism.—EDITORS.

metamorphism) of the ancient plutonic syenite porphyry of the
Schmalenberge near Harzburg, or the uralitization of the augite
diabase which predominates in the hornfels zone surrounding granite, identified today not only in the Harz, but by Philipps and All-
port in Cornwall in England and by Michel-Lévy in the Mâcon
area in France. . . .

Such observations acquire a real and positive worth in my judgment, not when they are described as isolated phenomena, as is
ordinarily done, but only if it is proved that they are characteristic
phenomena within a defined geological area and are considered as
relic traces of the geologic history of the segment of the earth
concerned.

It is in this geological respect that the proof, discovered in the
Harz and confirmed in Cornwall and the Mâconnais, that there
is a phase of the diabase characterized by definitely determined
properties at the granite contact, is thoroughly significant.   In
this phase, the uralitic hornblende which constitutes only one of its
characteristic properties appears clearly, along with the partial
preservation of the primary structure of the diabase. . . . With
more or less complete external alteration of their specific, eruptive-
rock characteristics, the rocks have definitely become diabase
hornstone.*   One type of these has the appearance of schist horn-

---

* I have used this term in the text of the Harzgerode sheet.   It seems to me
just as appropriate as schist, lime, or graywacke hornstone, as all these rocks,
the names of which are added to the word hornstone, are due to precisely
the same metamorphosing influence in the contact or hornstone ring surrounding the granite stocks.   Thus they have many mineral components in common,
as, for example, pyrrhotite, biotite, and uralitic or aluminous, actinolite-like
hornblende.   The name hornstone is a well-established expression, used for
the non-schistose rock of granite contact metamorphism.   The word diabase
hornstone can only seem ambiguous if the word hornstone is extended, as was
done recently by Herr Schenck in his able dissertation *Über die Diabase des
oberen Ruhrthals und deren Contacterscheinungen mit dem Lenneschiefer*, to
include the harder, metamorphosed sedimentary rocks, hitherto termed horn
schists, adinole schists, or adinole stone.   I cannot recommend such an extension because, so far as I have had opportunity to become acquainted with
them, granite contact rocks and diabase contact rocks seem to belong to two
entirely different categories of rock formation.   The former originate under
the action of granite on an essentially well-folded, sedimentary series with its
eruptive inclusions.   The latter, on the contrary, certainly do not owe their
present nature solely to the action of the diabase eruption before the folding.
Their regional variety in one and the same mountain is an objection to this, as

stone from the formation of numerous biotite flakes or from the accumulation of ferric pigments instead; another type has the appearance of lime hornstone from the separation of secondary lime silicate. Wherever seen, this diabase hornstone appears to be more massive than its mother rock, the diabase, as though it were impregnated with something which makes it denser, harder, and more finely splintery. . . . In diabase hornstone as in schist hornstone, the content of brown phyllitic mineral, instead of chlorite and sericitic mica, is always secondary, and is characteristic of contact metamorphism by granite. But just as characteristic of this metamorphism of crystalline as of bedded rocks is the fact that, in spite of the presence of so many mica plates, all tendency toward schistose structure* is lacking. Aside from the thermal effect and other conditions accompanying the eruption of granite, it is manifest that the ascending granite magma exerts an entirely different type of pressure action on its sheath during the formation of hornstone, from that which results in compacting, crushing, and tearing, under the friction-developing conditions of intermittent folding, or in compression within the still flexible masses of the earth crust, which manifests itself as dislocation metamorphism. In the literature mentioned earlier, I have furnished proofs from the Harz, the Rhenish-Westphalian Schiefergebirge, and the Alps that these last actions have transformed not only the sediments but also the eruptive rocks which solidified long ago. If I refer again to the Harz and . . . especially to the regional metamorphic zone of Wippra in the southeast of this mountain, I do so because the observations from a single geologic area may be compared better with each other than those from different areas. The still more conspicuous advantage of the Harz Mountains is that a narrowly limited area contains . . . a single sedimentary series with its diabase enclosures, first in the stage of granite contact metamorphism as hornstone and nodular schist with diabase hornstone,

---

is the increase in their crystalline development where, as in the southeast Harz, the shales of the region ordinarily show a phyllitic habit at the same time that the diabase shows an increase in the metamorphic phenomena. These are circumstances which indicate that the original contact metamorphism was followed by a dislocation metamorphism. In the future we shall have to distinguish more sharply between contact forms actively involved in mountain building and passively enfolded eruptive rocks.

* But not every tendency toward parallel bedding.

then as normal shale with normal diabase, although rich in chlorite and calcite, and finally in a state of strikingly marked* dislocation metamorphism as phyllite with streaked diabase and schist diabase.

Each of these three conditions, occurring in the same formation-members in different parts of the Harz, reflects a different act of the mountain-building process by virtue of various modes of action on the same geological substratum.   In the wide Rhenish-Westphalian Schiefergebirge, which first led me to recognize the mountain-building process as the origin of regional metamorphism . . . there is lacking that exposure of the granite or any other eugranitic eruptive rock playing the same geologic role, and therefore it is not possible to make a comparison between the results of granite

---

* The term "strikingly marked" seems necessary because the so-called normal shale, as exemplified in the Wieder shale of the Middle Harz, particularly in the south half, can in no way be considered simply as deposited in its present state or even completely formed before folding and mountain building.   It can only be considered as different in degree but not in kind from the phyllite or nodular clay schist.   It must be an effect of dislocation metamorphism which has left a greater or less part of the primary substrata of sedimentary material more or less unaltered.   The argument for the original simple formation of clay schists which F. Zirkel derived from the clay schist needles, since recognized as rutile, will probably be abandoned by the discoverer of the needles in the light of our present knowledge.   The reasoning brought forward in those pioneer treatises on the microscopic examination of these rocks was contradicted in the very beginning by the fact that the schist and slate, which were always sliced by Zirkel in the direction of the schist plane, were partly bedded schist and partly transverse schist—as, for instance, the Goslar slates— so that the observed crowding of the rutile needles in the plane of the section cannot be considered a result of an original sedimentary stratification.   The way the formation of the rutile microliths (in case such are present at all) keeps step in the sediments with the more or less crystalline organization of the rock bespeaks a secondary metamorphic origin for these.   In the phyllite of Wippra and in the phyllitic clay schist of Krebsbachthal near Magdesprung in the forecourt of the contact zone about the Rammberg granite, the microliths are more numerous and pronounced than in the ordinary Wieder shales near Harzgerode.   In the feldspar-bearing, gneisslike hornstone north of Rosstrappe Tavern and in hornstones from the vicinity of Ilsenstein they occur in much more compact little columns than in the nodular clay schists of the outer ring about Rammberg.   Their occurrence in the chlorite of metamorphose eruptive rocks also indicates this. . . . Interesting as the clay schist needles are, in percentage they are in a minority in comparison with the phyllitic minerals, mica and chlorite, the secondary nature of which, save for coarser fragments, is clearly shown by comparison with the graywacke, clay schist, phyllite, and hornstone of the Harz and other areas.

contact metamorphism and those of dislocation metamorphism. In the Harz, on the contrary, this not unimportant comparison is possible under relatively favorable circumstances.

The differences between the mineral composition of the material affected by granite contact metamorphism and that which has undergone regional or dislocation metamorphism display the most complete gradations in one and the same paleozoic terrain. Such a transition confirms the conclusion already derived from the broad study of the geologic role of forecourts and of atypical contact zones (chlorite-bearing hornstone near Friedrichsburnn, etc.), that on the whole no absolute geological distinction between granite contact metamorphism and dislocation metamorphism can be drawn.

This conclusion, however, does not annul the quite evident relative differences which have been mentioned. Indeed, additional differences should be noted. Other silicates of the hornstone, geologically and chemically related to garnet, such as vesuvianite and green augite, are lacking. The limestones of the regional metamorphic zones of the Harz do not form marble. Epidote is common to both types of metamorphism, and albite also. Cordierite, fluorite, pyrrhotite, tourmaline, titanite, and rutile, at least so far as known locally, play a conspicuously more dominant role in the contact zone, partly by their number, partly by the size of the individuals increased by accretion, than they do in regional metamorphic zones. Conversely, free iron oxide is much more characteristic for rocks of the latter zone, whereas frequently, although not always, it is of minor importance in the formation of hornstone. The same is true concerning admixtures of organic substances. All in all the observed differences of material indicate that in the Harz the granite contact metamorphism took place at a higher temperature than the dislocation metamorphism, and that it was consummated under special, locally varying conditions (emanations of boron and fluorine compounds) which, outside the contact zones or the surrounding forecourts, have left no trace of their activity, save in the ore veins.

BICKERTON

Alexander William Bickerton (1842–1914), professor of chemistry in Canterbury College, University of New Zealand, contributed many papers concerning the origin of the planets, most of which were published in New Zealand periodicals.

COLLISION AND ACCRETION HYPOTHESIS OF EARTH ORIGIN

From *Transactions and Proceedings of the New Zealand Institute,*Wellington, Vol. XIII, art. 14, pp. 154–159, and art. 13, pp. 149–154, 1880.

The order displayed in the structure of the Solar System strongly suggests the idea that it must have originated in some single event. Laplace has calculated that the probability of such a system having originated in a common cause, is not less than four millions to one. It is evident, therefore, that its origin is a legitimate subject for scientific speculation, and in order to account for the peculiarities of the motions of the planets and satellites, Laplace himself suggested the now well-known hypothesis of the release of nebular rings and their subsequent coalescence. This theory has found many supporters, but when it is examined in the light of the doctrine of the conservation of energy and the dynamical theory of gases, so many difficulties present themselves as to throw great doubt upon it, in fact, as Denison says, it has been so little accepted by English mathematicians that it has scarcely been discussed, and Faye has recently given his opinion that it must be given up. A modification of the theory is offered in Newcombe's "Astronomy,"—it is that the release of rings commenced on the inside.

I hope to examine a number of the difficulties of these theories in a future paper. Proctor has discussed the probability of the system having been formed by the coalescence of an immense number of meteorites, and to this hypothesis there appear to be fewer objections than to the other two. In fact it is highly probable that such an action has aided materially in giving symmetry to the system. But as the sole agent in the formation of the solar system these suggestions have two great objections:—the extreme slowness of the sun's present rotation; and the irregularities in the system, such as the eccentricity of the orbits, the inclinations of the axes and orbital planes, and retrograde motions. It cannot therefore be considered that a satisfactory solution of the problem has been given, and it is probable from its extreme difficulty, that nothing but a nearer and nearer approximation is to be expected. It is this opinion that must be my apology for bringing before the society the somewhat crude suggestions of this paper.

I recapitulate the more important of the points in what at present seems the most probable origin of the solar system.

Two rare bodies, moving with considerable velocity, rotating, and having revolving around them in all planes a large number of bodies, some of a large size, come within each other's attraction, are brought together by gravitation, and come into tangential collision. During the collision most of the accompanying bodies fly off in directions which are approximately in one plane; the component of the motion not in the plane being due to their original orbital rotation. The new orbit of all the bodies tends to be highly eccentric, but the general mass expands, and by its agency the orbit becomes nearly circular. Among the vast number of bodies thrown off during impact, the larger gradually collect the lesser up, also much of the matter that coalesces from the nebula, and many heavy molecules. Where this action is very considerable, the original mass forms so small a fraction of the final planet, that its original irregular motion almost disappears, and its axis is almost rendered perpendicular. The nebular resistance will tend to lessen the distance of the smaller bodies, and convert them into zodiacal light, or absorb them entirely into the sun, except the moons, which cannot escape the planet's attraction. All the smaller planets and those nearer the sun are robbed of their lighter molecules, and become very dense compared to the general mass of the system, but, as the nebulae contract within their orbit, they again pick up the lighter molecules which become the atmosphere.

In former papers it has been shown that the partial impact of cosmical bodies may not unfrequently produce a central mass and attendant bodies, which I have called respectively a sun or nebula, and planets. The sun is at a high temperature and rotates. The planets, in a solid, liquid, or gaseous state, revolve round in one general plane with orbits of varying area and of high eccentricity. All the motions, whether of sun or planets, have one common direction. Further it was shown that the planetary path is due to a portion of the original proper motion escaping conversion into heat at impact. For the same reason the temperature of the planet is lower than that of the sun, whose high molecular velocity, due to its temperature and comparatively small mass, may cause it to expand into a nebula.

The present paper requires that the central mass shall become a nebula, and shall expand beyond aphelion distance of the most remote planet. The forces acting on the planet will be the attraction of the nebula, gaseous adhesion while traversing the nebula, and at the same time exchange of molecules with those of the nebula. The heavier molecules will generally be attracted to the planet, while the lighter ones will leave it. The probability of such a system being formed, or the possibility of gaseous planets moving in a nebula, with its attendant effects on the size of the orbit and the change of apsides, is not treated in this paper. It is solely occupied with the change of eccentricity.

The following are five causes which are calculated to result in such a change:—

1st.   An alteration in the amount of the attractive force exerted on the planet by the nebula.

2nd.   The varying resistance and interchange of molecules incurred by the planet in its path.

3rd.   The gaseous adhesion to the planet revolving on its axis within a nebula.

4th.   The accretion of some of the vast number of small bodies which would exist in the nebula.

5th.   Some others which are too dependent upon the special character of the impact to be discussed at present.

———————

The general action of gaseous resistance is to convert the energy of the system into heat by gradually drawing the planet into the sun, or to the centre of attraction. It is maximum at perihelion, for there the density of the nebula is greater than at any other part of the orbit. Molecular exchange results from the varying densities of the different parts of the system. The planets are cooler than the central parts of the nebula, and will most likely be denser than the matter surrounding them in their path, and have sufficient attractive power to collect the heavy molecules in their vicinity. The temperature of the surface of the planet will be raised to an unknown extent by its immersion in the nebula and its progress towards perihelion. Its light molecules have their velocity so increased as to escape the planet, while the heavier molecules of the vicinity, with their lower velocity (though equal temperature), will be attracted, picked up, and become permanently part of the planet. A greater proportion of heavy molecules will be found towards perihelion, for at the centre of the nebula will probably be its greatest density, and the original expansion of the central mass into a nebula will result in the more rapid outward escape of the light molecules compared with the heavy, in obedience to the laws of gaseous diffusion. Thus the accretion of molecules to the planet will be maximum at perihelion distance. Its effect will be to retard the motion of the planet, as, in order to give its own velocity to a molecule, it will impart some of its energy. The escape of the light molecules will not affect the planet's orbit. We find therefore that gaseous resistance and molecular exchange act as resistances to planetary motion and are both maximum at perihelion, thereby decreasing aphelion distance and rendering the orbit more circular.

When the nebula has become stable, that is no longer expanding, much of its elementary matter will be at a sufficiently low temperature to combine; and these compound molecules will tend to aggregate into groups, and it may be shown that these masses may ultimately become so considerable as to form star clusters, associated by gravitation but not coalescing. I propose discussing this difficult question in a paper on star clusters; but in this paper I shall simply state that these little bodies will in all probability revolve in independent orbits at all eccentricities around the centre of the mass; these and other masses will be occasionally picked up by the planet. . . .

There can be but little doubt that this agency of accretion will be most important in giving regularity to any system. Proctor has discussed the influence of accretion of meteors, and it is certain he is right in giving it very great value. It probably played a great part in the formation of Jupiter.

LAPWORTH

Charles Lapworth (1842–1920), English geologist and paleontologist, professor of geology at the University of Birmingham, is best known for his work on graptolites. His suggestion for resolving peacefully the controversy between Sedgwick and Murchison has been generally accepted.

The Ordovician System

From *Geological Magazine*, Second Series, Decade II, Vol. VI, pp. 1–15, 1879.

At the present day it would be wholly superfluous to enter upon the discussion of the vexed question of the respective claims of Sedgwick and Murchison to the *Middle* and *Lowest* Divisions of the Lower Palæozoic Rocks. We may, however, without fear of contradiction, concede to Sedgwick the credit of having been the first to determine the limits and sequence of their larger subdivisions, and to Murchison and his followers the honour of having been the first to assign them their distinctive fossils.

For, amid all the confusion incident to this controversy, one grand fact stands out clear and patent to the most superficial student of Palæozoic geology—namely:—the strata included between the horizon marking the advent of *Paradoxides,* and the provisional line presently drawn at the summit of the Ludlow, imbed *three distinct faunas,* as broadly marked in their characteristic features as any of those typical of the accepted systems of a later age. . . .

So long as present systems of nomenclature survive, nothing can disturb the application of the title of Cambrian to the rocks of the *Primordial Series,* and that of Silurian to the strata of the *Third Fauna.* . . .

North Wales itself—at all events the whole of the great Bala district where Sedgwick first worked out the physical succession among the rocks of the intermediate or so-called *Upper Cambrian* or *Lower Silurian* system; and in all probability much of the Shelve and the Caradoc area, whence Murchison first published its distinctive fossils—lay within the territory of the *Ordovices:* a tribe as undaunted in its resistance to the Romans as the Silures. . . .

Here, then, have we the hint for the appropriate title for the central system of the lower Palæozoics. It should be called the ORDOVICIAN SYSTEM, after the name of this old British tribe.

THE SECRET OF THE HIGHLANDS

From *Geological Magazine*, Second Series, Decade II, Vol. X, pp. 120–128, 193–199, 337–344, 1883.

The final results of my investigations in the Durness-Eriboll region during August last seem to me to indicate most distinctly the probable truth of the theory which has long appeared to myself to be the only possible solution of the Highland difficulty. I believe that we have in the so-called metamorphic Silurian region of the Highlands of Scotland a portion of an old mountain system, formed of a complex of rock formations of very different geological ages. These have been crushed and crumpled together by excessive lateral pressure, locally inverted, profoundly dislocated, and partially metamorphosed. This mountain range, or plexus of ranges, which must have been originally of the general type of those of the Alps or Alleghanies, is of such vast geological antiquity, that all its superior portions have long since been removed by denudation, so that, as a general rule, only its interior and most complicated portions are preserved for us. In the area partly worked out by myself, the stratigraphical phenomena are identical in character with those developed by Rogers, Suess, Heim, and Brögger in extra-British mountain regions. They appear to me to account so naturally for the diverse views hitherto published by those who have personally studied the stratigraphy of the Northwest Highlands, and to indicate so clearly the common ground of accord upon which all parties may eventually meet, that I am emboldened to give them in outline in this place, in anticipation of a more detailed paper upon the subject, which I hope to publish elsewhere. It is for those who are interested in this great geological problem to test for themselves the truth of these conclusions, by their consonance with their own discoveries, or to point out those difficulties which at present stand in the way of their provisional adoption.

If the causes and results of overfolding of rocks under tangential thrust are correctly laid down in the preceding paragraphs, in so far as they affect the groups of strata involved in the *minor folds* upon the flanks of mountain chains, they must, theoretically, be equally true of the grander masses of strata enveloped in the *regional earthwaves,* each of which gives origin to a single chain in a mountain system.   When these mighty arches and troughs are brought closer together by the great crust-creep, the arches must rise and the intermediate troughs sink.   The great monoclinal folds defining the several mountain-flanks must gradually develope into magnificent overfolds, and these again in the process of time into overfaults, having the appearance of normal dislocations with tremendous vertical downthrow.   Only, however, in ranges geologically new, or composed of rock-formations of immense thickness, can these features be expected to be typically developed. The reader may collect abundant illustrations of this structure from the magnificent maps and sections of the Uinta and Wasatch areas in Western America, where both these conditions appear to obtain; and may observe how this theory brings into harmony many of the known facts of the interrelations of the folding, faulting, and remarkable physical geography of that wonderful region, and reconciles at the same time some of the apparently diverse interpretations of its geological development put forward by its enthusiastic describers.

In more easily convoluted regions, or in mountain chains of much higher antiquity, this inevitable relation of sunken trough to mountain arch is necessarily less conspicuous, and long eludes detection owing to the bewildering complexities of the strata involved.   But it is by no means impossible that the long straight (longitudinal or strike valleys) and so-called *anticlinal* valleys of the Scottish Highlands, such as those of the Great Glen and Loch Tay, walled in by steep hill-slopes and occupied by lakes of profound depth, are nothing more than greatly depressed intermont *synclinal troughs* owing their origin to the same causes which bring about the slow secular elevation and approximation of their flanking ranges.

GILBERT

Grove Karl Gilbert (1843–1918), American geologist, was a member of the Geological Survey of Ohio under J. S. Newberry, of the United States Surveys and Explorations West of the 100th Meridian under George M. Wheeler, and of the United States Geographical and Geological Survey of the Rocky Mountain Region under J. W. Powell. When the United States Geological Survey was established in 1879, he was appointed as geologist in that organization and continued in that association until his retirement from active service.

An Analysis of Subaerial Erosion

From *American Journal of Science*, Third Series, Vol. XII, pp. 85–103, 1876.

The deep gorges which so facilitate the examination of the strata and of their displacements, are themselves of interest as monuments of erosion. To account for their existence and unravel their history is to review the laws of erosion with great wealth of illustration. Results so extreme can have been produced only under conditions equally extreme; and natural laws are often best tested and exemplified by the consideration of their operation under exceptional circumstances. Already the problem of the cañons has been attacked, and I cannot better demonstrate its radical value than by presenting the present aspect of the case. For this purpose it is necessary to give a summary statement of the processes of erosion and of the conditions which determine its rate. The matter is so complex that this cannot be done briefly without the omission of the less important factors, and in undertaking it I shall take the liberty of either disregarding or slighting all considerations which have not an important bearing on the problem in question.

In order to analyse sub-aerial erosion, we must consider it (A) as consisting of parts, and (B) as modified by conditions.

A. All indurated rocks and most earths are bound together by a force of cohesion, which must be overcome before they can be divided and removed. The natural processes by which the division and removal are accomplished make up erosion. They are called disintegration and transportation.

Transportation is chiefly performed by running water.

Disintegration is naturally divided into two parts. So much of it as is accomplished by running water is called corrasion, and that which is not, is called weathering.

Stated in their natural order, the three general divisions of the process of erosion are (1) *weathering*, (2) *transportation*, and (3) *corrasion*. The rocks of the general surface of the land are disintegrated by *weathering*. The material thus loosened is *transported* by streams to the ocean or other receptacle. In transit it helps to *corrade* from the channels of the streams other material, which joins with it to be transported to the same goal.

(1.) In weathering the chief agents of disintegration are solution, change of temperature, the beating of rain, and vegetation.

(2.) A portion of the water of rains flows over the surface and is quickly gathered into streams. A second portion is absorbed by the earth or rock on which it falls, and after a slow underground circulation reissues in springs. Both transport the products of weathering, the latter carrying dissolved minerals, and the former chiefly undissolved.

(3.) In corrasion the agents of disintegration are solution and mechanical wear. Wherever the two are combined, the superior efficiency of the latter is evident, and in all fields of rapid corrasion the part played by solution is so small that it may be disregarded.

The mechanical wear of streams is performed by the aid of hard mineral fragments which are carried along by the current. The effective force is that of the current; the tools are mud, sand, and bowlders. The most important of them is sand; it is chiefly by the impact and friction of grains of sand that the rocky beds of streams are disintegrated.

Streams of clear water corrade their beds by solution. Muddy streams act partly by solution, but chiefly by attrition.

Streams transport the combined products of corrasion and weathering. A part of the debris is carried in solution, and a part mechanically. The finest of the undissolved detritus is held in suspension; the coarsest is rolled along the bottom; and there is a gradation between the two modes. There is a constant comminution of all the material as it moves, and the work of transportation is thereby accelerated. Bowlders and pebbles, while they wear the stream-bed by pounding and rubbing, are worn still more rapidly themselves. Sand grains are worn and broken by the continued jostling, and their fragments join the suspended mud.

Finally the detritus is all more or less dissolved by the water, the finest the most rapidly.

B. The chief conditions which affect the rapidity of erosion are (1) declivity, (2) character of rock, and (3) climate.

(1.) In general, *erosion is most rapid where the slope is steepest;* but weathering, transportation and corrasion are affected in different ways and in different degrees.

With increase of slope goes increase in the velocity of running water, and with that goes increase in its power to transport undissolved detritus.

The ability of a stream to corrade by solution is not notably enhanced by great velocity; but its ability to corrade by mechanical wear keeps pace with its ability to transport, or may even increase more rapidly. For not only does the bottom receive more blows in proportion as the quantity of transient detritus increases, but the blows acquire greater force from the accelerated current, and from the greater size of the moving fragments. It is necessary, however, to distinguish the ability to corrade from the rate of corrasion, which will be seen farther on to depend largely on other conditions.

Weathering is not directly influenced by slope, but it is reached indirectly through transportation. Solution and frost, the chief agents of rock decay, are both retarded by the excessive accumulation of disintegrated rock. . . . If, however, the power of transportation is so great as to remove completely the products of weathering, the work of disintegration is thereby checked; for the soil, which weathering tends to accumulate, is a reservoir to catch rain as it reaches the earth, and store it up for the work of solution and frost, instead of letting it run off at once unused.

(2.) Other things being equal, *erosion is most rapid when the eroded rock offers least resistance;* but the rocks which are most favorable to one portion of the process of erosion, do not necessarily stand in the same relation to the others. . . .

(3.) The influence of climate upon erosion is less easy to formulate. The direct influences of temperature and rainfall are comparatively simple, but their indirect influence, through vegetation, is complex, and is in part opposed to the direct influence of rainfall.

All the processes of erosion are affected directly by the amount of rainfall, and by its distribution through the year. All are accelerated by its increase and retarded by its diminution. When it is concentrated in one part of the year at the expense of the remainder, transportation and corrasion are accelerated, and weathering is retarded.

————————

Transportation and Comminution.   A stream of water flowing down its bed expends an amount of energy that is measured by the quantity of water and the vertical distance through which it descends.   If there were no friction of the water upon its channel the velocity of the current would continually increase; but if, as is the usual case, there is no increase of velocity, then the whole of the energy is consumed in friction.   The friction produces inequalities in the motion of the water, and especially induces subsidiary currents more or less oblique to the general onward movement.   Some of these subsidiary currents have an upward tendency, and by them is performed the chief work of transportation.   They lift small particles from the bottom and hold them in suspension while they move forward with the general current. The finest particles sink most slowly and are carried farthest before they fall.   Larger ones are barely lifted, and are dropped at once.   Still larger are only half lifted; that is, they are lifted on the side of the current and rolled over, without quitting the bottom. And finally there is a limit to the power of every current, and the largest fragments of its bed are not moved at all.

There is a definite relation between the velocity of a current and the size of the largest bowlder it will roll.   It has been shown by Hopkins that the weight of the bowlder is proportioned to the sixth power of the velocity.   It is easily shown also that the weight of a suspended particle is proportioned to the sixth power of the velocity of the upward current that will prevent its sinking.   But it must not be inferred that the total load of detritus that a stream will transport bears any such relation to the rapidity of its current. The true inference is, that the velocity determines the limit in coarseness of the detritus that a stream can move by rolling, or can hold in suspension.

. . . If, for the sake of simplicity, we suppose the whole load of a stream to be of uniform particles, then the measure of the energy consumed in their transportation, is their total weight multiplied by the distance one of them would sink in the time occupied in their transportation.   Since fine particles sink more slowly than coarse, the same consumption of energy will convey a greater load of fine than of coarse.

. . . As the energy expended in transportation increases, the velocity diminishes. If the detritus be composed of uniform particles, then we may also say that as the load increases, the velocity diminishes. But the diminishing velocity will finally reach a point at which it can barely transport particles of the given size, and when this point is attained, the stream has its maximum load of detritus of the given size. But fine detritus requires less velocity for its transportation than coarse, and will not so soon reduce the current to the limit of its efficiency. A greater per cent of the total energy of the stream can hence be employed by fine detritus than by coarse.

Thus the capacity of a stream for transportation is enhanced by comminution in two ways. Fine detritus, on the one hand, consumes less energy for the transportation of the same weight, and on the other, it can utilize a greater portion of the stream's energy.

Transportation and quantity of water. The friction of a stream upon its bed depends on the character of the bed, on the area of the surface of contact, and on the velocity of the current. When the other elements are constant, the friction varies directly with the area of contact. The area of contact depends on the length and form of the channel, and on the quantity of water. For streams of the same length, and the same form of cross-section, but differing in size of cross-section, the area of contact varies directly as the square root of the quantity of water. Hence, *ceteris paribus*, the friction of a stream on its bed, is proportioned to the square root of the quantity of water. But, as stated above, the total energy of a stream is proportioned directly to the quantity of water. And also, the total energy is equal to the energy spent in friction, plus the energy spent in transportation. Whence it follows, that if a stream change its quantity of water without changing its velocity or other accidents, the total energy will change at the same rate as the quantity of water, the energy spent in friction will change at a less rate, and the energy remaining for transportation will change at a greater rate.

It follows, as a corollary, that the running water which carries the debris of a district, loses power by subdivision toward its sources; and that, unless there is a compensating increment of declivity, the tributaries of a river will fail to supply it with the full load which it is competent to carry.

LAND SCULPTURE

From United States Geographical and Geological Survey of the Rocky Mountain Region, *Geology of the Henry Mountains*, pp. 115–141, 1877.

Erosion may be regarded from several points of view. It lays bare rocks which were before covered and concealed, and is thence called *denudation*. It reduces the surfaces of mountains, plateaus, and continents, and is thence called *degradation*. It carves new forms of land from those which before existed, and is thence called *land sculpture*. In the following pages it will be considered as land sculpture, and attention will be called to certain principles of erosion which are concerned in the production of topographic forms. . . .

Sculpture and Structure; the Law of Structure. We have already seen that erosion is influenced by rock character. Certain rocks, of which the hard are most conspicuous, oppose a stubborn resistance to erosive agencies; certain others, of which the soft are most conspicuous, oppose a feeble resistance. Erosion is most rapid where the resistance is least, and hence as the soft rocks are worn away the hard are left prominent. The differentiation continues until an equilibrium is reached through the law of declivities. When the ratio of erosive action as dependent on declivities becomes equal to the ratio of resistances as dependent on rock character, there is equality of action. In the structure of the earth's crust hard and soft rocks are grouped with infinite diversity of arrangement. They are in masses of all forms, and dimensions, and positions; and from these forms are carved an infinite variety of topographic reliefs.

In so far as the law of structure controls sculpture, hard masses stand as eminences and soft are carved in valleys.

The Law of Divides. We have seen that the declivity over which water flows bears an inverse relation to the quantity of water. If we follow a stream from its mouth upward and pass successively the mouths of its tributaries, we find its volume gradually less and less and its grade steeper and steeper, until finally at its head we reach the steepest grade of all. If we draw the profile of the river on paper, we produce a curve concave upward and with the greatest curvature at the upper end. The same law applies to every tributary and even to the slopes over which the freshly fallen rain flows in a sheet before it is gathered into rills. The nearer the water-shed or divide the steeper the slope; the farther away the less the slope. . . .

Under the *law of Structure* and the *law of Divides* combined, the features of the earth are carved. Declivities are steep in proportion as their material is hard; and they are steep in proportion as they are near divides. The distribution of hard and soft rocks, or the geological structure, and the distribution of drainage lines and water-sheds, are coefficient conditions on which depends the sculpture of the land.

Systems of Drainage. To know well the drainage of a region two systems of lines must be ascertained—the drainage lines and the divides. The maxima of surface on which waters part, and the minima of surface in which waters join, are alike intimately associated with the sculpture of the earth and with the history of the earth's structure; and the student of either sculpture or history can well afford to study them. In the following pages certain conditions which affect their permanence and transformations are discussed.

The Stability of Drainage Lines. In corrasion the chief work is performed by the impact and friction of hard and heavy particles moved forward by running water. They are driven against all sides of the channel, but their tendency to sink in water brings them against the bottom with greater frequency and force than against the walls. If the rate of wear be rapid, by far the greater part of it is applied to the bottom, and the downward corrasion is so much more powerful than the lateral that the effect of the latter is practically lost, and the channel of the stream, without varying the position of its banks, carves its way vertically into the rock beneath. It is only when corrasion is exceedingly slow that the lateral wear becomes of importance; and hence as a rule the position of a stream bed is permanent.

The Instability of Drainage Lines. The stability of waterways being the rule, every case of instability requires an explanation; and in the study of such exceptional cases there have been found a number of different methods by which the courses of streams are shifted. The more important will be noted.

Ponding. When a mountain uplift crosses the course of a stream, it often happens that the rate of uplift is too rapid to be equalled by the corrasion of the stream, and the uprising rock becomes a dam over which the water still runs, but above which there is accumulated a pond or lake. . . . As the uplift progresses the level of the pond is raised higher and higher, until finally it finds a new outlet at some other point. . . .

The disturbances which divert drainage lines are not always of the sort which produce mountains. The same results may follow the most gentle undulations of plains. It required a movement of a few feet only to change the outlet of Lakes Michigan, Huron, and Superior from the Illinois River to the St. Clair. . . .

Planation. It has been shown in the discussion of the relations of transportation and corrasion that downward wear ceases when the load equals the capacity for transportation. Whenever the load reduces the downward corrasion to little or nothing, lateral corrasion becomes relatively and actually of importance. The first result of the wearing of the walls of a stream's channel is the formation of a flood-plain. As an effect of momentum the current is always swiftest along the outside of a curve of the channel, and it is there that the wearing is performed; while at the inner side of the curve the current is so slow that part of the load is deposited. In this way the width of the channel remains the same while its position is shifted, and every part of the valley which it has crossed in its shiftings comes to be covered by a deposit which does not rise above the highest level of the water. The surface of this deposit is hence appropriately called the *flood-plain* of the stream. The deposit is of nearly uniform depth, descending no lower than the bottom of the water-channel, and it rests upon a tolerably even surface of the rock or other material which is corraded by the stream. The process of carving away the rock so as to produce an even surface, and at the same time covering it with an alluvial deposit, is the process of *planation.*

It sometimes happens that two adjacent streams by extending their areas of planation eat through the dividing ridge and join their channels. The stream which has the higher surface at the point of contact, quickly abandons the lower part of its channel and becomes a branch of the other, having shifted its course by planation.

The slopes of the Henry Mountains illustrate the process in a peculiarly striking manner. The streams which flow down them are limited in their rate of degradation at both ends. At their sources, erosion is opposed by the hardness of the rocks; the trachytes and metamorphics of the mountain tops are carved very slowly. At their mouths, they discharge into the Colorado and the Dirty Devil, and cannot sink their channels more rapidly than do those rivers. Between the mountains and the rivers, they cross rocks which are soft in comparison with the trachyte, but they can deepen their channels with no greater rapidity than at their ends. The grades have adjusted themselves accordingly. Among the hard rocks of the mountains the declivities are great, so as to give efficiency to the eroding water. Among the sedimentary rocks of the base they are small in comparison, the chief work of the streams being the transportation of the trachyte *débris*. So greatly are the streams concerned in transportation, and so little in downward corrasion (outside the trachyte region), that their grades are almost unaffected by the differences of rock texture, and they pass through sandstone and shale with nearly the same declivity.

The rate of downward corrasion being thus limited by extraneous conditions, and the instrument of corrasion—the *débris* of the hard trachyte—being efficient, lateral corrasion is limited only by the resistance which the banks of the stream oppose. Where the material of the banks is a firm sandstone, narrow flood-plains are formed; and where it is a shale, broad ones. In the Gray Cliff and Vermilion Cliff sandstones flat-bottomed cañons are excavated; but in the great shale beds broad valleys are opened, and the flood-plains of adjacent streams coalesce to form continuous plains.

River terraces as a rule are carved out, and not built up. They are always vestiges of flood-plains, and flood-plains are usually produced by lateral corrasion. There are instances, especially near the sea-coast, of river-plains which have originated by the silting up of valleys, and have been afterward partially destroyed by the same rivers when some change of level permitted them to cut their channels deeper; and these instances, conspiring with the fact that the surfaces of flood-plains are alluvial, and with the fact that many terraces in glacial regions are carved from unconsolidated drift, have led some American geologists into the error of supposing that river terraces in general are the records of sedimentation, when in fact they record the stages of a progressive corrasion. The ideal section of a terraced river valley which I reproduce from Hitchcock (Surface Geology, Plate XII, figure 1) [Fig. 32A] regards each terrace as the remnant of a separate deposit, built up from the bottom of the valley. To illustrate my own idea I have copied his profile [Fig. 32B] and interpreted its features as the results of

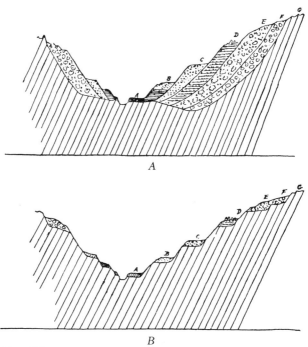

Fig. 32.—Gilbert's diagrams to illustrate alternative explanations of river terraces, 1877. A. Ideal section of a river valley, reproduced from Hitchcock. B. Gilbert's interpretation of the same features.

lateral corrasion or planation, giving each bench a capping of alluvium, but constituting it otherwise of the preëxistent material of the valley. . . .

There is a kindred error, as I conceive, involved in the assumption that the streams which occupied the upper and broader flood-plains of a valley were greater than those which have succeeded them. They may have been, or they may not. In the process of lateral corrasion all the material that is worn from the bank has to be transported by the water, and where the bank is high the work proceeds less rapidly than where it is low. A stream which degrades its immediate valley more rapidly than the surrounding country is degraded (and the streams which abound in terraces are of this character) steadily increases the height of the banks which must be excavated in planation and diminishes the extent of its flood-plain; and this might occur even if the volume of the stream was progressively increasing instead of diminishing.

Unequal and Equal Declivities. In homogeneous material, and with equal quantities of water, the rate of erosion of two slopes depends upon their declivities. The steeper is degraded faster. It is evident that when the two slopes are upon opposite sides of a divide the more rapid wearing of the steeper carries the divide toward the side of the gentler. The action ceases and the divide becomes stationary only when the profile of the divide has been rendered symmetric.

It is to this law that bad-lands owe much of their beauty. They acquire their smooth curves under what I have called the "law of divides," but the symmetry of each ridge and each spur is due to the law of equal declivities. By the law of divides all the slopes upon one side of a ridge are made interdependent. By the law of equal declivities a relation is established between the slopes which adjoin the crest on opposite sides, and by this means the slopes of the whole ridge, from base to base, are rendered interdependent.

One result of the interdependence of slopes is that a bad-land ridge separating two waterways which have the same level, stands midway between them; while a ridge separating two waterways which have different levels, stands nearer to the one which is higher.

It results also that if one of the waterways is corraded more rapidly than the other the divide moves steadily toward the latter, and eventually, if the process continues, reaches it. When this occurs, the stream with the higher valley abandons the lower part of its course and joins its water to that of the lower stream. Thus from the shifting of divides there arises yet another method of the shifting of waterways, a method which it will be convenient to characterize as that of *abstraction.* A stream which for any reason is able to corrade its bottom more rapidly than do its neighbors, expands its valley at their expense, and eventually "abstracts" them.   And conversely, a stream which for any reason is able to corrade its bottom less rapidly than its neighbors, has its valley contracted by their encroachments and is eventually "abstracted" by one or the other.

LACCOLITHS AND THEIR ORIGIN

From United States Geographical and Geological Survey of the Rocky Mountain Region, *Geology of the Henry Mountains*, pp. 19–21, 95–97, 1877.

It is usual for igneous rocks to ascend to the surface of the earth, and there issue forth and build up mountains or hills by successive eruptions.   The molten matter starting from some region of unknown depth passes through all superincumbent rock-beds, and piles itself up on the uppermost bed.   The lava of the Henry Mountains behaved differently.   Instead of rising through all the beds of the earth's crust, it stopped at a lower horizon, insinuated itself between two strata, and opened for itself a chamber by lifting all the superior beds.   In this chamber it congealed, forming a massive body of trap.   For this body the name *laccolite* (λάκκος, *cistern*, and λίθος, *stone*) will be used.   Figure 33A and Figure 33B are ideal sections of a mountain of eruption and of a laccolite.

The laccolite is the chief element of the type of structure exemplified in the Henry Mountains.

It is evident that the intrusion of a laccolite will produce upon the surface as great a hill as the extrusion of the same quantity of matter, the mass which is carried above the original surface being precisely equivalent to that which is displaced by the laccolite; and it is further evident that where the superior rock is horizontally stratified every stratum above the laccolite will be uplifted, and, unless it is fractured, will be upbent, and will portray, more or less faithfully, by its curvature, the form of the body it covers.

Associated with the laccolites of the Henry Mountains are *sheets* and *dikes.*

The term *sheet* will be applied in this report to broad, thin, stratified bodies of trap, which have been intruded along the partings between sedimentary strata, and conform with the inclosing strata in dip.   *Dikes* differ from sheets in that they intersect the sedimentary strata at greater or less angles, occupying fissures produced by the rupture of the strata.

The logical distinction between dike and sheet is complete, but in nature it not infrequently happens that the same body of trap is a sheet in one place and a dike in another. Between the sheet and the laccolite there is a complete gradation. The laccolite is a greatly thickened sheet, and the sheet is a broad, thin, attenuated laccolite.

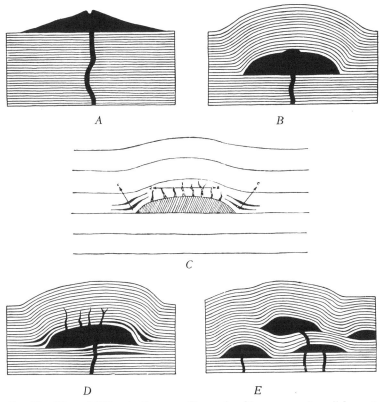

A   *B*

C

D   *E*

FIG. 33.—Five of Gilbert's diagrams illustrating his paper on laccoliths and their origin, 1877.

In the district under consideration the laccolite is usually, perhaps always, accompanied by dikes and sheets (see Figure 33C). There are sheets beneath laccolites and sheets above them. The superior sheets have never been observed to extend beyond the curved portion of the superior strata. Dikes rise from the upper surface of the laccolites. They are largest and most numerous about the center, but, like the superior sheets, they often extend nearly to the limit of the flexure of the uplifted strata. The larger often radiate from the center outward, but there is no constancy of arrangement. Where they are numerous they reticulate.

In the accompanying diagrams dikes are represented beneath as well as above the laccolites. These are purely hypothetical, since they have not been seen. In a general way, the molten rock must have come from below, but the channel by which it rose has in no instance been determined by observation.

The horizontal distribution of the laccolites is as irregular as the arrangement of volcanic vents. They lie in clusters, and each cluster is marked by a mountain. In Mount Ellen there are perhaps thirty laccolites. In Mount Holmes there are two; and in Mount Ellsworth one. Mount Pennell and Mount Hillers each have one large and several small ones.

Their vertical distribution likewise is irregular. Some have intruded themselves between Cretaceous strata, others between Jura-Triassic, and others between Carboniferous. From the highest to the lowest the range is not less than 4,000 feet. Those which are above not unfrequently overlap those which lie below, as represented in the ideal section, Figure 33E.

The erosion of the mountains has given the utmost variety of exposure to the laccolites. In one place are seen only arching strata; in another, arching strata crossed by a few dikes; in another, arching strata filled with a net-work of dikes and sheets. Elsewhere a portion of the laccolite itself is bared, or one side is removed so as to exhibit a natural section. Here the sedimentary cover has all been removed, and the laccolite stands free, with its original form; there the hard trachyte itself has been attacked by the elements and its form is changed. Somewhere, perhaps, the laccolite has been destroyed and only a dike remains to mark the fissure through which it was injected.

When lavas forced upward from lower-lying reservoirs reach the zone in which there is the least hydrostatic resistance to their accumulation, they cease to rise. If this zone is at the top of the earth's crust they build volcanoes; if it is beneath, they build laccolites. Light lavas are more apt to produce volcanoes; heavy, laccolites. The porphyritic trachytes of the Plateau Province produced laccolites.

The station of the laccolite being decided, the first step in its formation is the intrusion along a parting of strata, of a thin sheet of lava, which spreads until it has an area adequate, on the principle of the hydrostatic press, to the deformation of the covering strata. The spreading sheet always extends itself in the direction of least resistance, and if the resistances are equal on all sides, takes a circular form. So soon as the lava can uparch the strata it does so, and the sheet becomes a laccolite. With the continued addition of lava the laccolite grows in height and width, until finally the supply of material or the propelling force so far diminishes that the lava clogs by congelation in its conduit and the inflow stops. An irruption is then complete, and the progress of the laccolite is comparable with that of a volcano at the end of its first eruption. During the irruption and after its completion, there is an interchange of temperatures. The laccolite cools and solidifies; its walls are heated and metamorphosed. At the edges, where the surface of the laccolite is most convex, the heat is most rapidly dissipated, and its effect in metamorphism is least. A second irruption may take place either before or after the first is solidified. It may intrude above or it may intrude beneath it; and observation has not yet distinguished the one case from the other. In any case it carries forward the deformation of cover that was begun by the first, and combines with it in such a way that the compound form is symmetric, and is substantially the same that would have been produced if the two irruptions were combined in one. Thus the laccolite grows by successive accretions until at length its cooled mass, heavier and stronger than the surrounding rocks, proves a sufficient obstacle to intrusion. The next irruption then avoids it, opens a new conduit, and builds a new laccolite at its side. By successive shiftings of the conduit a group of laccolites is formed, just as by the shifting of vents eruptive cones are grouped. Each laccolite is a subterranean volcano.

The strata above the laccolite are bent instead of broken, because their material is subjected to so great a pressure by superincumbent strata that it cannot hold an open fissure and is *quasi-plastic*. But although quasi-plastic it is none the less solid, and can be cracked open if the gap is instantaneously filled, the cracking and the filling being one event. This happens in the immediate walls of the laccolite, and they are injected by dikes and sheets of the lava. The directions of the cracks are normal to the directions of the extensive strains (strains tending to extend) where they occur. From the top of the laccolite dikes run upward into the roof, marking horizontal strains. From the sides smaller vertical dikes run outward, marking horizontal, tangential strains. And parallel to the sides near the base of the laccolite, are numerous sheets, marking strains directed outward and upward. These last especially serve to show that the rigidity of the strata is not abolished, although it is overpowered, by the pressure which warps them.

Here we are brought face to face with a great fact of dynamic geology which though well known is too often ignored. The solid crust of the earth, and the solid earth if it be solid, are as plastic *in great masses* as wax is in small. Solidity is not absolute but relative. It is only a low grade of plasticity. The rigidity or strength of a body is measured by the square of its linear dimensions, while its weight is measured by the cube. Hence with increase in magnitude, the weight increases more rapidly than the strength; and no very large body is strong enough to withstand the pressure of its own weight. However solid it may be, it must succumb and be flattened. When we speak of rock masses which are measured by feet, we may regard them as solid; but when we consider masses which are measured by miles, we should regard them as plastic.

The same principle is illustrated by the limital area of laccolites. A small laccolite cannot lift its small cover, but a large laccolite can lift its correspondingly large cover. The strength or rigidity which resists deformation is overcome by magnitude.

HISTORY OF LAKE BONNEVILLE

From United States Geological Survey, *Second Annual Report*, 1881, pp. 169–200.

When, therefore, one views a slope from which an ancient lake has been withdrawn he recognizes its shore trace in a series of features which embody a horizontal line. Here the line is the meeting of a terrace and a cliff, there it is the crest of an embankment, and elsewhere it is the brow of a delta plain; its manifestations are diverse, but it is never wanting. To an eye placed at the proper height and distance all its elements blend together and it stands forth as a continuous, horizontal, indubitable *shoreline*.

Fig. 34.—The embankments and terraces of Lake Bonneville, near Reservoir Butte, drawn by H. H. Nichols, *Monograph I*, United States Geological Survey, 1890.

It is by records of this character that Lake Bonneville is chiefly known. All about the great basin of Utah the lower slopes of the mountains are skirted by these level tracings—not a single line merely at a single level, but a series of lines at many levels, testifying to a system of oscillations of an ancient lake.

The highest water line is 1,000 feet above the level of Great Salt Lake, and over every foot of the intervening profile can be traced evidence of the action of waves. There is, however, a great inequality of the record, and the most casual observation shows that the water lingered much longer at some stages than at others. One of the most conspicuous water lines is the highest of all, but its prominence is largely due to the fact that it marks the limit between the wave-wrought surface below and the rain-sculptured forms which rise above. Of the lower lines there is one lying about 400 feet below the highest, which is far more conspicuous than any other and has for this reason been given a special name—the Provo shoreline. The highest is distinctively known as the Bonneville shoreline. Between the Bonneville and the Provo four or five prominent lines can usually be seen, to which no individual names have been given, and it will be convenient to call these in this place the Intermediate shorelines. On the slopes below the Provo shore the water lines are less conspicuously drawn, and only a single one is so accentuated as to have been identified at numerous localities.

A lake without outlet cannot maintain a constant level, because its quantity of water depends upon the relation of the rainfall to the superficial evaporation, and these elements of climate are notoriously variable. A series of moist seasons increases its contents and causes its margin to advance upon the land, while a succession of dry seasons produces the reverse effect and makes it shrink from its borders. With a lake having a discharge the case is different, for every increase or diminution of supply, by slightly raising or lowering the surface, increases or diminishes the discharge, and thus a practical equilibrium is maintained. It can therefore rarely happen that the waters of a lake without outlet are held at one level for the time necessary to produce a strongly marked, individual shoreline, and for this reason it was early surmised that Lake Bonneville found an escape for its surplus waters at the times of the formation of the Bonneville and Provo shorelines. Search was therefore made for a point of outlet, and eventually it was discovered at the northern extremity of Cache Valley. The sill over which the water at first discharged was of soft material which yielded easily to the wear of the running water and permitted the lake level to be rapidly lowered by the mere corrasion of the outflowing stream, but eventually a reef of limestone was reached by which the erosion was checked and the lake was held at a nearly constant level until its outflow was finally stopped by climatic changes which diminished the water supply. The level of the soft sill first crossed by the outflowing water is the level of the Bonneville shoreline at the nearest point where it is visible; the level of the limestone sill is coincident with that of the Provo shoreline; and the discovery of these facts correlated in a satisfactory manner the history of outflow with the history of the most conspicuous shorelines. . . .

The detritus the waves bear away from one part of a coast is not all accumulated in the adjacent embankments; only the sand and gravel are there collected, and the finer matter, which is capable of being held in suspension by the water for a longer time, is borne to a greater distance from the shore and finally settles to the bottom in the form of mud. Moreover, the streams which flow from the land to a lake bring with them each its quota of mud too fine to subside quickly, and this, too, is deposited in the depths remote from the shore. Thus the whole surface of every lake bottom becomes coated with a fine mud derived from the demolition of the surrounding land by rains and rivers and by waves.

Nor is this all: the wear of the rains and even the wear of the waves is not limited to abrasion, but includes also solution. . . . A lake which receives the water of a river or rivers, and is itself drained by an outflowing river, catches in its sediments the mechanical detritus only, which settles to the bottom where the current is checked, and permits all or the greater part of the chemical contents of the water to escape; but a lake with no outlet accumulates the entire mineral contents of the tributary streams, storing the mechanical detritus as a sediment, and the dissolved minerals either as a precipitate or else, by the aid of animal life, in some form of organic débris. The record of a lake is therefore written, not only by its wave-wrought shorelines, but by its mechanical and chemical sediments, and the investigation of Lake Bonneville necessarily included a study of the beds deposited upon its bottom.

There can be no question of this history so far as it goes, and any changes which may be made in it must be of the nature of additions. The climate of the region was moist and Lake Bonneville was large for a period represented by 90 feet and more of Yellow Clay; the climate was dry and the lake was small for an unknown period represented by the intervening alluvia; there was a second epoch of moist climate and expanded lake represented by 15 feet of White Marl; and that was followed by the present period of shrunken waters and dry climate.

The Topographic Features of Lake Shores

From United States Geological Survey, *Fifth Annual Report,* pp. 75–123, 1884.

The Barrier. Where the sublittoral bottom of the lake has an exceedingly gentle inclination the waves break at a considerable distance from the water margin. The most violent agitation of the water is along the line of breakers; and the shore drift, depending upon agitation for its transportation, follows the line of the breakers instead of the water margin. It is thus built into a continuous outlying ridge at some distance from the water's edge. It will be convenient to speak of this ridge as a *barrier.*

The barrier is the functional equivalent of the beach. It is the road along which shore drift travels, and it is itself composed of shore drift. Its lakeward face has the typical beach profile, and its crest lies a few feet above the normal level of the water.

Between the barrier and the land a strip of water is inclosed, constituting a lagoon. This is frequently converted into a marsh by the accumulation of silt and vegetable matter, and eventually becomes completely filled, so as to bridge over the interval between land and barrier, and convert the latter into a normal beach.

The beach and barrier are absolutely dependent on shore drift for their existence. If the essential continuous supply of moving detritus is cut off, not only is the structure demolished by the waves which formed it, but the work of excavation is carried landward, creating a wave-cut terrace and cliff.

Cliffs. The sea-cliff differs from all others, first, in that its base is horizontal, and, second, in that there is associated with it at one end or other a beach, a barrier, or an embankment. A third valuable diagnostic feature is its uniform association with the terrace at its base; but in this respect it is not unique, for the cliff of differential degradation often springs from a terrace. Often, too, the latter is nearly horizontal at base, and in such case the readiest comparative test is found in the fact that the sea-cliff is independent of the texture and structure of the rocks from which it is carved, while the other is closely dependent thereon.

The sea-cliff is distinguished from the stream-cliff by the fact that it faces an open valley broad enough and deep enough to permit the generation of efficient waves if occupied by a lake. It is distinguished from the coulée edge by its independence of rock structure and by its associated terrace. It differs from the fault scarp in all those peculiarities which result from the attitude of its antecedent; the water surface concerned in the formation of the sea-cliff is a horizontal plane; the fissure concerned in the formation of the fault scarp is a less regular, but essentially vertical plane. The former crosses the inequalities of the preexistent topography as a contour, the latter as a traverse line.

The land-slip cliff is distinguished by the marked concavity of its face in horizontal contour. The sea-cliff is usually convex, or, if concave, its contours are long and sweeping. The former is distinguished also by its discontinuity.

Terraces. The only feature by which shore terraces are distinguished from all terraces of other origin, is the element of horizontality. The wave-cut terrace is bounded by a horizontal line at its upper edge; the delta is bounded by a horizontal line about its lower edge; and the wave-built terrace is a horizontal plain. But the application of this criterion is rendered difficult by the fact that the terrace of differential degradation is not infrequently margined by horizontal lines; while the inclination of the base of the stream terrace, though a universal and essential character, is often so small in amount as to be difficult of recognition. The fault terrace and land-slip terrace are normally so uneven that this character sufficiently contrasts them with all shore features.

The wave-cut terrace agrees with all the non-shore terraces in that it is overlooked by a cliff rising from its upper margin, and differs from them in that it merges at one end or both with a beach, barrier, or embankment. It is further distinguished from the terrace of differential degradation by the fact that its configuration is independent of the structure of the rocks from which it is carved, while the latter is closely dependent thereon. In freshly formed examples, a further distinction may be recognized in the mode of junction of terrace and cliff. As viewed in profile, the wave-cut terrace joins the associated sea-cliff by an angle, while in the profile wrought by differential degradation, the terrace curves upward to meet the overlooking cliff.

The wave-built terrace may be distinguished from all others by the character of its surface, which is corrugated with parallel, curved ribs. It differs from all except the stream terrace in its material, which is wave-rolled and wave-sorted. It differs from the stream terrace in that it stands on a slope facing an open basin suitable for the generation of waves.

The Recognition of Ancient Shores. The facility and certainty with which the vestiges of ancient water margins are recognized and traced depend on local conditions. The small waves engendered in ponds and in sheltered estuaries are far less efficient in the carving of cliffs and the construction of embankments than are the great waves of larger water bodies; and the faint outlines they produce are afterward more difficult to trace than those strongly drawn.

The element of time, too, is an important factor, and this in a double sense. A water surface long maintained scores its shore mark more deeply than one of brief duration, and its history is by so much the more easily read. On the other hand a system of shore topography, from which the parent lake has receded, is immediately exposed to the obliterating influence of land erosion and gradually, though very slowly, loses its character and definition. The strength of the record is directly proportioned to the duration of the lake and inversely to its antiquity.

It will be recalled that in the preceding description the character of horizontality has been ascribed to every shore feature. The base of the sea-cliff and the coincident margin of the wave-cut terrace are horizontal; and so is the crest of each beach, barrier, embankment, and wave-built terrace, and they not merely agree in the fact of horizontality, but fall essentially into a common plane—a plane intimately related to the horizon of the maximum force of breakers during storms. The outer margin of the delta is likewise horizontal, but at a slightly lower level—the level of the lake surface in repose. This difference is so small that for the purpose of identification it does not affect the practical coincidence of all the horizontal lines of the shore in a single contour. In a region where forests afford no obstruction, the observer has merely to bring his eye into the plane once occupied by the water surface, and all the horizontal elements of shore topography are projected in a single line. This line is exhibited to him not merely by the distinctions of light and shade, but by distinctions of color due to the fact that the changes of inclination and of soil at the line influence the distribution of many kinds of vegetation. In this manner it is often possible to obtain from the general view evidence of the existence of a faint shore tracing, which could be satisfactorily determined in no other way. The ensemble of a faintly scored shore mark is usually easier to recognize than any of its details.

It is proper to add that this consistent horizontality, which appeals so forcibly and effectually to the eye, cannot usually be verified by instrumental test. The surface of the "solid earth" is in a state of change, whereby the vertical relations of all its parts are continually modified. Wherever the surveyor's level has been applied to a fossil shore, it has been found that the "horizon" of the latter departs notably from horizontality, being warped in company with the general surface on which it rests. The level, therefore, is of little service in the tracing of ancient water margins, while the water margins afford, through the aid of the level, delicate measures of differential diastrophic movements. It might appear that the value of horizontality as an aid to the recognition of shores is consequently vitiated, but such is not the case. It is, indeed, true that the accumulated warping and faulting of a long period of time will so incline and disjoint a system of shore features that they can no longer be traced; but it is also true that the processes of land erosion will in the same time obliterate the shore features themselves. The minute elements of orographic displacement are often paroxysmal, but so far as observation informs us, the general progress of such changes is slow and gradual, so that, during the period for which shore tracings can withstand atmospheric and pluvial waste, their deformation is not sufficient to interfere materially with their recognition.

CHAMBERLIN

Thomas Crowder Chamberlin (1843–1928), American geologist and cosmologist, was called from the presidency of the University of Wisconsin in 1892 to become the first head of the Department of Geology in the new University of Chicago, a position which he occupied until his retirement in 1918. As geologist of the United States Geological Survey, his most notable work dealt with the Pleistocene glacial phenomena of the Mississippi Valley. During the last third of his life, his research was directed more generally toward cosmogony and the geologic problems which can be solved only by knowledge of the deeper structure of the earth. As a teacher of geology, he exerted a profound influence upon many American geologists of contemporary note.

THE METHOD OF MULTIPLE WORKING HYPOTHESES

From *Journal of Geology*, Vol. V, pp. 837–848, 1897.

There are two fundamental modes of study. The one is an attempt to follow by close imitation the processes of previous thinkers and to acquire the results of their investigations by memorizing. It is study of a merely secondary, imitative, or acquisitive nature. In the other mode the effort is to think independently, or at least individually. It is primary or creative study. The endeavor is to discover new truth or to make a new combination of truth or at least to develop by one's own effort an individualized assemblage of truth. The endeavor is to think for one's self, whether the thinking lies wholly in the fields of previous thought or not. It is not necessary to this mode of study that the subject-matter should be new. Old material may be reworked. But it is essential that the process of thought and its results be individual and independent, not the mere following of previous lines of thought ending in predetermined results. The demonstration of a problem in Euclid precisely as laid down is an illustration of the former; the demonstration of the same proposition by a method of one's own or in a manner distinctively individual is an illustration of the latter, both lying entirely within the realm of the known and old.

Creative study however finds its largest application in those subjects in which, while much is known, more remains to be learned. The geological field is preeminently full of such subjects, indeed it presents few of any other class. There is probably no field of thought which is not sufficiently rich in such subjects to give full play to investigative modes of study.

Three phases of mental procedure have been prominent in the history of intellectual evolution thus far. What additional phases may be in store for us in the evolutions of the future it may not be prudent to attempt to forecast. These three phases may be styled the method of the ruling theory, the method of the working hypothesis, and the method of multiple working hypotheses.

In the earlier days of intellectual development the sphere of knowledge was limited and could be brought much more nearly than now within the compass of a single individual. As a natural result those who then assumed to be wise men, or aspired to be thought so, felt the need of knowing, or at least seeming to know, all that was known, as a justification of their claims. So also as a natural counterpart there grew up an expectancy on the part of the multitude that the wise and the learned would explain whatever new thing presented itself. Thus pride and ambition on the one side and expectancy on the other joined hands in developing the putative all-wise man whose knowledge boxed the compass and whose acumen found an explanation for every new puzzle which presented itself. Although the pretended compassing of the entire horizon of knowledge has long since become an abandoned affectation, it has left its representatives in certain intellectual predilections. As in the earlier days, so still, it is a too frequent habit to hastily conjure up an explanation for every new phenomenon that presents itself. Interpretation leaves its proper place at the end of the intellectual procession and rushes to the forefront. Too often a theory is promptly born and evidence hunted up to fit in afterward. Laudable as the effort at explanation is in its proper place, it is an almost certain source of confusion and error when it runs before a serious inquiry into the phenomenon itself. A strenuous endeavor to find out precisely what the phenomenon really is should take the lead and crowd back the question, commendable at a later stage, "How came this so?" First the full facts, then the interpretation thereof, is the normal order.

The habit of precipitate explanation leads rapidly on to the birth of general theories. When once an explanation or special theory has been offered for a given phenomenon, self-consistency prompts to the offering of the same explanation or theory for like phenomena similar to the original one. . . . For a time these hastily born theories are likely to be held in a tentative way with some measure of candor or at least some self-illusion of candor. . . . It is in this tentative stage that the affections enter with their blinding influence. . . . Important as the intellectual affections are as stimuli and as rewards, they are nevertheless dangerous factors in research. All too often they put under strain the integrity of the intellectual processes. . . . Briefly summed up, the evolution is this: a premature explanation passes first into a tentative theory, then into an adopted theory, and lastly into a ruling theory.

The defects of the method are obvious and its errors grave. If one were to name the central psychological fault, it might be stated as the admission of intellectual affection to the place that should be dominated by impartial, intellectual rectitude alone.

So long as intellectual interest dealt chiefly with the intangible, so long it was possible for this habit of thought to survive and to maintain its dominance, because the phenomena themselves, being largely subjective, were plastic in the hands of the ruling idea; but so soon as investigation turned itself earnestly to an inquiry into natural phenomena whose manifestations are tangible, whose properties are inflexible, and whose laws are rigorous, the defects of the method became manifest and effort at reformation ensued. The first great endeavor was repressive. The advocates of reform insisted that theorizing should be restrained and the simple determination of facts should take its place. The effort was to make scientific study statistical instead of causal. Because theorizing in narrow lines had led to manifest evils theorizing was to be condemned. The reformation urged was not the proper control and utilization of theoretical effort but its suppression. We do not need to go backward more than a very few decades to find ourselves in the midst of this attempted reformation. Its weakness lay in its narrowness and its restrictiveness. There is no nobler aspiration of the human intellect than the desire to compass the causes of things. The disposition to find explanations and to develop theories is laudable in itself. It is only its ill-placed use and its abuse that are reprehensible. The vitality of study quickly disappears when the object sought is a mere collocation of unmeaning facts.

The inefficiency of this simply repressive reformation becoming apparent, improvement was sought in the method of the working hypothesis. This has been affirmed to be *the* scientific method. But it is rash to assume that any method is *the* method, at least that it is the ultimate method. The working hypothesis differs from the ruling theory in that it is used as a means of determining facts rather than as a proposition to be established. It has for its chief function the suggestion and guidance of lines of inquiry; the inquiry being made, not for the sake of the hypothesis, but for the sake of the facts and their elucidation. The hypothesis is a mode rather than an end. Under the ruling theory, the stimulus is directed to the finding of facts for the support of the theory. Under the working hypothesis, the facts are sought for the purpose of ultimate induction and demonstration, the hypothesis being but a means for the more ready development of facts and their relations.

It will be observed that the distinction is not such as to prevent a working hypothesis from gliding with the utmost ease into a ruling theory. Affection may as easily cling about a beloved intellectual child when named an hypothesis as if named a theory, and its establishment in the one guise may become a ruling passion very much as in the other. The historical antecedents and the moral atmosphere associated with the working hypothesis lend some good influence however toward the preservation of its integrity.

Conscientiously followed, the method of the working hypothesis is an incalculable advance upon the method of the ruling theory; but it has some serious defects. One of these takes concrete form, as just noted, in the ease with which the hypothesis becomes a controlling idea. To avoid this grave danger, the method of multiple working hypotheses is urged. It differs from the simple working hypothesis in that it distributes the effort and divides the affections. It is thus in some measure protected against the radical defect of the two other methods. In developing the multiple hypotheses, the effort is to bring up into view every rational explanation of the phenomenon in hand and to develop every tenable hypothesis relative to its nature, cause or origin, and to give to all of these as impartially as possible a working form and a due place in the investigation. The investigator thus becomes the parent of a family of hypotheses; and by his parental relations to all is morally forbidden to fasten his affections unduly upon any one. In the very nature of the case, the chief danger that springs from affection is counteracted. Where some of the hypotheses have been already proposed and used, while others are the investigator's own creation, a natural difficulty arises, but the right use of the method requires the impartial adoption of all alike into the working family. The investigator thus at the outset puts himself in cordial sympathy and in parental relations (of adoption, if not of authorship,) with every hypothesis that is at all applicable to the case under investigation. Having thus neutralized so far as may be the partialities of his emotional nature, he proceeds with a certain natural and enforced erectness of mental attitude to the inquiry, knowing well that some of his intellectual children (by birth or adoption) must needs perish before maturity, but yet with the hope that several of them may survive the ordeal of crucial research, since it often proves in the end that several agencies were conjoined in the production of the phenomena. Honors must often be divided between hypotheses. One of the superiorities of multiple hypotheses as a working mode lies just here. In following a single hypothesis the mind is biased by the presumptions of its method toward a single explanatory conception. But an adequate explanation often involves the coördination of several causes. This is especially true when the research deals with a class of complicated phenomena naturally associated, but not necessarily of the same origin and nature, as for example the Basement Complex or the Pleistocene drift. Several agencies may participate not only but their proportions and importance may vary from instance to instance in the same field. The true explanation is therefore necessarily complex, and the elements of the complex are constantly varying. Such distributive explanations of phenomena are especially contemplated and encouraged by the method of multiple hypotheses and constitute one of its chief merits. For many reasons we are prone to refer phenomena to a single cause. It naturally follows that when we find an effective agency present, we are predisposed to be satisfied therewith. We are thus easily led to stop short of full results, sometimes short of the chief factors. The factor we find may not even be the dominant one, much less the full complement of agencies engaged in the accomplishment of the total phenomena under inquiry. The mooted question of the origin of the Great Lake basins may serve as an illustration. Several hypotheses have been urged by as many different students of the problem as the cause of these great excavations. All of these have been pressed with great force and with an admirable array of facts. Up to a certain point we are compelled to go with each advocate. It is practically demonstrable that these basins were river valleys antecedent to the glacial incursion. It is equally demonstrable that there was a blocking up of outlets. We must conclude then that the present basins owe their origin in part to the preëxistence of river valleys and to the blocking up of their outlets by drift. That there is a temptation to rest here, the history of the question shows. But on the other hand it is demonstrable that these basins were occupied by great lobes of ice and were important channels of glacial movement. The leeward drift shows much material derived from their bottoms. We cannot therefore refuse assent to the doctrine that the basins owe something to glacial excavation. Still again it has been urged that the earth's crust beneath these basins was flexed downward by the weight of the ice load and contracted by its low temperature and that the basins owe something to crustal deformation. This third cause tallies with certain features not readily explained by the others. And still it is doubtful whether all these combined constitute an adequate explanation of the phenomena. Certain it is, at least, that the measure of participation of each must be determined before a satisfactory elucidation can be reached. The full solution therefore involves not only the recognition of multiple participation but an estimate of the measure and mode of each participation. For this the simultaneous use of a full staff of working hypotheses is demanded. The method of the single working hypothesis or the predominant working hypothesis is incompetent. . . .

A special merit of the use of a full staff of hypotheses coördinately is that in the very nature of the case it invites thoroughness. The value of a working hypothesis lies largely in the significance it gives to phenomena which might otherwise be meaningless and in the new lines of inquiry which spring from the suggestions called forth by the significance thus disclosed. Facts that are trivial in themselves are brought forth into importance by the revelation of their bearings upon the hypothesis and the elucidation sought through the hypothesis. The phenomenal influence which the Darwinian hypothesis has exerted upon the investigations of the past two decades is a monumental illustration. But while a single working hypothesis may lead investigation very effectively along a given line, it may in that very fact invite the neglect of other lines equally important. Very many biologists would doubtless be disposed today to cite the hypothesis of natural selection, extraordinary as its influence for good has been, as an illustration of this. While inquiry is thus promoted in certain quarters, the lack of balance and completeness gives unsymmetrical and imperfect results. But if on the contrary all rational hypotheses bearing on a subject are worked coördinately, thoroughness, equipoise, and symmetry are the presumptive results in the very nature of the case.

The loyal pursuit of the method for a period of years leads to certain distinctive habits of mind which deserve more than the passing notice which alone can be given them here. As a factor in education the disciplinary value of the method is one of prime importance. When faithfully followed for a sufficient time, it develops a mode of thought of its own kind which may be designated the habit of parallel thought, or of complex thought. It is contra-distinguished from the linear order of thought which is necessarily cultivated in language and mathematics because their modes are linear and successive. The procedure is complex and largely simultaneously complex. The mind appears to become possessed of the power of simultaneous vision from different points of view. The power of viewing phenomena analytically and synthetically at the same time appears to be gained. It is not altogether unlike the intellectual procedure in the study of a landscape. From every quarter of the broad area of the landscape there come into the mind myriads of lines of potential intelligence which are received and coördinated simultaneously producing a complex impression which is recorded and studied directly in its complexity. If the landscape is to be delineated in language it must be taken part by part in linear succession.

Over against the great value of this power of thinking in complexes there is an unavoidable disadvantage. No good thing is without its drawbacks. It is obvious upon studious consideration that a complex or parallel method of thought cannot be rendered into verbal expression directly and immediately as it takes place. We cannot put into words more than a single line of thought at the same time, and even in that the order of expression must be conformed to the idiosyncrasies of the language. Moreover the rate must be incalculably slower than the mental process. When the habit of complex or parallel thought is not highly developed there is usually a leading line of thought to which the others are subordinate. Following this leading line the difficulty of expression does not rise to serious proportions. But when the method of simultaneous mental action along different lines is so highly developed that the thoughts running in different channels are nearly equivalent, there is an obvious embarrassment in making a selection for verbal expression and there arises a disinclination to make the attempt. Furthermore the impossibility of expressing the mental operation in words leads to their disuse in the silent processes of thought and hence words and thoughts lose that close association which they are accustomed to maintain with those whose silent as well as spoken thoughts predominantly run in linear verbal courses. There is therefore a certain predisposition on the part of the practitioner of this method to taciturnity. The remedy obviously lies in coördinate literary work.

An infelicity also seems to attend the use of the method with young students. It is far easier, and apparently in general more interesting, for those of limited training and maturity to accept a simple interpretation or a single theory and to give it wide application, than to recognize several concurrent factors and to evaluate these as the true elucidation often requires. Recalling again for illustration the problem of the Great Lake basins, it is more to the immature taste to be taught that these were scooped out by the mighty power of the great glaciers than to be urged to conceive of three or more great agencies working successively in part and simultaneously in part and to endeavor to estimate the fraction of the total results which was accomplished by each of these agencies. The complex and the quantitative do not fascinate the young student as they do the veteran investigator.

The studies of the geologist are peculiarly complex. It is rare that his problem is a simple unitary phenomenon explicable by a single simple cause. Even when it happens to be so in a given instance, or at a given stage of work, the subject is quite sure, if pursued broadly, to grade into some complication or undergo some transition.   He must therefore ever be on the alert for mutations and for the insidious entrance of new factors.   If therefore there are any advantages in any field in being armed with a full panoply of working hypotheses and in habitually employing them, it is doubtless the field of the geologist.

On the Interior Structure, Surface Temperature, and Age of the Earth*

From *Science*, N.S., Vol. IX, pp. 889–901, and Vol. X, pp. 11–18, 1899.

With admirable frankness Lord Kelvin says (This JOURNAL, May 12, p. 672): "All these reckonings of the history of underground heat, the details of which I am sure you do not wish me to put before you at present, are founded on the very sure assumption that the material of our present solid earth all round its surface was at one time a white-hot liquid."   It is here candidly revealed that the most essential factor in his reasonings rests ultimately upon an *assumption,* an assumption, which, to be sure, he regards as "very sure," but still an assumption.   The alternatives to this assumption are not considered.   The method of multiple working hypotheses, which is peculiarly imperative when assumptions are involved, is quite ignored.   I beg leave to challenge the certitude of this assumption of a white-hot liquid earth, current as it is among geologists, alike with astronomers and physicists.

Is not the assumption of a white-hot liquid earth still quite as much on trial as any chronological inferences of the biologist or geologist?

It, of course, remains to be seen whether the alternative hypothesis of an earth grown up slowly in a cold state, or in some state less hot than that assumed in the address, would afford any relief from the limitations of time urged upon us.   At first thought it would, perhaps, seem that this alternative would but intensify the limitations.   Since the argument for a short history is based on the degree to which the earth is cooled, an original cold state should but hasten the present status.   But this neglects an essential factor.   The question really hinges on the proportion of

---

* Comments on Lord Kelvin's Address on the Age of the Earth as an Abode Fitted for Life.

potential energy convertible into heat which remained within the earth when full grown. There is no great difference between the alternative hypotheses so far as the amount of sensible heat at the beginning of the habitable stage is concerned. For, on the one hand, the white-hot earth must have become relatively cool on the exterior before life could begin, and, on the other, it is necessary to assume a sufficiency of internal heat coming from impact and internal compression, or other changes, to produce the igneous and crystalline phenomena which the lowest rocks present. The superficial and sub-superficial temperatures in the two cases could not, therefore, have been widely different.

So far as the temperatures of the deep interior are concerned there is only recourse to hypothesis. It is probable that there would be a notable rise of temperature toward the center of the earth in either case. In a persistently liquid earth this high central temperature would be lost through convection, but if central crystallization took place at an early stage through pressure, much of the high central heat might be retained. In a meteor-built earth, solid from the beginning, very much less convectional loss would be suffered, and the central temperature would probably correspond somewhat closely to the density. The probabilities, therefore, seem somewhat to favor a higher thermal gradient toward the center in the case of the solid meteor-built earth.

But if we turn to the consideration of potential energy, there is a notable difference between the two hypothetical earths. In the liquid earth, the material must be presumed to have arranged itself according to its specific gravity, and, therefore, to have adopted a nearly complete adjustment to gravitative demands; in other words, to have exhausted, as nearly as possible, its potential energy, i.e., its "energy of position." On the other hand, in an earth built up by the accretion of meteorites without free readjustment there must have been initially a heterogeneous arrangement of the heavier and lighter material throughout the whole body of the earth, except only so far as the partial liquefaction and the very slow, plastic, viscous and diffusive rearrangement of the material permitted an incipient adjustment to gravitative demands. A large amount of potential energy was, therefore, restrained, for the time being, from passing into sensible thermal energy. This potential energy thus restrained is supposed to have gradually become converted into heat as local liquefaction and viscous, molecular and massive movements permitted the sinking of the heavier material and the rise of the lighter material. This slow conversion of potential energy into sensible heat is thought to give to the slow-accretion earth a very distinct superiority over the hot liquid earth when the combined sum of sensible and potential heat is considered.

The phenomena of mountain wrinkling and of plateau formation, as well as the still greater phenomena of continental platforms and abysmal basins, seem to demand a more *deep-seated* agency than that which is supplied by superficial loss of heat. This proposition demands a more explicit statement than is appropriate to this place, but it must be passed by with this mere allusion. It would seem obvious, however, that an earth of heterogeneous constitution, progressively reorganizing itself, would give larger possibilities of internal shrinkage and that this shrinkage must be deep-seated as well as superficial. In these two particulars it holds out the hope of furnishing an adequate explanation for the deformation of the earth where the hypothesis of a liquid earth seems thus far to have failed.

But the essential question here is the possibility of sustained internal temperature. It is urged that the heterogeneous, solid-built earth is superior to the liquid earth in the following particulars: (1) It retains a notable percentage of the original potential energy of the dispersed matter, while in the liquid earth this was converted into sensible heat and lost in prezoic times; (2) it retains the conditions for a slow convection of the interior material, bringing interior heat to the surface, a function which was exhausted by the liquid earth in the freer convection of its primitive molten state; (3) it retains larger possibilities of molecular rearrangement of the matter and of the formation of new minerals of superior density, whereas the liquid earth permitted this adjustment in the prezoic stages. In short, in at least three important particulars, the slow-built meteoric earth delayed the exercise of thermal agencies until the life era and gradually brought them into play when they were serviceable in the prolongation of the life history, whereas the liquid earth exhausted these possibilities at a time of excessive conversion of energy into heat and thus squandered its energies when they were not only of no service to the life history of the earth, but delayed its inauguration until their excesses were spent.

Let it not be supposed for a moment that I claim that the alternative hypothesis of a slow-grown earth is substantiated. It must yet pass the fiery ordeal of radical criticism at all points, but it is the logical sequence of the proposition that a swarm of meteorites revolving about the sun in independent individual orbits and having any probable form of dispersion would aggregate slowly rather than precipitately. If the astronomers and mathematicians can demonstrate that the aggregation must necessarily have been so rapid as to crowd the transformed energy of the impacts into a period much too limited to permit the radiation away of the larger part of the heat concurrently, the hypothesis will have to be set aside, and we shall be compelled to follow the deductions from the white-hot liquid earth, or find other alternatives.

But I think I do not err in assuming that mathematical computations, so far as they can approach a solution of the exceedingly complex problem, are at least quite as favorable to a slow as to a rapid aggregation. If this be so, the problem of internal temperature must be attacked on the lines of this hypothesis as well as those of the common hypothesis before any safe conclusion can be drawn from it respecting the age of the earth.

Is present knowledge relative to the behavior of matter under such extraordinary conditions as obtain in the interior of the sun sufficiently exhaustive to warrant the assertion that no unrecognized sources of heat reside there? What the internal constitution of the atoms may be is yet an open question. It is not improbable that they are complex organizations and the seats of enormous energies. Certainly, no careful chemist would affirm either that the atoms are really elementary or that there may not be locked up in them energies of the first order of magnitude. No cautious chemist would probably venture to assert that the component atomecules, to use a convenient phrase, may not have energies of rotation, revolution, position and be otherwise comparable in kind and proportion to those of a planetary system. Nor would he probably feel prepared to affirm or deny that the extraordinary conditions which reside in the center of the sun may not set free a portion of this energy.

But assuming, as we are wont to do, that the limits of our present knowledge are a definition of the facts, has the evolution of the sun been worked out with such definiteness and precision as to give a determinate and specific history of its thermal stages from beginning to end? It is one thing to tell us, on the basis of the contractional theory, that the total amount of thermal energy originally potential in the system is only equal to so many million times the present annual output, but it is quite a different thing to give a specific statement of the *actual time occupied by the sun in the evolution and discharge of this amount of heat* and to define its successive stages. It is with this actual history that we are specially concerned. . . . It seems altogether necessary to determine specifically *the distribution of the heat in time* before any approach to a satisfactory application to geological history can be made. The period of 20 or 25 million years named can have little moral guiding force until this problem is solved. But the literature of the subject shows an almost complete neglect of this consideration. While certain of the physicists and astronomers have been instructing us "*e superiore loco*," they seem, with very rare exceptions, to have overlooked this vital factor in the case. Even in computing the sum-total of heat they have, for the most part, heretofore neglected the central condensation of the sun and in their computations have substituted a convenient homogeneity.

If we turn to the earth itself it may be remarked that the nature of its atmosphere very radically conditions the amount of heat requisite for the support of life. Dr. Arrhenius has recently made an elaborate computation relative to the thermal influence of certain factors of the atmosphere and has arrived at the conclusion that an increase of the atmospheric carbon dioxide to the amount of three or four times the present content would induce such a mild climate in the polar regions that magnolias might again flourish there as they did in Tertiary times. On the other hand, he concluded that a reduction of less than 50% would induce conditions analogous to those of the glacial period of Pleistocene times. The vast quantities of carbon dioxide represented in the carbonates and carbonaceous deposits of the earth's crust imply great possibilities of change in the constitution of the atmosphere of the earth in respect to this most critical element.

But there are more radical considerations that relate to the early thermal history of the earth. To be sure, if we are forced to adopt the hypothesis of a white-hot liquid earth, with all its extravagant expenditures of energy in the early youth of the earth, we can take no advantage of these possible resources, but under the supposition that the meteorites gathered in with measurable deliberation, it is theoretically possible to find conditions for a long maintenance of life on the earth, with little or no regard to the amount of heat which the early sun sent to it. In the earliest stages of the aggregation of the earth under this hypothesis, while it was yet small, it can scarcely be supposed to have been habitable, because its mass was not sufficient to control the requisite atmospheric gases, but when it had grown to the size of Mars, that is to a size representing about $\frac{1}{10}$ of its present aggregation, or, to be safe, when it had grown to twice the size of Mars, or about one-fifth of its present mass, it would have been able to control the atmospheric gases and water, and, so far as these essential items are concerned, it would have presented conditions fitted for the presence of life. At this stage the larger portion, four-fifths by assumption, of the matter of the earth would yet be in the meteoroidal form and doubtless more or less closely associated with the growing nucleus. If the infalling of this four-fifths of the material of the earth were duly timed, so as to be neither too fast nor too slow, it would give by its impact upon the atmosphere of the earth a sufficiency both of heat and of light to maintain life upon the surface of the earth. The plunging down of these meteorites upon the surface might be more or less destructive to the life, but only proportionately more so than the fall of meteorites to-day.

If astronomers, physicists and mathematicians will jointly attack the formational history of the solar system stage by stage, following each stage out into details of time and rate, and taking full cognizance of all the alternatives that arise at each stage, it will then be possible, perhaps, to decide whether the conditions of the early earth were such as to require a large or a small amount of heat from the sun for the sustenance of life, and whether the sun was wasting heat prodigally in those days or conserving it for later expenditure. The present measure of the earth's needs may be no measure of its early needs. The sun's present expenditure may be no measure of its early expenditure.

In view of all these considerations, I again beg to inquire whether there is at present a solid basis for any "sure assumption" with reference to the earth's early thermal conditions, either internal or external, of such a determinate nature as to place any strict limitations upon the duration of life.

THE PLANETESIMAL HYPOTHESIS

From Carnegie Institution, *Year Book*, No. 3, 1904, pp. 195–254, Washington, 1905.

Under the typical form of the planetesimal hypothesis it is assumed that the parent nebula of the solar system consisted of innumerable small bodies, planetesimals, revolving about a central gaseous mass, somewhat as do the planets to-day.   The hypothesis, therefore, postulates no fundamental change in the system of dynamics after the nebula was once formed, but only an assemblage of the scattered material.   The state of dispersion of the material at the outset and throughout, as now, was maintained *by orbital revolution*, or, more closely speaking, by the tangential component of the energy of revolution.   The planetesimal hypothesis by no means excludes gases from playing a part in the parent nebula or in its evolution, any more than it denies their presence in the sun or the atmosphere to-day, but it assigns to gaseous action a subordinate place in the evolution of the planetary system after the planetesimal condition had become established.

As the basis for developing the typical form of the planetesimal hypothesis, I have assumed that the parent nebula had a planetesimal organization from the outset.   The conception is a rather radical departure from the gaseous conception of the familiar nebular hypothesis, and from the meteoritic conception of Lockyer and Darwin, so far as fundamental dynamics and mode of evolution are concerned.   To develop the hypothesis as definitely and concretely as possible, I have further chosen a special case from among those that might possibly arise, viz, the case in which the nebula is supposed to have arisen from the dispersion of a sun as a result of close approach to another large body.   The case does not involve the origin of a star nor even the primary origin of the solar system, but rather its rejuvenation and the origin of a new family of planets. The general planetesimal doctrine does not stand or fall with the merits or demerits of this special phase of it, but to be of much real service in stimulating and guiding investigation, a hypothesis must be carried out into working detail so that it may be tested by its concrete and specific application to the phenomena involved, and hence the reason for developing a specific sub-hypothesis. This particular sub-hypothesis was selected for first development (1) because it postulates as simple an event as it seems possible to assign as the source of so great results, (2) because that event seems very likely to have happened, (3) because the form of the nebula supposed to arise in this way is the most common form known, the spiral, and (4) because spectroscopic observations seem at present to support the constitution assigned this class of nebulae, although it must be noted that spectroscopic observations have not reached such a stage of development as to demonstrate the motions of the nebular constituents. In future spectroscopic determinations lies one of the crucial tests which the hypothesis must yet undergo, for there is little doubt that spectroscopic work will in time reach such a degree of refinement as to demonstrate the motions of the constituents of the spiral nebulae.

All of the more familiar spiral nebulae have dimensions that vastly transcend those of the solar system, and they can not be taken as precise examples of the solar evolution. Because of these vast dimensions and of the probable feebleness of control of the central mass, which often appears to be itself quite tenuous, a rapid motion can not well be assigned to the arms. Seen from the immense distances at which these nebulae seem to be placed—no parallax having been as yet detected—changes of position must necessarily be slow in revealing themselves to observation. It is to be hoped, however, that the present rapid progress in the perfection of instruments and of skill will soon bring within the reach of successful study some of the smaller spiral nebulae that represent the solar system more nearly in mass and proportions.

With this much of knowledge and of limitation of knowledge relative to existing nebulae, the construction of a working hypothesis required not a little resort to supplementary deductive and hypothetical considerations. The inference that a spiral nebula is formed by a combined outward and rotatory movement implies a preëxisting body that embraced the whole mass. In harmony with this, an ancestral solar system has been postulated—a system perhaps in no very essential respect different from the present one.   My hypothesis does not, therefore, concern itself with the primary origin of the sun or of the stars, or with the ulterior questions of cosmic evolution.   It confines itself to a supposed episode of the sun's history in which the present family of planets had its origin, and in the initiation of which a possible previous family may have been dispersed, but no affirmation is made relative to this.   With some partiality, perhaps, this episode may be regarded as geologic, since it specially concerns the birth of the planet of which alone we have intimate knowledge.

With 100,000,000 or more known suns and an unknown number of dark bodies moving in various directions with various velocities, the possibility of collision is well recognized; but, owing to the vastness of the intervening spaces, the contingencies of collision for an individual sun are small.   However, no appeal is here made to collisions as a source of the parent nebula of the solar system, but only to an approach of the ancestral sun to another large body, and this approach is not assumed to have been very close.   This rather distant approach is a contingency that may fairly be assumed as likely to have been realized in fact.   I have elsewhere discussed the general effects of the close approach of celestial bodies* to one another, but the particular case of the supposed ancestral sun requires special consideration.

Our present sun shoots out protuberances to heights of many thousands of miles, at velocities ranging up to 300 miles per second and more.   If it were not for the retarding influence of the immense solar atmosphere, some of these outshoots would doubtless project portions of themselves to the outer limits of the present system, and perhaps in some cases quite beyond it, for the observed velocities sometimes closely approach the controlling limit of the sun's gravity, if they do not actually reach it.   The expansive potency of this prodigious elasticity is held in restraint by the equally prodigious power of the sun's gravity.   If with these potent forces thus nearly balanced the sun closely approaches another sun or body of like magnitude—suppose one several times the mass of the

* On the Possible Function of Disruptive Approach in the Formation of Meteorites, Comets, and Nebulae. Astrophys. Journal, vol. XIV, No. 1, July, 1901, pp. 17–40; Jour. Geol., vol. IX, No. 5, July–Aug., 1901, pp. 369–393,

sun, since it is regarded as a small star—the gravity which restrains this enormous elastic power will be *relieved along the line of mutual attraction*, on the principle made familiar in the tides. At the same time the pressure transverse to this line of relief is increased. Such localized relief and intensification of pressure must, it is believed, result in protuberances of exceptional mass and high velocity. According to the well-known tidal principle, these exceptional protuberances would rise from opposite sides, and herein lies the assigned explanation of the prevalence of two diametrically opposite arms in the spiral nebulae.

Nothing remotely approaching a general dispersion of the ancestral sun seems to be required. The present planets and their satellites altogether amount to about one seven-hundredth part of the mass of the system. Simply to supply the required planetary matter, the protuberances need include but this small fraction of the ancestral sun. However, some considerable part of the projected matter must probably have been gathered back into the sun, and some part may possibly have been projected beyond the control of the system. Making allowances for both these factors, the proportion of the sun's mass necessarily involved in the protuberances is still very small. Apparently 1 or 2 per cent of the sun's mass would amply suffice.

The protuberances, by hypothesis, would be thrust out as the sun was swinging about its temporary companion star in a sharp curve, and necessarily at a prodigious velocity. It is inferred that the projected protuberances would be differentially affected by the attraction of the companion star, and take different curves about it, out of which must spring rotatory motion. This seems logically clear, but the precise paths pursued by the parts of the protuberances would apparently vary widely with different cases. As each case constitutes an involved example of the problem of three bodies, the whole is beyond rigorous mathematical treatment, but special solutions seem to justify the inference that effective rotation would arise.

The distal portions of the protuberances would obviously be formed from the superficial portions of the sun, while the later portions of the ejections forming the proximal parts of the arms would doubtless come mainly from lower depths, and hence probably contain more molecules of high specific gravity. In this seems to lie a better basis for explaining the extraordinary lightness of the outer planets and the high specific gravities of the inner ones, than in the separation, from the extreme equatorial surface of a gaseous spheroid, of successive rings whose total mass only equaled one seven-hundredth part of the original nebula.

It seems consistent with the conditions of the case to assume that the protuberances would consist of a succession of more or less irregular outbursts, as the ancestral sun in its swift whirl around the controlling star was more and more affected by the latter's differential attraction; and hence the protuberances would be directed in somewhat changing courses, and would be pulsatory in character, resulting in rather irregular and somewhat divided arms, and in a knotty distribution of the ejected matter along the arms. These knots must probably be more or less rotatory from inequalities of projection.

It is thus conceived that a spiral nebula, having two dominant arms, opposite one another, each knotty from irregular pulsations, and rotatory, the knots probably also rotatory, and attended by subordinate knots and whirls, together with a general scattering of the larger part of the mass in irregular nebulous form, would arise from the simple event of a disruptive approach.

The ejected matter, at the outset, must have been in the free molecular state, since by the terms of the hypothesis it arose from a gaseous body; but the vast dispersion and the enormous surface exposed to radiation doubtless quickly reduced the more refractory portions to the liquid and solid state, attended by some degree of aggregation into small accretions; hence the continuous spectrum which this class of nebulae present.

As previously remarked, the verity of this particular mode of origin of spiral nebulae is not essential to the acceptance of the planetesimal hypothesis. It is merely necessary that two simple assumptions should hold good, viz, (1) that the nebular matter of the spiral can be in a finely divided solid or liquid condition, as the continuous spectrum implies, and (2) that the particles of this scattered material revolve in elliptical orbits about the central mass.

In attempting to follow the probable evolution of such a spiral nebula, three elements stand out conspicuously: (1) The central mass, obviously to become the sun; (2) the knots on the arms that are assumed to be the nuclei of the future planets and perhaps satellites; and (3) the diffuse nebulous matter to be added to the nuclei as material of growth. In the particular case of the solar nebula it is assumed (1) that the central mass was relatively very great; (2) that the knots were very irregular in size and placed at irregular distances from the center; and (3) that the nebulous portion was very small relative to the central mass and probably large relative to the knots.

It is assumed that the masses of matter in the knots were sufficiently large to hold themselves together in spite of the differential attraction of the central mass, otherwise they would soon have been scattered. They seem to have maintained themselves successfully in existing nebulae that appear to have undergone some notable degree of evolution.

On the other hand, it is presumed that the mutual attraction of the more tenuous nebular matter was insufficient to aggregate itself directly in the presence of the central attraction, for in the existing nebulae this portion seems to show a progressive tendency to a more general diffusion. The planetesimals of the diffused nebulous portion are assumed to be controlled essentially by the gravitation of the main mass and to revolve in individual orbits about it.

The irregularity of the forms of the knots seems to imply that their organization is also planetesimal, though the larger ones may be able to hold gases also. The direction of revolution of these knots is supposed to be usually the same as that of the rotation of the nebula as a whole, but subject to local and special influences that might lead to important variations.

While the knots of the solar nebula are regarded as the nuclei about which gathered the planetesimals to form the future planets, all such nuclei did not necessarily retain their independence and grow to planets, though no planet probably developed except from such a nucleus. Existing nebulae show clusters of knots and aggregates of irregular form susceptible of development into complex planetary systems, such as the large planets and their families of satellites. The earth-moon system is assigned to a couplet of companion nuclei of quite unequal sizes.

The collisions of isolated planetesimals with one another may be neglected, for it is uncertain whether the planetesimals would rebound from one another or would unite; probably the former when they were highly elastic, and the latter when inelastic; and probably much would also depend on their velocities and their modes of impact; but in any case the result would only affect the size and number of the planetesimals.   The important consideration is the impact of the isolated planetesimals upon the planetary nuclei.   In this case the usual result must apparently be the capture of the planetesimals by the nuclei; and with each capture the power of further capture would be augmented.

The rate of accretion is a matter of radical geological importance; indeed, it is, in some measure, the most critical feature of the whole nebular problem, for the rate of accretion determines whether the average temperature on the surface of the growing body will be high or low.   The surface temperature is not determined by the total heat produced by the collisions, but by *the heat produced in a given time*, which, in turn, is determined by *the frequency and force of the collisions on a given area*.   If the succession of collisions on a given square mile was not rapid enough to generate heat beyond the concurrent radiation from the square mile, a high average temperature for the whole could not be reached, however great the sum total of heat generated in the course of time.   It is to be noted that the heat generated after a solid nucleus was developed must have been superficial and hence readily radiated away.   While the nuclei remained assemblages of small bodies, perhaps gaseous in part in the larger ones, planetesimals from without may have penetrated to the interior and there developed heat not so readily lost.   But this state is only assignable to the early stages.

A further consideration bearing upon the critical subject of temperature is the manner of collision.   Since all the planetesimals and planetary nuclei were revolving *in the same direction* about the solar mass, the collisions were all overtakes, and could have been violent only to the extent of their differences of orbital velocity, modified by their mutual attractions.   These velocities are of a much lower order than the average velocities of meteoritic collisions.   Many of the overtakes would obviously be due to differences of velocity barely sufficient to bring about an overtake. When the relative mildness of impact is considered in connection with the intervals between impacts at a given spot, the conviction can scarcely be avoided that *the surface temperature would not necessarily have been high.* It seems probable that it would have been moderate throughout most of the period of aggregation, and certainly so in the declining stages of infall.

By graphical inspection of all probable cases, it may be seen that the possibilities of overtake favorable to forward rotation exceed those favorable to retrograde rotation. This holds true on the assumption of an equable distribution of planetesimals, which may fairly be assumed as an average fact, but not necessarily as always the fact; and hence the conclusion is not rigorous, and a backward rotation is not impossible. From the nature of the case, a varying rotation for the several planets is more probable than a nearly uniform one.

It is also obvious that the impacts on the right and left sides of a growing nucleus, as well as those on the outer and inner sides, might be unequal, and hence *obliquity* of rotation of varying kinds and degrees might arise. As the solar system presents these variations, the method of accretion here postulated seems to lend itself happily to the requirements of the case.

To bring out the geological bearings of the planetesimal hypothesis, I have given considerable time to a study of the probable stages of growth of the early earth, of the time and mode of introduction of the atmosphere and hydrosphere, and of the initiation of the great topographic features, together with the leading modern processes. . . .

Following the postulates of the previous sketch, a nebular knot is assumed to have been the nucleus of the growing earth. It has not been thought important to consider at much length the special state of organization of the material of this nucleus, since by assumption it constituted but a minor part of the grown planet, and its ultimate condition would probably be that of the dominant mass, or, if not, would be so deeply central as to have little geologic importance. Assuming that the nuclear mass was quite small, it is inferred that it was composed chiefly of matter of high molecular weight, since light molecules would be liable to escape because of their velocities. The nucleus is supposed to have been originally an assemblage of planetesimals grouped together by their mutual gravity, and to have passed gradually into a solid mass in connection with the capture of outside planetesimals. As the planetesimals were solid aggregates in the main, and only partially elastic, their collisions are assumed to have destroyed their orbital motions in a certain proportion of cases and to have led to their collection at the center. In other cases the orbital motions were doubtless increased, but any planetesimals which were thus temporarily driven away were subject to subsequent capture.

As the solid nucleus thus formed may not have been massive enough to control a gaseous envelope in its earlier stages, a possible atmosphereless stage is to be recognized. . . .

When the growing earth reached a mass sufficient to control the flying molecules of atmospheric material, there were two sources from which these could be supplied for the accumulation of an atmosphere, an external and an internal one.

By hypothesis, all the atmospheric and hydrospheric material of the parent nebula which has not gathered into the aggregated planetesimals remained as free-molecular planetesimals. While the planetary nucleus was small it probably could not gather and hold the lighter molecules, even when they collided with it, except as this was done by occlusion or surface tension, in which case they did not form an atmosphere; but when the growing earth reached the requisite mass these free atmospheric molecules were gathered about it and retained as an atmospheric envelope. This would be a more abundant source of supply during the nebular stages than afterward, but by hypothesis it continues to be a source of some supply even to the present time, for the very doctrine that postulates the loss of such high-speed molecules implies their presence in space, subject to capture by bodies capable of capturing them.

In the later stages of organization, and thence down to the present time, the molecules discharged from all the bodies of the solar system were possible sources of atmospheric accretion. Of these the most important were probably volcanic and similar discharges from the small bodies that could not hold gases permanently and discharges from the sun by virtue of the enormous explosive and radiant energies that are there resident.

As the planetesimals were gathered into the growing earth-nucleus they carried their occluded gases in with them, except as the superficial portion might be set free by the heat of impact. There was thus built into the growing earth atmospheric material. So, also, while the nucleus was growing it was subjected to the bombardment of free-molecular planetesimals of the atmospheric substances. In its early stages it might not be able to hold these as a free gaseous envelope, but to a certain extent it could hold, by virtue of capillary and subcapillary attraction, such molecules as were driven into the pores and other interstices of the fragmental surface arising from the infall of the solid planetesimals. . . .

The atmospheric material thus condensed within the growing earth could become a part of the atmospheric envelope only by extrusion. The assigned modes of extrusion will be considered presently; meanwhile it may be assumed that these internal gases were given forth progressively and fed the atmosphere.

The amount of oxygen in the early atmosphere is more uncertain from doubt as to a competent source of supply. Crystalline rocks and meteorites are not known to contain it in a *free* state. As above remarked, it occurs among volcanic gases, but it is not known that it comes from the deep interior. It is detected in the sun and not improbably existed in the nebula, from which it might have been gathered shortly after the accretion of carbon dioxide began. The safer inference seems to be that it was not very abundant relatively in the early atmosphere. This inference may be entertained the more freely because it seems to give the better working hypothesis, for the present large proportion of oxygen may be assigned to the reduction of carbon dioxide by plant action, and the present proportions and those of geologic history seem to come out best on this basis. For the primitive atmosphere there is theoretical need for only enough oxygen to support the primitive plant life until it could supply itself, after which it would produce a surplus. . . .

The problem of vulcanism assumes a quite new aspect under the planetesimal hypothesis, if very slow accretion without very high temperature be assumed. It has been taken for granted in the preceding statement that there was volcanic action. It is necessary, therefore, to consider how volcanic action may have arisen, and this involves the more radical question how the high internal temperatures of the earth may have arisen if the earth did not inherit its heat from a molten condition arising from a gaseous origin.

With the detailed conceptions now developed, the method of volcanic action deduced from the accretion hypothesis may be readily apprehended and the vital part assigned to it in earth history may be realized.   The chief portion of internal heat being assigned to compression, the temperature must have been highest at the center, because the compression was greatest there, and must have declined toward the surface.

Without attempting to fix the precise stage, it is not unreasonable to assume that surface waters had begun their accumulation upon the earth's exterior while yet it lay 1,500 to 1,800 miles below the present surface.   The present difference between the radii of the oceanic basins and the radii of the continental platforms is scarcely 3 miles, on the average; so that if the continental segments be assumed to be in approximate hydrostatic equilibrium with the oceanic segments today, as seems highly probable, the selective weathering process brought about a difference in depression of only 1 mile in 500 or 600 miles, or about one-fifth of 1 per cent. We appear, therefore, to be laying no heavier burden upon weathering than it is competent to bear.   It might well be thought to do much more, but the process of weathering is slow, and as new material was constantly falling in and burying the old, partial alteration was all that could take place; and, besides, a part of the basic material leached from the surface was redeposited beneath the ground water of the land and in landlocked basins and was not lost to the continental segments.

Not only is the evolution of the great abysmal basins and of the continental platforms thus assigned to a very simple and inevitable process, but there is therein laid the foundation for subsequent deformation of the abysmal and continental type.

The planetesimal hypothesis affords an undetermined lapse of time between the stage when conditions congenial to life were first possible and the stage when the first fairly legible record was made in the Cambrian period.   To this unmeasured period the whole pre-record evolution of life, whatever be its method, may be referred, with a strong presumption that the time was ample and that there is no occasion for an evasion of the profound problem of life genesis by referring it to some distant and unknown body; nor is the problem vexed by duress of severe time limits. A theoretical scantiness of time for a prolonged evolution previous to the Cambrian period has been deduced from a molten earth, but this does not apply to the planetesimal hypothesis. The supposed limitation of the sun's thermal endurance would apply if the arguments could be trusted, but their foundation has been cut away by recent discoveries. It is not the least of the virtues of the planetesimal hypothesis that it opens the way to a study of the problem of the genesis and early evolution of life free from the duress of excessive time limits and of other theoretical hamperings, and leaves the solution to be sought untrammeled, except by the conditions inherent in the problem itself, which are surely grave enough.

By hypothesis, volcanic action only began some time after the beginning of the earth's growth, for it was delayed (1) by the lack of sufficient compression in the central parts to give the requisite heat, and (2) by the time required for this central heat to move out to zones of less pressure, where it would suffice to melt the more fusible constituents. But, once begun, it is supposed to have gradually increased in actual and in relative importance until it reached its climax. This obviously came much later than the climax of growth, for it was dependent on the growth to give the increased compression from which arose the central heat on which the vulcanism depended. And so, owing to the sources of delay just cited, the maximum of volcanic action must have lagged much behind the accession of the material which remotely actuated it. It is therefore inferred that vulcanism continued to increase in activity long after growth had entered on its decline, and that there was an important period in which the dominant activity was volcanic.

It is conceived that in the late stages of the earth's growth the amount of material poured out on the surface in molten form or introduced into the outer parts of the earth from below was very much greater than the accessions from without. Still later, these declining accessions were so overwhelmed by the igneous extrusions that they became indistinguishable contributions. In this stage, too, it is held that the modifications wrought by the atmosphere, the hydrosphere, and organic life were also quite subordinate to the volcanic contributions. Disintegration is assumed to have gone little farther, usually, than to partially reduce rocks of the granitoid type to arkoses, and those of the basic type to wackes. Rather rarely, it is believed, was much pure quartzose sand, aluminous clay, or similar well-decomposed residuary materials accumulated; rarely, also, much carbonaceous shale. Arkoses and wackes, when metamorphosed later, took on such a similitude to igneous rocks as to be more or less unidentifiable.

The formations of this period of volcanic dominance, with very subordinate clastic accompaniment, are regarded as constituting the Archean complex, though perhaps only the later portions of the great volcanic series are represented by the *known* Archean.

BERTRAND

Marcel Alexandre Bertrand (1847–1907), French geologist, was professor of geology in the School of Mines and chief engineer of the Bureau of Mines as well as a member of the Service de la carte géologique détaillée de France. His work in the Jura and the Alps resulted in many significant contributions to the solution of the problems of orogeny and tectonics.

OVERTHRUSTING IN THE ALPS

Translated from *Bulletin de la Société géologique de France*, Third Series, Vol. XII, pp. 318–330, 1884.

A year ago Monsieur Lory presented before the Society a remarkable section, to which Escher de la Linth called attention in 1840. It shows the strongly folded Nummulitic Horizon (Eocene) covered by the normal and almost horizontal series of Permian and Triassic beds. The extraordinary nature of the occurrence is further enhanced by the intercalation between the two series of a very narrow band—often reduced to a few meters—of Jurassic and Triassic beds of varying age and composition nearly horizontal like the upper series, but always reversed in sequence.

Mr. Heim's excellent book,* where, in addition to theoretical views of great importance, observational data are grouped and described with remarkable clarity, leaves no room for doubt as to the accuracy of both the age determinations and the relative positions. The explanation proposed by Escher and developed by Mr. Heim has the great advantage of including all these singular anomalies in a very simple formula. It has made this phenomenon famous under the name of the "double fold of the Glarus Alps."

It would not be permissible for me to write this note about a region that I have not visited, had I not been struck, on studying the work of Mr. Heim, by the numerous structural resemblances which (regardless of surface relief) this part of the Alps bears to the Franco-Belgian coal basin. I have merely tried to extend the beautifully simple and rational explanation which Monsieur

---

* *Mechanismus der Gebildbildung [Gebirgsbildung].*

Gosselet has given for the North,* to the Alps. . . . I am repro-
ducing [Fig. 35] (fig. 1) one of Mr. Heim's sections, which is adequate to illustrate clearly the important points.   The dots indicate
the postulated continuation of the two great folds which slope,
one towards the north, the other towards the south, enclosing the
mass of Tertiary strata in a nearly closed ring.   The little band
of Upper Jurassic limestone which, with its intermittent accompaniment of reversed Dogger and Trias, separates the Nummulitic
from the Permian, should join in depth towards the north the
well-developed Jurassic beds which support the same Nummulitic.
Thus one would have a vast fold, the axis of which would be
inclined toward the south and of which the lower half, stretched
and compressed, would have nearly disappeared.

Let us now compare this with the coal basin of the North.   For
greater clarity, I reproduce [Fig. 35] (fig. 4) the diagrammatic section of the coal basin at Anzin according to Monsieur Gosselet. . . .

Monsieur Gosselet accounts for this complicated appearance as
follows: the Silurian shales of Condros were raised and folded
before the Devonian period, the deposits of which were therefore
laid down disconformably.   After the Carboniferous period, the
Condros fold was accentuated, then overturned to the north.   A
break formed along the axis, and the upper part, as though sliding
on a vast inclined plane, was thrust forward and upward, far to
the north, covering successive folds of the more northern region.
In this movement, the mass pushed along has carried with it fragments of the lower mass; in other words, there has been planing as
well as sliding. . . .

If, for the moment, the surface of separation of the Permian
with the little thinned band of reversed Trias and Jurassic in
Mr. Heim's section is called the south fault, and the surface of
separation of that band from the Nummulitic, the boundary fault,
one may unite in a single phrase, applicable to the two regions,
either the whole of the phenomena or the proposed explanation:

A vast, strongly folded region has been covered with a mantle
of more ancient strata, whose lower limit, parallel with their bedding planes, is but slightly inclined from the horizontal.   A band

---

* *Bulletin de la Société géologique*, Third Series, Vol. VIII, p. 505.

of strata of intermediate age, with varying thickness and always reversed, separates the two systems. The direction of the beds (at least, of the oldest of them) in this band is likewise parallel to the surface of separation. The explanation of this arrangement can be found in a general thrusting toward the north of the mass situated above the south fault, this mass having carried with it in its movement a "thrust-fragment."

In the North, an immense denudation, followed by the return of the sea, and the accumulation of new deposits have partially masked the original phenomena. In the Alps, this did not occur; the explanation, therefore, should be tied up more closely, or rather, followed further. It is not enough that it account for the abnormal superposition; it must explain the relationships with the structure of the rest of the country. In other words, it does not suffice to say that there has been a thrusting; it is necessary to ascertain where the mass of thrust strata has come from and also where it terminates.

The answer to the first of these questions seems easy to me. Let us take Mr. Heim's section [Fig. 35] (fig. 2) again, limiting our attention to the northern fold. This reclined fold can be considered the result of two distinct movements, the one corresponding to the formation of vertical folds, the other to the progressive inclination of their axes toward the north. Whether or not there is any reality in this succession of two distinct phases, the analysis of the movements is permissible; it is the simple application of one of the fundamental theorems of mechanics.

This done, let us replace the north fold in its vertical position, keeping the width postulated by Mr. Heim. Let us do the same, of course, for those folds which appear farther to the north and for those which can be imagined toward the south. We reach a diagram analogous to the dashed portion of fig. 2 [Fig. 35]. The second phase, that of the reclining of the folds is marked on the same figure by dotted lines. Let us suppose then that a break is formed along the axis $a'b'$ of the first reclined fold. With this all the elements of the final diagram, corresponding to the actual state of things, are available. It only remains to explain how the upper strip $a'b'k'$ could get down into $a''b''k''$. The distance is about fifteen kilometers.

It is clear at the start that the effect of weight is not to be invoked as an appreciable factor. The analogy with the Belgian basin,

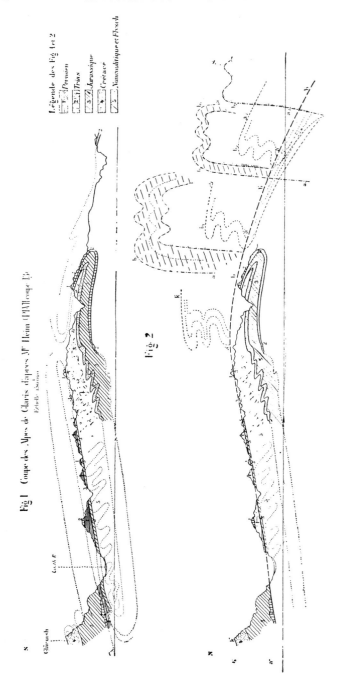

Fig 1 Coupe des Alpes de Claris d'après Mr Hein (Pl.II coupe B)

Fig. 2

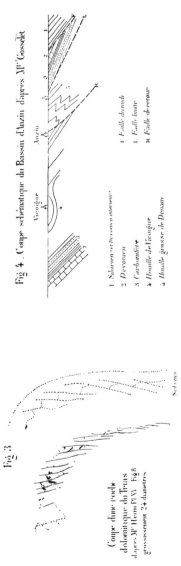

Fig. 4 . Coupe schématique du Bassin d'Anzin d'après M. Gosselet

Viroigne   Anzin

1 Silurien et Devonien inférieur

2 Devonien

3 Carbonifère

4 Houille de Viroigne

5 Houille grasse de Denan

t Faille d'amont

ι Faille limite

κ Faille de retour

Fig. 3

Coupe d'une roche
dolomitique du Trias
d'après M. Heim (Pl. XV. Fig.8
grossissement 24 diamètres

Schema

Fig. 35.—Diagrams used by Bertrand to illustrate his paper concerning overthrusting in the Alps, 1884.

where it would have acted on the contrary as a resistant force, suffices to prove this.

But one can conceive that the phenomenon of folding which I have tried to analyze may continue progressively towards the south with the same successive phases and that a series of folds a_1b_1, a_2b_2, all overturned and pushed towards the north as fast as they are formed, continues to transmit the impulse. From the same causes, the line of break, $a'b'$, continues in all the folds, doubtless changing in depth to a *zone of sliding*. According to the picture thus obtained, the idea of slow *flowing*, transformed near the surface to successive slidings, appears to sum up the phenomenon if not to explain it.

The sliding movement, being due to a thrust, could take place and be continued only if there had been no detachment from the central massif, that is, the overthrusting must have been continuous. Consequently, without the later action of denudation, a *regular* series from the Permian, perhaps even from the crystalline rocks, to the Cretaceous would be observed over a width of twenty kilometers. This series, over all this width, would cover another regular series, *complete* clear to the Eocene, which would be completely masked.

The study of the geological maps of Switzerland leads to the conclusion that the phenomenon of overthrusting is not limited to the Glarus Alps. The Sion sheet, published recently, gives, on the whole, the clear impression of a "base" of Flysch on which are spread great splotches of older rocks, usually forming the high peaks.

Finally, I return to the Glarus Alps. Adopting the method of diagrammatic representation which restores to certain folds all the phenomena of sliding, the structure of this region may be summarized thus: a series of folds, the axes of which, beginning from the central massif, are regularly inclined toward the Molasse plain but are raised more and more toward the vertical as they approach this plain, is covered by another fold, much more vast and with the axis inclined in the same direction in such a manner as even to exceed the horizontal and descend nearly to the plain. Fundamentally, it is simply the hypothesis of a "single fold" substituted for that of Escher's "double fold."

MARVINE

Archibald Robertson Marvine (1848–1876), American geologist, directed the work of the Middle Park Division during the early 1870's under the general supervision of F. V. Hayden, who at that time was in charge of the United States Geological and Geographical Survey of the Territories. His recognition of the efficacy of subaerial denudation in the planation of a large region is one of the earliest statements of its kind, although he ultimately appeals to "the encroaching ocean" to smooth the surface produced by stream erosion.

Processes of Erosion in the Colorado Front Range

From United States Geological and Geographical Survey ("Hayden Survey"), *Annual Report for 1873*, pp. 144–145, Washington, 1874.

Three causes combine to render the rapid study of the stratigraphy of the archaean rocks difficult and its results uncertain: First, their structure is not only often complex, but obscure, the evidence of it being at times nearly or wholly obliterated by the metamorphism, and often over large areas very difficult to find; second, this metamorphism renders lithological characters inconstant, so that a stratum that at one point may be characteristic among its neighbors, may, at another, become like them, or all may change so as to retain none of their geological features, becoming again like other series, so that lithological resemblances cannot often be taken as a guide to follow, and may even become misleading; third, the erosion producing the present surface features of the mountain region had the direction of its action determined by movements of the surface which were not closely connected with the extended plications of its rocks; and, moreover, since this erosion has not long been active among these rocks, there appears no well-defined connection between the topography and the structural geology. The ancient erosion gradually wore down the mass to the surface of the sea, and while previously to this it was no doubt directed by the structure, yet the mass was finally leveled off irrespective of structure or relative hardnesses of its beds by the encroaching ocean, which worked over its ruins and laid them down upon the smoothed surface in the form of the Triassic and other beds. The recent great uplift, while it probably added new plications to the accumulated plications of the past in the ancient rocks, was quite simple with respect to their total plication, and left the upper Triassic and other sedimentary beds comparatively simply structured, they having been affected alone by the later movements.

As the mass appeared above the sea and surface erosion once more commenced, but which now acts upon the recent rocks covering probably in greater part the complex underlying rocks, it was directed off from the line of greater uplift down the long slopes of the rising continent to the retiring sea. The channels of drainage started were directed solely by the structure and characters of the upper rocks, and when they gradually cut down through these and commenced sinking their cañons into the underlying complicated rocks, these cañons bore no relation whatever to their complications. It is but recently that the upper rocks have been completely removed from the summits of the mountain spurs, the ancient level of subaqueous erosion being still indicated by the often uniform level of the spurs and hill-tops over considerable areas, and large plateau-like regions which became very marked from certain points of view.

WHITE

Israel Charles White (1848–1927), American geologist, began his scientific career as a member of the Geological Survey of Pennsylvania. From 1877 to 1892, he was professor of geology in West Virginia University, a post which he relinquished in order to give his whole attention to his consulting practice and business interests. He was probably the first geologist to make practical use of the anticlinal theory of gas and oil accumulation. From 1897 until his death, he was state geologist of West Virginia, contributing his services to his native state without stipend.

THE ANTICLINAL THEORY OF GAS ACCUMULATION

From *Science*, Vol. V, pp. 521–522, 1885.

Practically all the large gas-wells struck before 1882 were accidentally discovered in boring for oil; but, when the great value of natural gas as fuel became generally recognized, an eager search began for it at Pittsburgh, Wheeling, and many other manufacturing centres.

The first explorers assumed that gas could be obtained at one point as well as another, provided the earth be penetrated to a depth sufficiently great; and it has required the expenditure of several hundred thousand dollars in useless drilling to convince capitalists of this fallacy which even yet obtains general credence among those not interested in successful gas companies.

The writer's study of this subject began in June, 1883, when he was employed by Pittsburgh parties to make a general investigation of the natural-gas question, with the special object of determining whether or not it was possible to predict the presence or absence of gas from geological structure. In the prosecution of this work, I was aided by a suggestion from Mr. William A. Earsenian of Allegheny, Penn., who had noticed that the principal gas-wells then known in western Pennsylvania were situated close to where anticlinal axes were drawn on the geological maps. From this he inferred there must be some connection between the gas-wells and the anticlines. After visiting all the great gas-wells that had been struck in western Pennsylvania and West Virginia, and carefully examining the geological surroundings of each, I found that every one of them was situated either directly on, or near, the crown of an anticlinal axis, while wells that had been bored in the synclines on either side furnished little or no gas, but in many cases large quantities of salt water. Further observation showed that the gas-wells were confined to a narrow belt, only one-fourth to one mile wide, along the crests of the anticlinal folds. These facts seemed to connect gas territory unmistakably with the disturbance in the rocks caused by their upheaval into arches, but the crucial test was yet to be made in the actual location of good gas territory on this theory. During the last two years, I have submitted it to all manner of tests, both in locating and condemning gas territory, and the general result has been to confirm the anticlinal theory beyond a reasonable doubt.

But while we can state with confidence that all great gas-wells are found on the anticlinal axes, the converse of this is not true; viz., that great gas-wells may be found on all anticlinals. In a theory of this kind the limitations become quite as important as, or even more so than, the theory itself; and hence I have given considerable thought to this side of the question, having formulated them into three or four general rules (which include practically all the limitations known to me, up to the present time, that should be placed on the statement that large gas-wells may be obtained on anticlinal folds), as follows:—

(a) The arch in the rocks must be one of considerable magnitude: (b) A coarse or porous sandstone of considerable thickness, or, if a fine grained rock, one that would have extensive fissures, and thus in either case rendered capable of acting as a reservoir for the gas, must underlie the surface at a depth of several hundred feet (five hundred to twenty-five hundred feet); (c) Probably very few or none of the grand arches along mountain ranges will be found holding gas in large quantity, since in such cases the disturbance of the stratification has been so profound that all the natural gas generated in the past would long ago have escaped into the air through fissures that traverse all the beds. Another limitation might possibly be added, which would confine the area where great gas-flows may be obtained to those underlaid by a considerable thickness of bituminous shale.

Very fair gas-wells may also be obtained for a considerable distance down the slope from the crest of anticlinals, provided the dip be sufficiently rapid, and especially if it be irregular, or interrupted with slight crumples. And even in regions where there are no well-marked anticlinals, if the dip be somewhat rapid and irregular, rather large gas-wells may occasionally be found, if all other conditions are favorable.

The reason why natural gas should collect under the arches of the rocks is sufficiently plain, from a consideration of its volatile nature. Then, too, the extensive fissuring of the rock, which appears necessary to form a capacious reservoir for a large gas-well, would take place most readily along the anticlinals where the tension in bending would be greatest.

HEIM

Albert Heim (1849–1937), Swiss geologist, professor at the University of Zurich, was the leader in modern work on Alpine structure.

ON ROCK DEFORMATION DURING MOUNTAIN BUILDING

Translated from *Untersuchung über den Mechanismus der Gebirgsbildung,*
Vol. 1, pp. 126–128; Vol. 2, pp. 32–75, Basel, 1878.

The Glarner Double Fold

As one proceeds southward from Walensee through the valley or on the ridges, massive Verrucano beds like those of the Sernf Valley are found as the base of mountains everywhere. Quartzite, red dolomite, Lias, and Brown Jura lie above this. Further, there is Malm at Mürtschenstock, and Cretaceous and nummulitic limestone above it at Kerezenberg. Therefore the bedding is entirely normal here. In general the strata beyond Walensee rise higher toward the south, while the covering of limestone formations becomes progressively thinner and more sporadic. At Gulderstock and Walenkamm, red dolomite and Lias rise highest in the south; then steep slopes formed by outcrops of Verrucano descend into the Sernf and Weisstannen valleys. There, deep in the valleys, we would expect to find rock of the central complex, crystalline schist, etc., under the Verrucano. Instead of this, the valleys lie in nothing but Eocene, which is separated from the Verrucano only by one or two, ordinarily thin, limestones. Some of the more southerly peaks carry a cap of Verrucano on their summit, encircled underneath by one or more limestones; while the remainder, clear to the foot, is Eocene.

From Limmernboden über Elm and the Foo Pass to Calfeuser Valley, there stretches a line of symmetry south of which the same phenomena are repeated with north and south reversed. Just as the base of Kärpfstock, Gulderstock, Walenkamm, and the Grauen-Hörner are formed of Eocene and the peaks of north-dipping Verrucano, covered, here and there, with limestone remnants—so Piz Dartgas, Vorab, Saurentstock, and Ringlekopf are crowned by south-dipping Verrucano beds covering the Eocene. A limestone horizon lies between the Verrucano and the Eocene, there as here. The north-dipping limestone formation lies on north-dipping Verrucano in normal succession at Walensee; similarly, in the Upper Rhine Valley, south-dipping limestone formations, red dolomite, *raubwacke*, Lias, and, at Neukirch in Obersaxon, even ironstone of the Dogger, rest on south-dipping Verrucano. In the north as in the south, the Verrucano rises in normal position toward the indicated line of symmetry, while the usual cover of secondary limestone is more and more contorted. But in the middle zone, between the two Verrucano sheets, there lies at depth what should be highest, namely Eocene. In general it dips rather steeply to the south and rests discordantly under the Verrucano cover. The line of symmetry itself is the prolongation of the median line of the central massif.

This whole structural condition appears at first sight to be the overthrust of both sides against the middle. We shall prove more definitely that the overthrusts on the two sides are not really faults but *extended recumbent folds*. The whole forms a *double fold*. (Baltzer called it a "loop"; Escher also frequently used this expression orally. I prefer the name "fold," because it indicates a surface corresponding to the spread of bent strata, while "loop" refers rather to the bending of a line.) The north flank of the double fold is a fold overturned toward the south; the south flank, one overturned toward the north. The unreversed limb of the arch of the north flank, or the north fold, may be seen on the south bank of Walensee; the unreversed limb of the south flank, or south fold, may be studied in the Upper Rhine Valley. The limestone layers in the middle zone between the Eocene and the overlying Verrucano represent the inverted *mid-limb* on both sides. The *trough limbs* of both folds unite at depth. The core of both troughs, opening towards each other, is filled by the Eocene beds, which the limbs of the trough almost entirely cover. Only on the side of the south fold is the mutual trough limb locally uncovered; and it is evident there that there are several little secondary folds which are, at the same time, extensions of the border chains of our area (Windgälle, Gemsfayrstock, Baumgartenalpen).

Going from west to east, the central massif and border chains are lost under and in the mutual trough limb, and their place is

taken by the overthrust coming from above on either side.   See
Profiles XIX and XX [Fig. 36]

The Deformation of Rock without Fracture

Whereas deformation by fracture overcomes the rigidity in the
arrangement of particles at only a few places, this is accomplished
by deformation without fracture in an almost infinitely large
number of places, though in somewhat different fashion.   Frac-

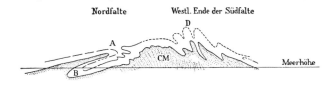

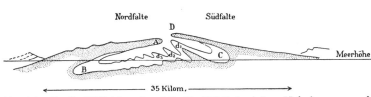

FIG. 36.—Profile XIX (above) and Profile XX (below) from Heim's monograph
on the mechanics of mountain building, 1878.

tureless deformation is a more efficient mechanical action than
deformation by fracture.

We may be amazed, when considering the many, enormous,
far-flung flexures of strata, by the thought of the force necessary
to produce such results.   However, the wide swing, the breadth
of the overthrust involved in a fold, only gives a measure for the
amount of compression produced, or for the route on which
the opposing resistances were overcome; it is not a measure of the
intensity of the force.   The physical difficulty of bending without
fracture is the greater, the less the radius of the bend.   The difficulty which opposes the deformation is the resistance of the rocks
to the plastic displacement of their particles.   These must change
their relative position all the more—i.e. the resistance must be
overcome in places all the more numerous and for a greater distance—the smaller the area in which the displacement has reached
a noticeable and measurable amount.   The narrow little folds
visible in a hand specimen, the stretching of a fossil to twice or

thrice its length, is a much greater performance, a more complete overcoming of the rigid cohesion, than the bending of a series of beds into a league-wide trough wherein the deformation of an enclosed fossil is too slight to be detected by measurement. In order to bend an enormous group of strata a little, more force is used, of course, than to crush a cubic meter of rock, but much less than to fold all parts of an equally enormous group of strata. From the local amount of deformation, readily measurable on a small scale, we estimate the locally effective intensity of the force, whereas the whole force is disseminated through the whole group of disturbed strata. If we have thus found a physical explanation for non-fractured deformation of brittle rocks visible even in hand specimens, the explanation for the wide flexures of strata will be gained even more readily, for these last are less intense molecular displacements or surmountings of cohesion.

We seek next to recognize some of the most important laws of these phenomena in order to reach by the route of inductive investigation conclusions which are to serve as a starting point for the explanation. We shall mention the earlier attempts at explanation and, as we are forced to consider them inadequate, shall offer a new explanation in their stead. Finally we shall see whether the conclusions deduced from our explanations are in agreement with nature. . . .

From the given examples, drawn mainly from our special research area, we obtain the following "Laws of the Phenomenon":

1. Fractureless deformations are found imprinted in very similar, often even identical, forms in rocks of extremely varied petrography—plastic as well as brittle. Fractureless deformation is not dependent on the type of the rock. . . .

2. The same rock types and rock varieties of the same beds which in one place have been deformed by fractures are found deformed in other places without fracture. . . .

3. The same piece of rock can undergo both kinds of deformation. . . .

4. Beds formed of different rocks, folded at the same time, frequently show, in the same locality, differences in the deformation which depend on the nature of the rocks. . . .

5. In the great and small folds associated with mountain building, the beds on the flanks of the folds are thinner, and in the axial areas, thicker. . . .

6. The microscopically ascertained cracks and structural openings, in spite of their frequency, suffice as little to illustrate the bending as the greater fissures. . . .

7. A schistose structure, slip cleavage, can develop through the accumulation of displacements which result in little folds—especially in plastic rocks. . . .

8. A second kind of transverse schistosity, microcleavage, originated thus—all rock particles are compressed into lamellar or columnar form in the slip direction, which is, for the most part, perpendicular to the direction of maximum pressure. . . .

9. A third type of cleavage arises thus—all lamellar and columnar mineral elements in a rock are moved into parallel position. . . .

10. In all cases where fossils have become deformed, a cleavage is set up in the rock;

and

11. the plane of schistosity coincides with the direction of elongation of the fossils. . . .

12. A linear direction of elongation is frequently recognized on the cleavage surfaces. . . .

13. The strike direction of cleavage generally coincides in the Alps with the strike direction of strata and chains, while the dip, for the most part, is rather steep and cuts the strata. . . .

14. Escher first postulated clearly what we wish to designate here as the fourteenth Law of the Phenomenon: Cleavage is ordinarily much more clearly developed in and near the areas of bending than in the more remote parts of the fold limbs. . . .

15. In general, the compacting or unfractured deformation of the beds increases with the depth under the mean mountain surface. . . .

At first glance those localities where marked fractureless deformation of the rock, elongation of fossils, cleavage bending, etc., occur seem to have no regular distribution. But, should we exclude all cases where the nature of the rock greatly aids the deformation and consider only those cases where the non-fractured deformation indicates much mechanical action in a hard, brittle material, it would be entirely different. In the eastern Swiss Alps, I do not find a single example in the Tertiary and none in the Cretaceous zone. All cases of fractureless deformation of very brittle rocks belong to the deeper sedimentary strata—the Jurassic and still older formations. All are found in inner sediment zones of the Alps, generally in the vicinity or on the border of the central massif (Frette de Saille, Windgälle, Pantenbrücke region) or are found in troughs squeezed between the rocks of the central massif. The sole exceptions known to me are formed by some of the Cretaceous and Eocene rocks of the Glarner double fold—for example, the folded quartzite back of Durnachthal (fig. 16) [Fig. 37]. On this rock lies, first about one hundred meters of Eocene formation, then the Lochseiten limestone, then the Verru-

FIG. 37.—A sketch, fig. 16, from Heim's monograph on the mechanics of mountain building, 1878.

cano, and over it—now of course removed by erosion—the entire Jurassic, Cretaceous, and Eocene formations. Although many rocks, folded there without fracture, belong to the upper formations, they were loaded in mountain building fully as much as though they belonged to the deepest strata. Taking the mechanical circumstances of mountain building into consideration, these exceptions are no more than apparent. Proof is given in the following section that we have to think of the central zone of the Alps as covered with an enormous thickness of sedimentary strata at the beginning of the folding process. In those mountains which were not formerly covered with much higher sediment— where loss from denudation accounts for some one hundred meters at the most, not for some one thousand meters of rock thickness— strata deformed by faults instead of bending are much more common than in the Alps. In the Jura this is very noticeable;

likewise in the Danish island of Möen, etc. We have hit upon a new law:

The sixteenth Law of the Phenomenon: Fractureless deformation in non-plastic rocks is found only at great depths under the original mountain surface.

But great depths mean great loading. The enormous horizontal pressure of mountain building alone does not suffice for fractureless deformation: loading, i.e. pressure from above, is needed. Of course, a corresponding back pressure enters into play with that. Fractureless deformation is best developed with high pressure as nearly as possible from all sides.

DAVIS

William Morris Davis (1850–1934), American geologist and geographer, was a member of the faculty of Harvard University throughout the greater part of his life, serving as Sturgis-Hooper professor from 1898 to 1912. As the great systematizer of the science of geomorphology, he was the acknowledged leader of the "American school" of physiographers.

THE EROSION CYCLE IN THE LIFE OF A RIVER

From "The Rivers and Valleys of Pennsylvania," *National Geographic Magazine*, Vol. I, pp. 15–71, 1889.

Rivers are so long lived and survive with more or less modification so many changes in the attitude and even in the structure of the land, that the best way of entering on their discussion seems to be to examine the development of an ideal river of simple history, and from the general features thus discovered, it may then be possible to unravel the complex sequence of events that leads to the present condition of actual rivers of complicated history.

A river that is established on a new land may be called an original river. It must at first be of the kind known as a consequent river, for it has no ancestor from which to be derived. Examples of simple original rivers may be seen in young plains, of which southern New Jersey furnishes a fair illustration. Examples of essentially original rivers may be seen also in regions of recent and rapid displacement, such as the Jura or the broken country of southern Idaho, where the directly consequent character of the drainage leads us to conclude that, if any rivers occupied these regions before their recent deformation, they were so completely extinguished by the newly made slopes that we see nothing of them now.

Once established, an original river advances through its long life, manifesting certain peculiarities of youth, maturity and old age, by which its successive stages of growth may be recognized without much difficulty. For the sake of simplicity, let us suppose the land mass, on which an original river has begun its work, stands perfectly still after its first elevation or deformation, and so remains until the river has completed its task of carrying away all the mass of rocks that rise above its baselevel. This lapse of time will be called a cycle in the life of a river. A complete cycle is a long measure of time in regions of great elevation or of hard rocks; but whether or not any river ever passed through a single cycle of life without interruption we need not now inquire. Our purpose is only to learn what changes it would experience if it did thus develop steadily from infancy to old age without disturbance.

In its infancy, the river drains its basin imperfectly; for it is then embarrassed by the original inequalities of the surface, and lakes collect in all the depressions. At such time, the ratio of evaporation to rainfall is relatively large, and the ratio of transported land waste to rainfall is small. The channels followed by the streams that compose the river as a whole are narrow and shallow, and their number is small compared to that which will be developed at a later stage. The divides by which the sidestreams are separated are poorly marked, and in level countries are surfaces of considerable area and not lines at all. It is only in the later maturity of a system that the divides are reduced to lines by the consumption of the softer rocks on either side. The difference between constructional forms and those forms that are due to the action of denuding forces is in a general way so easily recognized, that immaturity and maturity of a drainage area can be readily discriminated. In the truly infantile drainage system of the Red River of the North, the inter-stream areas are so absolutely flat that water collects on them in wet weather, not having either original structural slope or subsequently developed denuded slope to lead it to the streams. On the almost equally young lava blocks of southern Oregon, the well-marked slopes are as yet hardly channeled by the flow of rain down them, and the depressions among the tilted blocks are still undrained, unfilled basins.

As the river becomes adolescent, its channels are deepened and all the larger ones descend close to base level. If local contrasts of hardness allow a quick deepening of the down-stream part of the channel, while the part next up-stream resists erosion, a cascade or waterfall results; but like the lakes of earlier youth, it is evanescent, and endures but a small part of the whole cycle of growth; but the falls on the small headwater streams of a large river may last into its maturity, just as there are young twigs on the branches of a large tree. With the deepening of the channels, there comes an increase in the number of gulleys on the slopes of the channel; the gulleys grow into ravines and these into side valleys, joining their master streams at right angles (La Noë and Margerie). With their continued development, the maturity of the system is reached; it is marked by an almost complete acquisition of every part of the original constructional surface by erosion under the guidance of the streams, so that every drop of rain that falls finds a way prepared to lead it to a stream and then to the ocean, its goal. The lakes of initial imperfection have long since disappeared; the waterfalls of adolescence have been worn back, unless on the still young headwaters. With the increase of the number of side-streams, ramifying into all parts of the drainage basin, there is a proportionate increase in the rate of waste under atmospheric forces; hence it is at maturity that the river receives and carries the greatest load; indeed, the increase may be carried so far that the lower trunk-stream, of gentle slope in its early maturity, is unable to carry the load brought to it by the upper branches, and therefore resorts to the temporary expedient of laying it aside in a flood-plain. The level of the flood-plain is sometimes built up faster than the small side-streams of the lower course can fill their valleys, and hence they are converted for a little distance above their mouths into shallow lakes. The growth of the flood-plain also results in carrying the point of junction of tributaries farther and farther down stream, and at last in turning lateral streams aside from the main stream, sometimes forcing them to follow independent courses to the sea (Lombardini). But although thus separated from the main trunk, it would be no more rational to regard such streams as independent rivers than it would be to regard the branch of an old tree, now fallen to the ground in the decay of advancing age, as an independent plant; both are detached portions of a single individual, from which they have been separated in the normal processes of growth and decay.

In the later and quieter old age of a river system, the waste of the land is yielded slower by reason of the diminishing slopes of the valley sides; then the headwater streams deliver less detritus to the main channel, which, thus relieved, turns to its postponed task of carrying its former excess of load to the sea, and cuts terraces in its flood-plain, preparatory to sweeping it away. It does not always find the buried channel again, and perhaps settling down on a low spur a little to one side of its old line, produces a rapid or a low fall on the lower slope of such an obstruction (Penck). Such courses may be called locally superimposed.

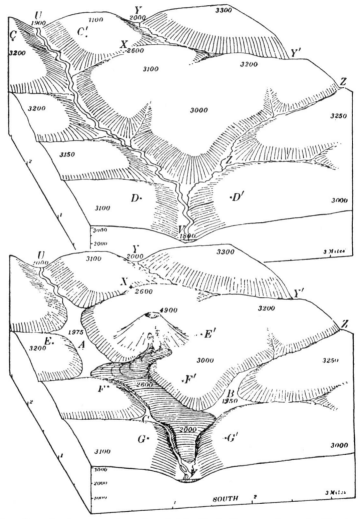

Fig. 38.—The first two of a series of eight block diagrams drawn by Davis to illustrate an ideal sequence of events in a volcanic region, c. 1900.

It is only during maturity and for a time before and afterwards that the three divisions of a river, commonly recognized, appear most distinctly; the torrent portion being the still young headwater branches, growing by gnawing backwards at their sources; the valley portion proper, where longer time of work has enabled

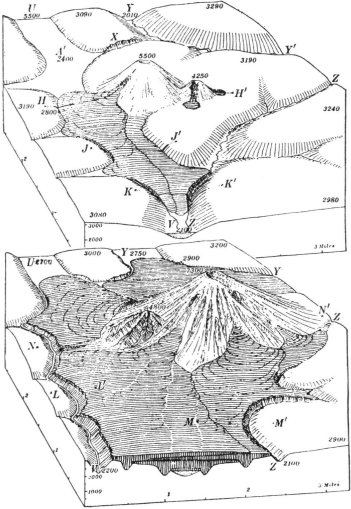

Fig. 39.—The third and fourth diagrams of the series beginning with Fig. 38.

the valley to obtain a greater depth and width; and the lower flood-plain portion, where the temporary deposition of the excess of load is made until the activity of middle life is past.

Maturity seems to be a proper term to apply to this long enduring stage; for as in organic forms, where the term first came into use,

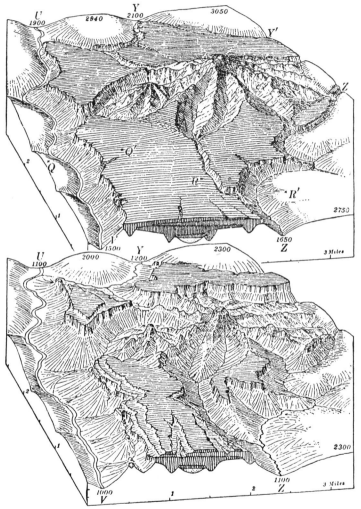

FIG. 40.—The fifth and sixth diagrams of the series beginning with Fig. 38.

it here also signifies the highest development of all functions between a youth of endeavor towards better work and an old age of relinquishment of fullest powers. It is the mature river in which the rainfall is best led away to the sea, and which carries with it the greatest load of land waste; it is at maturity that the regular descent and steady flow of the river is best developed,

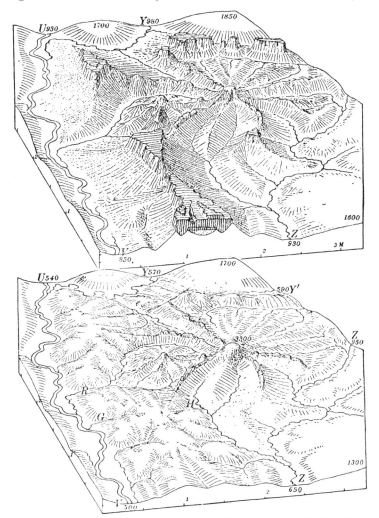

FIG. 41.—The last two diagrams of the series beginning with Fig. 38.

being the least delayed in lakes and least overhurried in impetuous falls.

Maturity past, and the power of the river is on the decay. The relief of the land diminishes, for the streams no longer deepen their valleys although the hill tops are degraded; and with the general loss of elevation, there is a failure of rainfall to a certain extent; for it is well known that up to certain considerable altitudes rainfall increases with height. A hyetographic and a hypsometric map of a country for this reason show a marked correspondence. The slopes of the headwaters decrease and the valley sides widen so far that the land waste descends from them slower than before. Later, what with failure of rainfall and decrease of slope, there is perhaps a return to the early imperfection of drainage, and the number of side streams diminishes as branches fall from a dying tree. The flood-plains of maturity are carried down to the sea, and at last the river settles down to an old age of well-earned rest with gentle flow and light load, little work remaining to be done. The great task that the river entered upon is completed.

Plains of Marine and Subaerial Denudation

From *Bulletin of the Geological Society of America*, Vol. VII, pp. 377–398, 1896.

Geologists today may be divided into two schools regarding the origin of regions of comparatively smooth surface from which a large volume of overlying rocks have been removed. These regions occur under two conditions: First, as buried "oldlands" on which an uncomfortable cover of later formations has been deposited, the oldlands being now more or less locally revealed by the dissection or stripping of the cover; second, as uplands or plateaus whose once even surface is now more or less roughened by the erosion of valleys.

The older school, now represented chiefly by English geologists, follows the theory of Ramsay, and regards these even oldlands as plains of marine denudation. The new school, represented chiefly by American geologists, but also by a number of continental European geologists, may be said to follow Powell, who first emphatically called attention to the possibility of producing plains by long continued subaerial denudation.

It is noteworthy that, with few exceptions, the more recent writers . . . do not discuss both processes by which smoothly abraded plains, whether buried or bare, may be produced, but directly announce their conclusion as to the origin—by marine or by subaerial agencies—of the surface under consideration. This, of course, implies that they regard the question as settled, just as for some time back it has been the habit of geologists on finding marine shells in stratified rocks to conclude, without reviving the discussions of earlier centuries, that the strata are of marine origin, and that their present position indicates a change in the relative attitude of the land and sea. But in this latter example all geologists are today agreed, while in the problem of the origin of plains of denudation each writer follows only the conclusion of his own school, not the conviction of the world. *It is chiefly to arouse attention to this aspect of the problem that the present review is undertaken.*

It is further noteworthy that, with few exceptions, the authors who discuss the matter at all do not attempt to discriminate between the two possible classes of denuded surfaces by searching for features peculiar to one or the other, but content themselves with *a priori* argument as to the possibility of producing plains by marine or subaerial agencies.

There is, however, a certain difference of attitude in the two schools regarding the doctrine of the other. The English school hardly considers at all the ability of subaerial agencies to produce smooth plains of denudation; their discussion of the question turned really on the possible origin of valleys by subaerial agencies. The American school does not, as far as I have read, deny the ability of marine agencies, but attributes greater ability, especially far in continental interiors, to subaerial agencies; their discussion of the question postulates the subaerial origin of ordinary valleys as a matter already proved, and goes on from this to the possible ultimate result of the valley-making processes. Again, the English school denies, tacitly or directly, the probability or even the possibility of a period of still-stand long enough for essentially complete subaerial denudation close to sealevel, but assumes the possibility of a period of still-stand or of slight depression continuous and long enough to allow the sea waves to plane off the sinking lands. The American school tacitly questions the occurrence of great erosive transgressions of the sea during either a still-stand or a slow depression of the land, but admits the possibility of essentially complete subaerial denudation to an average sealevel, above and below which the land long hovers in many minor oscillations before a new attitude is assumed by great depression, elevation or deformation. It should be borne in mind that the depressed and buried or the uplifted and dissected plains of denudation whose origin is in question are in no cases geometrical planes; they nearly always possess perceptible inequalities, amounting frequently to two or three hundred feet; but these measures are small compared to the inferred constructional relief of earlier date, or compared to the deep valleys often eroded beneath the plain if it has been uplifted. By whatever process the so-called "plain of denudation" was produced, an explanation that will account for a peneplain of moderate or slight relief is all that is necessary. Absolute planation is so rare as hardly to need consideration here.

In no respect is the contrast between the two schools more strikingly shown than in the beliefs concerning the cover of unconformable strata that lie or are supposed to have lain upon an oldland. The continental members of the English school generally regard these strata as an essential result of the process of marine denudation during slow depression; if such strata are absent from a dissected plateau, their absence is explained by denudation after uplift. The American school does not give the cover of unconformable strata an essential place in the problem; if present, it is generally ascribed to deposition following the submergence of a region already for the most part baseleveled by subaerial agencies.

In attempting to decide by arguing from effect to cause whether evenly denuded regions have been worn down by subaerial or marine agencies, let us try to stand on a provisional Atlantis, hoping that it may give steady support long enough for us to gain an unprejudiced view of the opinions that are so generally accepted on the lands to the east and west. From this neutral ground let us attempt to deduce from the essential conditions of each explanation of the problem as many as possible of its essential consequences, and then confront these consequences with the facts. The measure of accordance between consequences of theory and facts of observation will then serve as a measure of the verity of the theory from which the consequences are derived. No final decision can be reached in many cases; for, however clearly the consequences may be deduced, the facts with which they should be compared are often beyond the reach of observation. In such cases it is advisable to announce indecision as clearly as decision is announced in the others.

As far as I have been able to carry the analysis of the problems, it is more difficult to find positive criteria characteristic of plains of marine denudation than of plains of subaerial denudation; hence I will take up the latter class first. It should be remembered, however, that in each class of plains both classes of agencies may have some share, one preponderating over the other.

Consequences of Subaerial Denudation. Imagine a region of deformed harder and softer strata raised to a considerable elevation. Then let the land stand essentially still, or oscillate slightly above and below a mean position. The rivers deepen their valleys, the valleys widen by the wasting of their slopes, and the hills are slowly consumed. During this long process a most patient and thorough examination of the structure is made by the destructive forces, and whatever is the drainage arrangement when the rivers begin to cut their valleys a significant rearrangement of many drainage lines will result from the processes of spontaneous adjustment of streams to structures. This involves the adjustment of many subsequent streams to the weaker structures and the shifting of many divides to the stronger structures. Adjustment begins in the early stages of dissection, advances greatly in the mature stages, and continues very slowly toward old age, while the relief is fading away. Indeed, when the region is well worn down some of the adjustments of maturity may be lost in the wanderings of decrepitude, but this will seldom cause significant loss of adjustment except in the larger rivers. Now, if a region thus baseleveled or nearly baseleveled is raised by broad and even elevation into a new cycle of geographical life, the rivers will carry the adjustments acquired in the first cycle over to the second cycle. Still further adjustment may then be accomplished. The master streams will increase their drainage area in such a way that the minor streams will seldom head behind a hard stratum. In a word, the drainage will become more and more longitudinal and fewer and fewer small streams will persist in transverse courses. All this is so systematic that I believe it safe to assert that the advanced adjustments of a second cycle may in many cases be distinguished from the partial adjustments of a first cycle. It should be noted further that in the early stages of the second cycle the residual reliefs of the first will still be preserved on the uplands, and that they will be systematically related to the streams by which the dissection of the upland is in progress, as noted in the examples described by Darton and Hershey.

It is manifestly impossible to apply what may be called the river test to plains of denudation upon which a cover of unconformable sediments is spread; but, before assuming that such buried plains are of marine origin, their uppermost portion next beneath the cover should be examined to see if it presents indications of secular decay before burial; and, if so, a subaerial origin for the plain may be argued. Certain aspects of this division of the subject have been discussed by Pumpelly.* Another matter of importance is the character of the undermost layers of the cover. If these are fresh-water beds a subaerial origin for the plain on which they rest may be inferred. The Potomac formation offers an example of this kind.†

Consequences of Marine Denudation. Now suppose that a region of disordered structure is partly worn down by rain and rivers and is smoothly planed across by the sea during a time of still-stand or of gradual depression. The land waste gained in the later attack will be spread off-shore on the platform abraded in the earlier attack. The basal strata of the unconformable cover thus formed must indicate their marine origin and must be appropriately related in composition and texture to their sources of supply. The drainage systems of the land will be essentially extinguished by the encroaching sea. When the region rises, with the cover of new sediments lying evenly on its smoothed back, a new system of original consequent streams will take their way across it. If the elevation be sufficient, the streams will incise their valleys through the cover of new sediments and in time find themselves superposed on the "oldland" beneath. As time passes, more and more of the cover will be stripped off; at last it may disappear far and wide, although the stripped surface of the oldland may still retain a generally even skyline as a memorial of its once even denudation. Now, in this case, the rivers by which the dissected plateau is drained will have at most only a very slight adjustment to its structure. Their courses will have been inherited from the slope of the lost cover; they will at first run at random across hard and soft structures; a little later some adjustment to the discovered structures will be made, but as long as the even skyline of the upland is recognizable, only the incomplete adjustments appropriate to the adolescent stage of denudation can be gained.

---

* Bull. Geol. Soc. Am., vol. 2, 1891, p. 211.

† McGee: Am. Jour. Sci., vol. XXXV, 1888, p. 137; Fontaine: Monogr. XV, U.S. Geol. Survey, 1889, p. 61.

McGEE

W J McGee (1853-1912), American geologist and anthropologist, was for ten years in charge of the division of Atlantic Coastal Plain geology in the United States Geological Survey, resigning from that organization in 1893 to become ethnologist in the Bureau of American Ethnology. A year later he was placed in charge of that bureau, and most of his subsequent scientific studies dealt with anthropology. As one of the pioneers in geological research in the Coastal Plain, his work was however of such constructive nature that it occupies a permanent place in the history of geology.

THE COASTAL PLAIN AND THE FALL LINE

From United States Geological Survey, *Seventh Annual Report*, pp. 537-646, 1888.

Summarizing the leading pertinent considerations, it appears (1) that the true boundaries of the Middle Atlantic Coastal plain are the fall-line and the great submarine escarpment; (2) that the character of the downthrow of this plain is such as to produce superficial and structural landward tilting, culminating at the fall-line; (3) that the entire plain is undergoing loading by deposition of sediment, which attains a maximum near its inland margin; (4) that along the inland margin of the plain depression is far in excess of sedimentation, although the processes are more nearly and perhaps quite commensurate if the entire plain is regarded as a unit; (5) that it is uncertain whether or not the displacement extends to the crystalline subterrane, for the Virginia phenomena indicate that it does not so extend, while those of New Jersey indicate that it does; (6) that the mass of clastic deposits constituting the Coastal plain, together, possibly, with the crystalline subterrane, are in a condition of inequipotentiality; (7) that the congeries of phenomena of the great displacement are inexplicable save on the assumption either (a) that the entire region is affected by tangential tension or (b) that the vertical displacement is combined with some lateral movement; (8) that the assumption of general tangential force (or equivalent lateral movement) is such as the inequipotentiality of the mass tends to produce.

The several special hypotheses may be assembled for review as:

The hypotheses of
(1) Simple faulting
(2) Faulting along a curved space
(3) Compressional depression
(4) Compressional attenuation and elongation of hypogeal strata
(5) Settling of the clastic deposits, due to inequipotentiality
(6) Sliding of the clastic deposits, due to inequipotentiality
(7) Settling of clastic and crystallines, due to inequipotentiality
(8) Faulting along a curved surface, combined with inequipotential sliding and settling of clastics and crystallines.

Inspecting these hypotheses in connection with the conditions of the problem, it appears at once that the first is untenable: that the second consists fairly with the phenomena and upon certain postulates has strong elements of probability; that the third involves pertinent considerations but is alone incompetent; that the same is true of the fourth; that the fifth, like the second, is in harmony with many of the phenomena and has elements of probability without irrelevant postulates, though the sixth is in even closer accord with the phenomena and principles and appears to embody a true and perhaps adequate cause; that the seventh is free from the most serious objections to the possibly valid fifth hypothesis; that the eighth, which combines the plausible second and still more plausible sixth hypotheses, as well as the probable elements of the others, is not only consistent at the same time with the assemblage of phenomena and with established principles, but appears to afford a complete raison d'être for the displacement.

Briefly, the region is in an inequipotential condition, and the displacement is of precisely such character as the stresses arising from inequipotentiality combined with current sedimentation tend to produce; but it is impossible to demonstrate the quantitative sufficiency of these stresses, for although the inequipotentiality might be evaluated, the resistances opposed by cohesion and particle friction cannot be determined with present limited knowledge of the behavior of solids under such pressures as obtain at even inconsiderable depths beneath the terrestrial surface.

IDDINGS

Joseph Paxson Iddings (1857–1920), American petrologist, was connected with the United States Geological Survey from 1880 until 1895 and was professor of petrology at the University of Chicago until 1908.

PHENOCRYSTS

From *Bulletin of the Philosophical Society of Washington,* Vol. XI, pp. 71–113, 1892.

Lavas that reach the surface of the earth in a fluid condition consolidate upon cooling rapidly to a more or less perfect glass. The glass in most cases contains crystals of several minerals scattered uniformly through it. . . . Usually both large and small crystals occur in the same glass. The large ones that stand out prominently from the mixture of glass and small crystals are said to be *phenocrysts,** the remainder of the rock is called the *groundmass.*

Lavas may also cool in such a manner that the whole mass becomes crystalline and no glass is left. When there are porphyritical minerals (*phenocrysts*), they are scattered through a groundmass wholly made up of crystals. . . . But surface lavas seldom attain so high a degree of crystallization.

Upon studying these various structures under the microscope it is evident that in the case of porphyritical rocks, the porphyritical crystals (*phenocrysts*) in most instances were formed before those composing the groundmass, and belong to an earlier generation.

RELATION BETWEEN MINERAL COMPOSITION AND MODE OF OCCURRENCE OF IGNEOUS ROCKS

From *Bulletin of the Philosophical Society of Washington,* Vol. XI, pp. 191–220, 1892.

Comparing the two sets of analyses and the corresponding diagrams it is evident that the intrusive rocks of Electric Peak and

---

* These prominent crystals in porphyritic rocks are frequently termed *Einsprenglinge* by the Germans, which is a collective term without a good English equivalent. Realizing the advantages of such a word and the inconvenience of not having an equivalent in English, the writer, after consultation with Prof. J. D. Dana and others, suggests the term *phenocryst*, from φαίνω = conspicuous or eminent, and κρύσταλλος = crystal.

the volcanic rocks of Sepulchre Mountain [both are in Yellowstone National Park] have the same chemical character. . . . Furthermore the comparison demonstrates that the magmas that reached the surface of the earth in this place had exactly the same chemical composition as those which remained enclosed within the sedimentary strata. It proves with equal clearness that the different conditions attending the final consolidation of the extravasated and of the intruded magmas affected not only their *crystalline structure*, but their *essential mineral composition*. . . .

When we further consider the petrographical resemblance between the dike rocks of the two places, the correspondence of habit between the more acid members of both series, and the chemical identity of the magmas involved, we feel justified in concluding that:

I. The volcanic rocks of Sepulchre Mountain and the intrusive rocks of Electric Peak were originally continuous geological bodies.

II. The former were forced through the conduit at Electric Peak during a series of more or less interrupted eruptions.

III. The great amount of heat imparted to the surrounding rock was due to the frequent passage of molten lava through this conduit.

IV. Portions of the different magmas erupted found their way into vertical fissures and took the form of dikes; portions reached the surface and became lava-flows and breccias, while portions remained in the conduit.

V. The various portions of the magmas solidified under a variety of physical conditions, imposed by the different geological environment of each, the most strongly contrasted of which were the rapid cooling of the surface flows under very slight pressure, and the extremely slow cooling of the magmas remaining within the conduit under somewhat greater pressure.

It is to be remarked that the first of a series of eruptions would pass through a cooler conduit than the last magmas would, consequently the rate of cooling would be different in each case, and the pressures at which crystallization sets in would also differ. This difference would also obtain between the advance and rear ends of the magma of a single eruption. . . .

VI. . . . We find that the lower portion of the basic breccia is made up of andesites carrying phenocrysts of augite, hypersthene and plagioclase, and of other basic andesites without macroscopic crystals. The former correspond to the facies of the diorite in the conduit carrying the same kinds of phenocrysts; while the second modification of the andesite corresponds to the greater part of the basic diorite. For it is evident from a study of this diorite that much of its magma reached its position in the conduit in a completely fused condition, after which it crystallized into hypersthene, augite, hornblende, biotite, labradorite, alkali feldspars and quartz, with accessory magnetite, apatite and zircon. We feel justified, then, in the further conclusions:

VII. That the molecules in a chemically homogeneous fluid magma combine in various ways, and form quite different associations of silicate minerals, producing mineralogically different rocks.

VIII. In this region the greatest mineralogical and structural differences accompany the greatest differences in structure or degree of crystallization: hence, the causes leading to each are coëxistent.

IX. The causes of these mineralogical and structural differences must be sought in the differences of geological environment, and these affect the rate at which the heat escapes from the magmas, and the pressure they experience during crystallization.

Since it has been demonstrated by synthetical research that water- and other vapors are potent factors in the crystallization of quartz and other minerals that have not been produced artificially without their aid and as there is ample evidence both in the extravasated lavas and in the coarsely crystallized rocks in the conduit that water-vapor was uniformly and generally distributed through the whole series of molten magmas, and no evidence that there existed in the magmas which stopped within the conduit any more or different vapors than those which existed in the magmas that reached the surface, we may conclude that—

X. The efficacy of these absorbed vapors as mineralizing agents has been *increased* by conditions attending the solidification of the magmas within the conduit.

VAN HISE

Charles Richard Van Hise (1857–1918), American geologist, was associated throughout his life with the University of Wisconsin, where he attained the rank of professor in 1886 and served as president subsequent to 1903. He was a pioneer in the study of pre-Cambrian rocks and in the application of quantitative principles to the problems of metamorphism and diastrophism.

PRE-CAMBRIAN GEOLOGY

From "Principles of Pre-Cambrian North American Geology," United States Geological Survey, *Sixteenth Annual Report*, Pt. I, pp. 571–843, 1896.

By the term "pre-Cambrian rocks" is meant all formations which are older than those containing the Olenellus or Lower Cambrian fauna. As to the length of time represented by the pre-Cambrian, we have two points of view from which an approximate inference may be made. Accepting the nebular hypothesis, (1) we may look forward from the time when a crust first formed upon the globe, and (2) we may look backward from the broad domain of facts furnished us by the Lower Cambrian.

(1) After the first continuous solid crust of the earth formed there must have been an exceedingly long time before the conditions were such that life could exist. It is highly probable that this first outer crust was of an igneous character. After solidifying, this first shell must have thickened steadily by inward solidification. Much or all of the first crust may have been removed by subsequent erosion, but this is not true of its inward crystallization; that is, it can not be assumed that erosion has anywhere overtaken inward solidification. It is therefore to be expected that there exists in various parts of the earth the first outer crust or its inward continuation.

(2) The fauna of the Lower Cambrian is varied, complex, and abundant. It comprises all of the great classes of animals except the vertebrates, and it is by no means certain that these did not exist. One biologist says that if the differentiation of life since the Cambrian were represented by a line of a certain length the differentiation before the Cambrian would be represented by a

much longer line.   Another says that the amount of differentiation of life forms before the Cambrian is at least nine times as great as the amount of differentiation since the Cambrian.   Believing, as we now do, that this complex and varied life was produced by slow development under a definite order set in the universe, rather than by special creations, the Cambrian life implies that the time of pre-Cambrian life was very long, although it does not necessarily imply that it is nine times as long as post-Cambrian time, for it can not be assumed that the development of life was at a uniform rate.   So far as we know the laws of development, it appears to be true that higher forms differentiate more rapidly than lower forms.   If this law were applied it would follow that the time of pre-Cambrian life was more than nine times as long as Cambrian and post-Cambrian time, but unknown factors may enter into the problem of the earlier phases of development, and hence we can not assume any such time ratio as nine to one.   However, it seems reasonably certain that the time of pre-Cambrian life was as long as or longer, probably much longer, than all subsequent time.

If the conditions were such that life could exist, they were also such that ordinary sedimentary rocks could be deposited.   Therefore it is reasonable to think that in pre-Cambrian time there were eras in which vast groups of sedimentary rocks of essentially the same character as the Paleozoic sediments were deposited. Such rocks may have since been profoundly modified.

Between the time when all the rocks forming on the earth were igneous and the time when ordinary pre-Cambrian sediments began to form, there may have been a great length of time in which the conditions were materially different from those that we now know.   The waters of the ocean may have been at a high temperature, and the rocks deposited by this early ocean may have been different in character from ordinary sediments.   Thus a third possible group of rocks may have formed, which bridged the time interval between the early igneous rocks and the ordinary sediments.   If portions of such rocks still exist it is probable that transition phases unite them with the igneous rocks on the one hand and with ordinary sedimentary rocks on the other hand.

From our forward and backward views we then conclude that in pre-Cambrian time, at least two great classes of rocks were produced, and may, perhaps, now be found: first, those of igneous origin, representing the earliest outer crust of the earth or its inward crystallization, or both combined; second, those of sedimentary origin, deposited during the vast length of time in which life was growing from the first rudimentary form to the highly complex and varied life of the Cambrian; and, third, there may be a group of rocks intermediate in position between these two, in character and origin unlike any rocks which we now know to be forming, but connected by gradations with the later sediments. The first conclusion is a probable inference, based upon the truthfulness of the nebular hypothesis. The second conclusion is a certain one, based upon the known facts of the Cambrian life. Turning now to the field, do we find phenomena corresponding to these conclusions?

One who has worked long among the pre-Cambrian rocks in areas where the conditions are favorable for a structural study is usually impressed by the dual character of the pre-Cambrian. First, he finds a great series of gneissoid granites, gneisses, and schists, all completely crystalline, having the most intricate relations with one another, and showing the effects of repeated strong dynamic actions. These are injected by many undoubted igneous rocks of different characters and different ages. This group is always below the most ancient sedimentary rocks found. Without at present saying anything in regard to the origin of this Basement Complex, or as to the relations which obtain between it and the sedimentary rocks, it may be said that it is possible that it or some part of it corresponds either with the original crust of the earth and its inward crystallization or else with the transition rocks between these and the ordinary sediments which were deposited later. Second, he finds another great group of pre-Cambrian rocks, having all of the characteristics of ordinary sedimentary rocks. In many places these are divisible by unconformities into two or more different series. These rocks comprise conglomerates, sandstones, grits, shales, limestones, and their altered equivalents. These rocks may be reasonably regarded as occupying some part of the second great division of time, that in which life existed before the Cambrian, and as a matter of fact within these rocks we are not without evidence of the existence of life. This is demonstrated by the presence of undoubted fossils, which occur in Newfoundland, in the Grand Canyon region, in the

Lake Superior country, in Great Britain, in Belgium, in Brittany, and in other places. Further, within these rocks are great beds of carbonaceous and graphitic shales, from some of which hydrocarbons may be distilled, and are rarely so rich in nongraphitic carbon as to be combustible with difficulty. That these hydrocarbons were produced by any other than organic agencies is exceedingly improbable. Also, there exist great beds of limestone and gneiss through which are disseminated particles of graphite. Finally, beds of iron carbonate and other rocks which are by many regarded as products of life are abundant.

If the position be correct that abundant life existed for eras before Cambrian time, the rarity of recognizable fossils needs explanation. This Darwin mentioned as a serious difficulty in his theory of the origin of species. Recently Brooks has given a plausible answer to this difficulty. From his biological study, and particularly from the facts of embryology, he concludes that the great stems of animal life were established when all animals were free swimming and pelagic. When the conditions became favorable, certain of the forms of the different classes availed themselves of the advantages of location at the bottom of the sea. As a consequence, these animals were limited in their habitat to length and breadth rather than length, breadth, and depth. Hence there at once arose a far keener struggle for existence than had been known before. One of the great crises in life had come. There resulted the rapid development of the hard parts, such as are found in the Cambrian rocks. Professor Brooks's explanation* of free swimming, pelagic animals devoid of hard parts is the most plausible one yet offered.

In a treatment of the pre-Cambrian rocks the object of which is mainly to determine the principles of stratigraphy and rules to carry them out, the question arises in what respect these rocks differ from the post-Cambrian rocks, and what are the criteria which are especially applicable to them. Why should the pre-Cambrian rocks have a treatment separate from later formations? As has been seen, the first great point in which these rocks differ from later formations is in the apparent absence of abundant

---

* The origin of the oldest fossils and the discovery of the bottom of the ocean, by W. K. Brooks. Jour. of Geol., Vol. II, pp. 455–479, 1894.

fossils. Abundant life may, indeed must, have existed in some of them. Remains of this life, however, have not yet been found in sufficient quantities to serve for the purposes of correlation from district to district and region to region. Among post-Cambrian formations paleontology is the chief reliance in homotaxis and correlation. Upon paleontological data have been based nearly all inferences as to the intercontinental equivalence of strata, and even upon the same continent the equating of one set of strata with another set in a different region has been usually based upon their fossil contents. It then follows that in the pre-Cambrian there is an almost entire absence of the most important criterion upon which stratigraphical and structural work has been done among the post-Cambrian formations. We are thus driven to the use of physical data only. Being thus restricted, it becomes necessary to scrutinize them and to consider their limitations, in order that a judgment may be formed of the reliability of work done and correlations made. But the pre-Cambrian problem is still more complicated because the rocks are so old. They have been subjected to all of the constructive and destructive geological forces which have been at work during the whole of Cambrian and post-Cambrian time. Vast areas of pre-Cambrian rocks have been destroyed by the process of erosion. Other areas are hidden by later rocks. The major portion of the remnants at the surface have been profoundly modified by the processes of metamorphism.

The folding of the pre-Cambrian rocks is, upon the whole, of a more complicated character than that of the post-Cambrian, for not only have the former rocks been subjected to all of the earth movements which have affected the Cambrian and post-Cambrian sediments, but before Cambrian time they were affected by earlier movements; and if we are correct in supposing that some of these pre-Cambrian sedimentary rocks may be twice or thrice as old as those of the oldest Cambrian, the movements to which they have been subjected must have been upon the average twice or thrice as many, even if the law of uniformity applies, and it is wholly possible that the disturbing forces have increased in frequency and in power as we go backward in time, thus making the amount of folding even greater. This does not imply that there are no areas of post-Cambrian rocks which have been folded to an equal or greater degree than any pre-Cambrian rocks, but merely that in the same province the pre-Cambrian rocks are usually folded to a greater degree than later rocks; or, stated differently, the pre-Cambrian rocks have upon the average, been subjected to more periods of folding. Hence, in treating of pre-Cambrian stratigraphy we must consider the effects of this exceptional folding— what new structures have been developed, what interior alterations have occurred.

Following the same line of argument, it is plain that the jointing and faulting through which pre-Cambrian rocks have gone are, upon the whole, greater than in post-Cambrian rocks. Also, as a consequence of the folding, faulting, and jointing, brecciated or autoclastic rocks have been more extensively produced. A greater proportion of the rocks have become metamorphosed. Secondary structures have been more extensively developed in them. There have been intruded among them more numerous and abundant masses of igneous rocks; for the pre-Cambrian rocks have been subjected not only to all post-Cambrian intrusions, but to many pre-Cambrian intrusions and extrusions. It may also be that extrusive rocks were more widespread and abundant in pre-Cambrian than in post-Cambrian time. As a consequence of these various causes, producing profound effects upon the pre-Cambrian sedimentary rocks, it is inevitable that the problem of pre-Cambrian stratigraphy, even apart from the absence of fossils, is, upon the whole, a more difficult one than the problem of post-Cambrian stratigraphy; and the problem becomes increasingly difficult as we go back from the base of the Cambrian.

In the face of these difficulties, upon what criteria are we to depend in pre-Cambrian stratigraphy?

The first and most important of the criteria, as giving common horizons from which to work in comparison, is unconformity. If an unconformity is marked it must have a wide extent, for an unconformity implies an orogenic movement which raises the land above the sea; implies faulting or folding of the strata; implies truncation by means of epigene forces; and, finally, implies another orographic movement, which depresses the area below the sea. Later, evidence will be given to show that such a break can hardly be less than regional, while it may be continental or, perhaps, intercontinental in its effects. For instance, the great Appalachian revolution at the close of Paleozoic time not only caused a profound unconformity between the Paleozoic rocks and the post-Paleozoic rocks along the whole length of the Appalachian Mountains, but produced great physical changes throughout the central and western part of the continent.

Secondly, we may use the sequence of beds of the same character as guides in equivalency; that is, if a set of beds of peculiar lithological character occur in like order in different districts of the same geological province, it is probable that they are parts of a once-continuous series, and if such a series of beds is separated by an unconformity from a set of beds below and another set of beds above, the supposed equivalency has an increased degree of probability. Then, with proper restrictions, the lithological character of a single bed itself may have some value in comparative work. It is necessary that we consider carefully how an unconformity may be established, and in correlation what value may be placed upon unconformity, upon sequence of beds, and upon lithological character.

Finally, the very phenomena which make the pre-Cambrian stratigraphy a difficult problem may also give us assistance. If a lower series has been more extensively folded, or folded in a different way from a superimposed series, the question may be asked whether the two are not separated by a time interval; for if the upper series existed in the same district when the lower series was folded in the first instance, it should be folded in the same manner. Absence of folding, or a less complicated folding, indicates that the upper series was not present when the lower series was first folded. In using this criterion it must be certain that we have a case of superposition of the two series, not lateral position, for intense plications may die out rapidly transverse to the directions of the folds, and thus leave beds unaffected or little affected which were very close to those that are intensely folded. In the same way, faulting, jointing, brecciation, metamorphism, cleavage, fissility, and the relations of igneous rocks are criteria by which we may separate one series from another. For instance, in the case of igneous rocks, if a lower series is cut through and through by dikes which nowhere penetrate a superior series, the inference is that this superior series was not present at the time of the injection of the igneous rocks. Thus we see that the very causes which increase the difficulties of pre-Cambrian stratigraphy—that is, the greater amount of folding, faulting, jointing, brecciation, metamorphism, etc., to which they have been subjected—may be of assistance to us in particular cases in working out the structure.

THE MEANING OF ROCK FLOWAGE

From "Metamorphism of Rocks and Rock Flowage," *Bulletin of the Geological Society of America*, Vol. IX, pp. 269–328, 1898.

I have previously maintained that the rocks within the scope of our observation which have been deformed at considerable depths were deformed by rock flowage. . . . I shall take as a typical example of rocks which have been deformed in the zone of flow those laminated crystalline schists the mineral particles of which now show slight or no strain; for it is evident that these are the rocks which have nearly perfectly accommodated themselves to the deformation through which they have passed. The accommodation, as already explained, is accomplished by continuous solution and deposition, or by continuous recrystallization. While the adjustment during deformation at any moment was nearly as complete as though the rock were a magma, and while it nowhere shows even a microscopic space, it is evident that the flowage is wholly different from that of a liquid. At no time was the rock a liquid. On the contrary, it was at all times almost wholly a crystalline solid. At no time was more than an almost inappreciable fraction of it in a liquid form—that is, dissolved in water—yet at all times it was adjusting itself by means of this small percentage of water contained in the capillary and subcapillary spaces, this being the medium of solution and recrystallization. In order that such a continuous process shall be adequate to explain rock flowage, it is necessary only that it shall be sufficiently rapid to keep pace with the deformation. One's first thought is probably that it is not possible that the process can be sufficiently rapid to account for the phenomena. However, the experiments of Barus upon the solution of glass give us a basis upon which we can make a quantitative calculation.

Barus* has shown that a temperature of 180°C. is critical so far as the solution of glass by water is concerned. At temperatures lower than this the rate of solution by water is very slow. However, at temperatures of 185°C. and above, solution and crystallization of the silicates of glass go on with astonishing rapidity. In Barus' experiment, as already seen, water dissolved a sufficient

* Compressibility of liquids, by C. Barus: Bull. U.S. Geol. Survey, no. 92, 1892, pp. 78–84.

amount of glass and deposited the material as crystallized minerals to cause an apparent contraction of volume of the water amounting to 13 per cent of the water present in the capillary tubes in 42 minutes and 18 per cent in an hour. This shows that solution continued during the later stages of the experiment at about the same speed as during its earlier stages, for 13:42 is about as 18:60. This is apparent when the first terms of the ratios are made 1, for the proportion then stands $1:3\frac{3}{13}$ about as $1:3\frac{1}{3}$. From this proportion it appears that the action was apparently slightly more rapid during the last few minutes. Reasons which may be suggested for an increased rate of action, supposing the pressure, temperature, and composition of the glass to remain constant, are (1) the smooth glass at the beginning of the process was less readily attacked than the roughened surface produced by solution; (2) the roughening during the process gave the glass increased surface for action as the experiment continued. However, the apparent change of rate is so slight as to be possibly attributed to errors of observation.

During the experiment, unless hydrous minerals were produced, the water remained a constant quantity, and continued work. This could have been continued so long as the temperature and pressure were sufficient and glass was available for crystallization through solution, as a result of which the material is condensed. If no hydrated minerals are formed, there is no reason why a small amount of water cannot continue the process indefinitely.

If in this experiment we suppose the condensation of recrystallization to be 10 per cent, the amount of condensation in diabase in passing from the glassy to the crystalline condition, as shown by Barus,* this would mean (neglecting the condensation of the water) that in one hour, in order to have given an apparent volume contraction of 18 per cent, the water had dissolved 1.8 times its own volume of the glass, and deposited crystallized material with 10 per cent less volume. Therefore, for the water to dissolve a volume of glass equal to that of the water and deposit it in a crystallized form would require $33\frac{1}{3}$ minutes, or approximately one-half hour.

This illustrates the fact that the activity of water is amazing at a very moderate temperature, and one need not be surprised at its

---

* The contraction of molten rock, by C. Barus: Am. Journ. Sci., vol. 42, 1891, pp. 498, 499.

potency in the alteration of rocks deep below the surface of the earth. Temperatures higher than 180°C. exist at moderate depth, and therefore it is reasonable to suppose that a small amount of water may be the medium of rapid and most profound modification of the rocks.

We have already seen that during the process of deformation the material, if not dissolved, may be strained even to the point of granulation by the mechanical processes; also it has been seen that, so far as strain occurs, or the particles are small, the minerals are in a state in which solution is easier than for unstrained or larger mineral particles and the deposition of the material in an unstrained crystallized condition is considerably slower than that of amorphous glass, for it cannot be supposed that the same amount of energy is potentialized in the mineral particles as in the glass. But the further the strain goes before fracture the more energy is potentialized, or if fractures occur smaller particles are produced. Moreover, the contained water is in small capillary or subcapillary spaces, and therefore a given volume is acting upon a much larger surface than in the capillary tubes used by Barus in his experiments. In so far as granulation occurs, the surface of action is still further increased. All these conditions are favorable to solution and redeposition; therefore the greater the straining and resultant granulation, the more rapid the process of recrystallization; hence in the deep-seated zone mechanical disintegration never gets far in advance of solution and redeposition.

If it be supposed in the capillary and subcapillary spaces within the rocks that the speed of solution of minerals is .1 of that of glass, water would dissolve its own volume of minerals and redeposit the material in about five hours. If the deep-seated rocks be supposed to contain 2 per cent of water by volume—that is, less than 1 per cent by weight—the entire mass of rocks might be recrystallized in about 250 hours, or little more than 10 days. The percentage of water premised is known to be lower than the amount ordinarily found in the crystalline schists, and the rate hypothesized seems reasonable; but if this speed be decreased to .1 of that suggested or to .01 of that of glass, still the entire mass of the rocks might be dissolved and redeposited in about 100 days. Make the rate .1 of this or .001 of that of glass, and still recrystallization might be complete in about 1,000 days, or three years. If it be supposed that a mountain-making period occupied 150,000 years, and this is probably less rather than more than the time required for most mountain-making movements, during this period, at the slow rate suggested, the rocks could be recrystallized 50,000 times by 1 per cent of water, and this number certainly seems adequate to fulfill the requirement that at any given moment the crystalline rock shall exhibit but a slight strain effect. Of course, it is not thought probable that any rock has completely recrystallized 50,000 times. Indeed, it is well known that many of the rocks in which recrystallization is complete, in so far as the finer particles are concerned, contain many larger particles which have not been completely recrystallized or even granulated. Perhaps one of the best instances of this is furnished by the schist-conglomerates. The typical schist-conglomerates contain a crystalline schist matrix, embedded in which are numerous large fragments. In many of these the matrix is completely recrystallized. The fragments, unlike the matrix, show important strains, which not infrequently pass to the point of partial granulation or recrystallization.

Indeed, to explain the phenomena in the case of crystalline schists which have been developed during a continuous process of deformation it does not seem necessary to suppose that complete recrystallization of all of the material is necessary.

If the case of a large grain of quartz or feldspar in a recrystallizing rock be taken, we may suppose the process to go on somewhat as follows: Because of the lack of homogeneity of the rock the stresses are irregularly distributed. At the most exposed places upon the mineral particles the conditions are favorable for solution, for the following reasons: The particle is there greatly strained, perhaps to the point of granulation, and, so far as strain exists or small granules are formed, these conditions are favorable to solution. At the places of great strain the material is therefore taken into solution and transported to the parts of the particles less strained. At such places the conditions are favorable to deposition, on account of the relatively large size of the residual original grains as compared with the granules. The mineral where least strained separates from the solution material like itself, attaching it to itself, in orientation with the core in an unstrained or little strained condition. The process of growth is analogous to that of mineral growth by secondary enlargement. The entire process is similar in several respects to that of the continuous

solution and deposition of calcium carbonate in the chemical
laboratory when water is passed through a layer of this material
under pressure.   Where the pressure is greatest in the upper part,
the grains are taken into solution.   At the place of escape, where
the pressure is least, the material is deposited from solution, and the
grains increase in size or grow.

During the deformation of the rocks this process of solution and
deposition of a mineral particle is continuous.

If it be supposed to go on to a stage in which the original particle
is one-half or one-third as thick as it was originally, it is not
necessary to suppose that the central part of the mineral particle

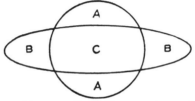

F<small>IG</small>. 42.—Diagram used by Van Hise to illustrate the possible relations of old
and new grains of recrystallized rocks, 1896.

has been recrystallized.   This is illustrated by figure 2 [Fig. 42].
The spherical grain is supposed to have changed to the superimposed spheroidal grain.   The common portion C may be an
uncrystallized part of the old grain, but the material AA has been
dissolved and added to the borders at BB.   Corresponding to this
explanation, some of the flat quartz grains of the mica-slates and
mica-schists of the Black Hills of Dakota show residual cores.

During the process of recrystallization, at any given moment
there will be the greatest shortening in the direction of greatest
stress, greatest addition in the direction of least stress, and there
may be shortening or addition in the direction of mean stress.
Consequently the shape of the modified particle may be that
which would be produced if a plastic grain were rolled out, the
sides being confined in one direction, but with liberty to elongate
in another direction; or it may be that which would be produced
if a roundish cake of dough were pressed between two boards, and
consequently elongated in all directions at right angles to the
direction of greatest pressure; or, finally, the mean stress may
approach so closely to the maximum stress that there is shortening
in two directions and elongation in a single one only, in which case

a fibrous structure is produced. However, from my study of the crystalline schists I am inclined to believe that shortening in one direction and unequal elongation in the directions at right angles to this is the most common case, though my thin sections give illustrations of all the cases.

In some cases the direction of greatest pressure varies within exceedingly short distances. The most common case of this is caused by the existence of large rigid particles, such as feldspar, garnet, or some other refractory mineral, which act as transmitters of pressure and deliver the stress nearly normal to themselves. In such cases the direction of greatest pressure adjacent to the rigid mineral particles is modified from point to point. The new particles there forming may curve about the rigid granules. . . . In such a case the direction of orientation for the quartz, in order to adjust itself to the pressure, would continuously vary, and the flat individuals might show undulatory extinction as originally developed without having been strained subsequently.

What is true of one mineral particle is true of all others, and therefore we conclude that, while recrystallization is constantly occurring in the deformation of rocks, at any time the majority of the mineral particles retain their integrity and are nuclei which at any moment orient the material being deposited. In many crystalline schists evidence may be seen that this has happened. The old mineral particles, represented by the cores, may have been slightly altered, and in consequence of this may be discriminated from the freshly added material, or the cores may show a border of iron oxide or other mineral, or the old and new material may have slightly different compositions, and this be discovered by a difference in the color, refraction, extinction, or in some other way. Finally, all of the old mineral particles may be regenerated or recrystallized.

Therefore a given portion of a definite mineral particle in a crystalline schist may not have been recrystallized at all, or, on the other hand, may have recrystallized several or many times. It is believed that ordinarily the recrystallization is far advanced or complete for all parts of a typical schist, although this is far from the case in the semicrystalline schists or imperfectly schistose rocks.

Of course, in this rearrangement it is not supposed that the identical molecules which are taken from the more severely stressed parts of a grain are necessarily deposited at the places of less stress upon the grain. Undoubtedly there is great interchange of material between the particles by means of the solutions. It is, however, thought probable that in many cases of deep-seated deformation, where the passage of solutions is difficult and slow, that much of the identical material which is taken from a grain at one place is added to it at another place.

When new individuals are produced in any way, as by granulation or by the deposition of new mineral particles, perhaps as different species from any originally in the rock, they are subject to the same laws as the original mineral particles. Many have a tendency to form with similar crystallographic orientation. However, it is only rarely that the orientation of the particles of a given mineral approximates exactness. One mineral—for instance, mica—may be well oriented, whereas such minerals as quartz or calcite may not be oriented.

In proportion as the minerals readily respond to the forces of recrystallization or are mobile, they do not gain or retain regularity of arrangement. After mass movement has ceased the temperature may be sufficiently high and the heat be held for a sufficient time, so that the solutions may completely recrystallize the minerals under mass static conditions, and therefore orientation may be lost. In proportion as minerals do not readily recrystallize or stubbornly resist the force of recrystallization, the minerals once oriented retain their regularity of arrangement.

The most mobile of the important minerals is calcite. Quartz is also somewhat mobile. Therefore these minerals in marble and in recrystallized quartzite frequently lack regularity of arrangement. However, in some cases even calcite may show well developed, similar crystallographic orientation. The usual almost complete lack of regularity for calcite is illustrated by most of the marbles from the Laurentide mountains to Alabama. The complete recrystallization of quartz to a coarse granolitic textured rock, the individuals wholly lacking orientation, is illustrated by the quartzites of the Wausau district of central Wisconsin.

In the process of recrystallization of rocks it is not supposed that every large mineral particle retains a nucleus for lateral growth; indeed, it is certain that some particles of a mineral may retain the modified integrity above described in a rock in which other particles of the mineral may be wholly destroyed. While as yet it has not been fully worked out, it is believed that orientation of the mineral particles in reference to the varying stresses may have an influence upon their preservation. If the original particles happen to be in such positions as they would develop as authigenic minerals under the differential stresses, this is believed to be favorable to the preservation of their nuclei and to lateral growth. In proportion as the orientation of the particles is removed from such positions, it is believed that the mineral particles are likely to be destroyed. The effect of orientation with reference to the principal stresses upon the persistence of a given particle is probably great in proportion as the mineral has a tendency to be influenced in its crystallographic orientation by the stress differences which exist during deformation. To illustrate : The position of the crystallographic axes in reference to the greatest pressure in mica and feldspar, which show a marked tendency to similar orientation, would be a more important factor in their preservation than in quartz, which only rarely shows similar orientation.

In the foregoing non-rotational distortion has been assumed. In case the deformation includes a rotational element, there would be no discoverable difference in the crystalline schist produced; but during the deformation and recrystallization each of the particles would be similarly rotated, as well as flattened or recrystallized, or both, and consequently the direction of shortening and elongation might momentarily change.

Of course, in proportion as the conditions are unfavorable for recrystallization—that is, as mineral particles are refractory, as they are coarse-grained, as the deformation is slight but rapid, as the depth is little, as the temperature is low, as the water content is small—residual strain or mechanical granulation will occur instead of recrystallization. . . .

It is therefore concluded from the foregoing that rock flowage, as deep as observation extends, is plastic deformation through continuous solution and deposition, or, in other words, recrystallization. During the adjustment all or only a part of the material may have passed through this change. However, if a matrix, plastic by recrystallization, be filled with rigid granules which are not recrystallized, the whole mass may be deformed by true flowage of the matrix and by slipping or shearing readjustment of the granules. So far as the average mass deformation is concerned, the result is substantially the same as though each rigid granule had not acted as a unit. Indeed, the same average mass deformation may be accomplished wholly by granulation and welding, as in Adams' experiments; but it may, perhaps, be doubted whether this is ever strictly the case with rocks in nature, for some small amount of water is always present, and probably, even in the cases of apparent perfect granulation, some degree of solution and recrystallization from solution has occurred. In the case of the imperfect crystalline schists, which are very widespread rocks, the adjustment to the new form is accomplished in part by the solution and redeposition. It is only in the case of the typical granulated rocks that we can suppose that the process of deformation is mainly accomplished by the movement of the solid particles over one another, and it is only in the perfect crystalline schists that we can suppose that the deformation is accomplished almost wholly by recrystallization.

GUIDE TO SUBJECT MATTER

The more significant contributions to certain important phases of geological
science are here listed in the chronological order of the dates of publication.

Age of the Earth

	PAGE
Buffon, 1778 (1807)	65
Thomson, 1864	472
Chamberlin, 1899	612

Cosmogony

Descartes, 1637	14
Leibnitz, 1680	45
Kant, 1755	88
Buffon, 1776	58
Laplace, 1796	143
Proctor, 1870	547
Bickerton, 1880	574
Chamberlin, 1904	618

Crystallography

Steno, 1669	33
de l'Isle, 1772	108
Hauy, 1801	129
Nicol, 1829	176
Sorby, 1858	488
Fouqué and Michel-Lévy, 1878	494

Diastrophism

Leibnitz, 1680	45
Hooke, 1705	28
von Buch, 1810	209
von Buch, 1825	210
Babbage, 1834 (1847)	237
Prévost, 1835	230
Gesner, 1839	280
H. D. and W. B. Rogers, 1840	338
Darwin, 1846	361
Dana, 1847	419
Élie de Beaumont, 1852	288
Airy, 1855	401

PAGE

Pratt, 1855.. 393
Pratt, 1858.. 398
Hall, 1859.. 406
Suess, 1875.. 503
Daubrée, 1878.. 445
Heim, 1878.. 642
Favre, 1880.. 448
Lapworth, 1883.. 579
Bertrand, 1884.. 631
McGee, 1888.. 661
Dutton, 1889.. 566
Le Conte, 1889.. 467
Shaler, 1895.. 551
Van Hise, 1898.. 673

Earthquakes

Hooke, 1705.. 28
Michell, 1760.. 80
Humboldt, 1810.. 179
Humboldt, 1819.. 182
Mallet, 1846.. 384
Dutton, 1889.. 560

Economic Geology (see also *Ore Deposits*)

Smith, 1815.. 201
H. D. Rogers, 1840.. 346
White, 1885.. 639

Experimental Geology and Laboratory Techniques

Saussure, 1787.. 114
Cavendish, 1798.. 103
Spallanzani, 1798.. 101
Hall, 1805.. 161
Cordier, 1816.. 214
Hall, 1826.. 165
Cordier, 1827 (1828).. 216
Sorby, 1859.. 492
Daubrée, 1862.. 440
Daubrée, 1878.. 445
Fouqué and Michel-Lévy, 1878.. 494
Fouqué and Michel-Lévy, 1878.. 495
Favre, 1880.. 448

Fossils

da Vinci, c.1500.. 1
Steno, 1669.. 33

PAGE

Woodward, 1695...................................... 49
Hooke, 1705... 28
de Maillet, 1748.................................... 47
Pallas, 1771.. 123
Schlottheim, 1813................................... 174
Cuvier, 1817.. 188
Studer, 1827.. 257
von Buch, 1830...................................... 212
Lyell, 1835... 268
Hitchcock, 1836..................................... 250
Dawson, 1855.. 462
W. B. Rogers, 1856.................................. 336
Bronn, 1858... 299
Miller, 1858.. 313
Logan, 1863... 296
Owen, 1864.. 317

Geomorphology (see also *Subaerial Erosion*)

Evans, 1755... 55
Gesner, 1839.. 280
Darwin, 1840.. 354
Logan and Hunt, 1855................................ 291
Lesley, 1869.. 457
McGee, 1888... 661
Shaler, 1895.. 553

Glacial Geology

Saussure, 1786...................................... 114
Playfair, 1802...................................... 131
Bernhardi, 1832..................................... 327
Agassiz, 1840....................................... 329
Ramsay, 1855.. 435
J. Geikie, 1871..................................... 549
Pumpelly, 1879...................................... 542

Ground Water

Agricola, 1546...................................... 7
Kircher, 1678....................................... 17
Gilmer, 1818.. 233
Reade, 1885... 509
Pošepný, 1893....................................... 537
Van Hise, 1898...................................... 673

Historical Geology

Steno, 1669... 33
Hooke, 1705... 28
Pallas, 1771.. 123

	PAGE
Desmarest, 1774	91
Buffon, 1778 (1807)	65
Soulavie, 1781	155
Dolomieu, 1789	152
Hutton, 1795	92
Werner, 1795	140
Schlottheim, 1813	174
Cuvier, 1817	188
Maclure, 1817	168
Omalius d'Halloy, 1822	220
Lyell, 1833	263
Emmons, 1842	284
W. B. Rogers, 1856	336
Bronn, 1858	299
Van Hise, 1896	666

Igneous Rocks

Desmarest, 1774	90
Raspe, 1776	111
Dolomieu, 1785	151
Saussure, 1786	114
Werner, 1788	138
Hutton, 1795	92
Spallanzani, 1798	101
Playfair, 1802	131
von Buch, 1802	208
Hall, 1805	161
Cordier, 1816	214
Macculoch, 1819	205
Scrope, 1825	274
Sedgwick, 1835	222
Gesner, 1839	280
Darwin, 1844	357
Dana, 1846	416
Bunsen, 1853	381
Sorby, 1858	488
Jukes, 1862	378
Richthofen, 1868	511
Fouqué and Michel-Lévy, 1878	495
Iddings, 1892	663

Internal Structure of the Earth

Kircher, 1678	17
Leibnitz, 1680	45
Descartes, 1681	14
Dolomieu, 1789	152
Mallet, 1846	384

PAGE

Airy, 1855 401
Pratt, 1858 395
Pratt, 1864 398
Thomson, 1864 472
Le Conte, 1889 467
Chamberlin, 1899 612

Isostasy

Airy, 1855 401
Pratt, 1855 393
Pratt, 1858 395
Dutton, 1889 566

Life Development

Steno, 1669 33
Cuvier, 1817 188
von Buch, 1830 212
Bronn, 1858 299
Darwin, 1859 363
Huxley, 1876 479

Marine Denudation

Varenius, 1672 24
Playfair, 1802 131
Prevost, 1835 230
Gesner, 1839 280
Ramsay, 1848 430
A. Geikie, 1868 523
Richthofen, 1882 511
Davis, 1896 656

Metamorphic Rocks

Arduino, 1779 76
Dolomieu, 1791 154
Hall, 1805 158
Macculoch, 1819 205
Studer, 1827 257
Sedgwick, 1835 222
Sharpe, 1847 321
Sorby, 1853 484
Hitchcock, 1861 252
Heim, 1878 642
Lossen, 1884 569
Van Hise, 1898 673

Mineralogy

Agricola, 1546 7
Steno, 1669 33

PAGE

Hauy, 1801... 129
Nicol, 1829... 176
Nicol, 1831... 177
Dana, 1846... 416
Sorby, 1858... 488
Daubrée, 1862....................................... 440
Fouqué and Michel-Lévy, 1878...................... 494

Ore Deposits

Agricola, 1546....................................... 7
Descartes, 1681...................................... 14
Werner, 1795... 140
Pošepný, 1893.. 537

Paleogeography

Buffon, 1778 (1807)................................. 65
Hall, 1857 (1882)................................... 413

Philosophical Geology

Hutton, 1795... 92
Chamberlin, 1897.................................... 604

Physics of the Earth

Cavendish, 1798...................................... 103
Cordier, 1827.. 216
Babbage, 1834.. 237
Scrope, 1825... 274
Mallet, 1846... 384
Élie de Beaumont, 1852.............................. 288
Thomson, 1864.. 472
Gilbert, 1877.. 592
Chamberlin, 1899.................................... 612

Sedimentary Rocks

Steno, 1669.. 33
Woodward, 1695....................................... 49
Strachey, 1719....................................... 53
de Maillet, 1748..................................... 47
Guettard, 1757....................................... 78
Saussure, 1786....................................... 114
Lavoisier, 1789...................................... 126
Hutton, 1795... 92
Hall, 1805... 158
Cuvier and Brongniart, 1811......................... 194
Scrope, 1825... 274
Hall, 1826... 165

PAGE

Darwin, 1840. 354
H. D. Rogers, 1840. 346

Sedimentation

da Vinci, c.1500. 1
Steno, 1669. 33
Varenius, 1672. 24
Boyle, 1673 (?). 27
Woodward, 1695. 49
Lavoisier, 1789. 126
Hutton, 1795. 92
Quoy and Gaimard, 1823. 235
Everest, 1832. 365
Darwin, 1840. 354
Dawson, 1855. 461
Ramsay, 1855. 435
Hall, 1857 (1882). 413
Humphreys and Abbott, 1861. 367
Sidell, 1861. 369
Newberry, 1874. 465
Pumpelly, 1879. 542
Brewer, 1885. 497
Ochsenius, 1888. 499

Shorelines and Shore Processes

Gilbert, 1881. 596
Gilbert, 1884. 600
Shaler, 1895. 553

Stratigraphy

Owen, 1603. 12
Steno, 1669. 33
Guettard, 1746. 77
Michell, 1760. 84
Cuvier and Brongniart, 1811. 194
Smith, 1815. 201
Maclure, 1817. 168
Omalius d'Halloy, 1822. 220
Studer, 1827. 257
Lyell, 1835. 268
Murchison, 1835. 244
Murchison, 1841. 247
Emmons, 1842. 284
Sedgwick, 1852. 227

Structural Geology

Michell, 1760. 84
Pallas, 1771. 123

PAGE

Saussure, 1786.. 114
Cuvier, 1817.. 188
Maclure, 1817.. 168
Sedgwick, 1835... 222
H. D. and W. B. Rogers, 1840....................... 338
Lesley, 1856.. 450
Le Conte, 1889... 467
Suess, 1897... 507

Subaerial Erosion

da Vinci, c.1500.. 1
Agricola, 1546.. 7
Hooke, 1705.. 28
Targioni-Tozzetti, 1752.................................. 74
Saussure, 1786... 114
Playfair, 1802.. 131
Gilmer, 1818... 233
Everest, 1832.. 365
Surell, 1841 (1870)...................................... 372
Dana, 1849.. 419
Oldham, 1854.. 371
Scrope, 1858... 278
Humphreys and Abbott, 1861......................... 367
A. Geikie, 1868.. 523
Lesley, 1869... 457
Marsh, 1874... 305
Marvine, 1874... 637
Powell, 1875... 518
Sternberg, 1875.. 477
Gilbert, 1876.. 581
Gilbert, 1877.. 586
Pumpelly, 1879.. 542
Dutton, 1881.. 555
Richthofen, 1882.. 512
Reade, 1885.. 509
de la Noë and de Margerie, 1888..................... 529
Davis, 1889.. 649
Davis, 1896.. 656

Systematic Classification and Nomenclature

Maclure, 1817.. 168
Omalius d'Halloy, 1822.................................. 220
Studer, 1827... 257
Lyell, 1835... 268
Murchison, 1835... 244
Murchison, 1841... 247
Emmons, 1842... 284

	PAGE
Hall, 1843	406
Sedgwick, 1852	227
Logan and Hunt, 1855	291
W. B. Rogers, 1856	336
Logan, 1857	295
Newberry, 1874	465
Lapworth, 1879	578
Van Hise, 1896	666

Vulcanism

Agricola, 1546	7
Kircher, 1678	17
Buffon, 1778 (1807)	65
Soulavie, 1781	155
Dolomieu, 1789	152
Spallanzani, 1798	101
Humboldt, 1810	179
Humboldt, 1819	182
Scrope, 1825	274
von Buch, 1825	210
Prevost, 1835	230
Darwin, 1844	357
Dana, 1849	423
Jukes, 1862	378
Gilbert, 1877	592

Index

The use of boldface numbers following certain names in this index indicates that there appear on such pages brief biographical notes concerning the individuals.

A

Abbot, Henry Larcom, **367**, 509
Abstraction of one stream by another, 591
Adams, Frank Dawson, 681
Adéhemar, Alphonse Joseph, 551
Agassiz, Louis Jean, 316, **329**, 336
Age determinations in a volcanic region, 91
Ages, geological, 299
Agricola, **7**
Airy, George Bedell, 396, **401**
Alluvial marine soils, 461
Alpine geology, 114
Alps, overthrusting in, 631
American sedimentary rocks, circles of deposition in, 465
Analysis of a general map of the middle British colonies in America (Lewis Evans), 55
Anhydrithut, 499, 502
Antecedent valleys, 520
Anticlinal dip, 341
Anticlinal mountain, 340
Anticlinal theory of gas accumulation, 639
Appalachian coal strata, limits of, 346
Appalachian folding, date of, 345
Appalachian structure, origin of, 421
and topography, 450
Arch, 341
Archeopteryx from Solenhofen, 317
Arduino, Giovanni, **76**
Arrhenius, Svante, 616

Ash, 379
Atmospheric denudation, 528
Auvergne volcanics, 152, 208
Azoic rocks, 295

B

Babbage, Charles, **237**, 392
Bad-lands, 591
Baltzer, Armin, 643
Barrier, 600
Barus, Carl, 673
Basalt, aqueous origin of, 138
volcanic origin of, 90
Basaltiform lavas, 101
Base level of erosion, 522, 558
Basin ranges, 470
Basin region, structure of, 467
Basin structure, 469
Bauer, Georg, **7**
Beaumont, Jean Baptiste Armand Louis Leonce Élie de, 213, **288**, 448, 490
Beche, Henry Thomas De la, **260**, 355, 379
Bernhardi, Reinhard, **327**
Bertrand, Marcel Alexandre, **631**
Bickerton, Alexander William, **574**
Bischof, Carl Gustav Christoph, 476
Blind coal, 161
Boulders and rolled pebbles, origin of, 115
Boyle, Robert, **27**
Brewer, William Henry, **497**
Brögger, Waldemar Christofer, 579

Brongniart, Alexandre, **194**
Bronn, Heinrich Georg, 213, **299**
Brooks, William Keith, 669
Buch, Leopold von, **208**, 232, 276, 330, 357
Buckland, William, 334, 350
Buffon, Comte de, George Louis Leclerc, **58**
Bunsen, Robert Wilhelm Eberhard von, **381**

C

Cambrian series, 228
Cambrian system, 293, 466
Canary Islands, 211
Caracas, earthquake of, 26th March, 1812, 182
Carboniferous fossils of Joggins section, 462
Catchment basin, 375
Cavendish, Henry, **103**
Cenchrites, 118, 119
Chalk, 196
Chamberlin, Thomas Crowder, **604**
Charleston earthquake of 1886, 561
Chronologie physique des volcans (Jean Louis Giraud Soulavie), 155
Circles of deposition in American sedimentary rocks, 465
Clarke, William Branwhite, 359
Classification, of igneous rocks, 1868, by Ferdinand von Richthofen, 513
and nomenclature of lower Paleozoic rocks of England and Wales, 227
of volcanic rocks, 511
Clay, deposition of, in salt water, 497
Cleavage, origin of slaty, 484
Cliffs, 600
Coal measures, character of, 350
Coal strata, origin of, 346, 351
"Coales, vaynes of," 13
Coastal plain and fall line, 661
Coasts, elevated, 361
Compleat system of general geography (Stephen Dugdale), 24

Conglomerates, fluviatile origin of, 78
Consequent valleys, 520
Consolidation of strata, role of heat in, 165
Contemporaneous and intrusive trap, distinction between, 378
Continents, mean height of, 524
and mountain ranges, origin of, 419
movement of, 30
Contraction of earth, effects of, 420
Cooling, of earth, secular, 472
of melted rock, slow, results of, 161
Coral islands, formation of, 235
Coral reefs and islands, origin of, 354
Cordier, Pierre Louis Antoine, **214**
Corgoloin, 118
Corrasion, 582
Correlation of Paleozoic rocks of New York with those of Europe, 406
Countries swallowed up into the earth, 32
Crest line of mountain, 455
Croll, James, 524
Crustal balance, hypothesis of, 401
Crystallization, laws of, 108
Crystals, formation of, 108
microscopical structure of, 488
settling in molten lava, 357
shape and growth of, 40
structure of, 129
Cuvier, Léopold Chrétien Frédéric Dagobert, **188, 194**, 213, 316
Cycle of erosion, 649

D

Dana, James Dwight, **416**
Darton, Nelson Horatio, 659
Darwin, Charles Robert, **354**, 669
Darwin, George, 566, 618
Daubrée, Gabriel August, **440**, 540
da Vinci, Leonardo, **1**
Davis, William Morris, **649**
Dawson, John William, **461**
De l'origine des fontaines (Pierre Perrault), 20
De re metallica (Agricola), 7

Degradation, 586
Deluc, Jean André, 448
Deluge, due to volcanic eruptions, 125
Denudation, 586
  atmospheric, 528
  marine, 527, 656
  now in progress, 523
  subaerial, 637, 656, 659
Deposition, of clay in salt water, 497
  of rock salt, 499
Descartes, René, 14
Description of Western Islands of Scotland (John Macculloch), 205
Desmarest, Nicolas, 90
Devonian system, 466
Dikes, 592
Discours de la méthode (René Descartes), 14
Dolomieu, Déodat Guy Sylvain Tancrède Gratet de, 151, 332, 387
Dolomite and limestone, distinctions between, 154
Dutton, Clarence Edward, 555

E

Earsenian, William A., 639
Earth, alterations on surface of, 29
  beds of, 43
  composition of, 14
  crust of, constitution, 398
  effects of changes in temperature in, 239
  density of, 103
  epochs in history of, 65
  fires at center of, 45
  forces which change, 7
  interior of, 16
  interior structure, surface temperature and age of, 612
  origin of, collision hypothesis of, 58
    collision and accretion hypothesis, 574
    meteoritic hypothesis of, 547
    naturalistic hypothesis of, 88
    planetesimal hypothesis of, 618
  originally molten, 45

Earth, secular cooling of, 472
  structural features of, 507
  structure of, 7, 95
  theory of, 92
Earth as modified by human action (George Perkins Marsh), 305
Earthquake of Caracas, 26th March, 1812, 182
Earthquake intensity shown by isoseismal lines, 560
Earthquake observatories, 390
Earthquakes, 11
  artificial, 390
  dynamics of, 384
  effects of, 28
  motion of earth in, 82
  nature and origin of, 80
  velocity of, 83
Eastern United States, physiographic provinces of, 457
Elasticity, modulus of, 389
Emmons, Ebenezer, 284
England, map of strata of, 201
Eocene period, 270
Eozoic system, 465
Eozoon canadense, 297
Epicentrum, 563
Epóques de la nature (George Louis Leclerc, Comte de Buffon), 65
"Eremacausis," 473
Erosion, analysis of, 557, 581
  in Appalachian belt, 459
  base level of, 522
  in Colorado front range, 637
  effects of unequal resistance to, 531
  importance of solution as factor in, 509
  influenced by rock character, 586
  rate of, 510, 524, 555, 583
Erosion cycle in life of river, 649
Erratic blocks of the Jura, 332
Essai de cristallographie (Jean Baptiste Louis Rome de l'Isle), 108
Essai sur la géographie minéralogique des environs de Paris (Alexandre Brongniart and Léopold Chrétien Frédéric Dagobert Cuvier), 194

Essai sur la Nouvelle Espagne (Friedrich Heinrich Alexander von Humboldt), 179
Essay towards a natural history of the earth (John Woodward), 49
Estuaries, origin of fiords and, 553
Evans, Lewis, 55
Evans, Nevil Norton, 507
Everest, Robert, 365
Ewing, James Alfred, 565
Expansion of rock with increase in temperature, 240

F

Fall line, 661
Faults, 341
  theory of, 467
Favre, Alphonse, 448
Faye, Hervé, 574
Feldspars, artificial production of, 494
Fiords and estuaries, origin of, 553
Flood-plain, 588
Floods, influence of forest on, 305
Folding and faulting of strata, experiments on, 445
Folds, and faults, origin of, 343, 406, 448
  recumbent, 643
Forest, influence on floods, 305
Formation, 195
Fossil footprints in Triassic rocks of Connecticut Valley, 250
Fossil organisms of sedimentary strata in natural series, distribution of, 299
Fossil shells, 31, 48, 51
  distortion of, 321
Fossils, of early Paleozoic age in New England, 336
  Carboniferous, of Joggins section, 462
  importance of, 212
  origin and meaning of, 3
  problem of, 33
  use of, in geological investigations, 174

Fouqué, Ferdinand André, 494, 565
Fractures resulting from torsional strains, 443

G

Gaimard, Joseph Paul, 235
Geikie, Archibald, 523
Geikie, James, 545, 549
Geocosm, 18
Geographia generalis (Bernhard Varenius), 24
Geography, physical, 24
Geologic formations, systematic classification of, 220
Geologic outline of Canada, 1855, 291
Geological sections, drafting of, 260
Geosynclines, 406, 505
Gesner, Abraham, 280
Gilbert, Grove Karl, 469, 471, 581
Gilmer, Francis Walker, 233
Glacial action, 544
Glacial deposits, inter-glacial beds in, 549
Glacial epoch, changes of climate during, 549
Glacial erosion and transportation, 136
Glacial period, theory of, 327, 329
Glaciation, Permian, 435
Glaciers, action of, 119
Glarner double fold, 642
Godwin-Austen, Robert Alfred Cloyne, 436
Gosselet, Jules, 632
Grand Manan, geology of, 280
Granite, experiments on, 114
Granite contacts, nature of, 359
Granitic veins, origin of, 133
Great Lake basins, origin of, 609
Ground water, deep circulation of, 539
  and deposition of ore bodies, 537
  vadose circulation of, 537
Grünstein, 163
Guettard, Jean Étienne, 77

H

Hall, Basil, 359
Hall, James, 406

Hall, Sir James, **158**, 334
Halloy, Jean Baptiste Julien d'Omalius d', **220**
Hauy, René Just, **129**
Hayden, Ferdinand Vandiveer, 637
Heat, operation of, 97
Heim, Albert, 579, 631, **642**
Hershey, Oscar H., 659
Himalaya mountains, 395
Hirsenstein, 118
Hitchcock, Charles Henry, 252
Hitchcock, Edward, **250**
Holmes, William Henry, 521
Hooke, Robert, **28**
Hoover, Herbert Clark, 7
Hoover, Lou Henry, 7
Hopkins, William, 421, 475
Horse, geological history of, 479
Humboldt, Friedrich Heinrich Alexander von, **179**, 276, 524
Humphreys, Andrew Atkinson, **367**, 509
Hunt, Thomas Sterry, **291**
Huronian system, 293, 295
Hutton, James, **92**, 161, 350
Huttonian theories, proofs of, 131
Huxley, Thomas Henry, **479**
Hypotheses, method of multiple working, 604

I

Ichthyolites, 316
Iddings, Joseph Paxson, **663**
Igneous origin of basalt, 208
Igneous rock, artificial production of, 495
Igneous rocks, analyses of, 382
  genetic relations of, 381
  origin of mineral constitution of, 416
  relation between mineral composition and mode of occurrence of, 663
Isoseismal lines, 560
Isostasy, 566
Isothermal surface, 242

J

Joggins section, Carboniferous fossils of, 462
Jointing, columnar, 111
Jukes, Joseph Beete, **378**
Jura valley, origin of, 116

K

Kant, Immanuel, **88**
Kelvin, Lord (Sir William Thomson), **472**
King, Clarence, 469
Kircher, Athanasius, **17**

L

Laccolite, 592
Laccoliths, 592
Lake Bonneville, history of, 596
Lake shores, topographic features of, 600
Land sculpture, 586
Laplace, Pierre Simon, **143**, 548, 574
Lapworth, Charles, **578**
Laurentian mountains, 291
Laurentian system, 292, 295
Lava, basaltic, 111
  fluidity of, 153
  rhyolitic, 151
Lavoisier, Antoine Laurent, **126**
Le Conte, Joseph, **467**
Leibnitz, Gottfried Wilhelm von, **45**, 59, 473
Lesley, Joseph Peter, **450**
Lévy, Auguste-Michel, **494**, 565, 570
Life development, geologic record of, 363
Life-principle, plastick faculty of, 30
L'Isle, Jean Baptiste Louis Rome de, **108**
Linth, Arnold Escher von der, 631, 636, 643
Lithoid pastes, 215
Lithophytes, 235
Littoral beds, 126
Lockyer, Joseph Norman, 618
Loess, 542

Logan, William Edmond, **291, 295,** 411
"Loose stones," problem of, 135
Lory, Charles, 631
Lossen, Karl August, **569**
Lyell, Charles, **263,** 330

M

Macculloch, John, **205**
McCurdy, Edward, 1
McGee, W J, **661**
Maclure, William, **168**
Maillet, Benoit de, **47**
Mallet, Robert, **384,** 565
Maps, geologic, value of, 77
Marble, and dolomite, origin of, 76
  from limestone in the laboratory, 158
Margerie, Emmanuel de, **529**
Marine abrasion and transgression, 512
Marine alluvial soils, 461
Marine denudation, 515, 527
  consequences of, 660
  in Southern Wales, 430
Marsh, George Perkins, **305**
Marvine, Archibald Robertson, 520, **637**
Mastozootic terrains, 221
Memoir to map of strata of England, 1815 (William Smith), 201
Mémoire sur les Îles Ponces (Déodat Guy Sylvain Tancrède Gratet de Dolomieu), 151
Mesozoic rocks distinguished from Paleozoic, 257
Metamorphic rocks, 223
Metamorphic Mesozoic rocks distinguished from Paleozoic, 257
Metamorphism, distinction between contact and regional, 569
  of Newport conglomerate, 252
  of sedimentary rocks, 222
Meteoric systems and origin of earth, 547
Method of multiple working hypotheses, 604

Meyer, Hermann von, 317
Michell, John, **80,** 103
Microscopical structure of crystals, 488
Middle British Colonies in America, physiography of, 55
Miller, Hugh, **313**
Milne, John, 565
Mineral composition, relation between, and mode of occurrence of igneous rocks, 663
Mineral constitution of igneous rocks, origin of, 416
Mineralogy, basis for system of, 129
Minerals, source of, 8
Miocene period, 270
Mississippi delta, geological phenomena of, 369
Mississippi River water, analysis of, 509
Molecules, integrant, 129
Monoclinal mountain, 341
Monoclinal ridges, 455
Monoclinal valley, 341
Moraines, glacial, 120
Mountain building, mechanics of, 644
  nature and cause of, 288
  rock deformation during, 642
Mountain ranges and continents, origin of, 419
Mountain structures, and forces which form them, 503
  reproduced experimentally by lateral compression, 448
Mountains, anticlinal, 453
  construction and destruction of, 10
  fault block, 469
  primitive, schistose, 123
  relation of, to regions of thick sedimentary accumulation, 406
  secondary, 123
  synclinal, 453
  Tertiary, 123, 124
  water-chambers of, 19
  zigzag, 457

Mundus subterraneus (Athanasius Kircher), 17
Murchison, Roderick Impey, **244**, 435, 466, 578

N

Natural bridge of Virginia, 233
Natural history of animals, vegetables, and minerals (Comte de Buffon, George Louis Leclerc), 58
Nature, tricks of, 31
Nebular hypothesis, 143
Newberry, John Strong, **465**
Newcomb, Simon, 574
Nicol prism, 176
Nicol, William, **176**
Noë, Gaston de la, **529**
Normal dips, 341
Normal faults, origin of, 467
Normal flexure, 341
North America, structure of continent of, 169
Note-books (Leonardo da Vinci), 1

O

Observations on geology of United States, 1817, 168
Observations sur la formation des montagnes (Peter Simon Pallas), 123
Ochsenius, Carl, **499**
Old red sandstone (Hugh Miller), 313
Oldenburg, Henry, 42
Oldham, Thomas, **371**
Oölites, inorganic nature of, 118
Ordovician system, 578
Ore bodies, ground water and deposition of, 537
Ore-deposits, origin of, 540
Origin of rolled pebbles and boulders, 115
Ornitholites, 250
Osservazioni (Giovanni Arduino), 76
Owen, George, **12**
Owen, Richard, **317**

P

Pacific Islands, origin of valleys in, 425
Paleozoic seas, physical conditions in, 413
Pallas, Peter Simon, **123**
Paris basin, stratigraphy of, 194
Pelagic beds, 126
Penbrokshire, description of (George Owen), 12
Penbrokshire, rock seams in, 12
Peneplain, 658
Permian glaciation, 435
Permian system, 247
Perrault, Pierre, **20**
Phenocrysts, 663
Phillips, John, 484, 570
Physiographic provinces of eastern United States, 457
Physiography of Middle British Colonies in America, 1755, 55
Pietre dure, 113
Plains of marine and subaerial denudation, 656
Planation by running water, 588, 637
Planetesimal hypothesis, 618
Planets, density of, 61
Playfair, John, **131**
Pliocene period, 269
Plumb-line, deflection of, in India, 393, 395
Poisson, Jules, 473
Pošepný, Ferencz, **537**
Powell, John Wesley, **518**, 558, 656
Pratt, John Henry, **393**, 401
Pre-Cambrian fossil, 296
Pre-Cambrian geology, 666
Prévost, Louis Constant, **230**
Primitive rocks, 170, 220
Primordial terrains, 220
Principles of geology (Charles Lyell), 263
Proctor, Richard Anthony, **547**, 574
Prodromus (Nicolaus Steno), 33
Protogaea (Gottfried Wilhelm von Leibnitz), 45

Pumpelly, Raphael, **542**, 660

Pyroid terrains, 221

Q

Quarry laborer as geological observer, 313

Quoy, Jean René Constant, **235**

R

Raised coasts of Sweden, 209

Ramsay, Andrew Crosbie, **430**, 527, 545, 656

Raspe, Rudolph Eric, **111**

Rate of erosion, 524

of Mississippi basin, 510

Reade, Thomas Mellard, **509**

Recrystallization, process of, 677

Relazioni d'alcuni viaggi fatti in diverse parti della Toscana (Giovanni Targioni-Tozzetti), 74

Revolutions and catastrophes in history of earth, 188

Richthofen, Ferdinand von, **511**, 543

River terraces, 589

Rock-basins, origin of, 545

Rock-disintegration, secular, 542

Rock flowage, meaning of, 673

Rock formation, three modes of, 277

Rock salt, deposition of, 499

Rockwood, Charles Greene, 560

Rocky mountains, origin of, 422

Rogenstein, 118

Rogers, Henry Darwin, **338, 346**, 414, 421, 579

Rogers, William Barton, **336, 338**, 421

Rolled pebbles and boulders, the origin of, 115

Rossi-Forel scale of earthquake intensity, 560

Russell, Israel Cook, 469

S

Saint-Germain, Bertrand de, 45

Salt, rock, deposition of, 499

Salt water, deposition of clay in, 497

Sandstone, old red, 313, 407

Saussure, Horace Benedicte de, **114**, 214, 448

Schlottheim, Ernst Friederich von, **174**

Scottish highlands, structure of, 579

Scrope, George Poulett, **274**, 358

Sea, action of, 131

calmness at bottom, 27

saltness of, 1

Sea-cliff, 600

Sea level, change of, 551

Seas, Paleozoic, physical conditions in, 413

Secondary formation, 172

Secondary rocks, 220, 221

Secular cooling of the earth, 472

Sedgwick, Adam, **222**, 334, 435, 466, 578

Sediment carried by Mississippi, amount of, 367

Sedimentary rocks, American, circles of deposition in, 465

metamorphism of, 222

Sedimentation, 555

Seismographs, 389, 566

Seismoscopes, 565

Selce, 113

Serapis, Temple of, 237

Shaler, Nathaniel Southgate, **551**

Shape of rock bodies, relation to composition, 275

Sharks' teeth, 36

Sharpe, Daniel, **321**, 484, 488

Sheet, 592

Shell fish, petrified, 6

Shorelines, Bonneville and Provo, 598

Shores, recognition of ancient, 602

Sidell, William H., **369**

Silicates, formation of, action of superheated water in, 440

Silurian series, 228, 244, 406, 465

Sky, trap veins of, 206

Slaty cleavage, 225, 321

origin of, 484

Slope limit, 375

Slopes, fashioning of, 529

profile of, 533

Smith, Andrew, 359
Smith, William, **201**
Soils, marine alluvial, 461
Solar system, creation of, 14
Solids naturally contained within solids, 33
Solution, importance of, as a factor in erosion, 509
Somersetshire coal mines, strata in, 53
Sorby, Henry Clifton, **484**
Soulavie, Jean Louis Giraud, **155**
South American coast, recent uplift of, 361
Spallanzani, Lazarro, **101**
Springs and rivers, source of, 20
Steno, Nicolaus, **33**
Sternberg, Hermann, **477**
Strabo, 551
Strachey, John, **53**
  structure section drawn by, 54
Strata, of earth due to deposits of fluid, 37
  formed beneath sea, 47
  regular and uniform, 84
  in Somersetshire coal mines, 53
Stream activity, 132
Stream deposits, 2
Stream erosion and transportation, 371
Stream gradient, 373
Stream profiles, 372, 477, 586
Stream transportation, quantitative study of, 365, 367
Streams and valleys, classification of, 518
Structural features of earth, 507
Structure, of Appalachian mountains, 338
  of basin region, 467
Structure section drawn by John Strachey, 54
Studer, Bernard, **257**
Subaërial denudation, 528
  consequences of, 659
Subaërial erosion, analysis of, 581
Submerged forests and marshes, 282
Subsequent streams, 659

Subterranean fire, 153
Subterranean fire cells, 17
Subterranean heat, 218
Suess, Eduard, **503**, 579
Superposed streams, 660
Superimposed valleys, 521
Surell, Alexandre, **372**
Surfaces of equal temperature within earth, 243
Synclinal dip, 341
Synclinal mountain, 340
Synclinal mountains, 453
Systematic classification of geologic formations, 220

T

Taconic system, 284
Targioni-Tozzetti, Giovanni, **74**
Telliamed (Benoit de Maillet), 47
Temperature gradient of earth's interior, 216
Terrace, wave-built, 601
  wave-cut, 601
Terraces, stream, 589
Terrains ammonéens, 221
Terrains pénéens, 221
Tertiary epoch, subdivisions of, 268
Theory of the earth (James Hutton), 92
Thin sections, preparation of, 177
  and their value to geologist, 492
Thomson, James, 476
Thomson, William, Lord Kelvin, **472**
Tillite, 436
Tivoli pills, 119
Topography, science of, 450
Torrential streams and their control, 372
Torrents, 372, 428
  control of, 375
Torsion balance, 103
Torsional strains, fractures resulting from, 443
Transition rocks, 245
Transportation by running water, 584
Transverse compression, 289
Trilobites, 336
Trough, 341

Troughs, synclinal, 580
Tufa, 379
Tuff, 379

U

Underground waters, 8
Uniformitarianism, 92, 132, 263, 472
Universal deluge, 51
Uplift, recent, of South American coast, 361
Usiglio, J., 500

V

Vadose underground circulation, 537
Valleys, antecedent, 520
    anticlinal, 580
    consequent, 520
    formed by stream erosion, 74
    origin of, 278
    superimposed, 521
Van Hise, Charles Richard, **666**
Varenius, Bernhard, **24**
Vegetation and stability of slopes, 535
Vein systems, 7
Veins, granitic, origin of, 133
    theory of formation, 140
Volcanic activity in Mexico, 179
Volcanic energy, source of, 274
Volcanic eruptions, nature of, 423
    role of gases in, 101
Volcanic island not upheaved, 230
Volcanic mountains, systems of, 186
Volcanic region, age determinations in, 91

Volcanic rocks, classification of, 511
    crystalline nature of, 214
    relation to ancient eruptive rocks, 512
Volcanic sources at great depths, 153
Volcanoes, upheaval of, 210
Voyages dan les Alpes (Horace Benedicte de Saussure), 114

W

Wacken, 163
Wagner, Andreas, 317
Water, circulation of, 29
    superheated, action of, in formation of silicates, 440
Weathering, 544, 582
Werner, Abraham Gottlob, **138**
Western Islands of Scotland, metamorphic and igneous rocks of, 205
Whinstone, 163
White, Israel Charles, **639**
Winter, John Garrett, 33
Wollaston, Francis John Hyde, 103
Woodford, Alfred Oswald, 477
Woodward, John, **49**
Works (Robert Boyle), 27
Works, posthumous (Robert Hooke), 28

Z

Zigzag mountains, 457
Zirkel, Ferdinand, 572